POLARIZATION AND INTENSITY OF LIGHT IN THE ATMOSPHERE

Studies in
Geophysical Optics and Remote Sensing

Series Editor: Adarsh Deepak

Published Volumes and Volumes in Preparation

Advances in Remote Sensing Retrieval Methods
Adarsh Deepak, Henry E. Fleming, and Moustafa T. Chahine (Eds.)

Aerosols and Climate
Peter V. Hobbs and M. Patrick McCormick (Eds.)

Aerosols and Their Climatic Effects
Hermann E. Gerber and Adarsh Deepak (Eds.)

Applications of Remote Sensing for Rice Production
Adarsh Deepak and K.R. Rao (Eds.)

Atmospheric Aerosols: Their Formation, Optical Properties, and Effects
Adarsh Deepak (Ed.)

Hygroscopic Aerosols
Lothar H. Ruhnke and Adarsh Deepak (Eds.)

IRS '84: Current Problems in Atmospheric Radiation
Giorgio Fiocco (Ed.)

Light Absorption by Aerosol Particles
Hermann E. Gerber and Edward E. Hindman (Eds.)

Microwave Remote Sensing
A. Chedin (Ed.)

Optical Properties of Clouds
Peter V. Hobbs and Adarsh Deepak (Eds.)

Proceedings in Atmospheric Electricity
Lothar H. Ruhnke and John Latham (Eds.)

Radiative Transfer in Scattering and Absorbing Atmospheres: Standard Computational Procedures
Jacqueline Lenoble (Ed.)

RSRM '87: Advances in Remote Sensing Retrieval Methods
Adarsh Deepak, Henry E. Fleming, and John S. Theon (Eds.)

Spectral Line Shapes, Volume 4
R.J. Exton (Ed.)

Tropical Rainfall Measurements
John S. Theon and Nobuyoshi Fugono (Eds.)

POLARIZATION AND INTENSITY OF LIGHT IN THE ATMOSPHERE

Kinsell L. Coulson
Professor Emeritus
University of California, Davis

A. DEEPAK Publishing 1988
A Division of Science and Technology Corporation
Hampton, Virginia USA

A. DEEPAK Publishing
A Division of Science and Technology Corporation
101 Research Drive
Hampton, Virginia 23666–1340 USA

Library of Congress Cataloging-in-Publication Data

Coulson, Kinsell L.
Polarization and intensity of light in the atmosphere / Kinsell L. Coulson.
p. cm. — (Studies in geophysical optics and remote sensing)
Includes bibliographies and indexes.
ISBN 0-937194-12-3
1. Polarization (Light) 2. Meteorological optics. I. Title.
II. Series.
QC976.P7C68 1988
551.5'6—dc19 88-25698
CIP

Printed in the United States of America

Table of Contents

FOREWORD

Since its discovery in the 17th century, polarization of electromagnetic radiation has been used to great advantage in laboratory investigations. Polarization of visible light in the natural environment was investigated by Arago, Brewster, and others in the early years of the 19th century, and an explanation of the blue color and polarization of the sunlit sky was provided by Lord Rayleigh in the latter part of that century. However, in spite of fundamental contributions by these and other "giants" of their times, a detailed understanding and meaningful application of this well-known optical phenomenon to planetary atmospheres has been far from being "sugar and spice, and everything nice." This is because the question is not "Why is the sky blue?" but is "Why is the sky not blue?" and, one may add, "anymore." The radiance and polarization characteristics of planetary atmospheres are a consequence of intricate interplays among phenomena of reflection, refraction, scattering, and absorption by molecules, manmade and natural pollutants, as well as by nonhomogeneous surfaces underlying the atmosphere. A detailed explanation of the various features has now become possible thanks to the valuable contributions by Chandrasekhar, Van de Hulst, Sekera, Rozenberg, Sobolev, and others over the past 40 years. Availability of modern digital computers and of advanced measurement technology has also played a very important role.

Remote sensing, the ability to perceive and study a feature from a distant location, is an area which has assumed great economic, political, and military importance in the last 30 years. It has now become possible to perform such activities from space, on a global basis in a routine manner. Until very recently all such observations have been restricted only to the radiance aspect of electromagnetic radiation, but the importance of polarization measurements in remote sensing is finally being recognized. Thus, publication of this book on polarization under the natural environment by Professor Coulson comes at a very opportune time. The author has not only participated in early measurements of polarization from space, but is one who understands the theory, has built his own equipment, and gathered very valuable observational data in the desert of California, in Antarctica, and also on the top of Mauna Loa in Hawaii. His objective evaluation and presentation of the subject matter will provide the reader with a solid foundation for future activities related to using polarization in remote sensing.

Jitendra V. Dave
August 1987
Palo Alto, California

FRONTISPIECE

Top photo: Solar depression angle 10.0°; Bottom photo: Solar depression angle 7.5°;

Photographs of the pre-sunrise sky taken by the author at the Mauna Loa Observatory on May 14, 1982. The colors are due to the scattering of sunlight

Top photo: Solar depression angle 5.3^0; Bottom photo: Solar depression angle 1.6^0.

by the stratospheric cloud from the eruption of the Mexican volcano El Chichon. Further descriptions of the cloud are given in Chapter 9.

LIST OF SYMBOLS

Symbol	Section	Definition
A	1.2	Complex amplitude
A	2.6	Indicator for aerosols
A	4.3.2	Average polarizability of particle
A	4.6.1	Albedo of surface
B	2.7	A constant
B	4.3.2	Average polarizability of particle
$\mathbf{B}$	1.2	Magnetic induction
B_i	5.6	$i=1$ to 3: Intensity measured at three positions of analyzer
CR	8.2	Color ratio
C	2.6	Indicator for clouds
C	2.7	A constant
C_T	3.2.8.1	Normalization factor in semi-empirical theory
$C_\parallel$, $C_\perp$	3.4	Coefficients of intensity reflected by a drop
$C'_\parallel$, $C'_\perp$	3.4	Coefficients of intensity refracted by a drop
DVI	9.1	Dust veil index of Lamb
D	3.4	Angular deviation of ray by water drop
D	4.1	Density of a particle
D	4.6.1	Indicator for direct transmission
D	8.3.3	Solar depression angle
$\mathbf{D}$	1.2	Dialectric displacement vector
$\mathbf{D}$	3.1	Scattering matrix for particles
D_m	3.4	Minimum deviation of ray by water drop
D'	4.1	Density of particle surroundings

$\mathbf{E}$	1.2	Electric field strength
F	–	Flux of energy
$\mathbf{H}$	1.2	Magnetic field strength
H_c	2.7	Critical supersaturation for hygroscopic nuclei
I	1.4	Intensity
I	1.5	A Stokes parameter
I_m	8.3.3	Intensity due to multiple scattering
I_p	8.3.3	Intensity due to primary scattering
I_e^*, I_r^*	4.6.1	Intensities due to surface reflection
J	11.2.1	Coefficient of birefrigence
K	4.1	Constant for material characteristics
K, L, M, N, W	4.4	Factors in Chandrasekhar scattering theory
L	6.1	Luminance of sky
$\mathbf{L}$	1.7	A rotation matrix
$\mathbf{L}$	11.8.2	Rotation matrix for rotation of analyzer
L_i	10.4.4	Portion of incident flux specularly reflected by leaves
M	3.2.1.2	Mass concentration
M	4.2	Molecular weight
$\mathbf{M}$	1.2	Magnetic dipole moment
$\mathbf{M}$	10.2	Reflection matrix for surface reflection
$\mathbf{M}_i$	11.7	Transmission matrix of *i*th component of optical train
$\bar{M}$	3.2.8.1	Parameter in EBCM method
N	3.2.8.1	Parameter in EBCM method
$\bar{N}$	4.6.1	Polarization matrix for reflected light
O	2.6	Indicator for ozone
P	1.4	Degree of polariztion
P	2.4	Phase angle
P	3.2	Legendre polynomial

Symbol	Section	Meaning
P	4.2	Pressure of gas
P	4.3.1	Phase function for scattered light
$\mathbf{P}$	2.4	Phase matrix for scattered light
$\mathbf{P}$	4.3.1	Dipole moment of a particle
$P_{\max}$	4.5.2.1	Polarization maximum
P_i	10.4.4	Probability of direct illumination of a leaf
P_{ij}	3.2.8.1	Elements of scattering matrix
P_m	8.3.3	Polarization due to multiple scattering
P_p	8.3.3	Polarization due to primary scattering
P_r	10.4.4	Probability of transmission of reflected ray by plant canopy
P_r	11.8.2	Polarization introduced by retardation plate
$\mathbf{P}^{(i)}$	4.4	Matrices in Chandrasekhar scattering theory
Q	1.5	A Stokes parameter
Q	4.3.5	Scattering efficiency of a particle for small volume
$\mathbf{Q}$	4.4	Diagonal matrix in Chandrasekhar scattering theory
Q_a	3.2	Efficiency factor for extinction by absorption
Q_s	3.2	Efficiency factor for extinction by scattering
Q_t	3.2	Efficiency factor for total extinction
R	2.6	Indicator for Rayleigh
$\mathbf{R}$	4.6.2	Reflection matrix
R_v	4.2	Gas constant for water vapor
R^*	4.2	Universal gas constant
$\mathbf{S}$	1.3	Poynting matrix
$\mathbf{S}$	4.4	Reflection matrix in Chandrasekhar scattering theory
$S_\parallel$, $S_\perp$	3.2.2	Amplitude functions in Mie theory
T	3.4	Fractional transmission of light by a drop
T	4.2	Absolute temperature

T	11.8.2	Total transmission of a retardation plate
$\mathbf{T}$	4.4	Transmission matrix in Chandrasekhar scattering theory
U	1.5	A Stokes parameter
V	1.5	A Stokes parameter
V	4.1	Volume of a particle
V_i	11.8.1	Signal values at different settings of analyzer
W	2.6	Indicator for water vapor
W	2.7	Molecular weight
W	10.3.3	Fractional coverage of sea surface by whitecaps
X	3.4.1	Distance along a wave front
X_i	4.4	Fundamental function in Chandrasekhar scattering theory
Y	3.4.1	Displacement of wave front
Y_i	4.4	Fundamental function in Chandrasekhar scattering theory
a	1.4	Semimajor axis of ellipse
a	3.2.7.3	A constant
a_n	3.2	A Mie coefficient
a_x	1.2	Component of amplitude in *x*-direction
a_y	1.2	Component of amplitude in *y*-direction
b	1.4	Semiminor axis of ellipse
b	3.2.7.3	A constant
b	3.2.8.1	A constant in semi-empirical theory
b_n	3.2	A Mie coefficient
c	1.2	Speed of light in a vacuum
c	3.2	Particle concentration
d	4.6.1	Indicator for diffuse transmission
e	4.4	Indicator for vertical axis
e	11.8.2	Instrumental constant of retardation plate
$\mathbf{e}$	2.1	Unit vector

f	2.7	Relative humidity
f_A	8.3.3	Normalized aerosol scattering function
f_R	8.3.3	Normalized Rayleigh scattering function
f_a	10.4.4	Probability density function for leaf orientation
f_ν	3.2.8.1	Expansion coefficient in EBCM method
g	4.2	Acceleration of gravity
g	4.6.2	Indicator for radiation at the ground
g_ν	3.2.8.1	Expansion coefficient in EBCM method
h	8.3.3	Height above the surface
i	3.2	Intensity
i	–	Square root of -1
i	–	An index
j	1.2	Current density
j_b	1.2	Current density, bound
j_f	1.2	Current density, free
k	1.2	Wavenumber
m	1.2	Index of refraction
m	2.7	Mass
n	–	An index
n	2.7	Real part of index of refraction
$\mathbf{n}$	1.2	Unit vector
n_a	4.2	Avagadro's number
n'	2.7	Imaginary part of index of refraction
p	–	A point
p_i	1.2	Dipole moment in given direction
q	4.6.2	Element of refiection matrix
r	–	A distance
r	–	Radius of a particle
$\mathbf{r}$	1.2	Vector distance
r_c	2.7	Critical radius of hygroscopic nucleus
r_d	2.7	Critical radius of dry nucleus

r_m	9.2.1.4	Modal radius of particles in corona
r_v	4.2	Mixing ratio: water vapor/dry air
$\bar{r}$	3.2.8.1	Distance from surface of a particle
s	–	A distance
$\bar{s}$	4.6.1	Function for computing effects of Lambert surface reflection
t	–	Time
t_f	11.8.2	Transmission along fast axis of retardation plate
t_s	11.8.2	Transmission along slow axis of retardation plate
$\mathbf{u}$	2.1	Unit vector
v	–	Speed or velocity
w	2.8	Attenuation coefficient for specular reflection
x	3.1	Size parameter for aerosols
z	–	A distance
Δ	–	A small quanitity
Δ	4.3.4	Depolarization factor
Δ	9.2.4.1	Angle from sun to first red border in Bishop's ring
Θ	2.1	Scattering angle
π	1.2	Electric polarization
π_n	3.2.2	Angular function in Mie theory
T_n	3.2.2	Angular function inMie theory
Φ	1.3	Flux of energy
Ψ	4.4	Characteristic function in Chandrasekhar theory
Ω	1.3	Solid angle
Ω	2.1	Indicator of angle pair ( )

α	3.2.7.2	A constant
α	4.3.1	Polarizability of a particle
$\bar{\alpha}$	4.3.4	Mean polarizability
β	1.4	Ellipticity
β	2.3	Volume attenuation coefficient
β	4.3.6	Volume scattering coefficient
γ	1.2	An angle
γ	3.2.7	Exponent in Junge particle size distribution
γ	3.2.8.1	Index in EBCM method for nonspherical particles
γ	4.2	Lapse rate of temperature
γ	4.3.4	Anisotropy factor for light scattering
γ	10.4.4	Phase angle between incident and reflected radiation
γ_e, γ_r	4.6.1	Gamma functions for Lambert ground reflection
δ	3.2.2	Phase difference
ϵ	1.2	Electric permittivity
ϵ	11.2.1	Index of refraction for extraordinary ray
η	10.2	Angle of plane of polarization of incident light
θ	2.1	Zenith angle
θ_c	10.4.5	Critical angle for total internal reflection
θ_o	2.1	Zenith angle of sun
κ	2.3	Mass attenuation coefficient
λ	1.2	Wavelength

μ	1.2	Magnetic permeability
μ	2.1	Cosine of zenith angle
μ_o	2.1	Cosine of zenith angle
$\bar{\mu}$	2.7	Linear mass increase coefficient
ν	1.2	Frequency
ξ_n	3.2.1	Riccati-Bissel function in Mie theory
π	–	Pi
ρ	1.2	Charge density
ρ	2.3	Mass density
ρ	3.2.8.1	Parameter in anomalous diffraction theroy
ρ	4.6.1	Bidirectional reflection of surface
ρ_b	1.2	Charge density, bound
ρ_f	1.2	Charge density, free
$\rho_\parallel$	11.4	Reflection coefficient for vibrtions parallel to surface
$\rho_\perp$	11.4	Reflection coefficient for vibrations normal to surface
σ	1.2	Conductivity
σ	1.3	Area
σ	4.3.5	Scattering cross section of particle
σ_a	2.7	Total extinction coefficient
σ_a	3.2.5	Absorption cross section of particle
σ_s	2.7	Scattering extinction coefficient
σ_s	3.2.5	Scattering cross section of particle
σ_t	2.7	Total extinction coefficient
σ_t	3.2.5	Total cross section of particle
τ	2.6	Normal optical thickness of atmosphere
τ	3.4	Angle of incidence of ray on water drop
τ^A	5.2.2	Optical thickness of atmospheric aerosols
τ^O	5.2.2	Optical thickness of atmospheric ozone
τ^R	4.3.7	Optical thickness of Rayleigh atmosphere

τ_D	3.4	Angle of incidence of Descartes ray on water drop
τ_n	3.2	Angular function in Mie theory
τ'	3.4	Angle of refraction of ray in water drop
ϕ	2.1	Azimuth
ϕ_o	2.1	Azimuth of sun or source
ϕ_x	1.2	Phase angle along x-direction
ϕ_y	1.2	Phase angle along y-direction
χ	1.4	Angle of the plane of polarization
χ_a	1.2	Electric susceptibility
χ_m	1.2	Magnetic susceptibility
ψ	11.2	Angle of rotation of axes of optical device
ψ_n	3.2.1	Ricatti-Bessel function in Mie theory
$\psi,\phi,\chi,\xi,\zeta,$ η,σ,θ	4.4.4	Scattering functions in Chandrasekhar scattering theory
ω	–	Elemental solid angle
ω	11.2.1	Index of refraction for ordinary ray in crystal
ω	1.2	Angular frequency
$\tilde{\omega}$	2.8	Albedo of single scattering

CHAPTER 1

THE NATURE OF POLARIZED LIGHT

1.1 Historical notes

Polarization of light was first discovered by a Danish professor, Erasmus Bartolinus (1625–1698), and described in a publication in 1669 on the optical properties of Iceland spar (a variety of calcite). He found that when a ray of light is refracted by such a crystal, two rays are produced. One, the "ordinary" ray, follows the "ordinary" laws of refraction, but the other, the "extraordinary" ray, follows a different law. In a series of detailed experiments, Bartolinus was able to show the main properties of the two types of rays. In some respects, however, history has not given Bartolinus proper credit for his discovery, undoubtedly because it was overshadowed about 20 years later by the brilliant work of celebrated Dutch physicist Christian Huygens (1629–1695) on double refraction by crystals. Contrary to the popular views of the time, Huygens did not subscribe to the corpuscular theory of light put forth by Isaac Newton (1642–1727), but postulated instead an undulation in an all-pervading elastic medium—the aether. Huygens made many of the same observations as Bartolinus, noting two rays with equal intensity but different properties. Furthermore, if either ray were passed through a second crystal, the emergent intensity was found to be strongly dependent on crystal orientation, in contrast to the situation with the first crystal. The observation of this "wonderful phenomenon" in Huygens' experiment established the existence of an asymmetry around the direction of propagation of certain light rays, a property we now know is due to polarization.

The subject of polarization seemed to be dormant throughout the 18th century, but it awakened, full of vigor, in the early years of the 19th century and has remained in a more or less healthy state ever since. Here we shall concentrate on aspects related to polarization of light in the natural environment, as opposed to the physics of the laboratory, although overtones from the laboratory are unavoidable.

The two main polarizing processes operating in the natural environment were discovered in France in 2 successive years early in the 19th century. In 1809 the astronomer D.F.J. Arago (1786–1853) first observed that the light which reaches the earth from the sunlit sky is partially polarized, and went on to establish the general magnitude and position of the polarization maximum for skylight and to discover the neutral point (point of zero polarization about 20° above the antisolar point) which bears his name. Both of these features will be discussed in detail in later chapters of this book.

In the year 1808 the Institute of France announced a contest for papers "To give a mathematical theory, verified by experiment, of the double refraction which light experiences when transmitted through different crystallized bodies." This was a restatement of the same problem which had occupied the attention of Huygens some 120 years previously. The prize was awarded 2 years later to a Colonel of the French Imperial Corps of Engineers, E.L. Malus (1775–1812). During his experiments on the specified subject, Malus looked through a calcite crystal at the light of the setting sun reflected by the windows of a Paris building, and was surprised to observe the disappearance of images of the windows as the crystal was rotated. A desire to understand this observation led to his discovery of the laws of reflection by which his name is immortalized. In fact, Malus was the first to use the term "polarization" as applied to light. He hypothesized that the corpuscles which made up light were aligned by the process of reflection in a manner similar to the way magnetic bodies are aligned by the poles of a magnet.

Though Malus was still thinking of corpuscles, the wave theory of light had already been reasonably well established through experiments by the English physician Thomas Young (1773–1829). From 1802 to 1804, Young studied diffraction bands and other properties of light, which convinced him of the validity of the wave theory. However, his unfortunate manner of presentation, his youth, and the tenacity with which Newton's statements were relied upon, prevented Young's ideas from being recognized and accepted. A decade later the wave theory was so clearly presented by the youthful French civil engineer Augustin Fresnel (1778–1827) that there was little question of its validity. By this time, of course, the phenomena of polarization discovered by Arago and Malus were well known. An equivalent of the prize awarded earlier to Malus was won by Fresnel in Paris in 1819, a prize which heralded the eventual demise of Newton's corpuscular theory, despite its continued popularity with some physicists.

Fresnel and other scientists accepting the wave theory at the time considered the waves to be longitudinal in nature, as sound waves are. They could not conceive of a fluid, such as the aether was thought to be, which could sustain transverse waves. Interference phenomena might be explained by longitudinal waves, but when Fresnel began experimenting with polarized light he was unable to obtain interference from the differently polarized ordinary and extraordinary rays, regardless of the phase difference he introduced. On the basis of careful experiments he was forced to adopt the concept originally postulated by Thomas Young in 1817, that the waves must be transverse. Only transverse waves could explain the lack of interference between the differently polarized components. Furthermore, polarized rays showed symmetry around two orthogonal planes oriented parallel to the direction of propagation, and there is no way such symmetry could be exhibited by longitudinal waves. This and other historical puzzles of light were brilliantly solved by Fresnel, before his death at the early age of 39.

In the context of polarization, however, the name of Fresnel lives in the Fresnel formulas, which give the relative intensities and states of polarization of the light reflected and refracted at the boundary between two transparent media in terms of the incident light and the relative index of refraction of the media. We shall have occasion to discuss the Fresnel formulas further in the section on the polarizing effects of surface reflection.

At the time of Fresnel's death, the concept of the aether remained a puzzle. The aether had to be an all-pervasive fluid which filled both solids and free space, it had to be perfectly elastic, and have the capability of sustaining transverse waves. Only much later did conclusive evidence show that the phenomena of light could be explained without invoking the necessity for this exotic fluid, but during the early 19th century the aether was a useful rack on which to hang the ideas of wave phenomena.

In a completely different theatre of experiments being pursued at a time of hot debate on the nature of light, the Danish physicist H.C. Oersted (1777–1851) was experimenting with electric currents and magnets. In 1819 he discovered a magnetic field surrounding a wire carrying an electric current, the first significant step in establishment of the concept we know as electromagnetism. A year later A.M. Ampere (1775–1836) of Paris demonstrated a physical interaction between two wires carrying electric currents. Michael Faraday (1791–1867), a self-taught English experimentalist, built on Ampere's findings and discovered the laws of induction on which electric motors and generators are based. This, of course, was before there was any notion of the relationship between electromagnetism and light.

The suspicion that there might be such a relationship led Faraday to an experiment in which polarized light was sent through a glass rod oriented along the lines of a strong magnetic field. It was observed that a rotation was introduced into the plane of polarization by interaction of the light with the magnetic field, a phenomenon we now term the Faraday effect. By this famous experiment of 1845, any doubts as to the existence of a unity between light and electromagnetism were laid to rest.

During this same period, studies of the polarization of light in its own right were continuing to progress. In France, the earlier decades of the century saw the names of M. Guerard and M. Bravais, and of course Arago, in measurements of skylight polarization. At the same time the Scottish physicist Sir David Brewster (1781–1868) began a long and extraordinarily fruitful career in the field of optics. His paper before the Royal Society of 1815 described his experiments on the angle of reflection at which reflected light is completely polarized, now termed the Brewster angle. For historical accuracy, however, it should be pointed out that Malus had recognized the existence of such an angle in the reflection of light from a water surface in his prize-winning paper of 1810. Brewster himself quotes Malus (*Journal of Science,* 1824) as saying, "In order to analyze this phenomenon completely, I have placed vertically the principal section of a crystal; and, after having divided a luminous ray by double refraction, I received two pencils on the surface of water, and at an angle of 52° 45′. The ordinary ray was partly refracted, and partly reflected, as if it had been a pencil of direct light, but the extraordinary ray penetrated entirely the water, and not one of its particles escaped refraction." Though this citation takes nothing from Brewster's contribution in mathematically describing the angle, the quotation by Brewster gives proper credit to its true discoverer.

Shortly before mid-century the interest in skylight was waning until the report to the Academie des Sciences, Paris, in 1840 of the discovery by the French meteorologist Jacques Babinet (1794–1872) of a second neutral point of skylight polarization situated about 25° above the sun, as the Arago point is at a like distance above the antisolar point. The lateness of the discovery of the Babinet point is surprising in view of the previous interest in skylight polarization and the fact that the Babinet point is not difficult to see with the Savart polariscope which was in use at the time. After this second neutral point was found, however, the discovery of a third was not long in coming.

From considerations of symmetry, Sir David Brewster thought that since a neutral point existed above the sun, another should be situated at a similar position below the sun. Being familiar with polarization by reflection, with polarizing effects of scattering unknown at the time, Brewster theorized that skylight polarization was due to reflection and refraction of sunlight by the atmosphere. Thus, he started looking for the third neutral point as soon as he heard about Babinet's discovery. However, at the latitude of Scotland the sun never gets very high in the sky and Brewster was "perplexed beyond measure by the feeble and uncertain indications of the polariscope" as he searched the sky below the sun. On a clear day in February 1842, he had a good view of the subsolar sky from inside a long dark passage, and the elimination of glare from the sun permitted him to see the neutral point very distinctly. Brewster communicated this information to Babinet, who, after several futile attempts, confirmed the existence of the Brewster point in July 1846.

With Brewster's discovery, the three principal neutral points, and the only ones now bearing the names of the discoverers, were known. They have been the subject of many investigations in both the 19th and 20th centuries. In addition to the three, however, some secondary neutral points have been observed from time to time. In fact, Brewster, himself observed a secondary neutral point located below the Arago point, an unusual phenomenon produced by distant sea haze as seen from the Scottish coast. At a later date, Cornu, the inventor of an early polarimeter, observed neutral points on either side of the Arago point. All of the neutral points will be discussed in a later section dealing with the entire subject.

From the standpoint of the theory of polarization, most of the work in the early part of the 19th century was devoted to an explanation of the many phenomena of reflection and refraction of light by surfaces, and with the wave versus particle theories of light. Even after the brilliant work of Fresnel, Young, and others, acceptance of the wave theory was by no means universal. For instance, Brewster (1832) wrote, "the theory of undulations, with all its power and all its beauty, is still burthened with difficulties, and cannot claim our implicit assent."

The experiments of Fresnel, Faraday, and others with polarized light showed the waves to always be transverse, and no longitudinal light waves had ever been detected. But the concept of the aether, while highly artificial in many ways, permitted, and even virtually required, the existence of longitudinal waves because of the elastic-solid nature of the medium. So the paradox persisted. With Faraday's demonstration of the relationship of light waves and magnetic fields, before the middle of the 19th century the state of knowledge had developed to the point where an understanding of the electromagnetic character of light was possible if only the proper scientist came along to perform the synthesis.

Just such a scientist appeared shortly after mid-century in the person of the Scottish theoretical physicist James Clerk Maxwell (1831–1879). Actually, Maxwell's original ideas came from careful studies of Faraday's experimental results in electricity and magnetism, but his mathematical genius permitted him to see the necessary quantitative relationships in terms of equations which describe completely the electromagnetic field, a term coined by Maxwell, previously described as lines of force by Faraday. In a continuum, the two concepts are synonymous, and they eventually led to the field equations which describe in all detail the electromagnetic character of light. The first formal publication of Maxwell's equations was in 1861 and 1862, with modifications in 1864 and the first edition of the great treatise on the subject in 1873. The essential validity of the wave theory, together with proof of the similarity of light and radio waves, came in 1886, a few years after Maxwell's death, when the German physicist Heinrich Hertz (1857–1894) detected the first radio waves, established their electromagnetic character, and showed that they could be reflected, refracted, and polarized in the same way as light waves. Still, the quantum theory, investigations of photons, and similar phenomena of a submicroscopic nature relating all of them to their atomic origins were all well in the future.

We end this brief history of polarization investigations with the work of Maxwell, and introduce additional historical notes at appropriate points in the text. It should be mentioned that Maxwell's equations (section 1.2) and the Poynting energy vector (section 1.3) are introduced here mainly for completeness. The material of these two sections is not necessary for understanding the other parts of the book.

1.2 Maxwell's equations

Maxwell (1893) showed that an electromagnetic wave consists of electric and magnetic vectors, **E** and **H**, respectively, vibrating in a plane normal to the direction of propagation of the wave. The wave results from a motion of an electric charge in a magnetic field. Although the existence of electrons was not known at the time of Maxwell, we use the concept here for descriptive purposes. Consider an electron (unit electrical charge) oscillating at a constant frequency about a nucleus which is at rest. The existence of the charge creates an electric field, its motion produces a magnetic field, and an electromagnetic wave consisting of sinusoidal vibrations of the electric and magnetic vectors is produced in the space surrounding the atom. The strength of the wave (i.e., the amplitudes of the electric and magnetic vectors) is proportional to the strength of the charge (in this case unity), the vibrations of the vectors are in the plane normal to the direction of propagation and the frequency of the wave is the frequency of oscillation of the electron. If the traveling wave happens to encounter another electric charge, this second charge is set into motion and causes its own electromagnetic wave to be set up. This is the phenomenon of light scattering, which is one of the main topics of this book.

In dealing with electromagnetic radiation of wavelengths typical of those of ordinary light, the smallest volumes in a medium which have relevance to the problem have dimensions in the same order of magnitude as the wavelength of light. This means that in solids, liquids, and gases at reasonably high pressures, the number of atoms in an elemental volume is very large and their behavior may be taken as the average over the elemental volume. Thus we may consider the macroscopic form of Maxwell's equations, and neglect the more complicated interactions which occur on a scale the size of individual atoms or molecules.

In a macroscopic system there are two different types of charges which have importance in electromagnetic interactions. The first are those electrons which are elastically bound to their nuclei, and the second are electrons which are free to move about in the medium. We shall designate them as bound and free, and denote the total charge density and current density due to their combination as $\rho=\rho_b+\rho_f$ and $j=j_b+j_f$, respectively. The bound electrons oscillate about their equilibrium positions on excitation by a quasi-monochromatic incident wave. It is possible (see Stone, 1963) to define an electric dipole moment per unit macroscopic volume of the medium as the electric polarization Π, and the charge density due to bound electrons is simply

$$\rho_b = -\text{div } \mathbf{\Pi} \tag{1.1}$$

From this we have the current density as

$$j_b = \partial \mathbf{\Pi}/\partial t \tag{1.2}$$

By analogy, it is possible to define the magnetic dipole moment per unit volume **M**, termed the macroscopic magnetization of the medium. If this is taken into account then Eq. (1.2) is modified to

$$j_b = \partial \mathbf{\Pi}/\partial t + c \text{ curl } \mathbf{M} \tag{1.3}$$

where c is the speed of light in a vacuum. However, in the absence of a strong magnetic field in the medium, the elementary magnetic moments are randomly oriented, thereby making $\mathbf{M}=0$ for the average macroscopic volume. Furthermore, thermal agitation of the molecules of the medium prevents a rapid response of the molecules in adjusting to whatever magnetic field which happens to exist, and thus **M** does not respond at the optical frequencies under discussion here. As a result we are justified in neglecting **M** for practical purposes in media without strong static or slowly changing fields, but for completeness we continue to carry **M** in the formalism.

With this approximation we can write the Maxwell equations for a macroscopic isotropic medium in the form (Stone, 1963)

$$\text{div } \mathbf{E} = -4\pi \text{ div } \mathbf{\Pi} + 4\pi \rho_f \tag{1.4}$$

$$\text{curl } \mathbf{E} = -\frac{1}{c}\frac{\partial \mathbf{B}}{\partial t} \tag{1.5}$$

$$\text{div } \mathbf{B} = 0 \tag{1.6}$$

$$\text{curl } \mathbf{B} = \frac{1}{c}\frac{\partial \mathbf{E}}{\partial t} + \frac{4\pi}{c}\frac{\partial \Pi}{\partial t} + 4\pi \text{ curl } \mathbf{M} + \frac{4\pi}{c} j_f \tag{1.7}$$

where **D** and **H** are given by

$$\mathbf{D} = \mathbf{E} + 4\pi\,\mathbf{\Pi} \tag{1.8}$$

and

$$\mathbf{H} = \mathbf{B} - 4\pi\,\mathbf{M} \tag{1.9}$$

In the usual terminology, **E** is the electric field strength, **H** the magnetic field strength, **D** the dielectric displacement vector, and **B** the magnetic induction.

On substituting these into Eqs. (1.5) and (1.7) we have

$$\text{div}\,\mathbf{D} = 4\pi\rho_f \tag{1.4a}$$

$$\text{curl}\,\mathbf{H} = \frac{1}{c}\frac{\partial \mathbf{D}}{\partial t} + \frac{4\pi}{c}\mathbf{j}_f \tag{1.6a}$$

These last four relations together with Eq. (1.5) and (1.7) constitute the full set of the macroscopic Maxwell equations for a medium. The vectors **E**, **H**, **B**, and **D** are termed electric field strength, magnetic field strength, magnetic induction, and dielectric displacement, respectively.

In order to relate Maxwell's equations to an actual medium, we consider the medium to be excited by a sinusoidally varying monochromatic field, and investigate the oscillation of the various vectors in response to the field. We define an orthogonal coordinate system, as shown in Fig. 1.1, with the direction of propagation and vectors **E** and **H** along unit vectors **n**, $\mathbf{e}_x$, and $\mathbf{e}_y$, respectively. If we denote the complex vector amplitudes of the oscillations by a subscript *a*, then the parameters of the medium relate the various amplitudes through the relations

$$\mathbf{\Pi}_a = \chi_a\,\mathbf{E}_a \tag{1.10}$$

$$\mathbf{M}_a = \chi_m \mathbf{H}_a \tag{1.11}$$

$$\mathbf{j}_a = \sigma\,\mathbf{E}_a \tag{1.12}$$

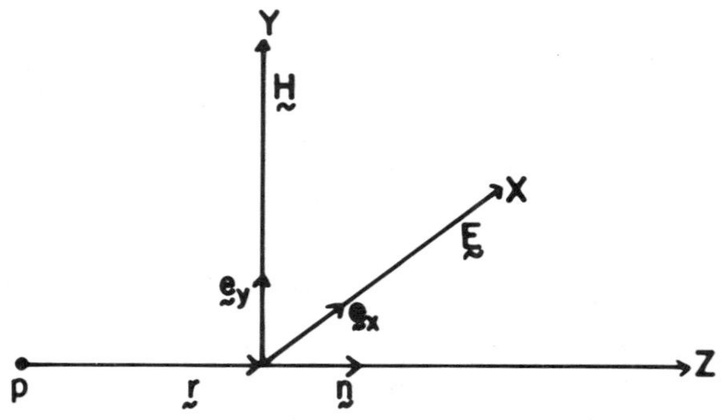

Figure 1.1 Coordinate system showing electromagnetic field vectors.

where χ_a, χ_m and σ are, respectively, the electric susceptibility, magnetic susceptibility, and conductivity of the medium (Stone, 1963). The first two of these parameters are related to the more commonly used electric permittivity ϵ and magnetic permeability μ of the medium by

$$\epsilon = 1 + 4\pi \chi_a \tag{1.13}$$

$$\mu = 1 + 4\pi \chi_m \tag{1.14}$$

In a vacuum $\epsilon = \epsilon_o = 8.85 \times 10^{-12}$ farad/meter and $\mu = \mu_o = 4\pi \times 10^{-7}$ henry/meter. The quantity ϵ/ϵ_o is the dielectric constant of the medium.

For the orthogonal coordinate system of Fig. 1.1, we have

$$\mathbf{E} \cdot \mathbf{n} = 0; \ \mathbf{H} \cdot \mathbf{n} = 0; \ \mathbf{H} = \mathbf{n} \cdot \mathbf{E} \tag{1.15}$$

Since the magnetic field is uniquely related to **E**, we need to consider only the electric vector for a complete specification of this radiation field.

If the source of electromagnetic radiation is a point or a very small volume, the radiation will leave the source as a spherical wave of a vibrating electric vector. We assume the source to be at p in Fig. 1.1. At a point in space at some vector distance **r** from the source, the electric vector is

$$\mathbf{E}(\mathbf{r}, \mathbf{n}, t) = \mathbf{A}(\mathbf{r}, \mathbf{n}) \ \frac{e^{\,i(kr-\omega t)}}{kr} \tag{1.16}$$

Here **A(r, n)** is a complex vector quantity, which will be discussed in detail below. The wave number k is related to the angular frequency ω and wavelength λ by the relations

$$k = 2\pi/\lambda = \omega/\mathrm{v} \tag{1.17}$$

where v is the speed of light in the particular medium. If the index of refraction of the medium is m, then

$$\mathrm{v} = c/m$$

where c is the speed of light in vacuo. Finally, the frequency of the wave ν in the medium is

$$\nu = \mathrm{v}/\lambda$$

where λ is the wavelength in the medium. For a vacuum, $\mathrm{v} = c/\lambda$

If the distance r of Eq. (1.16) is very large, a section of the original spherical wave is well approximated by an electric vector oscillating in a plane, which is oriented normal to the direction of propagation. This is the situation to be considered in this book. Eq. (1.16) then takes the form

$$\mathbf{E}(\mathbf{r},\ \mathbf{n},\ t) = \mathbf{A}(\mathbf{r},\ \mathbf{n})\ e^{i(kz-\omega t)} \tag{1.18}$$

where z is the distance from some arbitrary point along the direction of propagation.

The complex amplitude has components only in the $x-y$ plane, in which case we can write

$$\mathbf{A}(\mathbf{r},\mathbf{n}) = A_x\,\mathbf{e}_x + A_y\,\mathbf{e}_y \tag{1.19}$$

The components A_x and A_y are time-dependent complex functions which vary with the basic frequency ω. By the usual representation of a wave phenomenon, we can express them for a given time t in the form

$$A_x = a_x\, e^{i\phi x}\, e^{-ikz+i\omega t} \tag{1.20}$$

$$A_y = a_y\, e^{i\phi y}\, e^{-ikz+i\omega t} \tag{1.21}$$

where a_x, a_y are the real amplitudes and ϕ_x, ϕ_y are the corresponding phases of the oscillations.

1.3 Energy of electromagnetic waves

We know that in general, wave phenomena are associated with a flow of energy, and if the frequency is low enough, such as in long radio waves, we can measure the energy of individual waves. At optical frequencies, however, our sensing devices—the eye, photosensitive elements, etc.—are too crude to respond to the individual vibrations, but integrate the wave energy over a time interval corresponding to a large number of vibrations. Then we measure an intensity or flux of energy which is the average for many waves, which tells us little about the individual waves themselves. Thus we need a method of relating the energy of a wave to the electric and magnetic fields characterizing the wave.

Such a method was developed by the English physicist J.H. Poynting (1852–1914), and is expressed by the Poynting vector **S** written as

$$\mathbf{S} = \frac{c}{4\pi}\, Re[\mathbf{E} \cdot \mathbf{H}^*] \tag{1.22}$$

where *Re* indicates the real part of the quantity in brackets and $\mathbf{H}^*$ is the complex conjugate of **H**. The energy of the electromagnetic wave has dimensions of a flux density, which in the International System of units is watts per square meter (Wm^{-2}). From Eq. (1.15) and the knowledge that **n** is a real unit vector we have

$$\mathbf{H}^* = \mathbf{n} \cdot \mathbf{E}^* \tag{1.23}$$

in which case Eq. (1.21) reduces to

$$\mathbf{S} = \frac{c}{4\pi}\, Re\,[\mathbf{E} \cdot (\mathbf{n} \cdot \mathbf{E})] \tag{1.24}$$

or since **n** and **E** are orthogonal vectors, the vector calculus yields

$$\mathbf{S} = \frac{c}{4\pi} Re[(\mathbf{E} \cdot \mathbf{E}^*)\mathbf{n}] \tag{1.25}$$

This shows that the flow of energy is along the direction of propagation of the wave (i.e. in the direction of **n**), but its interpretation in terms of geometry requires some further consideration.

We reconstruct Fig. 1.1 to define the elemental solid angle $d\Omega$ and elemental area $d\sigma$ in Fig. 1.2. For convenience we assume $d\sigma$ is oriented normal to the direction of propagation from p. If, however, the angle of incidence is some angle $\gamma \neq 90°$, then $d\sigma$ must be replaced by its projected area $d\sigma \cos\gamma$ in the relevant expressions.

The rate at which energy Φ passes through $d\sigma$ from a point source at p is

$$\frac{d\Phi}{dt} = |\mathbf{S}|\, d\sigma \tag{1.26}$$

where the vertical bars on **S** indicate absolute magnitude. On substituting the complex amplitude of the electric vector from Eq. (1.16) at the distance r from the point into Eq. (1.25) we obtain

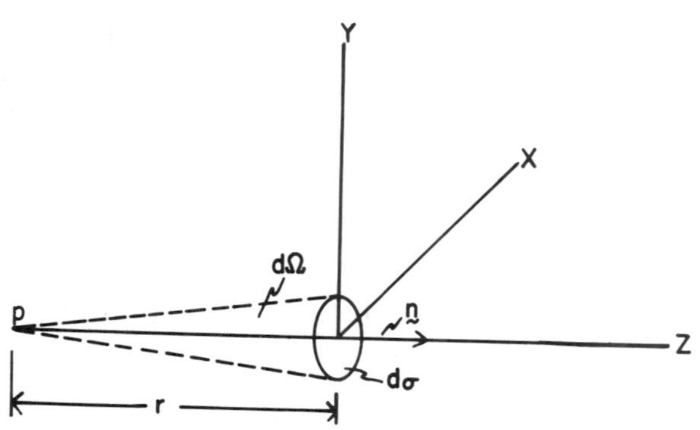

Figure 1.2 Geometry for determining the intensity of radiation from a point source.

$$\mathbf{S} = \frac{c}{4\pi} \frac{\mathbf{A} \cdot \mathbf{A}^*}{k^2 r^2} \mathbf{n} \tag{1.27}$$

and on introducing the elemental solid angle $d\Omega$ from its definition

$$d\Omega = \frac{d\sigma}{r^2}$$

into Eq. (1.26) we have scalar intensity of radiation, per unit frequency interval and unit area as

$$I = \frac{d\Phi}{dt d\Omega} = \frac{c}{4\pi} \frac{\mathbf{A} \cdot \mathbf{A}^*}{k^2} d\Omega \tag{1.28}$$

Neglecting the constant factor $c/4\pi k^2$ for the moment, we obtain the proportionality relation

$$\mathbf{I} \sim \mathbf{A} \cdot \mathbf{A}^* \tag{1.29}$$

Since we can measure the intensity, we can get some information about the complex amplitude of the waves.

Expansion of relation (1.29) into components in the x and y direction gives

$$\mathbf{I} \sim [A_x \, \mathbf{e}_x + A_y \, \mathbf{e}_y] \cdot [A_x^* \, \mathbf{e}_x + A_y^* \, \mathbf{e}_y] \tag{1.30}$$

or, since $\mathbf{e}_x$ and $\mathbf{e}_y$ are orthogonal,

$$\mathbf{I} \sim A_x A_x^* + A_y A_y^* \tag{1.31}$$

But for these complex quantities we have

$$\begin{aligned} A_x &= a_x \, e^{i\varphi_x} \qquad & A_x^* &= a_x \, e^{-i\varphi_x} \\ A_y &= a_y \, e^{i\varphi_y} \qquad & A_y^* &= a_y \, e^{-i\varphi_y} \end{aligned} \tag{1.32}$$

where a_x and a_y are the real amplitudes. On substitution of these into (1.31) we see that

$$I \sim a_x^2 \, e^{i(\varphi_x - \varphi_x)} + a_y^2 \, e^{i(\varphi_y - \varphi_y)} = a_x^2 + a_y^2 \qquad (1.33)$$

This important relationship shows that the intensity, and thus the amount of energy in a given stream of radiation, is proportional to the sum of the squares of the amplitudes of the orthogonal components. This fact will find many applications in the discussions which follow. It should be realized, however, that the actual waves in the radiation stream of principal interest here may undergo incessant variations of amplitude which are too rapid to detect with present instrumentation, and that the effective amplitudes a_x and a_y are only the time average over the real variations which may actually exist. It is interesting, however, that the phases of the components, or at least the phase differences, do not undergo such variations, as is clear from the polarization characteristics.

1.4 Types of polarization

If a stream of light waves shows complete symmetry around the direction of propagation, the stream is said to be unpolarized. Other terms such as natural or neutral are sometimes used for this case. No way has ever been found to detect a change of intensity or any other property with direction in the transverse plane for an unpolarized light beam. For instance, the energy transmitted by an analyzer at one orientation is the same as that transmitted at any orientation. This means that the electric vector is vibrating at all possible angles in the transverse plane in such a manner that the time average of the vibrations is independent of direction. Observations show that radiation direct from the sun is essentially unpolarized, as is thermally emitted radiation from diffuse surfaces. Otherwise, completely unpolarized light is relatively rare in nature and difficult to produce in the laboratory. Most light beams exhibit partial polarization of some type.

Polarization is produced by some type of phase relationship and/or magnitude difference in the orthogonal components of the vibration of the electric vector. The most general type of polarization is that in which the end point of the electric vector describes an ellipse of some type in the transverse plane, which is equivalent to a spiral around the direction of propagation as the wave travels through space. An intuitive idea of the behavior of the electric vector for elliptic polarization can be obtained from Fig. 1.3. It should be realized, of course, that the vector is only a graphical representation of the electric field strength, which is actually the quantity that changes. However, a diagram such as Fig. 1.3 is a useful visualization aid. The features necessary to characterize the ellipse are the ratio between the minor and major axes of the ellipse, which translates into the ellipticity (through the relation $\tan\beta = b/a$), and the angle χ which determines the orientation of the ellipse in the x–y plane. The ratio b/a is limited to $0 \leq b/a \leq 1$, and χ is unconstrained. The angle of the plane of polarization is, by convention, defined as the angle between the maximum of the electric vector (i.e., the a axis of the ellipse) and the reference axis.

Elliptic polarization results from an arbitrary phase difference $\varphi_x - \varphi_y$ between the orthogonal components of the electric vector. Plane and circular polarization are special cases of elliptical polarization. The ellipse degenerates to a plane under the following three circumstances:

1. $\varphi_x = \pm\varphi_y$: the plane of polarization (maximum of the electric vector) in this case makes equal angles with the two axes.

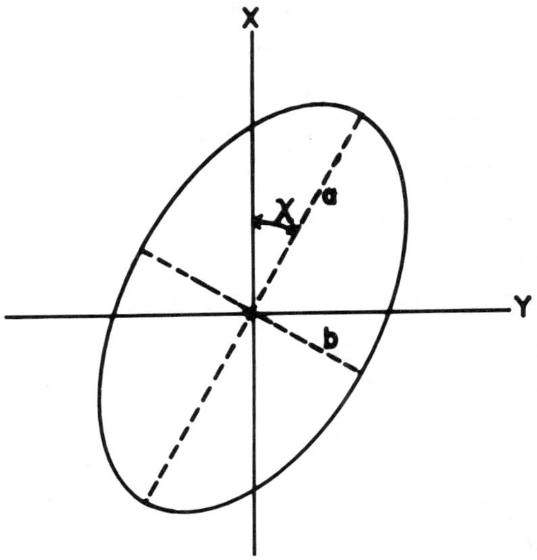

Figure 1.3 Simplified representation of the vibration of the electric vector for the case of elliptic polarization.

2. $\varphi_x - \varphi_y$ is an integral multiple of π

3. Either of the amplitudes a_x or a_y is zero. The plane of polarization in this case is parallel to one of the axes. The ellipse becomes a circle, resulting in circular polarization, if $\varphi_x - \varphi_y$ is an odd multiple of $\pi/2$ and $A_x = A_y$. If the former of these criteria is met but the latter is not, the light is elliptically polarized. Graphical representations of plane and circularly polarized light are shown in Figures 1.4a and 1.4b. Depending on the sign of the phase difference $\varphi_x - \varphi_y$, the sense of rotation for elliptic and circular polarization may be that of either a right-handed or left-handed helix. This concept introduces no ambiguity with respect to the observer's viewpoint, as a right-handed helix, such as that in the threads of a machine screw, appears right handed regardless of the direction from which it is viewed.

Complete polarization of any type rarely occurs in nature, although it is not impossible to closely approximate. For instance, light reflected at the Brewster

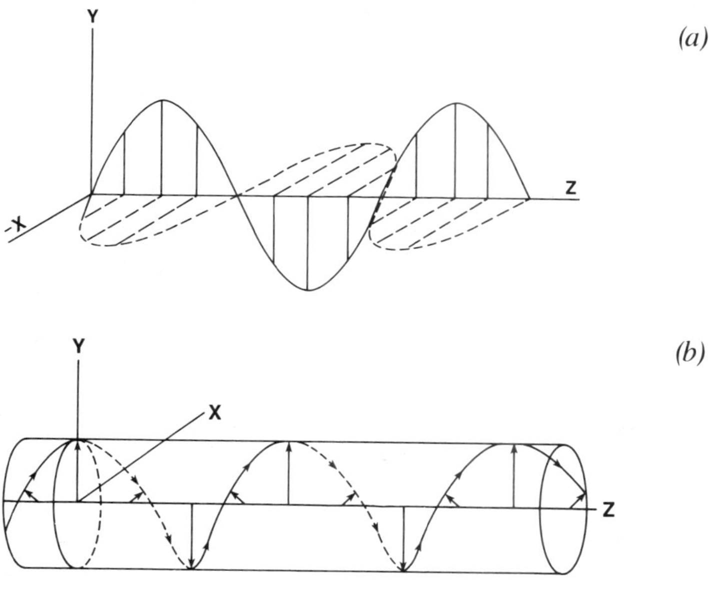

Figure 1.4 Schematic representation of (a) plane polarized wave vibrating in the y–z plane (solid curve) and another vibrating in the x–z plane (dashed curve) (b) right-handed circular polarization.

angle (approximately 56°) from glass or other dielectric material exhibits complete polarization, but the usual situation is that of partial polarization. In other words, the radiation streams generally consist of a combination of both polarized and unpolarized waves. The fraction of the total intensity which is polarized is termed the degree of polarization. If I is the total intensity and I_p that which is represented by completely polarized light, then the degree of polarization is simply $P = I_p / I$.

To summarize the above, the most general state of polarization is partial elliptic polarization. Linear and circular polarization are special cases of elliptic polarization. A light beam may be characterized completely by its total intensity I, degee of polarization P, ellipticity β, and angle χ of the plane of polarization with respect to the reference axis.

It is not immediately obvious how such diverse quantities as an intensity, a ratio, a pure number, and an angle, all necessary to characterize polarized light, can best be represented for analytical work. Various methods have been devised, but the one most commonly used is that of the Stokes parameters.

1.5 The Stokes parameters

A very powerful and convenient representation of the four parameters necessary to characterize polarized light was developed in 1852 by the English physicist Şir George Stokes (1819–1903). The method developed by Stokes was forgotten in the literature for nearly a century until it was resurrected and used in the treatise of Chandrasekhar (1950). The Stokes polarization parameters, as they are now termed, are particularly well suited to analytical work because they all have the dimensions of intensity and are therefore additive. This means that the Stokes parameters of a stream of light which is made by adding any number of streams with different states of polarization is characterized by the simple sum of the Stokes parameters of the individual streams. Thus they are completely general and by their use we can describe a radiation field in all details, regardless of its state of polarization.

Returning to Eq. (1.19), we can define the Stokes parameters in terms of the complex amplitudes of the wave as

$$I = A_xA_x^* + A_yA_y^* = A_x^2 + A_y^2 = I_x + I_y$$

$$Q = A_xA_x^* - A_yA_y^* = A_x^2 - A_y^2 = I_x - I_y \tag{1.34}$$

$$U = A_xA_y^* + A_yA_x^* = 2A_xA_y \cos(\varphi_x - \varphi_y)$$

$$V = i(A_xA_y^* - A_yA_x^*) = 2A_xA_y \sin(\varphi_x - \varphi_y)$$

These are the the Stokes parameters for a general radiation field. They can be combined with the parameters of the ellipse described by the electric vector, shown previously in Fig. 1.3, by use of the relations

$$A_x^2 = A^2(\cos^2\beta \cos^2\chi + \sin^2\beta \sin^2\chi)$$

$$A_y^2 = A^2(\cos^2\beta \sin^2\chi + \sin^2\beta \cos^2\chi) \tag{1.35}$$

$$\tan(\varphi_x - \varphi_y) = \tan 2\beta / \sin 2\chi$$

to give alternate forms of parameters

$$Q = I \cos 2\beta \cos 2\chi$$

$$U = I \cos 2\beta \sin 2\chi = Q \tan 2\chi \tag{1.36}$$

$$V = Q \tan 2\beta \sec 2\chi = I \sin 2\beta$$

For convenience of reference, we rewrite the parameters in two alternate forms referred to axes x and y, as they have been used by various authors in the past and as they will be used in the remainder of this book. Of course the axes will be shifted as necessary, but the form of the relations will remain the same.

$$
\begin{aligned}
I &= I_x + I_y \\
Q &= I_x - I_y \\
U &= Q \tan 2\chi \\
V &= I \sin 2\beta
\end{aligned}
\tag{1.37}
$$

$$
\begin{aligned}
&I_x \\
&I_y \\
U &= (I_x - I_y) \tan 2\chi \\
V &= (I_x + I_y) \sin 2\beta
\end{aligned}
\tag{1.38}
$$

It is instructive to see the behavior of the various parameters for different polarization conditions of a light stream. It should be emphasized that the values are the averages over many vibrations of the electric vector.

1.5.1 Case of unpolarized light

Unpolarized light is characterized by symmetry of intensity around the direction of propagation and random phases of the electric vector. This means that $I_x = I_y$ in the average, and $\varphi_x - \varphi_y = \Delta\varphi$ undergoes incessant variations. Since in the average of many vibrations, $\Delta\varphi$ will be negative as frequently as positive, we have $\cos\Delta\varphi = \sin\Delta\varphi = 0$. On introducing these relations into Eq. (1.34), the set of parameters for unpolarized light is seen to be $(I,0,0,0)$.

1.5.2 Case of complete linear polarization

Vibration is in one plane, the angle of which is arbitrary. For the vibration to remain in one plane the orthogonal components must vibrate with the same phase and frequency, reaching their maxima, minima, and zero points at the same time and making $\Delta\varphi = 0$. This means that $V = 0$ but $U \neq 0$. In general, the orthogonal scalar amplitudes a_x and a_y, which are responsible for the angle χ of the plane of polarization, are of different magnitudes, making $Q \neq 0$. In the singular case of $\Delta\phi = n\pi/4$, n being an odd integer, $a_x = a_y$ and $Q = 0$. A second singular case occurs for linear polarization if the plane of polarization is parallel to one of the axes. For instance, if $\chi=0$ (plane in x direction), the expression for U in Eq. (1.37) gives $U=0$ while $Q\neq 0$, and similarly for $\chi=\pi/2$. Thus linear polarization may have parameter sets $(I,Q,U,0)$, $(I,0,U,0)$ or $(I,Q,0,0)$, the last two of which are for special values of χ.

1.5.3 Case of complete elliptic polarization

If there is a phase difference between the orthogonal components (i.e., $\Delta\varphi\neq 0$), then in Eq. (1.34), $V\neq 0$ and some form of elliptic polarization results. As seen in the previous section, if $\Delta\varphi=0$ then $\beta=0$ and the ellipse degenerates to a plane, giving linear polarization. In general, however, for elliptic polarization $\Delta\varphi$ has a constant non-zero value, and $\beta\neq 0$. The parameter set for complete elliptic polarization is in general the full set (I,Q,U,V). Only for the singular case of $\chi=\mu\pi$, m being an integer, does $U=0$, as can be seen from Eq. (1.36). In that case, the parameters are (I,Q,O,V). If the sign of $\Delta\varphi$ is positive, the rotation of the end point of the electric vector is to the right, and the reverse if $\Delta\varphi$ is negative. A second particular case of elliptic polarization occurs when the ellipse becomes a circle, as discussed in the next section.

1.5.4 Case of circular polarization

If $\Delta\varphi$ is an odd multiple of $\pi/2$ and $a_x=a_y$, the elliptic polarization has the singular form of circular polarization. This is another case of symmetry around the direction of propagation, as was depicted in Fig. 1.4b. For circular polarization $\tan\beta=1$, and the criterion for direction of rotation is the same as for the general case (+: right; −: left). The Stokes parameters for circular polarization are (I,O,U,V), the zero value of Q resulting from $a_x=a_y$.

1.5.5 Case of partial (elliptic) polarization

As mentioned previously, the most general type of polarization is that of partial elliptic polarization, partial linear and partial circular polarization being singular cases of the general type. The concept of partial polarization is made clearest by considering the addition of two beams of light, of which one is unpolarized and the other is completely polarized. The additive character of the individual Stokes parameters for any number of individual streams is particularly convenient for representing this general case. The streams must be mutually incoherent, however, which means that there can be no permanent phase relationships between the vibrations of the two. For two such independent streams, if subscripts 1 and 2 represent the unpolarized and completely polarized streams, respectively, then the Stokes parameters for the composite stream are

$$I = I_1 + I_2;\; Q = Q_1 + Q_2;\; U = U_1 + U_2;\; V = V_1 + V_2 \tag{1.39}$$

The set of parameters of the composite stream are $(I_1 + I_2, Q_2, U_2, V_2)$, and the degree of polarization is

$$P = \frac{I_2}{I_1 + I_2} \tag{1.40}$$

Eq. (1.39) is a special application of the general principle for the addition of n independent streams of light of arbitrary states of polarization written as

$$I = \sum_{i=1}^{n} I_i;\; Q = \sum_{i=1}^{n} Q_i;\; U = \sum_{i=1}^{n} U_i;\; V = \sum_{i=1}^{n} V_i \tag{1.41}$$

While a partially polarized beam is conveniently achieved in the laboratory by adding individual light streams, various processes in the natural environment as well as in the laboratory produce partial polarization in a single stream of light such as that from the sun. These processes will be discussed fully in later sections. In the context of adding light streams, it should be mentioned that the addition of two streams of equal intensity and complete but opposite elliptic polarization (i.e., the ratio of the major to minor axes is the same, the major axes are normal to each other, and the sense of rotation of one stream is the reverse of that of the other) results in a composite which has all the properties of a completely unpolarized stream. This apparent anomaly is discussed by Chandrasekhar (1950).

1.6 Relationships among the Stokes parameters

By performing some algebraic manipulation from Eq. (1.34), it is seen that for completely elliptically polarized light we can write

$$I^2 = Q^2 + U^2 + V^2 \tag{1.42}$$

But it has already been established that for unpolarized light

$$Q = U = V = 0 \tag{1.43}$$

Thus for partially elliptically polarized light

$$I^2 > Q^2 + U^2 + V^2 \tag{1.44}$$

and for the most general situation

$$I^2 \geq Q^2 + U^2 + V^2 \tag{1.45}$$

Furthermore, from the definition of the degree of polarization, we have

$$P = \frac{\sqrt{Q^2 + U^2 + V^2}}{I} \tag{1.46}$$

Obviously, for complete polarization $P=1$ and for unpolarized light $P=0$, but in general light in the natural environment is characterized by $0<P<1$.

For convenience, we summarize the relations among the parameters as follows:

$$\begin{aligned}
I &= I_x + I_y \\
Q &= I_x - I_y \\
U &= Q \tan 2\chi \\
V &= I \sin 2\beta \\
I_x &= \tfrac{1}{2}(I + Q) \\
I_y &= \tfrac{1}{2}(I - Q) \\
I^2 &\geq Q^2 + U^2 + V^2 \\
P &= \frac{\sqrt{Q^2 + U^2 + V^2}}{I}
\end{aligned} \tag{1.47}$$

It may be noted that alternate and often more useful forms for the Stokes parameters are (I_x, I_y, U, V).

1.7 Transformation of Stokes parameters for a rotation of axes

For certain problems in radiative transfer in the atmosphere, it is convenient to utilize an orthogonal pair of axes which are rotated from the fixed x and y pair we have used so far. From Eq. (1.35) and the definition of intensity, it is clear that I is invariant to such a rotation and V, being a function of only I and the ratio of the two axes of the ellipse, is similarly invariant to an axis rotation. Q and U, however, change with an axis rotation. We consider a rotation of the axes by angle γ in the clockwise direction from the original pair and ask what the Stokes parameters referred to the new axes will be. We designate the new pair by subscripts e and r as shown in Fig. 1.5. The plane of polarization of the ellipse makes angle $(\chi - \gamma)$ with the e axis, in which case the parameters Q' and U' for the new axes are, in analogy with Eq. (1.36), expressed as

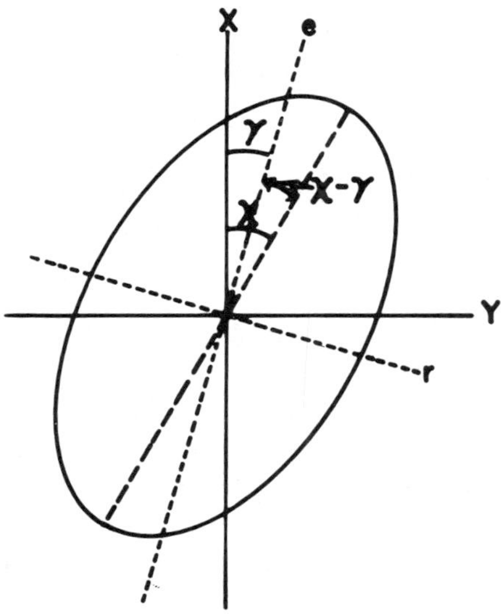

Figure 1.5 Diagram showing a rotation of axes from (x,y) through angle γ to (e,r).

$$\begin{aligned} Q' &= I \cos 2\beta \cos 2(\chi - \gamma) \\ U' &= I \cos 2\beta \sin 2(\chi - \gamma) \end{aligned} \tag{1.48}$$

On expanding these relations by trigonometric identities and rearranging, we see that for angle γ

$$\begin{aligned} Q' &= Q \cos 2\gamma + U \sin 2\gamma \\ U' &= -Q \sin 2\gamma + U \cos 2\gamma \end{aligned} \tag{1.49}$$

Following Chandrasekhar (1950), we adopt the component form of the Stokes parameters (I_x, I_y, U, V) shown in Eq. (1.38), and similarly for the parameters (I_e, I_r, U', V') applicable to the rotated axes. Since I and V are invariant to such a rotation we have

$$\begin{aligned} I_e + I_r &= I_x + I_y \\ V &= V' \end{aligned} \tag{1.50}$$

With these and Eq. (1.49), we find, after some algebraic manipulation that

$$\begin{aligned} I_e &= I_x \cos^2\gamma + I_y \sin^2\gamma + \tfrac{1}{2} U \sin 2\gamma \\ I_r &= I_x \sin^2\gamma + I_y \cos^2\gamma - \tfrac{1}{2} U \sin 2\gamma \\ U' &= -(I_x - I_y) \sin 2\gamma + U \cos 2\gamma \end{aligned} \tag{1.51}$$

1.8 Vector representation of the Stokes parameters

We can consider the Stokes parameters to constitute a generalized vector intensity **I** with four components, and thus use the convenience and power of vector algebra for analysis of the radiation stream in any arbitrary state of polarization. Then the vector intensity **I** assumes the form of either of the column matrices

$$\mathbf{I} = \begin{vmatrix} I \\ Q \\ U \\ V \end{vmatrix} \qquad (1.52) \qquad\qquad \mathbf{I} = \begin{vmatrix} I_x \\ I_y \\ U \\ V \end{vmatrix} \qquad (1.53)$$

1.9 Axis rotation matrix

With the vector notation it is a simple matter to represent the vector intensity $\mathbf{I}_{e,r}$ referred to the (e,r) axes by the linear transformation $\mathbf{L}(\gamma)$ operating on the intensity $\mathbf{I}_{x,y}$ referred to the x,y axes, where $\mathbf{L}(\gamma)$ is obtained directly from Eq. (1.50) and written as (Chandrasekhar, 1950).

$$\mathbf{L}(\gamma) = \begin{vmatrix} \cos^2\gamma & \sin^2\gamma & \tfrac{1}{2}\sin 2\gamma & 0 \\ \sin^2\gamma & \cos^2\gamma & -\tfrac{1}{2}\sin 2\gamma & 0 \\ -\sin 2\gamma & \sin 2\gamma & \cos 2\gamma & 0 \\ 0 & 0 & 0 & 1 \end{vmatrix} \qquad (1.54)$$

Thus

$$\mathbf{I}_{e,r} = \mathbf{L}(\gamma)\, \mathbf{I}_{x,y} \qquad (1.55)$$

It should be emphasized that the rotation represented by operation $\mathbf{L}(\gamma)$ is in the clockwise direction from the (x,y) axes to the (e,r) axes.

The application of the above relations to problems of polarization of light in the environment will be an integral part of the discussions in the remaining chapters of this book.

References

Brewster, David (1832): Report on the recent progress of optics, British Assoc. Report, 1831–32, 308–322.

Chandrasekhar, S. (1950): *Radiative Transfer,* Clarendon Press, Oxford.

Maxwell, J.C. (1893): *Treatise on Electricity and Magnetism,* see *Electricity and Magnetism,* 3rd ed., Oxford Univ. Press, New York (1904).

Stokes, G.G. (1852): On the composition and resolution of streams of polarized light from different sources, *Trans. Cambridge Phil. Soc., 9,* 399.

Stone, J.M. (1963): *Radiation and Optics: An Introduction to the Classical Theory,* McGraw-Hill, New York.

CHAPTER 2

BASIC PROBLEMS OF RADIATIVE TRANSFER IN THE ATMOSPHERE

2.1 General Problem

The general problems of radiative transfer in a medium that absorbs, emits, and scatters radiation are extremely complex and no general solution has been obtained. Every approach developed so far has involved simplifying assumptions of some type. For present purposes we shall restrict the discussions mainly to the case of solar radiation in the atmosphere-surface system of the earth, but of course the basic principles are applicable to other planets and to media in general. In addition, the emphasis on the solar radiation regime permits us to effectively neglect the whole problem of long-wave or terrestrial-type radiation due to emission from material at temperatures typical of those in the natural environment of the earth. Furthermore, the energy of solar radiation is so dominant over other sources of visible or near-visible radiation that for practical purposes we can neglect sources such as airglow, starlight, and aurorae, although we shall have occasion to mention such sources from time to time. Emphasis will be on the field of polarization in all its various aspects, but the intensity will be included where appropriate.

Although we confine our attention to our own planet, there are several aspects that are applicable to other planets, as will be evident in due course. The atmosphere may be considered as an homogeneous or inhomogeneous medium, and as spherical or plane parallel (no variation in the horizontal direction) in geometry. The problems may involve efforts to determine the radiative fields resulting from the interaction of sunlight with the medium, or an inversion of the problem to deduce the properties of the medium from measurements of the radiation. Since the atmosphere is illuminated from the bottom by light reflected from the underlying surfaces, be they soils, sands, water, snow, vegetation, etc., the reflection properties of natural surfaces are an important part of the overall radiative problem. In fact, an attractive possibility in remote sensing is a deduction of surface properties from measurements of the intensity and polarization of their reflected radiation.

The features of the general problem are shown in a schematic fashion in Fig. 2.1. Radiation of flux $\mathbf{F}\,(\infty, -\mu', \phi')$ from the sun is incident at the top of the atmosphere from a direction specified by $(-\mu', \theta')$ where $-\mu' = -\cos^{-1}\theta'$, θ' being zenith angle and the negative sign indicating downward-directed radiation, and ϕ is azimuth angle with respect to an arbitrary reference. A certain portion of the original beam is attenuated by absorption and scattering, but the remainder $\mathbf{F}(0; -\mu', \phi')$, still from the original direction, reaches the surface. That which is absorbed goes into heating the atmosphere, and that which is scattered, represented by intensity $\mathbf{I}(z;\mu,\phi)$, may undergo various fates before being absorbed or leaving the system. The intensity $\mathbf{I}\,(\infty, +\mu,\ \phi)$ is returned to space, while $\mathbf{I}(0; -\mu, \phi)$ is scattered down to the ground. Multiple scattering, represented in the diagram by $I(z_1;\mu_1,\phi_1)$ and $I(z_2;\mu_2,\phi_2)$, occurs inside the atmosphere and $I(0; +\mu, \phi)$ is radiation reflected from the ground surface. Obviously, any of these quantities may undergo further absorption and scattering as they traverse the medium.

As indicated in the diagram, each of these radiation streams may be represented as a Stokes vector of the forms noted in the previous chapter. For

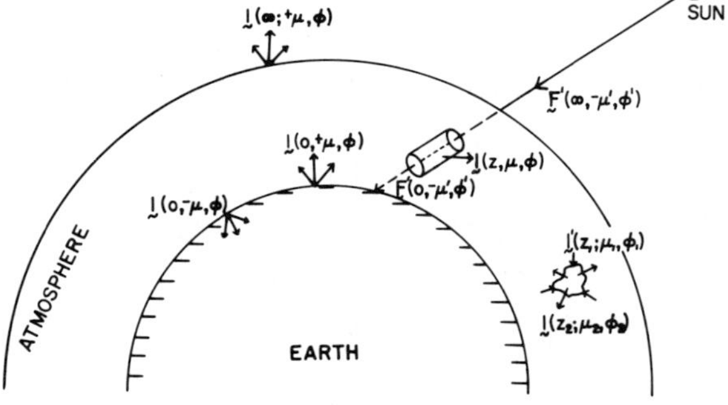

Chapter 2.1 Schematic diagram of radiative parameters in the earth-atmosphere system. The primed symbols represent incident radiation and those without primes are scattered or reflected radiation. A positive sign of μ is for radiation directed into the upward hemisphere, and the negative sign indicates the radiation directed downward.

instance, we can represent the unpolarized incident solar radiation by the equivalent forms

$$\mathbf{F}' = \begin{vmatrix} F \\ 0 \\ 0 \\ 0 \end{vmatrix} \qquad \mathbf{F}' = \begin{vmatrix} \tfrac{1}{2}F \\ \tfrac{1}{2}F \\ 0 \\ 0 \end{vmatrix} \tag{2.1}$$

The scattering process generally introduces polarization into the scattered light, in which case we have the generalized intensity vectors

$$I = \begin{vmatrix} I \\ Q \\ U \\ V \end{vmatrix} \qquad I = \begin{vmatrix} I_e \\ I_r \\ U \\ V \end{vmatrix} \tag{2.2}$$

where the orthogonal axes are in the e and r directions.

2.2 Coordinate systems and notation

From the variety of coordinate systems used by various authors in the past, we shall select two for use here. Both are polar coordinates, the only difference between them being the orientation of the orthogonal axes in the polar coordinate system. They are shown schematically in Figs. 2.2 and 2.3.

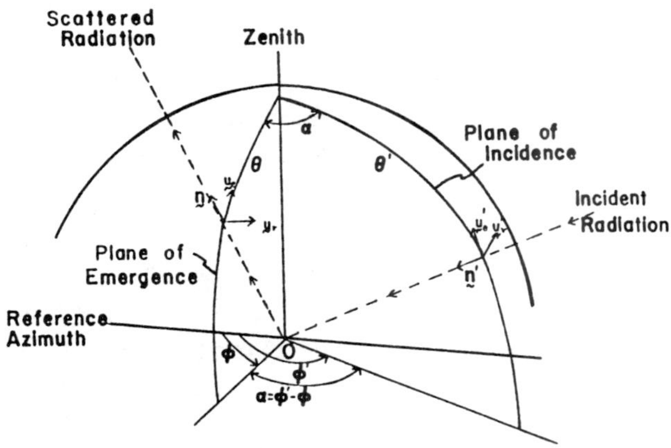

Figure 2.2 Geometry of the radiative transfer problem for the case in which the orghogonal axes (e,r) and (e',r') are referred to the meridonal planes through the zenith and the directions of propagation of the radiation.

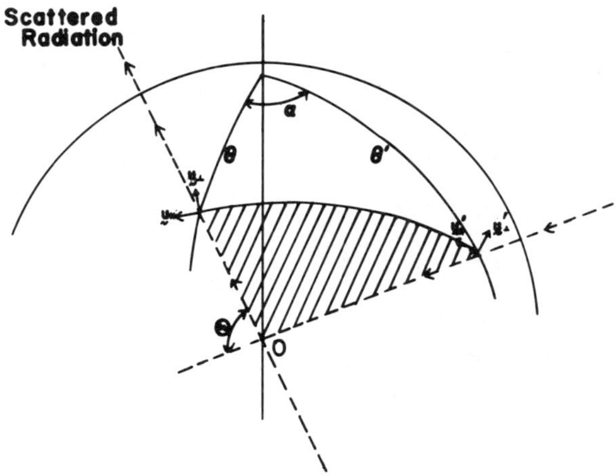

Figure 2.3 Same as the previous diagram except here the orthogonal axes (‖, ⊥) and (‖ ′, ⊥ ′) are referred to the scattering plane (the plane shown shaded defined by the directions of propagation of incoming and outgoing radiation.) Angle Θ is termed the scattering angle.

In Fig. 2.2 the scattering or reflecting volume is at point 0, and the reference planes are meridional planes defined by some azimuth ϕ or ϕ' and the zenith direction. (Primed symbols are for incident radiation and unprimed for scattered or reflected radiation.) Angles from the zenith are θ or θ', and $\alpha=\phi'-\phi$. Radiation proceeds in the direction of unit vector $\mathbf{n}$ or $\mathbf{n}'$. The unit vector pairs ($\mathbf{u}_e,\mathbf{u}_r$) or ($\mathbf{u}_e',\mathbf{u}_r'$) denote directions along the orthogonal (e,r) or (e',r') axes, the e and e' axes being parallel to the reference planes and the r and r' axes being normal to those planes. As before, we shall define $\pm\mu=\pm\cos\theta$ and $\pm\mu'=\pm\cos\theta'$, where the + sign denotes radiation toward space and the − sign is for that directed toward the surface.

Before proceeding further, it is worth mentioning an ambiguity regarding directions considered in radiative transfer discussions. Some authors (e.g., Chandrasekhar, 1950) take the various angles as referring to the direction to which the radiation is propagated, whereas other authors (e.g., Sekera,1956) refer to the direction from which the radiation is propagated. The two cases are exemplified by (θ,ϕ) and (θ',ϕ') in Fig. 2.2. Angles (θ,ϕ) specify the direction toward which the radiation is propagated and (θ',ϕ') specify that from which it comes. The ambiguity is so deeply embedded in the literature that an effort to eliminate it here would probably prove confusing. Furthermore, for orthogonal coordinates the angles are simply supplements of each other, and the reference direction will usually be clear from the context.

The second coordinate system that will be used for some problems is shown in Fig. 2.3. The only difference between this system and that of the previous diagram is that here the orthogonal axes are referred to the plane defined by the two directions of propagation. This plane, which is shaded in the diagram, is usually termed the plane of scattering or scattering plane. The axes designated by parallel are represented by unit vectors $\mathbf{u}_{\parallel}$ and $\mathbf{u}'_{\parallel}$ situated parallel to this plane, and the perpendicular axes, with unit vectors $\mathbf{u}_{\perp}$ and $\mathbf{u}'_{\perp}$, are normal to it. The method of modifying the Stokes vector for a change of axis orientation was given previously in section 1.9. It will be remembered that parameters I and V are invariant in such a rotation, but Q and U are not.

Following Sekera (1968), we shall simplify the notation by the following convention for the angle pairs (θ, ϕ) and (θ', ϕ'). The pairs will be designated by Ω and Ω', respectively, with the direction of upward-directed radiation being $+\Omega$, or $+\Omega'$ and that of downward-directed radiation $-\Omega$ or $-\Omega'$. This is the same sign convention as was mentioned previously for $\pm\mu=\pm\cos\Omega$, but the azimuth angle ϕ is taken as always positive.

2.3 Equation of radiative transfer in a realistic medium

The equation of transfer for a medium that absorbs and scatters radiation from a monodirectional incident beam is readily derived in the following manner. We do not include true emission since we are concentrating on the solar radiation regime, and atmospheric emission at the wavelengths relevant to the solar regime may be neglected for practical purposes. We consider a cylindrical volume of unit cross section $d\sigma=1$ and length ds, as shown in Fig. 2.4, to be embedded in a medium such as the atmosphere. A pencil of monochromatic radiation of intensity $I_\nu(\Omega)$ traverses the cylinder and emerges with intensity $I_\nu(\Omega)+dI_\nu(\Omega)$ after having undergone attenuation by absorption and scattering in the medium. It is clear that in general I_ν is a vector intensity, but for convenience we carry the scalar notation. We can define a mass attenuation coefficient κ such that for a cylinder of unit cross section

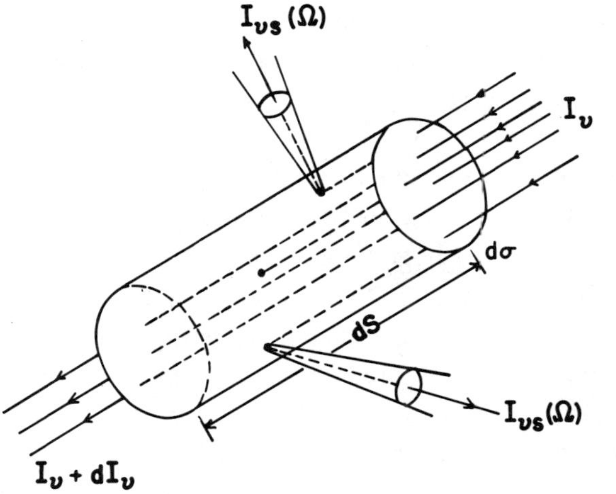

Figure 2.4 Diagram of the change of radiant energy over a pathlength ds in a medium which exhibits scattering and absorption.

$$dI_\nu = \kappa\rho I_\nu ds \tag{2.3}$$

where ρ is the density of material in the cylinder. Similarly, a volume attenuation coefficient for the medium is $\beta_\nu = \kappa_\nu\rho$, and

$$dI_\nu = -\beta_\nu I_\nu ds \tag{2.4}$$

It should be noted that β is a scalar coefficient, which means that the polarization properties of the incident beam would not be affected by its passage through the volume. This is not true of course for the scattered radiation emerging from the volume, since the scattering process is very effective in changing the state of polarization of the scattered light.

Since attenuation is due to both scattering and absorption, we can write

$$\beta_\nu = \beta_{\nu s} + \beta_{\nu a} \tag{2.5}$$

where subscripts s and a indicate the two processes. As we shall see, scattering is produced by various types of particles, principally gaseous molecules, aerosols, and cloud droplets. Absorption plays a less dominant role in some problems of polarization, but it is crucial in studies of the effects of multiple scattering.

Radiation of a given stream may consist of various components arising from different orders of multiple scattering, and the relative magnitudes of the components are affected by their history of absorption as they traverse the medium. Furthermore, as will be shown later, the polarizing characteristics of atmospheric aerosols depend on their index of refraction, the imaginary part of which is responsible for absorption by the particles.

On combining Equations (2.4) and (2.5), we can write the equation of transfer for the simple geometry as

$$\frac{dI_\nu}{I_\nu} = d(ln\ I_\nu) = -(\beta_{\nu s}+\beta_{\nu a})\ ds \tag{2.6}$$

If the incident radiation is not monodirectional but is confined to a small solid angle $d\omega'$, then

$$dI_\nu = -\beta_\nu I_\nu ds\, d\omega' \tag{2.7}$$

2.4 The diffuse radiation field

So far we have considered the change of intensity during the direct transmission of a beam through the cylinder, but the radiation which is scattered from the beam also emerges from the cylinder, albeit in other directions, and contributes to the general radiation field. Since the radiation is scattered in all directions, we shall characterize its angular distribution as it emerges from the cylinder into an elemental solid angle $d\omega$ oriented in direction Θ with respect to the original beam by the phase function $p\,(\cos\Theta)$. (In the vector case necessary for defining the polarization of scattered radiation, $p(\cos\Theta)$ becomes a phase matrix $\mathbf{P}(\cos\Theta)$, but otherwise its form is the same.) The intensity $I_{\nu s}(\Omega)$ due to scattering into direction Ω is

$$dI_{\nu s}(\Omega) = -\frac{1}{4\pi}\ \beta_{\nu s} I_\nu(\Omega)\ P(\cos\ \Theta)\ ds \tag{2.8}$$

Since the phase function (or matrix) simply characterizes the angular distribution of the scattered light and does not add or subtract from the total energy, we have the constraint that its integral over the entire solid angle surrounding the volume cannot exceed unity. Thus we have the normalization relation for conservative scattering

$$\frac{1}{4\pi} \int_{4\pi} P(\cos\Theta)\, d\omega = 1 \tag{2.9}$$

If some absorption occurs during the process, then we have

$$\frac{1}{4\pi} \int_{4\pi} P(\cos\Theta)\, d\omega < 1 \tag{2.9a}$$

In either case, this is one definition of the albedo of single scattering (see section 2.8).

Radiation which is scattered from an incident beam and contributes to the surrounding field as diffuse radiation appears as if it were emitted by the volume. In our case it is virtual emission only, as we have arbitrarily eliminated true atmospheric emission from consideration as being negligible at wavelengths of the solar regime. The basic equation of transfer for scattering has exactly the same form as Eq. (2.3) for total attenuation and, indeed, for a purely scattering atmosphere, κ is a scattering coefficient. In accord with Eq. (2.8), the intensity scattered into direction Ω due to radiation incident from each direction Ω' surrounding the volume is

$$dI_{\nu s}(\beta_{\nu s}, \Omega) = \frac{1}{4\pi}\, \beta_{\nu s}\, ds \int_{4\pi} P(\Omega, \Omega')\, I_\nu(\Omega')\, d\Omega' \tag{2.10}$$

where the integration is taken over the entire solid angle. One of the most demanding tasks in radiative transfer studies in the atmosphere is a realistic representation of the phase function. The phase matrix formulation is even more difficult, and the entire matrix has been determined only for simplified cases such as very small particles (Rayleigh scattering), spherical particles (Mie scattering), and a few others.

In summarizing, there are two types of radiation emerging from the reference volume under discussion. The first is the incident beam from direction Ω' which undergoes the change of intensity $-\beta_\nu I_\nu ds$ on its passage, but is still in the same direction. The second type is a radiation field emerging from the volume arising from scattering of radiation incident on the volume from all directions Ω' into direction Ω. The sum of these two is

$$dI_\nu(\Omega) = -\beta I_\nu(\Omega)\, ds + \frac{\beta}{4\pi}\, ds \int_{4\pi} P(\Omega, \Omega')\, I_\nu(\Omega')\, d\Omega' \qquad (2.11)$$

2.5 Plane parallel model of the atmosphere

For many purposes the actual atmosphere can be well approximated by a model in which the properties (density, composition, etc.) vary only in the vertical direction. In such a plane parallel model, the atmosphere is of infinite horizontal extent and horizontally homogeneous, and any curvature of the layers is neglected. This much-simplified case has been found to be quite realistic for a wide range of radiative problems in the actual atmosphere. It breaks down for major horizontal inhomogeneities of atmospheric properties or surface reflection and for cases of very low sun elevations or measurement directions. It is inapplicable to the case of twilight studies, but even here some authors have adapted the model by empirical arguments. We shall utilize the plane parallel approximation extensively in later sections.

For the plane parallel model, we can replace ds in Eq. (2.3) by the equivalent distance $\sec\theta\, dz = (1/\mu)\, dz$, where z is altitude taken in the vertical direction, θ is angle from the zenith (or nadir for outward directed radiation), and $\mu = \cos\theta$. We shall generally restrict θ to $0° \leq \theta \leq 90°$, in which case μ is always a positive quantity. The differential form of the equation of transfer for the plane parallel model is then

$$dI_\nu(\beta_\nu, \Omega) = \beta_\nu I_\nu(\Omega)\, \frac{dz}{\mu} \qquad (2.12)$$

If we define two levels z_1 and z_2 in the atmosphere and let the equivalent intensities be I_1 and I_2, then on integration of Eq. (2.7) from z_1 to z_2 we have

$$I_2(\Omega) = I_1(\Omega) \exp\left(-\frac{1}{\mu}\int_{z_1}^{z_2} \beta(z)\, dz\right) \qquad (2.13)$$

2.6 Normal optical thickness

It is most convenient to express the attenuation function in terms of a nondimensional parameter termed the normal optical thickness τ, defined for any level z by the relation

$$\tau_\nu(\beta_\nu, z) = -\int_\infty^z \beta_\nu(z)\, dz \tag{2.14}$$

Here ∞ represents the top of the atmosphere and the sense of the integration is always from the top downward for z increasing upward. On reversing the limits we have the useful form

$$\tau_\nu(\beta_\nu, z) = \int_z^\infty \beta_\nu(z)\, dz \tag{2.15}$$

In the real atmosphere, β_ν is due mainly to scattering by gaseous molecules (Rayleigh scattering), scattering and absorption by dust, haze, and other atmospheric aerosols, absorption by water vapor and ozone, and scattering and absorption by liquid or frozen water clouds. Thus we have

$$\tau_\nu = \tau_\nu^R + \tau_\nu^A + \tau_\nu^W + \tau_\nu^O + \tau_\nu^C \tag{2.16}$$

where the superscripts represent the components listed. All of these components are strong functions of altitude, all except τ^R are extremely variable in time and geographical location, and all are spectrally dependent. These variabilities make τ a very complicated function in any realistic case, and simplifying assumptions are usually necessary for applications to the real atmosphere.

The magnitudes τ^R, τ^A, and τ^O due to the Rayleigh atmosphere, aerosols, and ozone, respectively, and the sum of these, according to the atmospheric model derived by Elterman (1968), are shown as a function of wavelength in Fig. 2.5. The curves represent normal optical thicknesses (i.e., optical thicknesses taken in the zenith direction) at sea level, and thus are for the whole atmosphere. Their variations with altitude are included in the original tabulations of Elterman. The atmospheric model was developed from a synthesis of observations of various types by many observers, and was meant to represent a relatively clear case in which the meteorological range at sea level is about 25 km at wavelength

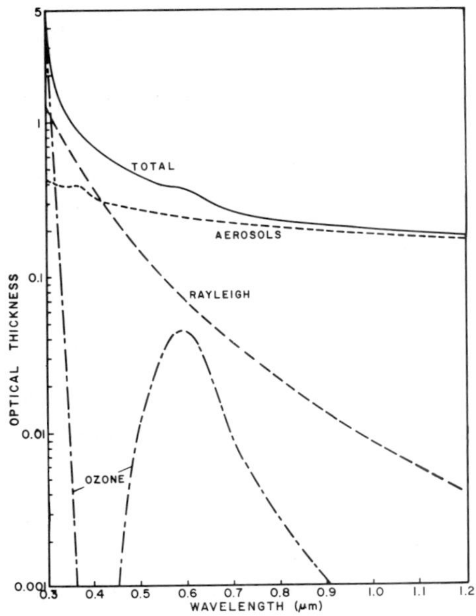

Figure 2.5 Normal optical thickness as a function of wavelength for aerosols, ozone, and the Rayleigh atmosphere, and for the total, according to the atmospheric model of Elterman (1968).

$\lambda = 0.5\ \mu$m. The basic Rayleigh component was assumed to be that of the U.S. Standard Atmosphere, 1962, to which an amount of ozone of 0.35 atm cm, typical of a mid-latitude location, was added. The aerosol concentration as a function of altitude was derived as the mean of 79 aerosol profiles measured by the searchlight technique during the April 1964–April 1965 time period. Although this was shortly after the eruption of the Agung volcano, the data showed that the effects of the volcano had been abated and that atmospheric conditions had stabilized. Thus the model is taken to represent an unperturbed and reasonably clear mid-latitude atmosphere.

The simplest curve of Fig. 2.5 is that for the Rayleigh atmosphere. However, we postpone a discussion of τ^R for Chapter 4, in which many aspects of the Rayleigh atmosphere, including τ^R, are covered. The optical thickness of ozone is due only to absorption, and its spectral dependence is thus similar to the ozone absorption spectrum. Elterman used the ozone absorption coefficients as determined by Vigroux (1953) for the computations of τ^O. Within the wavelength range of the diagram, the influence of the very strong Hartley-Huggins band is seen at and slightly above $\lambda = 0.30\ \mu$m, in which region τ^O is larger than those of the other components combined. It reaches the very high values of 35.8 and 71.0 at $\lambda = 0.28$ and $0.27\ \mu$m, respectively. It dies off very quickly, however, with increasing wavelength, being down by more than three orders of magnitude at $\lambda = 0.35\ \mu$m. It picks up again in the relatively weak Chappuis absorption band to the point that τ^O is nearly as large as τ^R in the center of the band at $\lambda = 0.60\ \mu$m. Aside from these two regions, τ^O is of negligible magnitude in relation to τ^R and τ^A.

The dominant role throughout most of the range of wavelengths of Fig. 2.5 is played by τ^A. Its magnitude for this model is slightly less than that of τ^R in the ultraviolet region, but beyond $\lambda = 0.42\ \mu$m, τ^A/τ^R progressively increases and becomes a factor of 10 at $\lambda = 80\ \mu$m and 50 at $\lambda = 1.2\ \mu$m. Thus the aerosol optical thickness is clearly dominant over that of the Rayleigh atmosphere, including ozone, throughout the near-infrared region of the spectrum. Elterman (1968) pointed out that the aerosol content of the model used for these computations is merely representative, and that there are wide variations in particle concentrations and properties in the real atmosphere. There may be some locations, such as the Antarctic or other remote regions, in which the aerosol content is much lower than that of the model, in which case the relative magnitude of τ^A would be less than that of Fig. 2.5. There are urban locations, of course, in which τ^A would be much increased over that shown, and in such cases τ^A would dominate the picture even in the ultraviolet.

The optical thicknesses τ^W and τ^C listed in Eq. (2.16) were not included in Elterman's model, perhaps because they are so extremely variable. For instance, measurements by Aufm Kampe (1950) gave cloud attenuation coefficients of between 0.0267 m^{-1} and 0.195 m^{-1} for different types of clouds. If we take 0.10 m^{-1} for an example, we see that τ^C reaches a value of 2 in only 20 m of cloud thickness, and of 10 in 100 m of cloud. In addition, the relatively large-sized droplets of which clouds are constituted make τ^C almost independent of wavelength. Thus it becomes obvious that τ^C is much greater than all of the other optical thicknesses combined for the wavelength range from the near ultraviolet to the near infrared when a significant cloud cover is present.

The magnitude of τ^W due to the absorption of radiation by water vapor in the atmosphere is critically dependent on wavelength, as well as on the total amount of water vapor present. The absorption spectrum of water vapor is very rich in detail, as shown by Goody (1964). The wavelengths of the electronic absorption bands are all too short to be of interest here, the long-wave limit being only 0.186 μm. The strongest fundamental of the vibration-rotation spectrum of water vapor is at 6.25 μm, which is too far out to be of much interest in the solar energy regime. The other two fundamentals, however, are at 2.66 and 2.74 μm, which is a region in which the solar spectrum has significant amounts of energy. In addition to these strong bands of the fundamentals, there is a large number of bands which varies from relatively strong to very weak spaced throughout the near-infrared spectrum. For instance, the six bands usually designated as δ, σ, τ, ϕ, ψ, Ω fall in the range $\lambda=0.90$ to 1.91 μm and are strong enough in combination to absorb significant amounts of solar energy. An even larger number of bands, all very weak, arising from ground state transitions of the water vapor molecule occur in the region $\mu=0.544$ to 0.846 μm, but they are not strong enough to be of much significance for present purposes.

In summary, we see that τ^W has a very complicated spectral distribution and varies from high values to negligibly small values in the various bands. The strongest peaks of τ^W are at the wavelengths of the fundamentals (2.66, 2.74, and 6.25 μm), in which case τ^W would be dominant over the other optical thicknesses whenever the atmosphere has a normal amount of humidity. At most other wavelengths, however, the weakness and narrowness of the bands prevent them from being of much practical significance in comparison to other sources of attenuation.

2.7 Indices of refraction of aerosol particles

One of the main parameters determining the radiative effects of atmospheric aerosols is their complex index of refraction $m=n-in'$, where n is the real part and n' is the imaginary part of m. Unfortunately, m is not well known for the aerosols which exist in the real atmosphere, although the measurements of Volz (1954), Fischer (1973), Hanel (1968), and others have established the strong variations which occur. The indices for water droplets, however, are well established, thanks to the work of Hale and Querry (1973) for the predominantly solar radiation regime, and to that of Irvine and Pollack (1968) for the infrared wavelengths. Mineralogists have measured the indices of refraction of many mineral species, including quartz, graphite and other forms of carbon, sea salt, calcite, metals, etc., which make up portions of the natural aerosol, but such mineralogical determinations have been for the materials in their dry state. The extent to which the natural humidity and impurities of the atmosphere alter the indices for the natural aerosol from those of the dry materials has been studied in some detail by Hanel (1976).

The following discussion of the complex index of refraction will be divided into that for the dry materials or particles, that for liquid water, and that for materials either dissolved in water or surrounded by a layer of liquid water.

2.7.1 Index of refraction for dry materials

Although many aerosol particles in the natural atmosphere are known to be either liquid or have a liquid coating, aerosols derived from non-hygroscopic mineral soils maintain their dry state until the ambient humidity reaches at least 70% to 90%. The constituents comprising such mineral particles are obviously dependent on geographic location and wind regime, but airborne dust over deserts is known to have a high quartz content with some other mineral species in lesser amounts. Even sea-salt particles derived from the evaporation of sea spray are in the dry state in conditions of low humidities, as are particles of limestone, gypsum, and numerous other common materials. Most of these require humidities near saturation before they collect a water coating, but of course when they do so, the index of refraction of the composite is strongly modified. The effect of humidity on the radiative properties of aerosols is the subject of section 2.7.3.

The refractive indices of various materials in the dry state, as given by the indicated sources, are shown in Table 2.1. For birefringent materials such as quartz and calcite the tabulated values are for the ordinary ray. Equivalent data for the longer wavelengths applicable to the terrestrial radiation regime are given by Peterson and Weinman (1969) and Volz (1972).

The data for the aerosol models of Hanel (1976) require some explanation. Model 3 was developed from aerosol collections taken from a research vessel in the Atlantic, Model 5 was from a wintertime aerosol in the urban area of Mainz, Germany, and Model 6 was from a collection taken in summer from

TABLE 2.1 Refractive indices versus wavelength of certain mineral species known to be constituents of atmospheric aerosols and mean values for various aerosol models considered to be in the dry state. (n is the real and n' the imaginary parts, respectively.)

Material	$\lambda(\mu m)$	n	n'	Source	Model	$\lambda(\mu m)$	n	n'	Source
Quartz	0.340	1.57	10^{-7}	(1)	Marine	0.30	1.61	0.089	(4)
(SiO_2)	0.589	1.55	"	"	aerosol	0.40	1.57	0.082	"
	0.760	1.54	"	"	"	0.50	1.56	0.066	"
	0.768	1.539	"	(2)	"	0.55	1.55	0.055	"
	1.159	1.53	"	"	"	0.60	1.55	0.047	"
	1.681	1.53	"	"	"	0.70	1.54	0.045	"
	3.50	1.48	"	"	"	0.80	1.54	0.050	"
Calcite	0.340	1.70	—	(1)	"	1.0	1.54	0.066	"
($CaCO_3$)	0.589	1.66	—	"	"	2.0	1.47	0.132	"
	0.760	1.65	—	"	"	2.5	1.47	0.132	"
	1.23	1.64	—	"	"	3.0	1.34	0.264	"
Rock salt	0.589	1.54	—	"	Industrial	0.30	1.61	0.037	"
(NaCl)	0.884	1.53	—	"	aerosol	0.40	1.57	0.042	"
	1.179	1.53	—	"	"	0.50	1.55	0.046	"
	2.357	1.53	—	"	"	0.60	1.55	0.050	"
NH_4Cl		1.64	—	"	"	0.70	1.54	0.055	"
NH_4NO_3		1.60	—	"	"	0.80	1.54	0.059	"
$(NH_4)_2SO_4$		1.52	—	"	"	1.0	1.53	0.068	"
$MgCl_2$		1.54	—	"	"	2.0	1.47	0.136	"
$NaNO_4$		1.59	—	"	"	2.5	1.47	0.136	"
$CaSO_4$		1.57	—	"	"	3.0	1.33	0.272	"
KCl		1.49	—	"	Continental	0.30	1.57	0.012	"
Na_2SO_4		1.48	—	"	aerosol	0.40	1.53	0.013	"
SiO_2		1.49	—	"	"	0.50	1.51	0.014	"
K_2SO_4		1.49	—	"	"	0.60	1.50	0.015	"
Carbon	0.30	1.84	0.70	(3)	"	0.70	1.50	0.016	"
(C)	0.40	1.88	0.69	"	"	0.80	1.50	0.018	"
"	0.50	1.94	0.66	"	"	1.0	1.49	0.020	"
"	0.60	1.99	0.64	"	"	2.0	1.47	0.040	"
"	0.80	2.07	0.61	"	"	2.5	1.47	0.040	"
"	1.00	2.12	0.59	"	"	3.0	1.43	0.080	"

Sources: (1) Handbook of Chemistry and Physics (1964–65) (2) Peterson and Weinman (1969) (3) Bergstrom (1972) (4) Hanel (1976)

a 1000-m elevation in the foothills of the Alps. Determination of the real part of the refractive index was made by immersing the aerosol sample in a liquid of known index and measuring the change of index due to the aerosol (Hanel, 1968). The imaginary part was measured by Fischer (1973) by use of an integrating sphere. In applying the results, it was assumed that the mean values obtained applied to a hypothetical particle typical of the aerosol as a whole. These same aerosol samples were used by Hanel (1976) for studying the effects of relative humidity on particle characteristics (see section 2.7.3).

The values of the real part of the index of refraction of most dry components of atmospheric aerosols lie in the 1.45 to 1.60 range for the wavelengths of most significance in the solar radiation regime, with the imaginary part between 0.001 and 0.1. Carbon, of course, has higher values, particularly for the imaginary part, and the real part for calcite is considerably higher. The measurements for the aerosol collections show generally decreasing values with increasing wavelength, but as a whole they are in the 1.55 region for the real part. The imaginary part for the oceanic and urban aerosols is generally higher than that for the Alpine continental aerosol, probably because of higher concentrations of carbonaceous material in the lower layers of the atmosphere.

The index of refraction of volcanic ash, particularly the imaginary part, is of considerable current interest because of the possible effects of volcanoes in promoting changes of the climate. The eruption of Mount St. Helens in the western United States on May 18, 1981, provided an opportunity for a concentration of studies on the properties of the ash emitted from the volcano. Patterson (1981) measured the imaginary part n' of the index of refraction for ash samples taken from the surface various distances from the volcano and for samples taken by aircraft from 15- and 18-km altitudes. The results of Patterson's measurements are shown in Fig. 2.6 for wavelengths of 0.30 to 0.70 μm. The samples from the surface at Bozeman, Montana, and Boulder, Colorado, were, according to Patterson, a uniform gray color, with that from Cheney, Washington a lighter gray. The imaginary index is by roughly a factor of two less for the light-colored sample than for the darker material, but even in the latter case n' is consistently less than 0.01. For the longer part of the visible region, $n' \approx 0.004$ for the darker sample and less than 0.002 for the lighter one. The stratospheric samples, particularly that from 18 km taken 3 days after the eruption, show a strong spectral dependence, n' being more than an order of magnitude less at 0.70 μm than at 0.30 μm for the more extreme case. Both stratospheric samples, however, show similarly small values of n', indicating relatively modest amounts of absorption of solar radiation and heating of the atmospheric layers containing the volcanic material.

Not all volcanic ash has such small absorptive capacity as that from Mount St. Helens, as shown by the upper curve of Fig. 2.6 for the Fuego volcanic

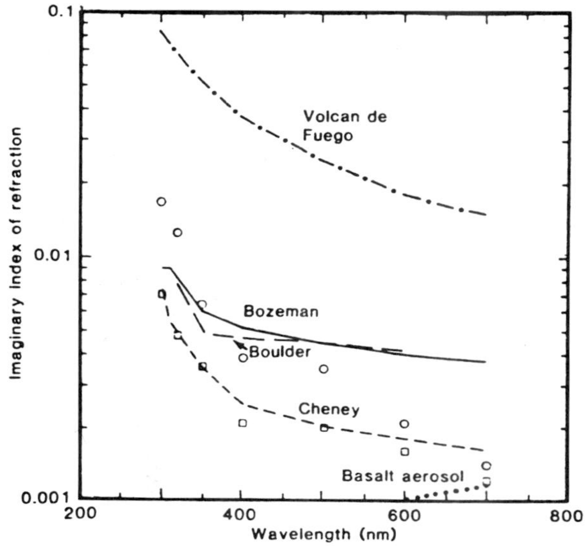

Figure 2.6 Imaginary part of the refractive index for five samples of volcanic ash from the eruption of Mount St. Helens, a sample from the eruption of Fuego, and a sample of aerosol derived from basalt. The curves are for surface samples from Bozeman, Montana; Boulder, Colorado; and Cheney, Washington; squares are for a sample from 15 km altitude on May 20, and circles for one from 18 km on May 21 (after Patterson, 1981).

material, in which n' is about an order of magnitude higher than that of the ash from Mount St. Helens. The wavelength dependence, however, is about the same for the two.

2.7.2 Index of refraction of water

Many measurements have been made to establish the index of refraction of water, both in its liquid and solid phases. Irvine and Pollack (1968) have given data for the infrared region, and Hale and Querry (1973) have given very complete tabulations for the shortwave region of the spectrum. Both the real and imaginary parts of the refractive index of liquid water, as given by Hale and Querry (1973) for many different wavelengths, are shown in Table 2.2. The real part is very close to 1.33 throughout the visible range, with a slight decrease with increasing wavelength. A significant minimum of 1.2 occurs at $\lambda = 2.9$ μm, with a relatively

TABLE 2.2 Real and ccmplex parts of the index of refraction ($m = n - in'$) of liquid water at various wavelengths (Hale and Querry, 1973)

$\lambda(\mu m)$	n	n'	$\lambda(\mu m)$	n	n'
0.200	1.396	$1.1x10^{-7}$	0.875	1.328	$2.91x10^{-7}$
0.250	1.362	$3.35x10^{-8}$	0.900	1.328	4.86
0.300	1.349	$1.6x10^{-8}$	0.925	1.328	$1.06x10^{6}$
0.325	1.346	$1.08x10^{-8}$	0.950	1.327	2.93
0.350	1.343	$6.5x10^{-9}$	0.975	1.327	3.48
0.375	1.341	$3.5x10^{-9}$	1.00	1.327	2.89
0.400	1.339	1.86	1.2	1.324	9.89
0.425	1.338	1.3	1.4	1.321	$1.38x10^{-4}$
0.450	1.337	1.02	1.6	1.317	$8.55x10^{-5}$
0.475	1.336	$9.35x10^{-10}$	1.8	1.312	$1.15x10^{-4}$
0.500	1.335	$1.00x10^{-9}$	2.0	1.306	$1.1x10^{-3}$
0.525	1.334	1.32	2.2	1.296	$2.89x10^{-4}$
0.550	1.333	1.96	2.4	1.279	9.56
0.575	1.333	3.60	2.6	1.242	$3.17x10^{-3}$
0.600	1.332	$1.09x10^{-8}$	2.65	1.219	6.7
0.625	1.332	1.39	2.70	1.188	0.019
0.650	1.331	1.64	2.75	1.157	0.059
0.675	1.331	2.23	2.80	1.142	0.115
0.700	1.331	3.35	2.85	1.149	0.185
0.725	1.330	9.15	2.90	1.201	0.268
0.750	1.330	$1.56x10^{-7}$	2.95	1.292	0.298
0.775	1.330	1.48	3.00	1.371	0.272
0.800	1.329	1.25	3.2	1.478	0.0924
0.825	1.329	1.82	3.3	1.450	0.0368
0.850	1.329	2.93	3.4	1.420	0.0195

pronounced increase beyond that point. The imaginary part is very low for $\lambda < 1.8$ μm, oscillates around 0.001 in the 1.8- to 2.4-μm region, and reaches a maximum of almost 0.30 in the 2.9- to 3.0-μm absorption band. There is, of course, relatively little energy in solar radiation at these longer wavelengths.

2.7.3 Effects of ambient humidity on aerosols

The radiative properties of aerosol particles change with the relative humidity of the air due to condensation of water vapor on the particles with increasing humidity and its evaporation with decreasing humidity. The condensed water changes the mean density, size, and index of refraction of the particles, and thereby their fall velocities and radiative properties. In addition, the differing rates of growth of different sizes in a distribution tend to broaden the size-frequency distribution, another factor affecting the radiative properties of a real aerosol population.

The changes of an aerosol particle which result from changes of relative humidity are strongly dependent on the type of material of which the particle in its dry state is composed, as well as on the total mass of dry material in the particle. A hygroscopic material such as sodium chloride, for example, begins growing at lower relative humidity and achieves a complete water coating more readily than does a quartz particle of the same original size. Seldom are the natural aerosol compositions as simple as these extreme cases, but are instead likely to be a mixture of soluble and insoluble materials of diverse characteristics. The growth or shrinking of the amount of water condensed on a particle is entirely dependent on the difference in vapor pressure over the liquid of the particle and in the surrounding air. In a stable-sized particle the vapor pressure of the air is the same as that over the liquid, the latter of which is lowered by the salts dissolved in the solution (solute effect) and raised by the surface curvature of the particle (curvature effect). The two effects can be seen from Fig. 2.7

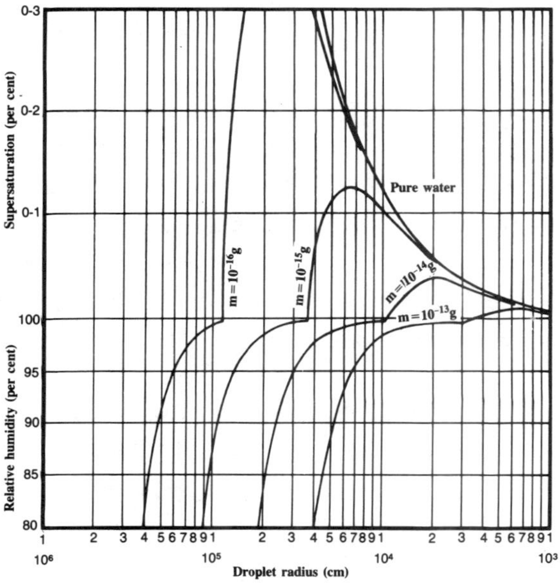

Figure 2.7 Equilibrium relative humidity as a function of droplet radius for droplets containing the indicated masses of sodium chloride (after Mason, 1971).

(Mason, 1971), in which the radii of droplets resulting from the condensation of water on four different masses of sodium chloride are plotted as a function of relative humidity. The solute effect resulting from the smallest mass of salt (10^{-16} grams) is very weak, whereas the curvature effect for the small droplet is strong. This means that the droplet remains small until a relatively high supersaturation of 0.42% in the surrounding air is reached, after which the increase in droplet size decreases the curvature effect and permits the droplet to grow under conditions of decreasing supersaturation. The critical radius of the droplet which corresponds to the critical value of supersaturation in this case is approximately 2.5×10^{-5} cm. On the other hand, for the largest mass of salt (10^{-13} g), the strong solute effect causes the droplet to be relatively large even at 80% relative humidity, and the curvature effect is weak for the large droplet. Thus the droplet requires only a very small supersaturation to overcome the curvature effect and permit continued growth of the droplet. In this case the critical radius is about 7×10^{-4} cm and the critical supersaturation only about 0.01%. Droplets for intermediate masses of salt remain between these two extremes. Eventually so much water accumulates that the solute concentration approximates that of pure water and the surface of the large drop approximates a plane water surface, so that the drop grows or shrinks depending on whether the relative humidity of the surrounding air is slightly above or slightly below 100%.

Particles which are insoluble but have a wettable surface may also serve as condensation nuclei. Due to the absence of the solute effect for droplets that form on insoluble nuclei, their equilibrium vapor pressure is higher than that of droplets containing dissolved salts. This is shown by the curve for pure water in Fig. 2.7. Such insoluble particles will generally not be activated into forming droplets as long as there are a sufficient number of soluble nuclei of appropriate sizes in the air taking up the available water. The initial size of the insoluble nucleus must be much larger than that of a hygroscopic nucleus for the start of condensation. Mason (1971) has given the critical radii and supersaturation for hygroscopic nuclei of various sizes listed in Table 2.3. These values were computed from the relations given by Mason for hygroscopic nuclei

$$r_c = (3C/B)^{4} \tag{2.17}$$

$$H_C/100 = 1 + (4B^3/27C)^{1/2} \tag{2.18}$$

TABLE 2.3 Radii of dry particles r_d, critical radii r_c, and values of critical supersaturation H_c for hygroscopic (sodium chloride) nuclei of various masses as given by Mason (1971). Numbers in parentheses are the powers of 10 by which the entry is multiplied.

log m(g)	−16	−15	−14	−13	−12	−11	−10	−9	−8
$r_d(\mu m)$	0.022	0.048	0.103	0.22	0.48	1.03	2.2	4.8	10.3
$r_c(\mu m)$	0.20	0.62	2.0	6.2	20	62	200	620	2000
$H_c(\%)$	0.42	0.13	4.2(−2)	1.3(−2)	4.2(−3)	1.3(−3)	4.2(−4)	1.3(−4)	4.2(−5)

where B and C are constants applicable to the given nucleus of mass m, molecular weight W, and temperature T. Their values are $B=3.2\times10^{-5}\,T$ and $C=8.6$ m/W.

The natural aerosol in the atmosphere is known to include hygroscopic nuclei, insoluble nuclei, and combinations with an insoluble core surrounded by a hygroscopic substance (such as H_2SO_4). Thus the effects of humidity on the radiative properties of natural aerosols are not so simple as those for the idealized cases shown above.

In an extensive investigation, Hanel (1976) determined the effects of humidity on the physical properties (density, index of refraction, etc.) of aerosol samples collected under six different types of atmospheric conditions of continental and marine locations. The three samples considered by him to be most representative of general conditions were a marine aerosol from a ship in the Atlantic, an urban aerosol from an industrial city in winter, and a continental aerosol taken at an altitude of 1000 m in the foothills of the Alps in summer (Hanel's models 3, 5, and 6). The effects of humidity on these samples were investigated in the laboratory. The mean complex indices of refraction measured for these samples were shown previously in Table 2.1, and the effects of relative humidity on their radiative properties as given by Hanel (1976), are outlined below.

By considering the entire aerosol sample to be represented by a typical particle or sample of typical particles, the real part n of the complex index of refraction can be computed from the relation

$$n = n_w + (n_o - n_w)\left(1 + \frac{\rho_o}{\rho_w}\,\bar{\mu}\,\frac{f}{1-f}\right)^{-1} \tag{2.19}$$

where n and ρ_o are the index and density, respectively, for the dry particle, n_w and ρ_w are the same parameters for pure water, f is relative humidity, and $\bar{\mu}$ is a quantity termed the linear mass increase coefficient which is dependent on the composition of the particular aerosol sample. It is defined by

$$\bar{\mu} = \frac{m_w}{m_o} \frac{1 - f}{f} \tag{2.20}$$

where m_o and m_w are the masses of the dry material and the collected water, respectively. In Hanel's study, $\bar{\mu}$ was determined by weighing the samples at different humidities. The relationship for the complex part of the index is the same as Eq. (2.19) except that n, n_w and n_o are replaced by n', n'_w, and n'_o. Values of $\bar{\mu}$ for the marine, urban industrial, and continental aerosol samples as a function of relative humidity f are shown for increasing and decreasing humidities in Table 2.4. A larger hysteresis effect in aerosol response is clearly

TABLE 2.4 Values of the linear mass increase coefficient $\bar{\mu}$ as a function of relative humidity f for marine, urban, industrial, and continental aerosols (Hanel's models, 3, 5, and 6) for use in computing the complex index of refraction of the aerosol samples by Eq. (2.19). The left columns are for increasing and the right for decreasing relative humidity (after Hanel, 1976).

Model		Model		Model		Model		Model		Model	
f	3	f	5	f	6	f	3	f	5	f	6
0.204	0.032	0.195	0.033	0.231	0.025	0.701	0.428	0.860	0.194	0.827	0.212
0.305	0.026	0.304	0.030	0.301	0.026	0.599	0.085	0.820	0.214	0.716	0.096
0.399	0.027	0.398	0.029	0.394	0.027	0.516	0.071	0.789	0.230	0.582	0.067
0.497	0.030	0.493	0.030	0.485	0.032	0.400	0.057	0.704	0.174	0.517	0.067
0.557	0.031	0.556	0.042	0.544	0.037	0.317	0.046	0.597	0.115	0.362	0.069
0.604	0.042	0.605	0.068	0.590	0.040	0.202	0.056	0.513	0.101	0.264	0.079
0.653	0.077	0.655	0.091	0.625	0.042			0.401	0.079		
0.710	0.155	0.704	0.109	0.677	0.045			0.317	0.068		
0.772	0.369	0.772	0.180	0.731	0.056			0.200	0.061		
0.834	0.328	0.829	0.187	0.768	0.120						
0.900	0.280	0.886	0.180	0.827	0.160						
0.950	0.243	0.952	0.171	0.885	0.150						
0.962	0.234	0.975	0.177	0.950	0.105						
0.975	0.225	0.980	0.180	0.972	0.103						
0.986	0.211	0.986	0.186	0.986	0.115						
0.990	0.206	0.990	0.192	0.990	0.117						
0.9969	0.207	0.9969	0.213	0.9969	0.131						
10.0	0.256	10.0	0.256	10.00	0.163						

evident from the larger masses of particles in the decreasing humidity regime than in times of increasing humidity.

The effects of humidity on the radiative characteristics of natural aerosols are evidenced by the change of the coefficients of extinction, absorption, and scattering as a function of ambient relative humidity f to which the aerosols are subjected. The ratios of extinction coefficient $\sigma_e(f)$ at $f \neq 0$ to those for the dry aerosol $\sigma_e(o)$ for $\lambda = 0.55\ \mu$m are shown as a function of the quantity $(1-f)$ for three aerosol models by the upper set of curves in Fig. 2.8. Selected values of relative humidity are shown at the top of the diagram. To a good approximation (Hanel, 1976), the ratio $\sigma_e(f)/\sigma_e(o) = Q_e(f)/Q_e(o)$ where Q_e is the extinction efficiency of the aerosol particles (see section 2.9). The data of Fig. 2.8 show that the particles are relatively insensitive to changes of relative humidity at $f \leq 65\%$ to 70%, but in the marine and urban industrial aerosols the hygroscopic component of the aerosol grows very rapidly at $f \approx 70\%$. The continental aerosol appears to be less hygroscopic than the others, the change of growth rate being less and occurring in the region $f = 75\%$ to 80%. As the humidity exceeds 80%, the particles of all three aerosols grow very rapidly and reach values of 16 to

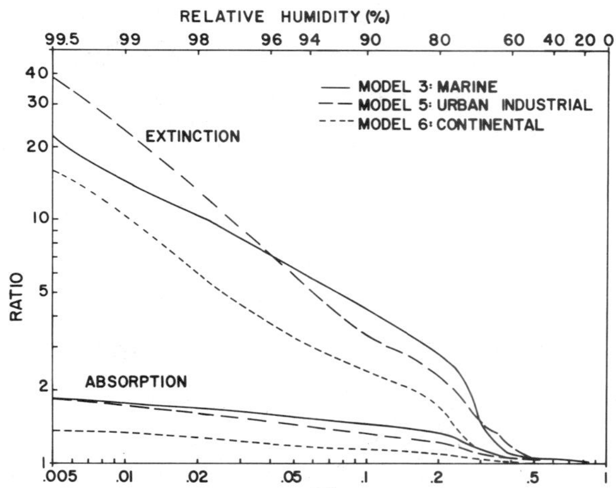

Figure 2.8 Ratios of the extinction and absorption cross sections of aerosol particles to those of the dry aerosol as a function of $(1-f)$ for three aerosol models, as computed by Hanel (1976). (f is relative humidity.) The top scale shows selected values of f.

40 times their dry cross section at $f=99.5\%$. No data on the critical radii or critical supersaturation values for these aerosol collections are available, but presumably the hygroscopic nature of at least the marine and urban industrial aerosols would require relatively low supersaturations for continued growth of the particles.

Curves for the ratio of absorption coefficients $\sigma_a(f)/\sigma_a(o)$ for the same aerosol collections are shown also in Fig. 2.8. The equivalent ratios of the scattering coefficients $\sigma_s(f)/\sigma_s(o)$ are represented by the differences between the respective pairs of curves. The absorption coefficients are not very sensitive to relative humidity. This is because the increased absorption caused by increaseing size is largely offset by the accumulation of pure water, which has a small value for the complex part of the index of refraction, on the dry particle. Thus the index for the composite is decreased, and at the higher humidities the particles are mostly water. The particles in this case are inefficient absorbers, and the extinction is mostly due to the efficient scattering typical of water droplets.

2.8 Albedo of single scattering

In the most general case, a particle or assemblage of particles attenuates a beam of radiation by both absorption and scattering. If we define a quantity $\tilde{\omega}$, usually termed the albedo of single scattering, to characterize the fraction of the attenuation which is due to scattering alone then we have

$$\tilde{\omega}(\nu) = \frac{\beta_{\nu s}}{\beta_{\nu s} + \beta_{\nu a}} \tag{2.21}$$

where $\beta_{\nu s}$ and $\beta_{\nu a}$ are the monochromatic volume scattering and absorption coefficients, respectively. The fraction of the attenuation due to absorption is $1-\tilde{\omega}$. The concept and terminology of the albedo of single scattering originally arose from considerations of scattering and absorption by single particles, and its applicability to a volume encompassing many particles, with possible multiple scattering among the particles, is not so obvious. However, if the elemental volume is taken small enough, multiple scattering effects may be made arbitrarily small, in which case the concept is valid for a multiple-particle volume, though it is most directly appropriate for single particles such as atmospheric aerosols.

For an absorbing medium we have, in analogy with Eq. (2.8), the inequality

$$dI_{\nu s}(\Omega) < \frac{\pi}{4} \beta_\nu I_\nu P(\cos\Theta) \, d\Omega \tag{2.22}$$

A convenient method of restoring the equality is to demand that the normalization of the phase function be less than unity. Thus

$$\frac{\pi}{4} \int_\Omega P(\cos\Theta) \, d\Omega = \tilde{\omega} < 1 \tag{2.23}$$

In fact, this is an alternate definition of $\tilde{\omega}$. The magnitude of $\tilde{\omega}$ is constrained to $0 \leq \tilde{\omega} \leq 1$, the upper limit for a particle which exhibits either scattering only (a dielectric) or pure reflection (a metal : $m=\infty$) and the lower limit is for absorption only. For a large particle which absorbs all the energy falling on it (a classic blackbody), then $\tilde{\omega}=0.5$, because one-half of the incident energy is diffracted around the particle and thus does not get absorbed.

The relationship between $\tilde{\omega}$ and the complex part of the refractive index n' is not a simple one. Usually, an increase of n' corresponds to a decrease in $\tilde{\omega}$, but this is not always the case. Furthermore, the relationship is a function of particle size. For example, $\tilde{\omega}$ is low for small particles (generally less than 0.2 for all $x=2\pi r/\lambda < 0.5$), even for small values of n', whereas particles with $x>5$ generally have $\tilde{\omega}>0.5$. In the limit as $n'=\infty$, the particle becomes a perfect reflector, giving $\tilde{\omega}=1$. The scattering efficiencies of various types of particles are shown as a function of particle size in Chapter 3.

There are two asymptotic forms of interest for setting limits on the single scattering albedo of different sizes of particles. For very large dielectric particles the scattering efficiency Q_s (see Chapter 3) is 2.0, one-half of the scattered radiation being reflected and refracted by the drop and the other half being due to diffraction around the edges. For the general cases of $n' \neq 0$, we have (Kerker, 1969)

$$Q_s = 1 + w \tag{2.24}$$

where w is due to the specular reflection of the large drop and may be computed from the original Mie formulation (Van de Hulst, 1957). Kerker (1969) gives values of w for several combinations of n and n' for large spheres. These data have been used for computing $\tilde{\omega}$ from the relation

$$\tilde{\omega} = \frac{1}{2}(1 + w) \tag{2.25}$$

the results of which are shown as a function of n' for six values of n in Fig. 2.9. The scattering albedo is essentially independent of n' for $n' < 0.1$, the case most applicable to large spheres in the atmosphere. Only for $n' > 0.5$ is there a strong dependence of $\tilde{\omega}$ on n', and only as the spheres assume a metallic nature with $n' > 4$ do the spheres approach perfect reflection ($\tilde{\omega} = 1$). The strong dependence of $\tilde{\omega}$ on the real refractive index n for small n' is due to the increase of refraction of the incident light for n increasing from unity. For n=1.0, a case not shown in the diagram, refraction would be zero. Scattering would then be due to diffraction ($\tilde{\omega} = 0.5$) for small n' and specular reflection plus diffraction for large n'.

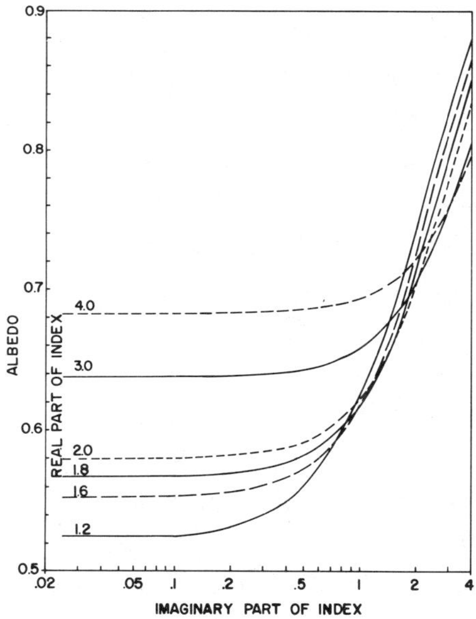

Figure 2.9 Single scattering albedo for large spheres as a function of the imaginary part n' of the refractive index for six different values of the real part n.

In the small particle limit, $\tilde{\omega}$ is given by the relation (Kerker, 1969)

$$\frac{1}{\tilde{\omega}} = 1 - \frac{3}{2x^3} \frac{Im(m^2 - 1)/(m^2 + 2)}{[(m^2 - 1)/(m^2 + 2)]^2} \tag{2.26}$$

For very small particles of $x<1$, the absorptivity is very high for even small values of n', and the single scattering albedo is correspondingly low. Curves of $\tilde{\omega}$ as a function of x, as computed from the relation (2.26), for values of the complex index of refraction representative of natural aerosols in the atmosphere are shown in Fig. 2.10. The range of relative sizes shown corresponds to a range

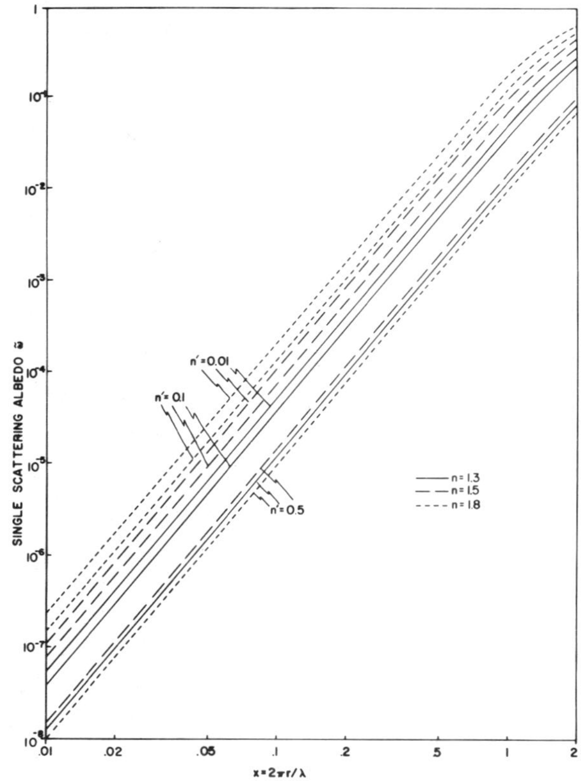

Figure 2.10 Single scattering albedo as a function of relative particle size as computed under the small particle approximation in Eq. (2.26) for a number of complex indices of refraction representative of those of natural atmospheric aerosols.

of radii from 0.008 to 0.16 μm for radiation of wavelength $\lambda = 0.50$ μm. The curves are closely linear on this double logarithmic plot from $x = 0.01$ to $x = 1$, making extremely low values of $\tilde{\omega}$ for the smallest particles. Van de Hulst (1957) has shown that the scattering cross section of a very small particle is proportional to r^2, V being volume, and since $x = 2\pi r/\lambda$, $V \sim x^3$. Thus the scattering cross section is proportional to r^6, a relation which explains the very strong dependence of $\tilde{\omega}$ on x, as shown in the diagram.

For particles of moderate sizes, which satisfy neither of the limiting cases described above, $\tilde{\omega}$ generally, but not always, decreases with increasing values of n', as shown by Fig. 2.11 (Kerker, 1969). The very strong dependence of

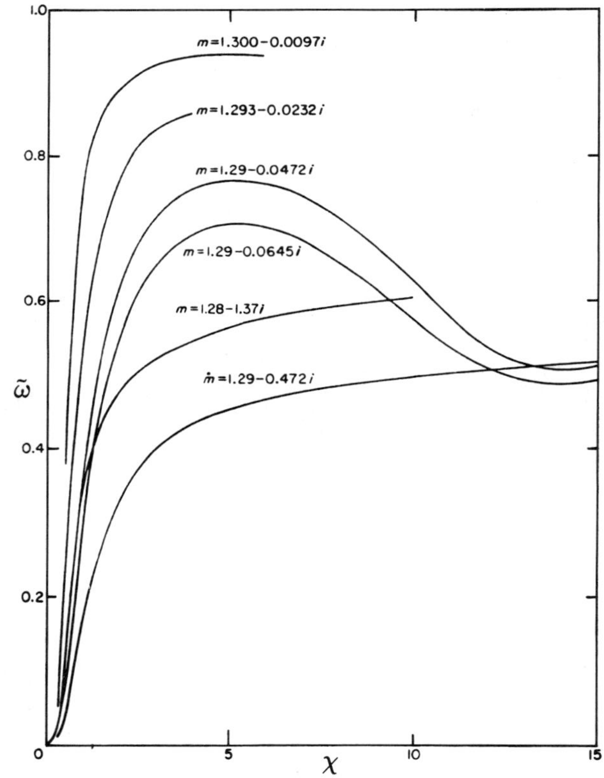

Figure 2.11 Single scattering albedo as a function of particle size parameter for various complex indices of refraction (after Kerker, 1969). The coordinates have been relabeled to correspond to our notation.

$\tilde{\omega}$ on x for small particles, seen above in the small particle approximation, is shown here as well, but the dependence is much less pronounced for $x>2$. In fact, for $n'=0.0472$ and 0.0645, $\tilde{\omega}$ decreases with increasing x for $x>5$, and there is an indication of a similar behavior for small values of n'. This is not the case for large values of n', however. As the particle becomes more metallic in nature, the specular component of reflection increases and the energy entering the particle decreases, thereby causing an increase of $\tilde{\omega}$. This effect is responsible for $\tilde{\omega}$ being greater for $n'=1.37$ than for $n'=0.472$ in Fig. 2.11. As $n\rightarrow\infty$ the particle becomes pure metallic, no energy enters the particle to be absorbed, and $\tilde{\omega}\rightarrow 1$.

References

Anon. (1964–65): *Handbook of Chemistry and Physics,* 45th ed., Chemical Rubber Co., Cleveland, Ohio.

Aufm Kampe, H.J. (1950): Visibility and liquid water content in clouds and in the free atmosphere, *J. Meteorol.*, *7,* 54–57.

Bergstrom, R.W. (1972): Predictions of the spectral absorption and extinction coefficients of an urban air pollution aerosol model, *Atmos. Environ.*, *6,* 247–258.

Chandrasekhar, S. (1950): *Radiative Transfer,* Clarendon Press, Oxford.

Elterman, L. (1968): *UV, Visible, and IR Attenuation for Altitudes to 50 km, 1968,* Env. Res. Pap. No. 258, Air Force Cambridge Research Laboratories, Bedford, Mass.

Fischer, K. (1973): Mass absorption coefficient of natural aerosol particles in the 0.4–2.4 μm wavelength interval, *Contrib. Atmos. Phys.*, *46,* 89–100.

Goody, R.M. (1964): *Atmospheric Radiation, I: Theoretical Basis,* Clarendon Press, Oxford.

Hale, G.M., and M.R. Querry (1973): Optical constants of water in the 200 nm to 200 μm wavelength region, *Appl. Opt.*, *12,* 555–563.

Hanel, G. (1968): The real part of the mean complex refractive index and the mean density of samples of atmospheric aerosol particles, *Tellus, 20,* 371–379.

Hanel, G. (1976): The properties of atmospheric aerosol particles as functions of the relative humidity at thermodynamic equilibrium with surrounding moist air, *Adv. in Geophys.*, *19*, 73–188.

Irvine, W.M., and J. B. Pollack (1968): Infrared optical properties of water and ice spheres, *Icarus*, *8*, 324–360.

Kerker, M. (1969): *The Scattering of Light and Other Electro-magnetic Radiation*, Academic Press, New York.

Mason, B.J. (1971): *The Physics of Clouds*, Clarendon Press, Oxford.

Patterson, E.M. (1981): Measurements of the imaginary part of the refractive index between 300 and 700 nanometers for Mount St. Helens ash, *Science*, *211*, 836–838.

Peterson, J.T., and J.A. Weinman (1969): Optical properties of quartz dust particles at infrared wavelengths, *J. Geophys. Res.*, *74*, 6947–6952.

Sekera, Z. (1956): Recent developments in the study of the polarization of skylight, *Advan. Geophys.*, *3*, 43–104.

Sekera, Z. (1968): Theory of Radiative Transfer, Lecture Notes, Meteorology 228, University of California, Los Angeles (courtesy of Larry Stowe, Jr.).

Van de Hulst, H.C. (1957): *Light Scattering by Small Particles*, John Wiley & Sons, New York.

Vigroux, E. (1953): Contribution a l'etude experimentale de l'absorption de l'ozone, Doctoral Thesis, Fac. Sci., Univ. of Paris.

Volz, F. (1954): Die Optik und Meteorogie der atmospharischen Trubung, *Ber. Dtsch. Wetterdienstes*, *13*, 1–47.

Volz, F. (1972): Infrared absorption by atmospheric aerosol substances, *J. Geophys. Res.*, *77*, 1017–1031.

CHAPTER 3

SCATTERING BY SINGLE PARTICLES

3.1 Scattering domains for different types of particles

It is to be expected that particles of different sizes and indices of refraction will have different scattering properties. This is indeed the case, and certain approximations may be used to make the scattering theory more tractable. The two parameters most useful for this purpose are the size parameter x defined as the ratio between circumference of the particle and wavelength of the radiation, $x=2\pi r/\lambda$, and the index of refraction of the particle, $m=n-in'$, where n and n' are the real and imaginary parts. On the basis of these parameters, a number of different scattering domains determined by types of particles have been defined.

3.1.1 Rayleigh scattering: $x \ll 1$ but m arbitrary

This very useful approximation applies to the molecules of atmospheric gases and, as will be seen in detail in later sections, it explains many of the features of polarization and intensity of light in the earth's atmosphere. It is also applicable to scattering of radar and other microwaves by cloud and rain droplets. Rayleigh scattering in the atmosphere is the subject of Chapter 4.

3.1.2 Rayleigh–Gans scattering: $(m-1) \ll 1$ and $2x(m-1) \ll 1$

This is more stringent than the Rayleigh case because of the requirement that the index of refraction of the particle not be much different from that of the surrounding medium. However, particle size is less restrictive than for the pure Rayleigh case. This approximation is not very useful in atmospheric problems because natural aerosol particles generally have m considerably higher than that for air.

3.1.3 Anomalous diffraction: $(m-1) \ll 1$ but x arbitrary

Since the radiation field in this case is due to the straight transmission of the incident wave by the particle plus a diffraction field according to Huygens' principle, the scattered light is concentrated in the region of small scattering angles. Thus, the method is most applicable to studies of the solar aureole and similar problems. It is similar to the W.K.B. approximation (Saxon, 1955).

3.1.4 Geometric optics: $x \to \infty$ and m real

This is the classical case for dealing with refraction, reflection, and diffraction in explanations of optical phenomena such as rainbows. Ray tracing techniques are applicable.

3.1.5 Mie scattering: x and m both arbitrary, but for spheres only

The Mie theory covers the entire size range from Rayleigh scattering to geometric optics, and in this respect is completely general. Since it is directly applicable to (assumed spherical) aerosol particles in the atmosphere, it has been widely applied to both polarization and intensity of light in planetary atmospheres, and particularly that of the earth. We shall devote considerable attention to it.

A complete discussion of single particle scattering is given by Van de Hulst (1957) and others, and will not be repeated here. We shall outline the main principles, however, for eventual application to scattering in the earth's atmosphere. It is possible to develop a general form for scattering by particles of arbitrary shapes (Saxon, 1955), but the scattering matrix is known in sufficient detail for computations only for some simplified shapes such as spheres and ellipsoids, and for the case of very small particles that scatter according to Rayleigh's law in which the shape is of secondary importance. A popular concept in the past has been that random orientation of the many particles that are responsible for scattering in the atmosphere would cause an aggregate of particles with different orientations to scatter as if the particles were spheres, but more recent work has shown the fallacy of such an argument (Chylek et al., 1976; Asano and Sato, 1980). However, the extensive computations of scattering by aerosol particles has been confined mainly to the spherical case.

3.2 Mie scattering by spherical particles

The law of scattering by spherical particles was derived first by the German physicist Gustav Mie (1868–1957) in 1908. Later versions have been developed by Stratton (1941) and Van de Hulst (1957), and various sets of tables of the functions have been issued (Lowan, 1949; Shifrin and Zelmanovich, 1966, 1968; Zelmanovich and Shifrin, 1968; Deirmendjian, 1969). As mentioned previously, the Mie theory is particularly useful for atmospheric problems, as it is applicable to all particle sizes from gaseous molecules (Rayleigh case) through intermediate sizes of aerosol particles to the largest rain drops (geometric optics).

For outlining the Mie theory, we assume an oscillating electromagnetic field of radiation of wavelength λ incident on a homogeneous dielectric sphere of radius r and dielectric constant ϵ. The incident wave may consist of parallel radiation or radiation confined to an elemental solid angle dw. Energy may be scattered or absorbed by the particle. Absorbed energy goes into heat. In the space surrounding the particle and at a distance that is large compared to the particle radius, the scattered energy has the form of a diverging spherical wave with oscillations in the plane normal to its direction of propagation and at the same frequency as that of the incident wave. For unpolarized incident radiation and spherical particles, the scattered field has axial symmetry around the direction of propagation of the the incident beam, the only variation being in scattering angle Θ between the directions of propagation of incident and scattered radiation. This is not true, of course, if the incident radiation is partially or completely polarized, as will be seen in later sections.

We now define an arbitrary plane through the center of the particle and along the two directions of propagation, as shown in Fig. 3.1, with scattering angle Θ and unit vectors $(u_\parallel, u_\perp)$ and $(u'_\parallel, u'_\perp)$ parallel and normal to the scattering plane for the scattered and incident radiation, respectively. A linear relationship exists between the electric vectors of the scattered and incident radiation such that

$$\begin{aligned} \mathbf{E}_\parallel &= D_\parallel(\cos\Theta)\, \mathbf{E}'_\parallel \\ \mathbf{E}_\perp &= D_\perp(\cos\Theta)\, \mathbf{E}'_\perp \end{aligned} \tag{3.1}$$

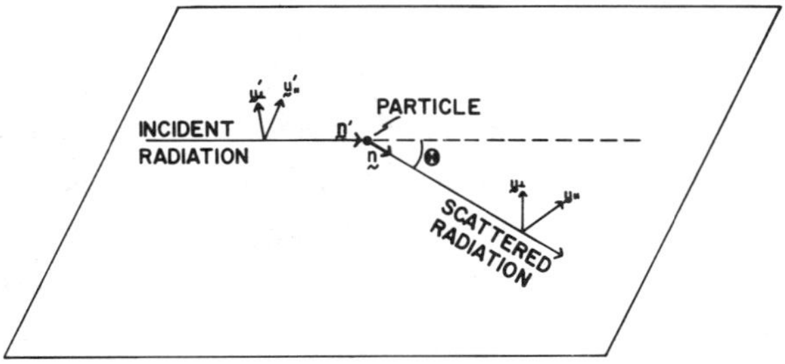

Figure 3.1 Representation of the scattering plane, with scattering angle Θ and orthogonal unit vectors $u_\parallel$ and $u_\perp$ for scattered radiation and $u'_\parallel$ and $u'_\perp$ for incident radiation. This is a different representation of, but equivalent to, the scattering plane shown previously in Fig. 2.3.

A convenient feature of these relations is that for a given particle the scattering matrices $D_{\parallel,\perp}$ are functions only of scattering angle.

The fundamentals of the Mie theory are described briefly in the following terms. The relations required are a pair of coefficients, the Mie coefficients a_n and b_n (which are independent of scattering angle but are functions of size and relative index of refraction of the particle and of wavelength of the radiation), and a pair of functions $\pi_n(\cos\Theta)$ and $\tau_n(\cos\Theta)$ (which are dependent on only scattering angle). Functions π_n and τ_n are real, but a_n and b_n are complex.

3.2.1 The Mie coefficients

The Mie coefficients are defined by the following relations

$$a_n = \frac{\psi_n'(y)\,\psi_n(x) - m\psi_n(y)\,\psi_n'(x)}{\psi_n'(y)\,\xi_n(x) - m\psi_n(y)\,\xi_n'(x)}$$

$$b_n = \frac{m\psi_n'(y)\,\psi_n(x) - \psi_n(y)\,\psi_n'(x)}{m\psi_n'(y)\,\xi_n(x) - \psi_n(y)\,\xi_n'(x)} \tag{3.2}$$

Here, x is given by $x=2\pi r/\lambda$, where r is particle radius and λ is wavelength of the radiation; m is relative index of refraction of the particles; and $y=mx$. The functions ψ_n and ξ_n are Riccati–Bessel functions having the form of infinite series. Subscript n is a sequential integer index and primes denote derivatives.

The Mie coefficients show an interesting behavior with increasing size of the particles, although for present purposes the interest is mainly from a mathematical standpoint. Plots of a_1 and b_1 in the complex plane as functions of relative size parameter x for a dielectric sphere with $m=1.29$ and for a partially absorbing sphere with $m=1.29-0.0472i$, as determined by Deirmendjian (1969), are shown in Fig. 3.2. Since we are dealing with atmospheric scattering, we emphasize the case of nonabsorbing or slightly absorbing spheres. For nonabsorbing spheres, $a_n(x)$ and $b_n(x)$ always fall on a circle of radius 0.5 centered at point 0.5 on the real axis and 0 on the imaginary axis of the complex plane, as shown by curve A in the diagram (Van de Hulst, 1957). The addition of some

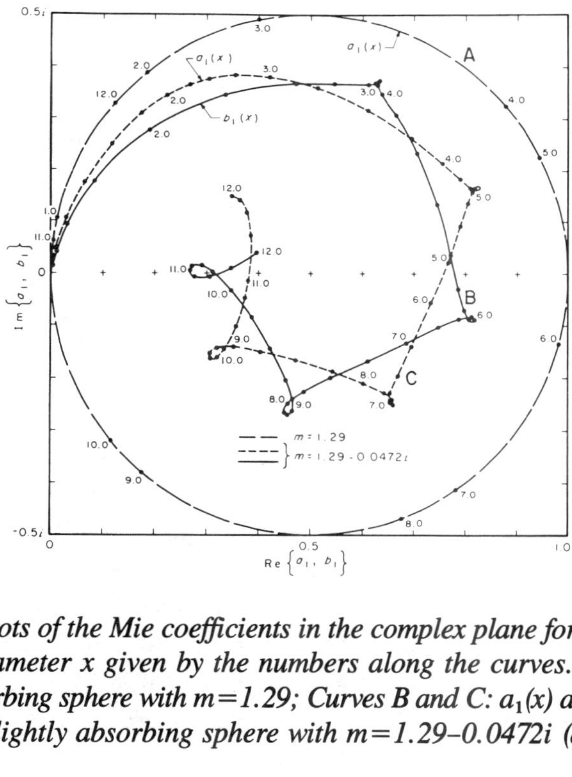

Figure 3.2 Plots of the Mie coefficients in the complex plane for increasing particle size parameter x given by the numbers along the curves. Curve A: $a_1(x)$ for a nonabsorbing sphere with m=1.29; Curves B and C: $a_1(x)$ and $b_1(x)$, respectively, for a slightly absorbing sphere with m=1.29–0.0472i (after Deirmendjian, 1969).

absorption creates a spiraling behavior such as that shown by curve B for $a_n(x)$ and curve C for $b_n(x)$. As x gets very large, both curves approach the circle of convergence, shown by the dashed circle near the center of the diagram. The radius of the circle of convergence is ½ $(m-1/m+1)$. It is known (e.g., Hansen and Travis, 1974) that the coefficients a_n and b_n can be interpreted, as a multipole expansion of the scattered light, a_n being for electric multipole radiation and b_n for magnetic multipole radiation. Rayleigh scattering, for instance, results from small particles with small or moderate values of refractive index, in which case only electric dipole radiation is important. Resonance phenomena result from small particles with large refractive indices. Large particles in general exhibit the full multipole behavior, but Van de Hulst (1957) has given a number of limiting cases for simplification of the computational requirements. For moderately large particles with moderate values of refractive index, however, no significant simplifications for alleviating the computational task for the full Mie expansion have been devised.

3.2.2 The angular functions

The angular dependence of the scattered radiation field is introduced by a pair of functions $\pi_n(\cos\Theta)$ and $\tau_n(\cos\Theta)$ involving Legendre polynomials and their derivatives written as

$$\begin{aligned} \pi_n(\cos\Theta) &= \frac{1}{\sin\Theta}\, P_n'(\cos\Theta) \\ \tau_n(\cos\Theta) &= \frac{dP_n'(\cos\Theta)}{d(\cos\Theta)} \end{aligned} \tag{3.3}$$

Expressions for the various parameters in a form suitable for machine computation are given by Deirmendjian (1969). The behavior of these angular functions $\pi_n(\cos\Theta)$ and $\tau_n(\cos\Theta)$ is of interest from a computational standpoint. Errors of computation are more easily detected at this stage than after they have been incorporated into the amplitude functions. Fairly regular patterns of π_n and τ_n are observed, at least for $n=1$ to 6, as shown in Figs. 3.3 and 3.4 over the range of angles $\Theta=0°$ to $90°$ (Van de Hulst, 1957). Except for changes of sign, symmetry around $\Theta=90°$ is to be expected from the nature of the Legendre

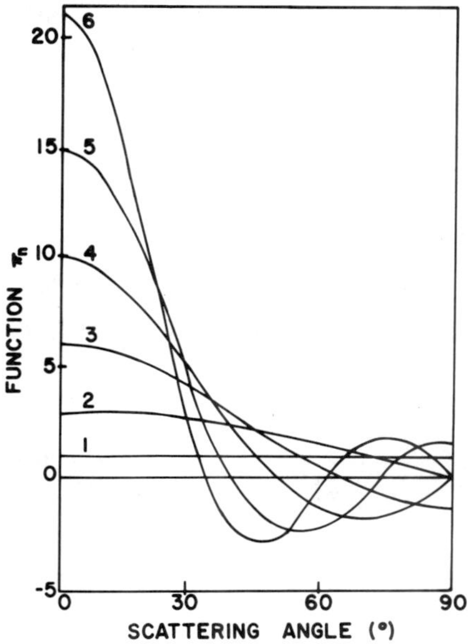

Figure 3.3 Behavior of the angular function $\pi_n(\cos\Theta)$ as a function of scattering angle for n=1 to 6 (adapted from Van de Hulst, 1957).

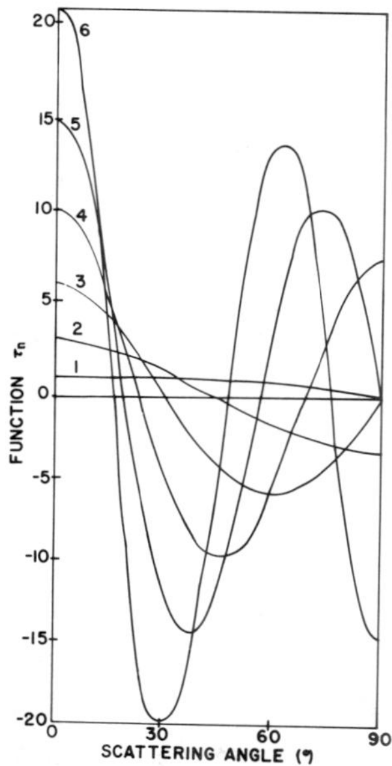

Figure 3.4 Behavior of the angular function $\tau_n(\cos\Theta)$ as a function of scattering angle for n=1 to 6 (adapted from Van de Hulst, 1957).

polynomials, and the curves will be continuous throughout the angular domain for all values of n.

3.2.3 The amplitude functions

Once a_n, b_n, π_n, and τ_n are determined for a sufficient number of values of n for convergence of the series, we can immediately determine the complex amplitude functions for scattered radiation field. They are given by the relations

$$S_{\perp}(\Theta) = \sum_{n=1}^{\infty} \frac{2n+1}{n(n+1)} \left[a_n \pi_n(\cos\Theta) + b_n \tau_n(\cos\Theta)\right]$$

$$S_{\parallel}(\Theta) = \sum_{n=1}^{\infty} \frac{2n+1}{n(n+1)} \left[b_n \pi_n(\cos\Theta) + a_n \tau_n(\cos\Theta)\right] \tag{3.4}$$

3.2.4 The intensity components

For present purposes we shall be more interested in the intensity components represented by the squares of the amplitude functions than in these functions themselves. In later sections, emphasis will be on the amplitude matrix necessary for characterizing the state of polarization of the scattered light.

The intensity components perpendicular and parallel to the scattering plane are given by the relations

$$i_{\perp}(\Theta) = |S_{\perp}(\Theta)|^2 = S_{\perp} S_{\perp}^*$$

$$i_{\parallel}(\Theta) = |S_{\parallel}(\Theta)|^2 = S_{\parallel} S_{\parallel}^* \tag{3.5}$$

plus a phase difference $\delta(\Theta)$ between them. For the scattering of unpolarized incident radiation, there is no permanent phase relation between the components, in which case the scattered field is defined completely by the magnitudes of the two. For the scattering of polarized incident light, however, there is a phase difference between them, which is the same as saying that some elliptic polarization is introduced by the scattering process. Chu and Churchill (1955) have derived expressions for computing $\delta(\Theta)$. The subject will be discussed further in relation to the polarization field. In anticipation of that discussion, we point out that if a particle is illuminated by an incident flux F', the intensity per unit solid angle in any direction is

$$I(\Theta) = F'(i_\perp + i_\parallel)/2k^2 = F'(i_\perp + i_\parallel)\lambda^2/8\pi^2 \tag{3.6}$$

where $k=2\pi/\lambda$.

If the incident radiation is plane polarized with the electric vector vibrating normal to the scattering plane, the scattered intensity is:

$$I(\Theta) = I_\perp(\Theta) = F'_\perp i_\perp \lambda^2/4\pi^2 \tag{3.7}$$

In this case $i_\parallel=0$, and the scattered light is plane polarized with a plane of polarization $x=90°$. If the incident flux F' has the electric vector in the scattering plane, we have the opposite case, or

$$I(\Theta) = I_\parallel(\Theta) = F'_\parallel i_\parallel \lambda^2/4\pi^2 \tag{3.8}$$

and $x=0°$. If the plane of polarization of F' is between $0°$ and $90°$, say x', then the scattered intensity in that same plane is

$$I(\Theta,x) = F'(x')\frac{\lambda^2}{4\pi^2}(i_\perp \sin^2 x + i_\parallel \cos^2 x) \tag{3.9}$$

In this case, the scattered light has some elliptic polarization if the scattering particle is partially absorbing (i.e., if the index of refraction is complex). The ellipticity is introduced by the two components having undergone different phase retardations in the scattering process.

3.2.4.1 Intensity in the forward and backward directions

The intensity components in the immediate forward ($\Theta=0°$) and backward directions ($\Theta=180°$) have special significance, the former being related to transmission or attenuation efficiency and the latter to the backscatter of radar or lidar beams in the atmosphere. In both of these directions, the scatter of an unpolarized incident beam results in $i_{\perp} = i_{\parallel}$.

The character of $i_{\perp,\parallel}$ in the forward and backward directions for scattering of unpolarized incident radiation of unit flux by a single dielectric and absorbing sphere is shown as a function of particle radius in Fig. 3.5. The points for these curves were calculated from the tabulations of the real and complex parts of $S_{\perp,\parallel}$ given by Deirmendjian and Classen (1962). The scale of the abscissa has been changed from the tabulated size parameter $x = 2\pi r/\lambda$ by assigning

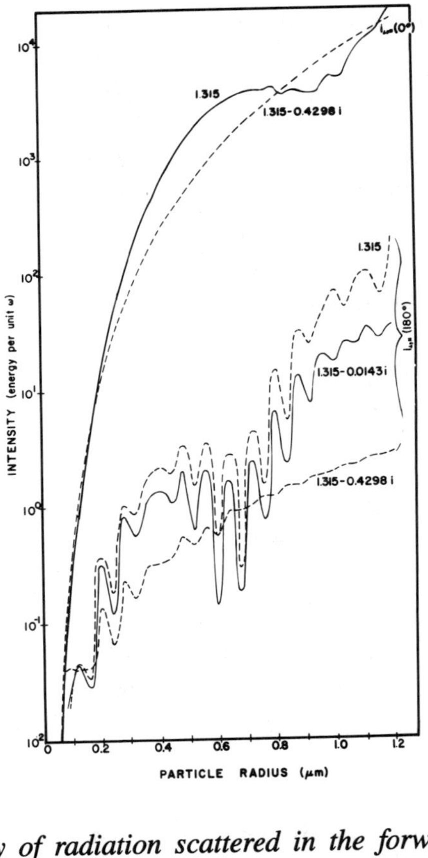

Figure 3.5 Intensity of radiation scattered in the forward (top curves) and backward directions (bottom curves) as a function of particle radius for non-absorbing and partially absorbing spherical particles. The incident radiation is assumed to be unpolarized and of unit flux at a wavelength of 0.50 μm.

$\lambda=0.50$ μm to make the curves conceptually more specific, the original tabulations being for $0.5 \leq x \leq 15$. The great sensitivity of forward scattering to particle size is shown by the curves, the intensity changing by more than a million times for a range of particle radius from 0.06 to 1.2 μm. Although absorption by the particles tends to smooth out the irregularities exhibited for purely dielectric particles, there is relatively little influence by absorption in the overall intensity or its dependence on particle size.

These characteristics have two important consequences. First, the total intensity of light that would be observed in the near-forward direction for a collection containing both large and small particles would be due mainly to effects of the larger end of the particle spectrum. Thus, the intensity of the solar aureole is predominantly attributable to scattering by the large particles in the atmosphere, and very little to the smaller particles, so conclusions on atmospheric aerosols derived from studies of the solar aureole are subject to a strong bias toward large particles. Second, the radiation field in the forward direction is largely insensitive to the index of refraction of the particles. This is much less true for other scattering angles. For instance, the maximum difference in the forward scattered intensity of Fig. 3.5 is only about a factor of 2, whereas in the backward direction, represented by the second set of curves in the figure, the difference reaches nearly two orders of magnitude. Since we know relatively little about the index of refraction for aerosol particles in the actual atmosphere, these differences may prove useful in studies of atmospheric aerosols.

The curves for the backward intensity in Fig. 3.5 also show a large change of overall intensity with particle size, the difference over particle radii from 0.5 to 1.2 μm being fully four orders of magnitude. This means that the backscattered light from a lidar would tend to be mainly due to the larger end of the aerosol spectrum, but of course the increase of particle number with decreasing size would tend to modify this tendency somewhat. At the small radii, the approximate equivalence of forward and backward intensities is obvious. (It will be remembered that the two are identical for single particle Rayleigh scattering.) Strong absorption tends to damp out the large intensity oscillations and causes the backward intensity difference between strongly absorbing and nonabsorbing particles to reach nearly two orders of magnitude.

3.2.4.2 Intensity at other scattering angles

The intensity components scattered at angles other than 0° and 180° are more complicated. In general, $i_\perp \neq i_\parallel$, a fact which indicates the existence of linear polarization, even for unpolarized incident light. For polarized light incident on a particle which exhibits some absorption, the scattering process may introduce phase differences in the components, giving some elliptic or circular polarization. These aspects are discussed in a later section. For the present, we assume the incident light to be unpolarized with unit energy flux.

The distribution of $i_\perp$ and $i_\parallel$ for light scattered by a particle of size parameter $x=3$, typical of the smaller range of atmospheric aerosol particles for visible light, is shown for two different indices of refraction in Fig. 3.6 (Deirmendjian and Classen, 1962). The asymmetry of the scattering pattern between backward and forward directions is obvious for both types of particles.

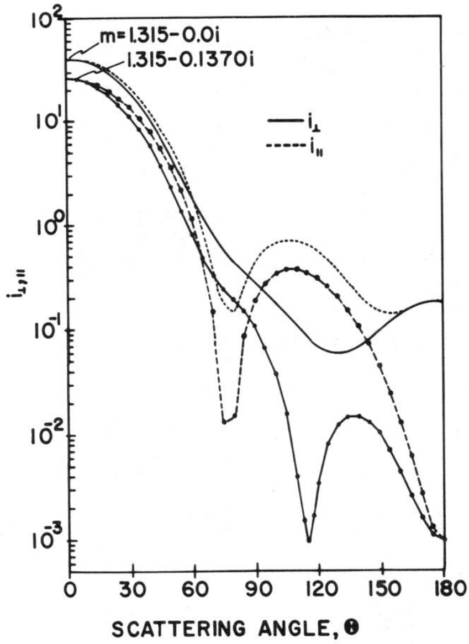

Figure 3.6 Intensity components $i_\perp$ and $i_\parallel$ as functions of scattering angle for spheres of $x=3$ and two different indices of refraction (adapted from Deirmendjian and Classen, 1962).

It is particularly pronounced for scattering by partially absorbing spheres, in which case the difference between maximum and minimum reaches more than four orders of magnitude. Absorption decreases the intensity in all directions for this case (although of course this is not a universal characteristic), and it introduces details into the curves that do not appear for purely dielectric particles. For instance, the behavior of $i_\perp$ in the region Θ=100° to 180° is quite different for $m = 1.315 - 0.01370i$ from that for $m = 1.315 - 0.0i$, whereas, aside from a difference of magnitude, the curves for $i_\parallel$ are very similar. These effects show up most clearly in the degree of polarization of the scattered light, a subject which is covered in later sections.

3.2.5 Cross sections and efficiency factors

Some of the most useful indices of the ability of a particle to abstract energy from an incident beam are the scattering, absorption, and attenuation cross sections of the particle. If we let these be represented by σ_s, σ_a, and σ_t, respectively, then we have

$$\sigma_t = \sigma_s + \sigma_a \tag{3.10}$$

Here σ_s represents the cross section for scattering in all directions. We can also specify the scattering cross section $\sigma_s(\Theta)$ applicable to a given scattering angle and leading to the phase function of the scattered radiation. No similar concept applies for σ_a because of the nature of the absorption, and σ_t is normally considered to apply to total attenuation, as opposed to an angular distribution. Since scattered radiation proceeds into the entire 4π steradians surrounding the particle, we have

$$\sigma_s = \int_0^{4\pi} \sigma_s(\Theta)\ dw \tag{3.11}$$

All of these cross sections have dimensions of area, except $\sigma_s(\Theta)$ is area/steradian. They each represent the amount of energy abstracted from an incident beam of unit flux. The concepts apply to radiation in any state of polarization.

It is emphasized that these are effective cross sections, and not the geometrical cross section (πr^2) of the particle. The two are related by the corresponding efficiency factors

$$Q_s = \sigma_s / \pi r^2 \tag{3.12}$$

and similarly for $\sigma_s(\Theta)$, σ_a, and σ_t. In analogy with Eq. (3.10), we have

$$Q_t = Q_s + Q_a \tag{3.13}$$

The efficiency factors are given in terms of the original Mie coefficients by the relations

$$Q_t = \frac{2}{x^2} \sum_{n=1}^{\infty} (2n + 1) \, Re(a_n + b_n)$$

$$Q_s = \frac{2}{x^2} \sum_{n=1}^{\infty} (2n + 1) \, (|a_n|^2 + |b_n|^2) \tag{3.14}$$

$$Q_a = Q_t - Q_s$$

We have used $x = 2\pi r/\lambda$.

Computed values of the efficiency factors for single particles and for various combinations of particles have been given by many authors, including Gumprecht and Sliepcevich (1951), Penndorf (1957), Van de Hulst (1957), Kerker (1969), Kattawar and Plass (1967), and others.

The scattering coefficient Q_s is shown as a function of the size parameter for $x \leq 28$ in Fig. 3.7 for four different real indices of refraction. Several features of the diagram are of interest. For $x < 2$ the Q_s is critically dependent on x. In fact, $Q_s \sim x^6$ in this small particle region, as was seen for the single scattering albedo in section 2.8. The curves reach maxima at relatively small values of x, after which they have the form of a damped oscillation. For large particles (see inset) they settle down to a value of 2. The position of the first maximum

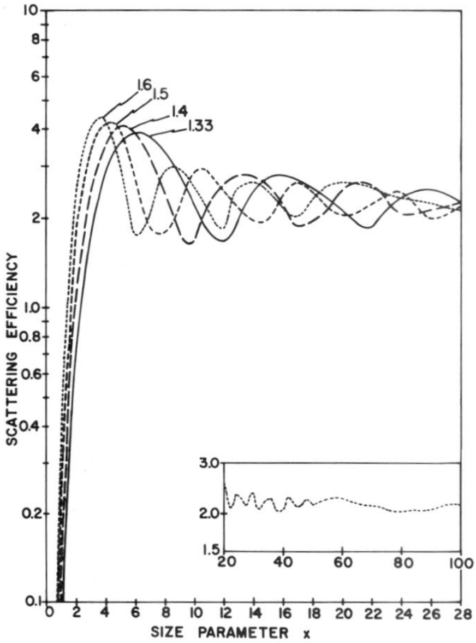

Figure 3.7 Single particle scattering efficiency as a function of size parameter for four different real indices of refraction for $x \leq 28$ (main curves) and for $20 \leq x \leq 100$ for index of 1.6 (inset).

and the frequency of the oscillations are dependent on index of refraction m, increasing values of m corresponding to smaller values of x at the maximum and somewhat higher maximum efficiencies.

The apparent paradoxical nature of the curves showing that a particle can abstract as much as four times as much energy from a beam as would be incident on the area of the cross section of the particle, has been explained by many authors (e.g., Van de Hulst, 1957) and has a sound physical basis. It comes about because of the wave nature of light. The extinction is due to a combination of a blocking of the wave by the particle and an interference between the incident and scattered waves. Some of the energy beyond the dimensions of the particle is scattered also. According to Babinet's principle, if the particle is very large compared to the wavelength ($r \gg \lambda$), an amount of energy proportional to the geometric cross section is abstracted from the beam by scattering and absorption, but a like amount is diffracted out of the beam. Thus, for very large particles $Q=2$, as indicated by the curves. The diffracted energy is confined to the near forward direction and is difficult to measure separately from the incident beam.

For particles with a nonzero imaginary part of the refractive index ($m = n - in'$, $n' \neq 0$), true absorption occurs and modifies the dependence of scattering efficiency Q_S on particle size from that of Fig. 3.7. The results of computations of Q_S for $n = 1.5$ and various values of n' typical of those for atmospheric aerosols and visible light are shown in Fig. 3.8. As n' increases from 0, the principal maximum of the curves decreases in magnitude and shifts toward smaller values of x, and the amplitudes of the secondary oscillations are correspondingly reduced. In fact, for $n' = 0.50$, the oscillations are completely damped out. The very steep slope of Q_S in the small particle range of this plot somewhat masks the actual effects of absorption in this size range.

The absorption efficiency Q_a of single particles is shown as a function of size parameter x for various values of n' in Fig. 3.9. Although generally Q_a increases with n', this is not universally true. Reflection, such as that from metals, plays an important role at $n' = 0.50$ at the larger values of x. The very important absorption efficiency of the small particle range is shown more advantageously in this diagram than in the previous one. It points out the fact that the attenua-

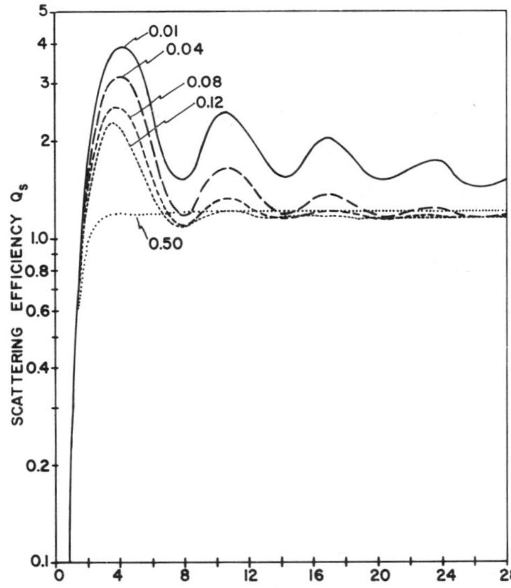

Figure 3.8 Scattering efficiency for partially absorbing particles with a real index of refraction $n = 1.5$ and various values of the complex part n' as a function of size parameter of the particles.

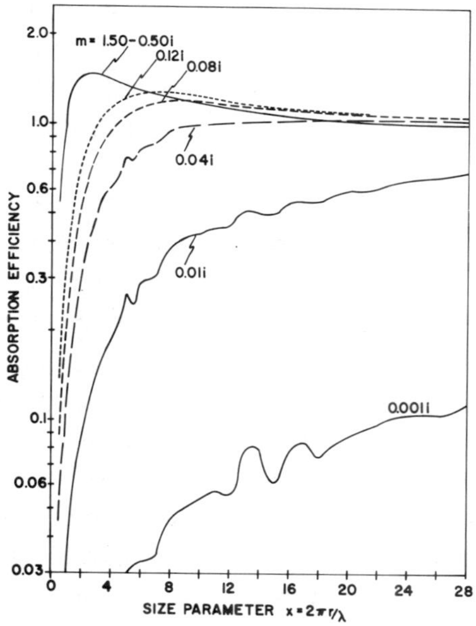

Figure 3.9 Absorption efficiency of particles with a real index of refraction n=1.5 and various values of the complex part n' as a function of size parameter of the particles.

tion of a beam of light traversing a cloud of very small and partially absorbing particles is caused more by absorption than by scattering by the particles. For instance, for the 0.1 μm diameter particles, which are very numerous in urban pollution, n' is often found to be in the range 0.01 to 0.1. Representative values of x for visible light for this case are $x=1.0$, for which case Fig. 3.9 shows Q_a to be at least 0.1 for $n'=0.04$, and as much as 0.4 for $n'=0.12$. These are comparable to the scattering efficiencies for this size particle shown in the previous diagram. Such high absorption efficiency explains the often-observed dark appearance of clouds of urban pollution.

3.2.6 Resonances in the scattering pattern

On a high resolution plot of scattering efficiency versus size parameter, a large number of small ripples are superimposed on the major oscillations, such as those shown in Fig. 3.8. Chylek (1976) and others have shown these ripples to be due to resonances in the partial wave expansion of the scattering amplitudes a_n and b_n, but it is only relatively recently that the theory has been brought together with experiment to define the details of the resonances. In an experiment utilizing optical levitation to suspend a single particle at a given point in space, Ashkin and Dziedzic (1977) observed resonance phenomena in radiation pressure. By the levitation technique, the radiation pressure of a vertically directed laser beam is used to support a particle against gravity for studies of particle growth in clouds, evaporation of liquid particles, light scattering by single particles, and other purposes. During such an experiment on a small droplet of oil ($m=1.42$), these authors observed the radiation pressure showing several sharp spikes as the tunable laser was scanned over a small range of wavelengths. Other authors (Irvine, 1965; Shipley and Weinman, 1978) have computed the fine structure of Mie scattering for water droplets over the size parameter $x=50$ to $x= 4520$ at a resolution of $\Delta x=10^{-2}$ and found a periodic ripple structure of $\Delta x=0.83$. From these calculations it was assumed that all details of Mie scattering had been resolved. Such was not the case, however, as shown by later computations by Chylek and his colleagues.

Returning to the levitation experiment, the radiation pressure responsible for suspending the particle is given in terms of the Mie coefficients a_n and b_n of Eq. (3.2) by the relation

$$Q_{pr} = Q_t - \frac{4}{x^2}\left\{ \sum_{n=1}^{\infty} \frac{n(n+2)}{n+1} Re(a_n^* a_{n+1} + b_n^* b_{n+1}) + \sum_{n=1}^{\infty} \frac{2n+1}{n(n+1)} Re(a_n^* b_n) \right\} \tag{3.15}$$

where Q_t is the extinction efficiency of Eq. (3.14). Resonances of a_n and b_n are responsible for the observed spikes in radiation pressure, according to the computations of Chylek et al. (1978). The curves of both observation and computations are shown in Fig. 3.10, in which the backscattered intensity is plotted as a function of wavelength for the experimental curve (top) and as a function of size parameter x for the Mie calculations (middle and bottom diagrams). The

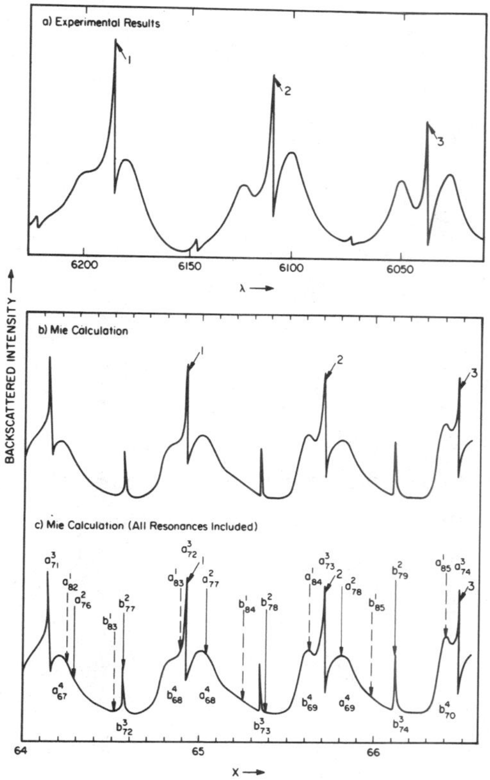

Figure 3.10 Experimental measurement of backscattered intensity of laser light from a droplet of oil (top curve) and that computed by the Mie theory for a comparably sized particle (middle and bottom curves). The peaks in the computations are due to different orders in the Mie coefficients a_n and b_n (after Chylek et al., 1980).

experimental curve is plotted on a scale of wavelength, because the radius of the droplet was known to only $\pm 5\%$, which is not sufficient to correspond to the high resolution of the computations, but the wavelength of the laser beam was known very precisely. It is obvious from the diagram that the two curves are describing the same phenomenon.

It will be remembered that a_n and b_n are complex quantities. They can be expressed in the form

$$a_n = \frac{a_n^{(1)}}{a_n^{(1)} - i a_n^{(2)}}$$
$$b_n = \frac{b_n^{(1)}}{b_n^{(1)} - i b_n^{(2)}} \tag{3.16}$$

If m is real then all of these quantities are real. In this case a_n and b_n reach maxima at the values of x at which the imaginary values of a_n and b_n are zero. These maxima are the resonance peaks of Fig. 3.10. By matching the numbered resonance peaks in top and middle curves, it is evident that the values of x used for the computations are slightly in error with respect to the precise wavelengths of the top curve. This means that the measured size of the particle was similarly in error. In fact, this technique would yield very precise measurements of the size of particles suitable for levitation in a laser beam.

The diagram shows that the resonances are grouped in families of three or four members. The bottom curve of Fig. 3.10, being computed at the very high resolution of $\Delta x/x$ of the order of 10^{-9}, shows the entire complement of resonances in this small range of x. If the particle has a significant amount of absorption the resonances tend to be wiped out (Chylek et al., 1978).

One of the features of the scattering problem, which has been subject to some question, is the concept of surface waves. Van de Hulst (1957) thought them to be waves that travel around the surface of the body, which, by interference with the forward-diffracted wave, could explain the ripple superimposed on the general curve of extinction efficiency versus size parameter. Similarly, interference with the backward-reflected wave could explain the phenomenon of the glory. While an intuitive explanation of surface waves must be considered open, the unambiguous connection of both the ripples in extinction efficiency and details of the glory with resonances of the Mie coefficients require at least a new look at surface waves.

3.2.7 Scattering by a polydispersion of particles

So far in this chapter we have concentrated on single-particle scattering. While the scattering properties of single particles or groups of similar particles are of interest in many applications (studies of biological systems, calibration of light-scattering instruments, etc.), the scattering and absorption of light by volumes containing a dispersion of particles of different sizes, shapes, and indices of refraction is more applicable to problems of radiative transfer in planetary atmospheres (including the atmosphere of the earth), hydrosols, and other turbid media. Thus we include a brief discussion of this more general and interesting problem, even though it does not meet the single particle criterion. The term polydispersion is normally applied to such systems of varying particle characteristics, as opposed to monodispersion as a system of particles with uniform properties.

The polydispersion of main interest in the present context is the natural aerosol in the earth's atmosphere. The atmospheric aerosol is made up of dust, haze, pollution, pollen, spores, and many other types of particulate material in the air. Even clouds, fog, and rain fall under the category of polydispersions, but are not normally thought of as atmospheric aerosols. Since the natural aerosol consists of such a disparate conglomeration of particles, with their diversity of sizes and chemical makeup, the problem of scattering in the atmosphere is a complicated one. The problem is compounded by both a lack of knowledge of the particle characteristics themselves and our inability to handle the theory of light scattering by particles of a general nature. Simplifying assumptions have to be made in both of these areas to achieve results that even approximate radiation fields in the actual atmosphere.

In addition to the radiative effects of aerosols there are those of atmospheric gases—oxygen, nitrogen, ozone, carbon dioxide, water vapor, and numerous trace gases. These gases exhibit both scattering and absorption. Consisting as they do of molecular size particles, they constitute the aerosol-free component, which is termed the Rayleigh atmosphere. The properties and radiative effects of the Rayleigh atmosphere are the subjects of Chapter 4 of this book.

Because of the widespread nature and radiative importance of atmospheric aerosols, an extensive body of literature on their properties and effects has been developed. Relatively recent monographs devoted to aerosols and their effects are those of Hidy (1972), Schuerman (1980) and McCartney (1976). These are in addition to the standard references of a decade or two ago, which continue to be in widespread use, such as Van de Hulst (1957), Junge (1963), Kerker

(1969), Deirmendjian (1969), Tricker (1970), Cadle (1966), and Sobolev (1972), to name a few. Bullrich (1964) and Hansen and Travis (1974) developed reviews of the scattering of radiation by atmospheric aerosols, and Rozenberg (1966) gave a thorough discussion of the twilight problem. Some of the more classical publications in the field are Pernter and Exner (1910), Humphreys (1940), Stratton (1941), and Minnaert (1940).

In order to compute the scattering and absorption of radiation by an aerosol, the size distribution, number concentration, spatial distribution, and complex index of refraction of the aerosol particles must be known. Data on the index of refraction of substances composing aerosols in the atmosphere were given in Chapter 2, but since the data are still incomplete and the aerosols themselves are extremely variable, the index of refraction applicable to the aerosol at a given location and a given time remains one of the major unknowns for computations of the radiative effects of natural aerosols. The sizes of aerosol particles are also widely variable, covering at least four orders of magnitude and extending from large molecules or Aitken nuclei of radii 10^{-3} μm to large dust particles of 10 μm or more in radius. The smallest particles are produced mainly by the process of gas-to-particle conversion. The gases undergoing such conversion in the atmosphere are normally from combustion, although other sources make their contributions. For instance, the most persistent particles in the stratosphere are the very small sulfuric acid droplets resulting from the conversion of gases injected into the stratosphere by volcanic eruptions. These very small particles are not very important for light scattering because of their size and relatively low concentrations. The lower size limit for particles of the natural aerosol which are radiatively effective is usually taken at about 0.04 μm. Even though each particle of this size is not very effective in light scattering, their concentrations are high and their radiative effects may be significant under certain conditions. The large particle limit is usually taken as a radius of about 10 μm, although there may be an occasional particle larger than this. Such large particles fall out rapidly, however, and since they consist mostly of dust picked up from the surface they are confined to low levels and contribute little to the overall radiation regime of the atmosphere. Thus, for practical purposes, the bulk of the natural aerosol, in terms of total mass, number density, and radiative effects, is in the size range 0.04 μm to 10 μm radius.

As seen above, the various parameters of the radiation field are critically dependent on particle size. For computations of radiative transfer in the natural atmosphere, it is convenient to represent the size versus frequency by a statistical distribution function of some type. A number of different distributions have been proposed and used for computations of aerosol effects, three of the principal ones being shown in Fig. 3.11.

3.2.7.1 The Junge distribution

Perhaps the most widely used distribution, and the one most firmly based on aerosol observations from different parts of the world, is that of Junge (1963). The Junge distribution is a simple power law written as

$$dn(r) = cr^{-\gamma}\, d(\log r) \tag{3.17}$$

where n is number, r is radius, c represents the total particle concentration, and γ is a fitting parameter that generally lies between 2.5 and 4.0 for the natural

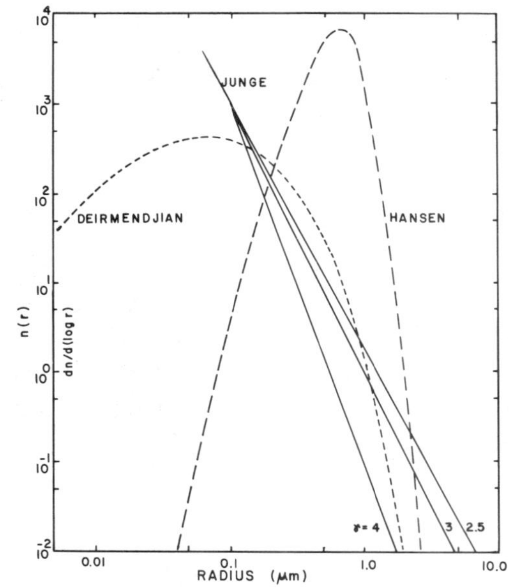

Figure 3.11 Particle size distributions for the power law of Junge for γ=2.5, 3.0, and 4.0 (inside ordinate scale), the modified gamma distribution of Haze L of Deirmendjian, and the standard distribution of Hansen.

aerosol. As seen in the diagram, γ represents the slope of the curve on this type of plot. A change of γ corresponds to a change of the relative proportions of small and large particles in the aerosol. Observations have shown that continental-type aerosols have $\gamma \approx 3.0$, and the generally smaller particles of maritime aerosols give $\gamma = 3.5$ to 4.0. Typical results from use of the Junge distribution are given in the review by Bullrich (1964).

3.2.7.2 The gamma distribution

This well-known statistical distribution was used by Zelmanovich and Shifrin (1971) for providing tabulations of total cross sections for scattering, attenuation, and radiation pressure for 124 wavelengths, and angular cross sections and degree of polarization at 37 scattering angles from 0° to 180° for 12 wavelengths. The gamma distribution takes the form

$$n(r) = ar^{\alpha} \exp(-br) \tag{3.18}$$

where a, α, and b are constants to be determined by observation. Constant a is a measure of the number concentration of particles, and α and b are parameters for fitting the theoretical curve to observations.

3.2.7.3 The modified gamma distribution

The gamma distribution has been modified by Deirmendjian (1969) and by Hansen and Travis (1971) to give slightly different distributions of particle sizes. The choice between them is made difficult by the lack of adequate data on actual distributions, and is probably best made on the basis of convenience. Deirmendjian (1969) introduced an additional exponent γ into the exponent of Eq. (3.18) to yield

$$n(r) = ar^{\alpha} \exp(-br^{\gamma}) \tag{3.19}$$

The constants a, α, and b are the same as for the gamma distribution, but the additional constant γ provides more flexibility in determining the shape of the distribution. The two distributions are the same for $\gamma = 1$. Although an infinite number of distribution curves are possible, the three which have been applied to aerosols by Deirmendjian and have been found most useful are the haze models *H, L,* and *M*. Their general configuration is shown by the curve of Deirmendjian in Fig. 3.11, which is for the haze *L* model. The haze *M* model has the largest numbers of both very large and very small particles, while haze *H* is the converse of this. Haze *L* has more moderate numbers of both large and small particles. Tabulations of the elements of the scattering phase matrix as computed for these models are given by Deirmendjian (1969).

3.2.7.4 The standard distribution

This distribution, so termed by its developers (Hansen and Travis, 1971), is a second modification of the basic gamma distribution. It takes the form

$$n(r) = cr\left(\frac{1-3b}{b}\right)\exp\left(-\frac{r}{ab}\right) \tag{3.20}$$

where c is again particle concentration. A curve of this distribution, as given by Hansen and Travis (1971) for $a=1$ and $b=0.10$, is shown in Fig. 3.11. The numbers of both small and large particles are very low for this case, with the bulk of the particles falling at about 1.0 μm radius. Of course, the curve can be shifted or changed in shape by appropriate changes of the parameters. The extent to which this distribution corresponds to that of the natural aerosol is a subject for further investigation, but its simplicity is appealing for analysis and use in computations. For instance, the effective radius typical of the entire distribution is the parameter a, and b is the effective variance for the particles of the distribution.

3.2.7.5 Light scattering by polydispersions

The intensity of light scattered by the standard and Junge distributions of particles is shown as a function of scattering angle in Fig. 3.12 according to the computations of Hansen and Travis (1974). The effective radius for the standard distribution was taken as $a=1.0$ μm and the effective variance as $b=0.25$. The size limits for the Junge distribution were taken as 0.362 μm and 2.14 μm,

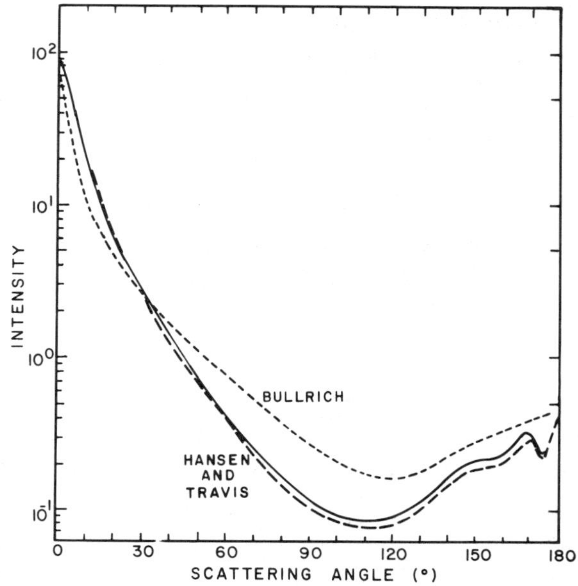

Figure 3.12 The intensity as a function of scattering angle of light scattered by a polydispersion of spherical particles according to the computations of Hansen and Travis for the Junge (dashed curve) and standard (solid curve) size distributions, and of Bullrich for the Junge distribution. (Data for Hansen and Travis computations: $m=1.33-0.0i$; $\lambda=0.55$ μm; effective radius and variance for standard distribution 1 μm and 0.25, respectively; range of radii for Junge distribution $r=0.362$ μm to 2.14 μm. Data for Bullrich computations: $m=1.50-0.0i$, $\lambda=0.55$ μm, $\gamma=3.0$, range of radii $r=0.04$ μm to 10 μm.)

values quite different from the 0.04 μm and 10.0 μm mentioned previously. The index of refraction was $m=1.33-0.0i$, the wavelength was $\lambda=0.55$ μm, and the incident light was assumed to be unpolarized. As can be seen, the scattered intensity varied by more than three orders of magnitude over the range of scattering angle Θ from 0° to 180°, with a minimum at about $\Theta=110°$. The intensity differences between the two size distributions are very minor, as might be expected from the small range of particle radii centered at about 1.0 μm for the Junge distribution and an effective particle radius of 1.0 μm for the standard distribution. Bullrich (1964) performed equivalent computations, also for the Junge distribution, but with a range of radii $r=0.04$ μm to 10.0 μm, and obtained the third curve of Fig. 3.12. The wider particle size spectrum, combined with the difference of refractive index, washed out the details of phase function near the backward direction and considerably decreased the importance of the major minimum. These and many other computations of a similar nature show that the phase function is sensitive to the size spectrum of the scattering particles, which means that, in principle, measurements of the intensity distribution of radiation scattered by an aerosol may be inverted to yield information on the aerosol itself. Unfortunately, at the present time, efforts in this direction have met with only modest success.

Computations have shown that the polarization distribution of the light scattered by an aerosol is even more sensitive to particle characteristics than the intensity distribution. This fact is shown very graphically in Fig. 3.13. The aerosol models for these polarization curves are the same as those for the intensities of the previous diagram, and in fact they both result from the same computations. The pronounced negative branch of the curves of Hansen and Travis is replaced by positive polarization in that of Bullrich, the positive polarization

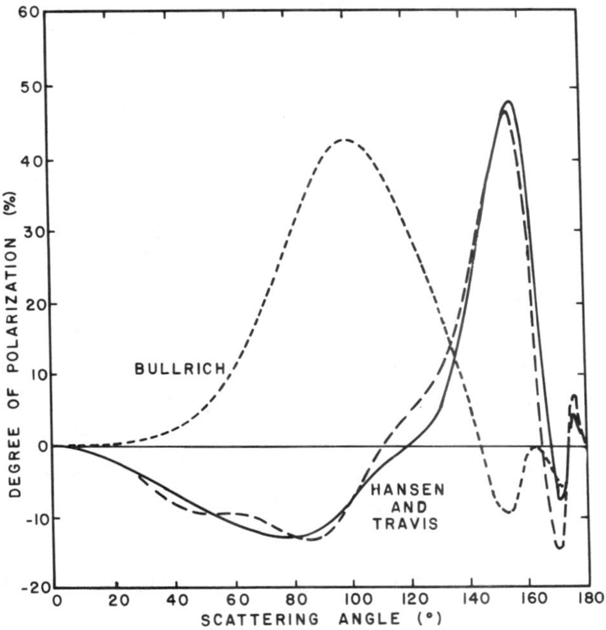

Figure 3.13 Same as Fig. 3.12 except for degree of polarization instead of intensity.

no doubt being due to the large number of particles in the $r = 0.04\ \mu m$ to $1.0\ \mu m$ region which were not included in the computations of Hansen and Travis. The most important point of this diagram is that it illustrates the sensitivity of the polarization distribution of the scattered light to the size spectrum of the scattering particles.

3.2.8 Scattering by nonspherical particles

The Mie theory has found extensive application in many fields, as spherically shaped particles are frequent in nature. Cloud, fog, and all but the largest rain drops are spherical in shape, and some atmospheric aerosols collect water on their surfaces in a thick enough layer to end up as a sphere, regardless of their initial configuration. Many types of particles of the atmospheric aerosol, however, retain their nonspherical form. These irregularly shaped particles exert a strong influence on the transfer of radiation in the atmosphere, and thereby on the energy balance of the atmosphere and illumination provided by the sun. They also affect methods of probing the atmosphere by laser and other types of electromagnetic radiation, remote sensing of atmosphere and surface, and many other uses of naturally and artificially produced radiation. As shown below, nonspherical particles are particularly important in influencing the polarization of light scattered by atmospheric aerosols.

There is presently great concern among people everywhere that the detonation of nuclear weapons might inject clouds of dust and smoke into the atmosphere in sufficient quantities to blot out the sun and bring about a dramatic decrease in temperature over much of the earth. This is the essence of the phenomenon termed "nuclear winter." With a sufficiently dense and persistent cloud, it is unlikely that shapes of the particles can either promote or eliminate the possibility of a drastic modification of the energy balance of the earth-atmosphere system leading to a "nuclear winter," but radiative calculations based on the assumption of spherical particles can yield erroneous estimates of the overall effects of the cloud. In addition, the use of measured radiation from the cloud to deduce the physical characteristics of cloud particles, the essence of the inverse problem, will unavoidably yield erroneous results unless proper account is taken of the nonspherical shapes of the particles.

Nonspherical shapes are likely to be more predominant among particles resulting from nuclear explosion than in particles of the natural aerosol for at least two reasons. First, the numerous particles of soil and sand that would be injected into the atmosphere at the site of the explosion are naturally irregular and jagged in shape, and most particles of this type, being composed largely of compounds of silicon, are nonhygroscopic. Thus they do not readily collect a layer of liquid water during their residence in the atmosphere, but maintain their original irregular shape and exhibit radiative characteristics commensurate with nonspherical particles. Second, smoke particles produced by the conflagrations feeding on carbonaceous materials of cities are known to be not only irregular in shape, but they may also form into long chains of individual particles by agglomeration in the dense cloud. An additional factor of radiative importance in the burning of carbonaceous materials are the holes, interstices, and projections which are observed to form in and on the surface of particles produced by different types of fires. For instance, under more normal burning situations, Ahlberg and Fangmark (1983) found smoke particles from an oil-fired power plant to be hollow cenospheres with a porosity of 88%. In contrast, Del Monte et al. (1983) found more than 90% of particles from a similar plant which was burning coal to be spherical in shape with a smooth external surface, with a second class of much less numerous particles being in the form of porous spheres. Chippett and Gray (1978) observed a great change of particle characteristics within less than 2 mm from the burning head for soot particles formed by propane burning in a deficiency of air. While the optical properties of smoke from the widespread burning of cities are (and must remain) unknown, there is little doubt that numerous particles would be nonspherical in shape.

3.2.8.1. Methods of computing scattering by nonspherical particles

Since the Mie theory gives the complete solution for scattering and absorption of radiation by spheres, it is of interest to compare scattering computed by the Mie theory with scattering resulting from nonspherical particles with the same sizes and indices of refraction. Chylek (1976) has proved that the extinction cross section of a randomly oriented nonspherical particle is always larger than that of a spherical particle of equal volume. Measurements by Holland and Gagne (1970) on silicon dioxide particles of flat plate-like shape with jagged edges and random orientation gave angular distributions of certain of the elements of the scattering matrix which were similar to those computed for spheres over scattering angles $\Theta = 0°$ to $80°$, but the results were widely divergent at larger scattering angles, and for some matrix elements at all scattering angles. Greenberg (1972) states that needles of iron of length/diameter ratio of 25 would absorb ten times more efficiently than equivalent volume iron spheres, and that rough particle extinction is 50% larger than that for smooth particles in the size range $x \geq 3$.

Several methods of computing the scattering of radiation by particles of various shapes have been developed. The main ones are outlined below.

3.2.8.1.1 The Mie theory for spheres: x and m arbitrary

This method was discussed in detail in section 3.2.1.

3.2.8.1.2 The Rayleigh-Gans approximation: particles of $|m-1| \ll 1$ and $(2\pi r/\lambda)\,|m-1| \ll 1$

Chapter 4 is devoted largely to this type of particle.

3.2.8.1.3 The method of geometric optics: very large particle sizes of $x \geq 2000$, m arbitrary

This method is included in discussions of the rainbow and halo in sections 3.4 and 3.5.

3.2.8.1.4 The extended boundary condition method (EBCM)

This is a powerful method for computing scattering by particles of arbitrary shape for sizes ranging from small particles in the Rayleigh-Gans region to the large particles of geometric optics (Barber and Yeh, 1975). It was originally developed by Waterman (1971) and has more recently been summarized and applied to scattering by homogeneous oblate and prolate spheroids and a finite circular cylinder by Barber and Yeh (1975) and Barber (1977). Barber et al. (1981) have developed the necessary programs for performing the calculations on a microcomputer without excessive running time requirements. The method is based on the concept of surface currents being generated on the surface of the particle by the incident radiation field, with these surface currents being the source of the scattered radiation. Barber and Yeh (1975) break the analysis down into five steps which may be summarized as follows:

1. An equivalence theorem is applied that, as far as the external fields are concerned, effectively replaces the scattering object by a set of surface currents over a surface s which is coincident with the surface of the original object. These surface currents are the source of the scattered field outside s.

2. An expression for the unknown surface currents which result in radiation outside the volume is derived from the fact that they radiate a scattered field which cancels the incident field throughout the interior of the volume. The equations are put in a form suitable for numerical solution by expansion of the various field quantities in vector spherical wave functions.

3. The internal fields are expanded in regular vector spherical wave functions with coefficients to be determined.

4. The boundary conditions at the surface are applied, and the continuity of the tangential fields leads to a linear system of integral equations for the coefficients of the unknown internal fields in terms of the expansion coefficients of the incident field.

5. The scattered far field is finally determined by evaluating the internal field at the surface and then solving the scattered field in terms of the surface fields. This gives the differential scattering cross section as proportional to the square of the far field amplitude.

The formalism for the EBCM requires expansion of incident and scattered fields in spherical wave functions, which Barber et al. (1981) denote as M and N. The incident field is given in the notation of Barber (1977) by the relation

$$\bar{E}^{(i)}_{(\bar{r})} = \sum_{\nu=1}^{\infty} D_\nu \left[a_\nu \bar{M}^{(1)}_\nu (k,\bar{r}) + b_\nu \bar{N}^{(1)}_\nu (k,\bar{r}) \right] \tag{3.21}$$

Here $k=2\pi/\lambda$ is wave number, $\bar{r}$ is distance from the surface, D is a normalization constant and ν is a combined index incorporating the spherical harmonic indices σ, m, and n. The expansion coefficients a_ν and b_ν are assumed known for the incident wave. The scattered field is given by

$$\bar{E}^{(s)}_{(\bar{r})} = \sum_{\nu=1}^{\infty} 4D_\nu \; [f_\nu \, \bar{M}_\nu^{(3)} \; (k,\bar{r}) \; + \; g_\nu \, \bar{N}_\nu^{(3)} \; (k,\bar{r})] \qquad (3.22)$$

where the expansion coefficients f_ν and g_ν are initially unknown and are to be determined. Superscript (1) in Eq. (3.21) indicates M and N are finite at the origin, and superscript (3) in Eq. (3.22) is used to indicate that the wave is outgoing. Functions $\bar{M}^{(1)}$ and $\bar{N}^{(1)}$ of Eq. (3.21) have a Bessel function radial dependence, and $\bar{M}^{(3)}$ and $\bar{N}^{(3)}$ have a Hankel function radial dependence.

The steps necessary for performing the computations are summarized by Barber (1977), and the development of a FORTRAN computer program for use on a microcomputer is discussed by Barber et al. (1981). Basically, the requirement is to first choose a value of N, which is a maximum value of index ν, solve a set of simultaneous linear equations $2N \times 2N$ in size, and then iterate on N until convergence to the required accuracy is achieved. The maximum value of N is dependent on size and index of refraction of the object, and on the extent to which the shape deviates from a sphere. Small spheres with m near unity require only small values of N, whereas a large cylinder with high index of refraction demands large N. Even with possible shortcuts in numerical techniques, computer requirements tend to be very demanding. For instance, for a small and mildly prolate spheroid ($x=5.6$, ratio of axes$=2.0$) with random orientation, an overnight run time of an LSI-11/2 microcomputer was required for computations at 5° increments from 0° to 180° scattering angle. With a more powerful computer, of course, time requirements could be greatly reduced, but for relatively large and highly asymmetric particles the computer requirements are formidable. For instance, the largest equivalent size parameter x for the computations of Barber (1977) for demonstration of the method using a UNIVAC 1108 computer was x=2.5, and the particles were all axially symmetric (spheroids and a cylinder). Latimer and Barber (1978) considered x up to 4.3 for indices of refraction m=1.05 and 1.20, and Barber and Yeh (1975) performed sample calculations for x as large as 9.5 for prolate and oblate spheroids. However, none of these computations is particularly applicable to sizes and indices of refraction appropriate to the natural aerosol in the atmosphere, and particularly for those aerosol particles which are not well approximated by axial symmetry. As far as is known, no computations for these more realistic cases have been performed by the EBCM, but with adequate computing power they no doubt could be. Barber and Yeh (1975) conclude that the method is suitable for determinations of either the radiation field resulting from scattering by given particles (the direct problem) or of the characteristics of particles producing a given radiation field (the inverse problem).

Computations performed by the EBCM reported so far support the following conclusions:

1. Scattering in the equatorial plane by cylinders is independent of cylinder length for cylinders of length/diameter ratios of 2 or 3 or greater.

2. Even for the same overall dimensions, differential scattering characteristics are different for capped cylinders and prolate spheroids.

3. The addition of some absorption to a particle causes much greater decrease of intensity in the backward than in the forward direction.

3.2.8.1.5 Method of anomalous diffraction

This method, developed by Van de Hulst (1957), is applicable in cases in which the index of refraction is arbitrarily close to 1: $|m-1| \ll 1$, but $x=2\pi r/\lambda$ is large. The case of x small and $|m-1| \ll 1$ is that of Rayleigh scattering. The controlling parameter for the anomalous diffraction theory is $\rho=2x(m-1)$. For a given ρ, the condition $m \rightarrow 1$ implies that $x \rightarrow \infty$. In this domain the intensity surrounding the particle is due to a straight transmission of the incident wave plus diffraction according to Huygens' principle. This means that the scattered light is concentrated in the region of small scattering angles. Van de Hulst (1957) shows that the extinction efficiency Q_t describes series of maxima and minima when plotted as a function of ρ. In fact, Q_t is given by

$$Q_t = 2 - \frac{4}{\rho} \sin \rho + \frac{4}{\rho^2}(1 - \cos\rho) \tag{3.23}$$

a formula which Van de Hulst calls the most useful in the whole domain of the Mie theory. Its usefulness arises from its ability to describe the important features of the extinction efficiency not only for m close to 1 but for values of m as large as 2. Thus it has application to the natural aerosols of the atmosphere for computations of Q_t. It is to be emphasized, however, that the anomalous diffraction theory gives the scattered intensity mostly in the forward direction.

Latimer and Barber (1978) compared results of computations of the anomalous diffraction method with those from the EBCM for an index of refraction $m=1.05$ and $m=1.20$ with $x=2.37$ and 4.31 over the range of scattering angles 0° to 180° for ellipsoids of revolution. The anomalous diffraction method failed badly at large scattering angles, but reasonably good agreement with EBCM results was obtained by repeated trials in which the long axis of the prolate spheroid was oriented at different angles with respect to the incident beam to change the projected cross-sectional area of the particle. The use of this artifact was not pursued beyond the exploratory stage. As would be expected, better results were obtained by the anomalous diffraction method for $m=1.05$ than for $m=1.20$. Latimer (1984a) used this method for scattering by homogeneous spheres with holes of various sizes passing through them for $m=1.05$ and a range of x from 0.1 to 1200. The results of this study are mentioned below.

3.2.8.1.6 Method of suppression of resonances

Chylek (1977) and Chylek et al. (1976) have shown that resonances in the Mie coefficients a_n and b_n are responsible for the ripple structure superimposed on the general curve of extinction efficiency versus size parameter of spherical particles. The resonances are equivalent to the so-called surface waves, which were thought by Van de Hulst (1957) to be waves traveling along the surface of the sphere. In this case, the ripple structure of the extinction efficiency could be due to interference of the surface wave with the forward diffracted wave, and the intensity variations in the glory could be explained by interference of surface waves with the backward reflected wave. While the intuitive physical model is still a matter of conjecture, the work of Chylek and his colleagues has shown unambiguously that resonances are responsible for both the glory and the ripple structure in extinction efficiency.

This being the case, Chylek et al. (1976) conjectured that surface waves probably do not exist on strongly nonspherical particles, so a good correction for nonsphericity would be the removal of resonances in the Mie theory. This requires a modification of a_n and b_n in the resonance regions. By setting $a_n=1/2$ and $b_n=1/2$, good agreement between theory and experiment was obtained for particles in the range of size parameter $3 \leq x \leq 30$. According to these authors, the method is applicable to medium-sized particles of arbitrary shapes and random orientations, but not to uniformly oriented particles or to those of $x>30$.

Welch and Cox (1978) have applied the resonance suppression method to randomly oriented and arbitrarily shaped particles of different sizes and indices of refraction and found significant changes of scattering, absorption, and extinction efficiencies due to nonsphericity. The extinction efficiency is increased by nonsphericity in the size range $x \gg 1$, but it is decreased in the range $3 \leq x \leq 5$. The maximum absorption occurs at $5 \leq x \leq 10$ for all values of the complex part of the index of refraction n'. Nonsphericity has little effect on attenuation for $x=2$ and $x=30$ for the real part of the refractive index $n-1.3$. Absorption efficiency is particularly sensitive to nonsphericity. For instance, for $n'=10^{-7}$, absorption can be increased by a factor of 10^5 by nonsphericity, and by several orders of magnitude for $n'=10^{-5}$ and $n \leq 2.0$.

3.2.8.1.7 The statistical approach

Nevitt and Bohren (1984) state that an exact solution to scattering by nonspherical particles, even if available, would give more detail than is needed. They chose instead to analyze the scattering statistics of randomly oriented anisotropic oscillators, which they term Rayleigh ellipsoids. This means that the particles have $x \ll \lambda$. They make use of the fact that such oscillators have resonances near-infrared absorption bands in minerals such as ammonium sulfate, quartz, and other silicates. In these cases, small irregularly shaped particles have absorption or extinction cross sections up to an order of magnitude greater than those of equivalent spheres. The results are particularly applicable, according to these authors, to remote sensing of aerosols at wavelengths of about 10 μm.

3.2.8.1.8 The expansion of vector wave equations in spheroidal coordinates

This method, developed in full by Asano and Yamamoto (1975), separates the vector wave equations in spheroidal coordinates and expands them in terms of the spheroidal wavefunctions. The unknown coefficients in the expansion are determined by a system of equations derived from the continuity of the tangential components of electric and magnetic vectors across the surface of the spheroid. The solutions are applicable to either prolate or oblate spheroids, but only to spheroids. The five parameters involved are (1) size parameter of particle, (2) eccentricity of the spheroid, (3) refractive index, (4) angle of incidence with respect to the rotational axis of the particle, and (5) angle of scattering. Asano (1979) extended the computations up to a size parameter of 30, at which point instabilities in numerical computations prevented computations for still larger sizes of particles. In a later paper, Asano and Sato (1980) applied the method to assemblages of randomly oriented prolate and oblate spheroids. The results obtained are shown further below.

3.2.8.1.9 The semi-empirical theory of Pollack and Cuzzi (1980)

For particles of size parameter $x \leq 5$, the Mie theory is used by these authors for nonspherical particles. For larger particles, however, the semi-empirical theory considers three separate components: diffraction, external reflection, and transmission. Physical optics theory is used for diffraction, geometrical optics theory for reflection, and a parameterization is used for transmission. The parameterization is based on observations, and takes the form

$$I_T = C_T \exp(-b\theta) \tag{3.24}$$

where C_T is a normalization factor derived from

$$\int I_T \frac{d\Omega}{4\pi} = 1 \tag{3.25}$$

and b is an empirically derived constant. The semi-empirical theory yields good fits to laboratory measurements on particles of various types.

3.2.8.2 Results of computations for nonspherical particles

While the theoretical problem of scattering by spherical particles and those of certain other shapes (mainly ellipsoids and cylinders) has been solved, such is not the case for most other particle shapes. For this latter case, two approaches are possible. First, approximate methods directed toward a theoretical solution are not only feasible but have been applied by many authors for particular types of particles. Some of the approximate methods were outlined above. A second approach in the absence of an exact theoretical solution is that of measurement. In the final analysis, measurements are required for validation of theoretical results, exact or approximate, but unfortunately this aspect has not received the emphasis commensurate with its importance. It may be remarked that the possibility of major contributions to solution of the radiative transfer problem lies more in the physical measurement aspect than in any other element of the problem. Computations of the radiative properties of nonspherical particles have generally been confined to particles, such as cylinders, with symmetry around one axis, those with symmetry around two or three axes (oblate and prolate spheroids), rough particles made by stacking cylinders together, ice crystals, and hypothetical assemblages of particles of random shape and orientation. The results of the computations for these generally nonspherical particles are normally compared to those for spherical particles of either equal cross-sectional area or equal volume, the differences being ascribed to the effects of nonsphericity.

3.2.8.2.1 Prolate and oblate spheroids

More computations have been made for spheroids than for any other type of nonspherical particles, due, no doubt, to the availability of theoretical methods for exact representations of their scattering properties. While details vary with particle characteristics and orientation, generally a comparison of radiative properties of spheroids with those of spheres show the following.

The scattering properties of spheroids are more sensitive to the size parameter than those of spheres. For instance, the first and largest peak of scattering cross section, which occurs at $x \approx 6$ for spheres, is displaced to larger sizes and is generally greater for randomly oriented spheroids than for spheres. This is shown by the computations of Asano and Sato (1980) in Fig. 3.14, in which normalized cross sections for extinction, scattering, and absorption are shown as functions of size parameter for prolate and oblate spheroids of index of refraction $m=1.33-0.05i$ and ratio of axes $a/b=5$. The extinction is mainly due to scattering and prolate spheroids scatter more efficiently than do oblate spheroids.

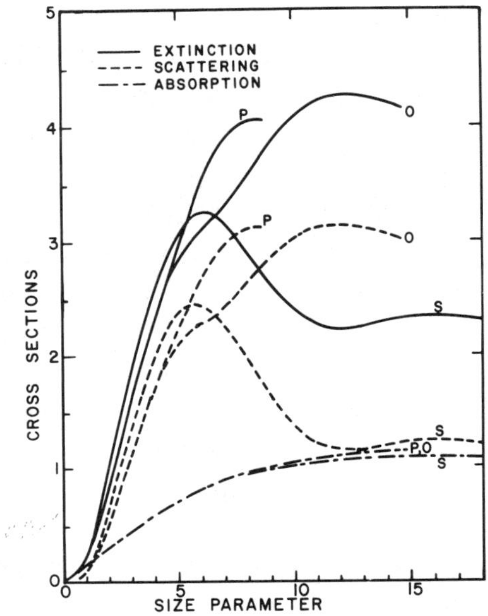

Figure 3.14 Cross sections for extinction, scattering, and absorption of randomly oriented prolate (P) and oblate (O) spheroids compared to those of equal volume spheres (S) as functions of equivalent size parameter. Index of refraction m=1.33–0.05i and ratio of spheroid axes a/b=5 (after Asano and Sato, 1980).

The spheroids have about the same absorption cross sections as equivalent spheres if they are randomly oriented. For prolate spheroids which are oriented with the long axis in the direction of the incident light, the scattering cross section shows strong differences from that of equal volume spheres, as shown by Fig. 3.15 (after Asano, 1979). The primary maximum is very pronounced for these uniformly oriented ellipsoids, and is shifted to smaller values of x. For values of x from 4 to 7 the efficiency of scattering is very low for ellipsoids, and the oscillations of scattering efficiency versus x are much more pronounced than for spheres. Since, however, there is probably no efficient way for the ellipsoids to become highly oriented if they did exist in significant numbers in the atmosphere, this case is of less interest in the present context than that of randomly oriented spheroids.

For studies oriented toward large amounts of carbonaceous smoke in the atmosphere, the absorption by nonspherical carbon particles is of paramount

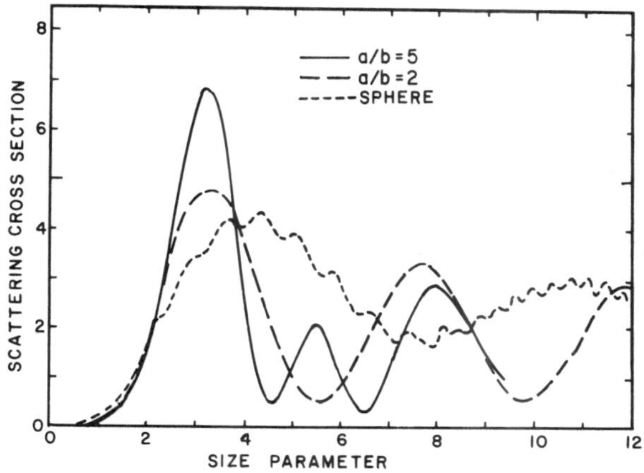

Figure 3.15 Scattering cross sections of prolate spheroids oriented with the long axis parallel to the incident radiation with axial ratios a/b=2 and 5 as functions of size parameter of equal volume spheres. A similar curve for spherical particles is given for comparison (adapted from Asano, 1979).

importance. While the smoke particles are not likely to be in the form of spheroids, it is of interest to see how the absorption by carbon spheroids compares with that of equal volume spheres. Such a comparison was made by Roessler et al. (1983) by the EBCM for randomly oriented spheroids of index of refraction $m=2.0–1.0i$ typical of carbon. The results are shown in Fig. 3.16 for a wavelength of $\lambda=0.55\ \mu$m. Included also are the results of computations for very small cylinders and discs. Parameter Z is the ratio of axes, and Pro. and Obl. indicate prolate and oblate, respectively. The diagram shows that carbon spheroids of equivalent diameters of less than about 0.12 and axial ratio of 3 are considerably more absorbent than equivalent spheres, but there is not much difference for larger sized particles. Very small and greatly elongated ($Z=10$) carbon spheroids are nearly twice as effective in absorbing radiation as are carbon spheres, and very flat oblate spheroids are over twice as effective. The highest absorption is by small carbon discs.

The angular distribution of the P_{11} elements of the scattering matrices for prolate and oblate spheroids of $m=1.33$ and axial ratio $a/b=5$ for size parameter $x=15$ is shown in Fig. 3.17 according to the computations of Asano and Sato (1980). The dashed curves give the elements for equivalent spheres. Element

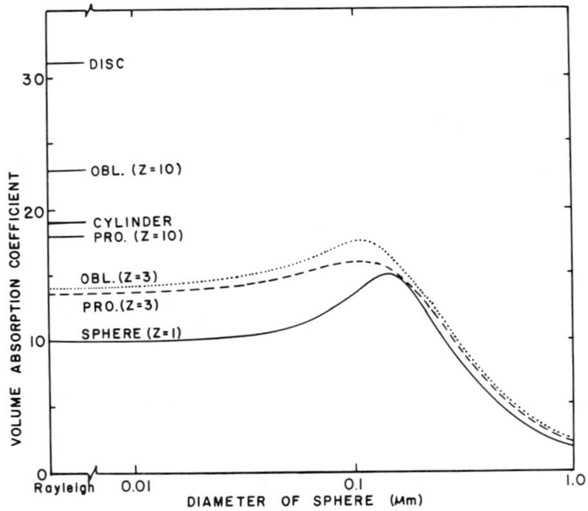

Figure 3.16 Volume absorption coefficient for spheres (Z=1), prolate and oblate spheroids with ratio of axes a/b=3 and 10, and for disks and cylinders as functions of the diameter of equal volume spheres for radiation of wavelength λ=0.55 μm and index of refraction m=2.0−1.0i (after Roessler et al., 1983).

P_{11} is the scalar intensity. While the intensity in the region of forward scattering is not much different for spheres and spheroids, its change with angle is significantly affected by nonsphericity, particularly for oblate spheroids. Prolate spheroids give less intensity at angles throughout the whole backward hemisphere, and some of the variation in the glory region is destroyed. Oblate spheroids give slightly more forward scattering, but much less intensity through angles 20° to 180°. The difference reaches more than an order of magnitude through a wide range of angles in the backward direction.

The degree of linear polarization given by the ratio of matrix elements $-P_{12}/P_{11}$ and shown in Fig. 3.18 appears to be particularly sensitive to nonsphericity. In fact there is little relation between the polarization curves for spheres and spheroids. Measurements of the phase function for polarization might be a method of characterizing nonspherical particles if they were known to be ellipsoids or some other well-defined shape. While the other matrix elements certainly show differences between spheres and ellipsoids, the differences are not as pronounced as for the degree of polarization.

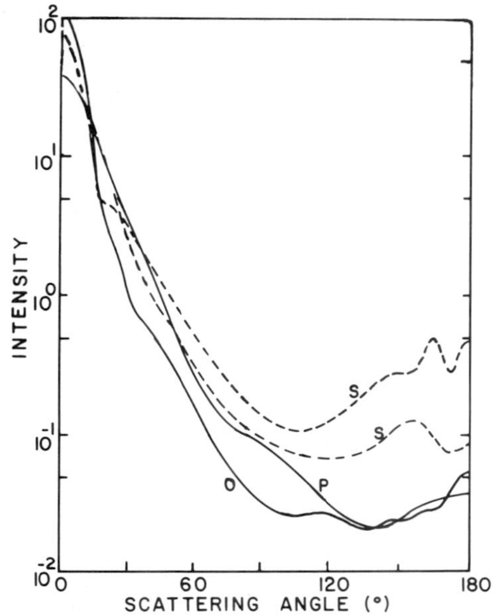

Figure 3.17 Angular distribution of the scalar intensity (scattering matrix element P_{11} of radiation scattered by randomly oriented prolate (P) and oblate (O) spheroids with ratio of axes a/b=5, size parameter x=15, and index of refraction m=1.33 compared to that of equal area spheres (adapted from Asano and Sato, 1980).

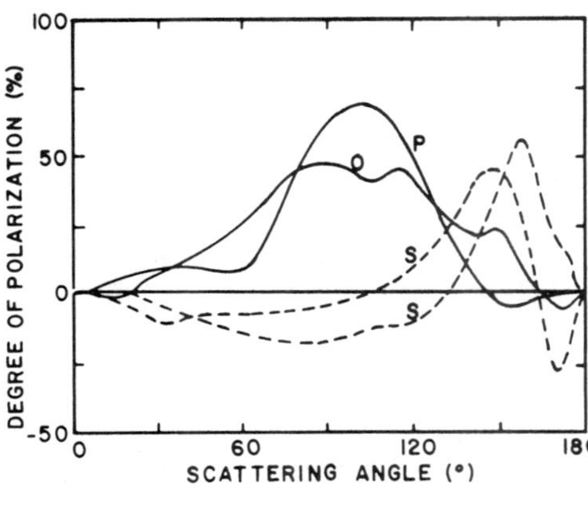

Figure 3.18 Angular distribution of the degree of linear polarization (scattering matrix elements $-P_{12}/P_{11}$) of radiation scattered by randomly oriented prolate (P) and oblate (O) spheroids with ratio of axes a/b=5, size parameter x=15, and index of refraction m=1.33 compared to that of equal area spheres (adapted from Asano and Sato, 1980).

The angular distribution of intensity for three sizes of oblate spheroids with axial ratio $a/b=5$, as given by Asano and Sato (1980), is shown in Fig. 3.19. As with spheres, the phase function shows an increasingly pronounced forward peak for the larger sizes of ellipsoids, but the decrease of intensity with increasing scattering angle is much more pronounced for the spheroids. It is noteworthy that the largest spheroids show a backward intensity maximum, which is almost absent for moderate sizes.

3.2.8.2.2 Homogeneous sphere with holes

Many particles of a natural aerosol have internal structure such as interstices or holes in their three-dimensional makeup. In fact, observation shows that most fly ash particles from coal-fired and oil-fired power plants are at least approximately spherical in general outline, but they have just such internal voids with resulting surface irregularities. An attempt to model the light-scattering proper-

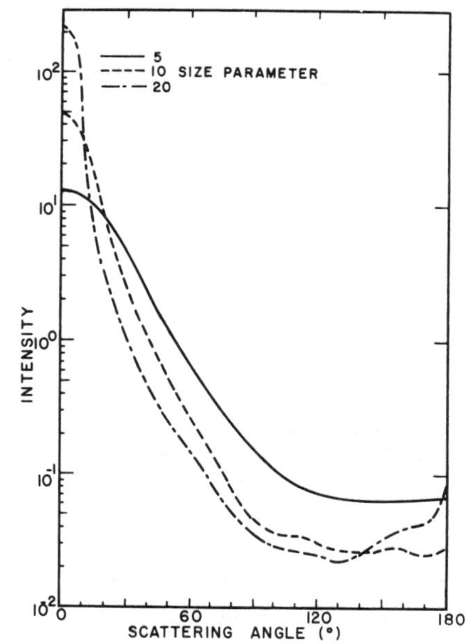

Figure 3.19 Angular distribution of the intensity (matrix element $P_{\|}$) of radiation scattered by three sizes of oblate spheroids, size parameter 5, 10, and 20, with ratio of axes a/b=5 and index of refraction m=1.33 (after Asano and Sato, 1980).

ties of particles with cylindrical-shaped holes of various sizes passing through the particle along the three coordinate axes has recently been made by Latimer (1984a). This type of structure is more applicable in the present context than is that of a spherical particle with spines projecting radially from the surface, a model also studied by Latimer (1984b). The approach taken for accounting for the effects of holes was to compute the scattering by smooth spheres with the same volume of material, the effect of holes being to simply increase the size of the sphere by an amount sufficient to include the volume of the holes. In two cases considered by Latimer, holes contributed 4.4% and 18.9% to the total volume of the sphere. His results show that the main factors controlling total scattering are the total volume of the particle and its index of refraction. The geometrical distribution of the volume is significant but not dominant. Latimer's computations of the effects of holes and spines indicate that the radial redistribution of 5% of the mass by hole or spine formation causes a change of total scattering of 3% to 10%. He made the encouraging conclusion from these numbers that the useful domain of homogeneous particle theory is greater than is usually supposed.

3.2.8.2.3 Rough particles made of stacked cylinders

The microwave analog method was used by Greenberg (1972) to measure the extinction efficiency of rough particles made by stacking together a large number of small cylinders with an appropriate index of refraction, and measuring the radiative properties. The results of the measurements for the simulated rough particles are compared with Mie calculations for equivalent size sphere in Fig. 3.20. The extinction efficiency is plotted against the dimensionless phase parameter ρ, which is given in terms of the size parameter $x=2\pi r/\lambda$ as $\rho=2x(m-1)$. The vertical lines are from microwave measurements for stacked cylinders and the curves are from Mie calculations. The results are quite comparable for $\rho\leq 4$, which is the primary resonance maximum for Mie scattering. For the larger particles, however, the extinction efficiency of rough particles is well above that of spheres. Pollack and Cuzzi (1980) point out that this condition should continue to exist for ever larger particles. The extinction efficiency Q_t is defined as the ratio of the extinction cross section to the geometrical cross section, a quantity which approaches 2.0 as particle size increases. Since a sphere has the minimum geometrical cross section for a given volume, the above ratio

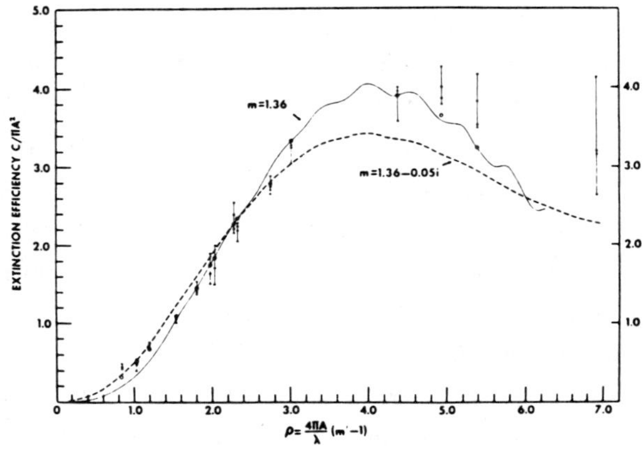

Figure 3.20 Extinction efficiency as measured in the microwave region by rough particles made up of stacked cylinders (vertical lines) as a function of the dimensionless phase parameter ρ. The curves are from Mie calculations with two different indices of refraction (after Greenberg, 1972).

of cross sections will be larger for an equal volume rough particle, and for the large rough particle $Q_t > 2$.

3.2.8.2.4 Irregular particles of various shapes

The microwave analog technique has been applied very successfully by Zerull and Giese (1974) and Zerull (1976) to studies of light scattering by rough spheres, cubes, octahedrons, convex particles, concave particles, and fluffy particles. They have compared the intensity and polarization of radiation scattered from 44 cubes and concave and convex particles with that computed for equivalent volume spheres in the size range $1.9 \leq x \leq 17.8$ and obtained the results shown in Fig. 3.21 for intensity and Fig. 3.22 for degree of polarization. In accord with the results of other workers they found that nonsphericity had little effect on scattering in the immediate forward direction, but very significant differences occur at intermediate and large scattering angles. Nonspherical particles give an almost isotropic intensity distribution in the backward hemisphere, whereas spheres produce a pronounced intensity minimum at about $\Theta = 120°$ and an intensity increase of an order of magnitude from $\Theta = 120°$ to $\Theta = 180°$. Nonsphericity eliminates the possibility of the glory phenomenon or the rainbow in the back-

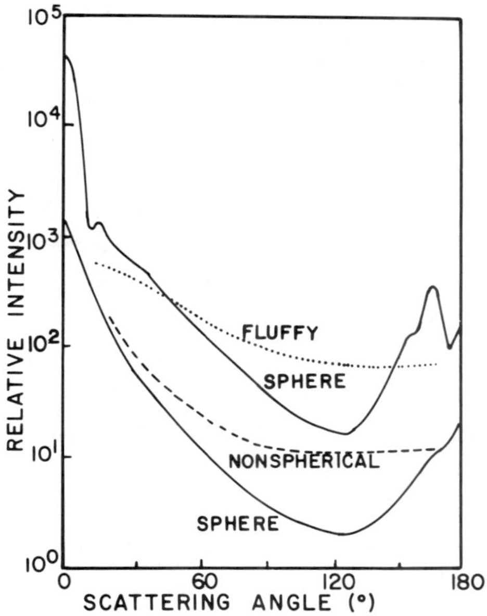

Figure 3.21 Angular distribution of intensity of microwave radiation scattered by a combination of rough spheres, cubes, octahedrons, convex and concave particles, and fluffy particles, as given by the measurements of Zerull (1976) and Zerull and Geise (1974) compared to that computed for equal volume spheres.

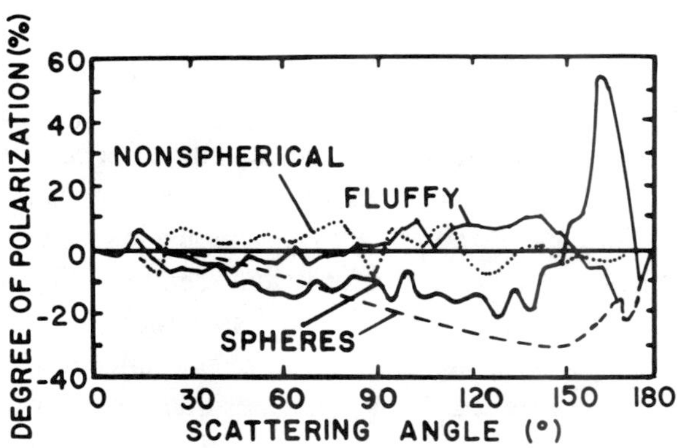

Figure 3.22 Angular distribution of the degree of linear polarization of microwave radiation scattered by a combination of rough spheres, cubes, octahedrons, convex and concave particles, and fluffy particles, as given by the measurements of Zerull (1976) and Zerull and Geise (1974) compared to that computed for equal volume spheres.

scattered light. Nonsphericity also washes out much of the polarization of the scattered light, as shown by the diagram for nonspherical particles in Fig. 3.22. Measurements on polydispersions, cubes, octahedrons, and concave and convex particles taken separately show essentially the same characteristics as those given in the diagram.

Also included in Figs. 3.21 and 3.22 are curves resulting from measurements on a polydispersion of fluffy particles. The term fluffy is used to describe a loose agglomeration of many smaller particles with open interstices among them. An example of such a particle might be a large particle of soot made up of many small smoke particles clinging together. The low intensity in the diffraction peak for fluffy particles indicates that the small individual component particles maintain their small particle lack of diffraction, but at the larger scattering angles the nonsphericity of the particle as a whole produces the isotropic intensity and deficit of polarization seen previously for the mixture of cubes, convex, and concave particles.

Compact particles, in the terminology of Zerull (1976), are generally convex bodies without appreciable voids or interstices in which no small particle structure is evident. The intensity and polarization patterns for scattering by a monodispersion of such compact particles are shown in Figs. 3.23 and 3.24 respectively, along with similar patterns computed by the Mie theory for single spheres.

The oscillations of the Mie curves are typical of single sizes of particles, whereas the smooth curves of previous diagrams result from the overlap of individual curves for different sizes of particles. The absence of any small particle character of the compact particles brings back the diffraction peak in the intensity phase function, and the approximately isotropic intensity in the backward hemisphere for compact particles is similar to that seen for fluffy and other types of nonspherical particles. The polarization distribution of Fig. 3.24, being somewhat similar to that resulting from pure Fresnel reflection but greatly different from that for pure Mie scattering, indicates that Fresnel reflection makes a strong contribution to the scattered field at all angles outside the diffraction peak.

3.2.8.2.5 Particles of carbonaceous smokes

Probably the most important component of dense clouds which might be produced by nuclear weapon detonations, leading to a nuclear winter would

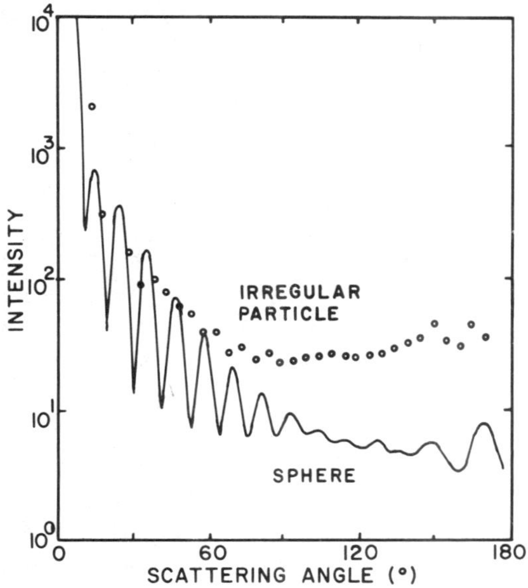

Figure 3.23 Angular distribution of intensity of microwave radiation scattered by a monodispersion of compact, irregularly shaped particles compared to that computed by the Mie theory for single spheres, as given by Zerull (1976).

be carbonaceous smoke emitted by fires raging in devastated cities. Thus any data pertinent to the radiative effects of smoke have direct applicability in the context of nuclear winter. Jennings and Pinnick (1980) have utilized the available data on smoke from coal-fired and oil-fired power plants, diesel engines, and the burning of acetylene to derive a linear relation between the extinction coefficient for visible and near-infrared radiation and mass concentration of the smoke. The relation is shown to be nearly independent of particle size distribution. A similar relation is shown to apply to the absorption coefficient. The concept is summarized in Fig. 3.25, in which the efficiencies of scattering and absorption are shown as a function of size parameter x for an index of refraction appropriate for carbon ($m = 2.0 - 1.0i$). The relation between volume extinction coefficient Q_t and mass concentration M is given by the relation

$$\frac{Q_t}{M} = \frac{\int \pi r^2 Q_t(m,x)\ n(r)\ dr}{\rho \int (4/3)\ \pi r^3\ n(r)\ dr} \tag{3.26}$$

where r is particle radius, $n(r)$ represents the size frequency distribution of particles, and ρ is particle density. If the efficiency factor can be approximated

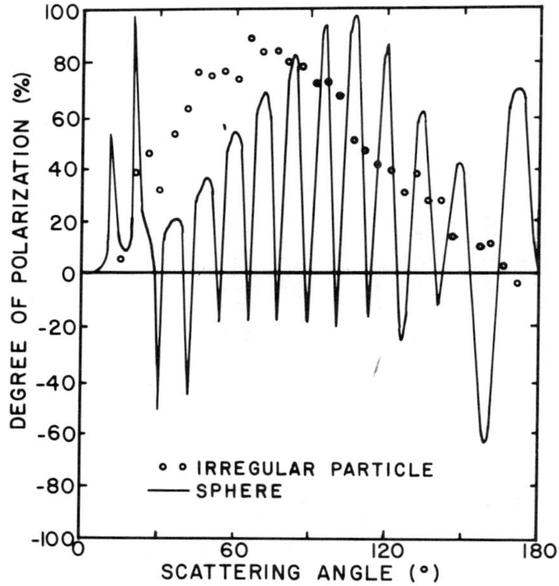

Figure 3.24 Angular distribution of the degree of linear polarization of microwave radiation scattered by a monodispersion of compact, irregularly shaped particles compared to that computed by the Mie theory for single spheres, as given by Zerull (1976).

by a linear function of size parameter $Q_t = cx$, then numerator and denominator of this relation both contain the factor $\int r^3 n(r)dr$. This eliminates the dependence on $n(r)$ and the relation reduces to

$$Q_t = \frac{3\pi c}{2\lambda\rho} M \tag{3.27}$$

The extent to which the approximation is valid for all $x < x_m$, where x_m is a specific size parameter equal to 1.4 for attenuation in the diagram can be seen from the fit of the straight line $Q_t = cx$ to the Mie curve. The fit is not perfect, but even for the worst case of a monodisperse aerosol of $x \approx 0.3$ for $x_m \approx 1.0$, the discrepancy is only about a factor of 2. For a more realistic case of a polydispersion of particles the error would be smaller than this, and it could be negligible for an appropriate size distribution of particles.

The equivalent case for absorption efficiency Q_a is shown by the dashed curve of Fig. 3.25. The fit here is considerably better than that for extinction

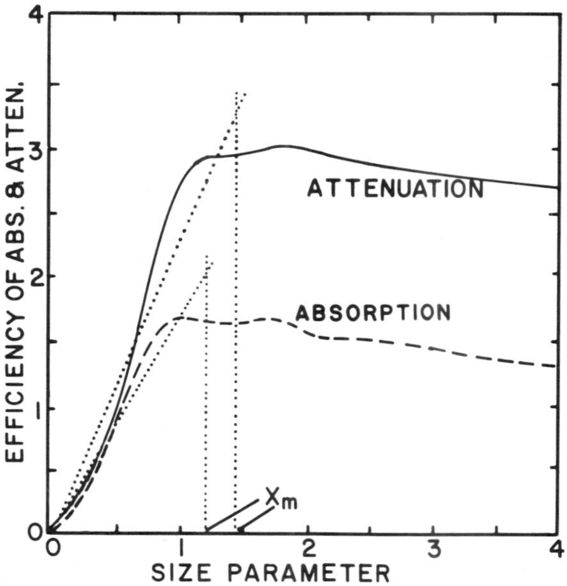

Figure 3.25 Efficiency of attenuation and absorption of carbonaceous particles as a function of size parameter as computed from the Mie theory and as approximated by linear relationships for all $x<x_m$ by Jennings and Pinnick (1980).

efficiency, and the accuracy should be adequate for practical purposes. It should be noted that for absorption $x_m=1.2$, as opposed to 1.4 for extinction.

A pertinent question for the application of these results to an actual aerosol is the validity of the values of x_m for the aerosol. No firm answer to the question is feasible, of course, but it is germane to note that in general, combustion processes yield small sizes of particles. It is likely that the majority of carbonaceous smokes would fall under the restrictions of x_m at the time they are injected into the atmosphere. Later processes of agglomeration and coagulation, however, may well change the size distribution to the range beyond the limits of x_m given here.

3.2.8.3 Conclusions on nonspherical particles

The following conclusions regarding nonspherical particles are supported by the above results:

1. Nonspherical particles have generally different scattering and attenuation properties for radiation in and near the visible range of wavelengths than those of their spherical counterparts.

2. The scattering and extinction efficiencies are generally larger for nonspherical particles than for equivalent spheres. The only exceptions to this are in very restricted particle size ranges, and in any real atmospheric aerosol the natural dispersion of sizes would obliterate the effects of these restricted ranges.

3. The magnitude of the effects of nonsphericity are generally not more than a factor of 2 in efficiencies of attenuation and scattering, and often considerably less than that. The only exceptions that have been reported are for very flat oblate spheroids of carbon and for iron needles. These are extreme cases, however, which have little applicability to aerosols in the atmosphere.

4. The angular scattering pattern is normally more strongly affected by nonsphericity of particles than are the total scattering efficiencies. While the intensity distribution at small scattering angles is relatively insensitive to particle shape, the intensity at scattering angles beyond 30° to 60° is often modified quite significantly for spheroidal or irregularly shaped particles from that for equivalent spheres. The backward intensity maximum typical of spheres is largely washed out by nonsphericity, as is the strong intensity minimum in the region around 120° scattering angle. These changes result in a distribution for nonspherical particles with a strong forward peak of intensity with a more or less isotropic intensity distribution in the backward hemisphere.

5. While information on the effects of nonsphericity on polarization of the scattered light is still scanty, the available evidence indicates that the polarization field is affected more by nonsphericity than is the field of intensity. From an overall energy standpoint, however, the effects of polarization, while not entirely negligible, are not of major significance. It is likely that polarization will find most application in remote sensing of aerosols, a problem to which it is well adapted.

6. In the context of the nuclear winter, nonspherical shapes of particles have the potential of adding a certain increment to the scattering and attenuation of incoming solar radiation, and thus may further exacerbate an already serious problem. In and of itself, however, the nonsphericity of particles could hardly either trigger a nuclear winter or add greatly to its severity.

3.3 Scattering by large water drops

The light-scattering properties of large water spheres have been studied by many authors, originally in efforts to explain by geometric optics the intensity and colors of light in the rainbow, and more recently to better understand intricacies of the scattering process. Perhaps the most detailed computations are those of Dave (1969). Dave assumed unpolarized light was incident on water droplets of various sizes and computed the intensity and degree of polarization of the scattered light. The results of the computations are summarized below. It is well to point out that the computing task for such large spheres is indeed formidable, particularly for the case in which adequate angular resolution of the scattered radiation field is achieved. As mentioned previously, the expressions for Mie scattering involve series of spherical Bessel functions of complex arguments and the first and second derivatives of Legendre Polynomials. The number of terms required to obtain convergence of the series is generally somewhat larger than the size parameter $x=2\pi r/\lambda$. For instance, for a cloud droplet with radius $r=10$ μm and a wavelength $\lambda=0.40$ μm, $x=157.1$. For a relatively small 1-mm radius rain drop and $\lambda=0.40$ μm, $x=15{,}710$. Obviously, computing this number of terms of a series for each of the functions for the intensity for a large number of scattering angles becomes a Herculean task for even the largest computers, and the task is magnified fourfold when the polarization is included. Fortunately, the much simpler approach of ray tracing is applicable to the larger spheres. Van de Hulst (1957) suggests that for a value of $x=2000$ or greater, ray tracing techniques yield generally reliable results, and spot checks can be made by comparison with the full Mie computations for specific angles. Although the majority of computations available have been limited to x much less than 200, Fahlen and Bryant (1968) reported data for x up to 3000 at the single scattering angle of $\Theta=180°$. Dave (1969) compared detailed Mie results for water spheres of $x=785.4$ with similar results obtained by ray tracing and found close agreement in the very forward lobe. The discrepancy increases rapidly with scattering angle, however, due to the increasing contributions of reflection and transmission by the sphere. The strongest of the oscillations are for nearly forward scattering, and are mainly the result of diffraction of light by the edges of the drop. In fact, the patterns in the first 2° correspond well with that predicted purely on the basis of diffraction theory. In the region $\Theta = 2°$ and 3°, discrepancies of nearly an order of magnitude are obtained, and the discrepancies continue to increase with increasing Θ. Dave's data for $x = 785.4$ are insufficient to characterize the ray tracing discrepancies further. As far as is known, no similar comparisons have been made for the polarization field of the scattered light.

The intensity pattern typical of the scattering of visible light by large water droplets, as given by the computations of Dave (1969) for $\lambda = 0.40$ μm and droplet radius 6.25 μm, is shown in Fig. 3.26. Numerous relatively strong oscillations are superimposed on a general decrease of intensity amounting to more than four orders of magnitude as Θ increases from 0° to 90°, with a recovery of roughly one order of magnitude between right angle scattering $\Theta = 90°$ and backward scattering. Strong oscillations near the forward direction are evident in the diagram. As mentioned above, they are mainly a result of diffraction by the drop. Outside of the diffraction region, radiation which has passed through the sphere, undergoing any number of internal reflections in the process, and that which is reflected from the outside surface of the sphere are the main contributors to the intensity, and the various streams may cause interference phenomena that

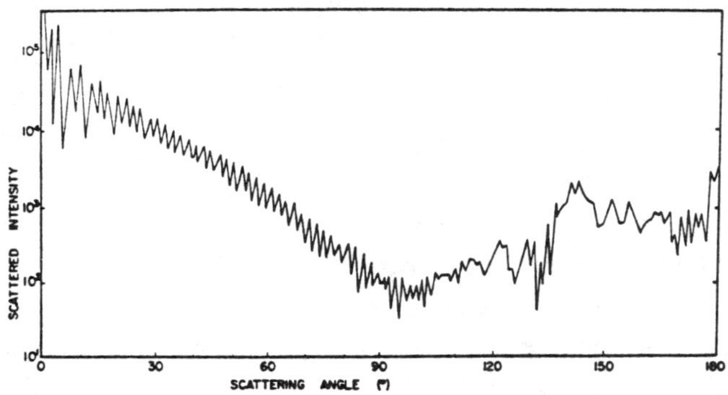

Figure 3.26 Intensity of scattered radiation at wavelength $\lambda = 0.40$ μm as a function of scattering angle for spheres of radius 6.25 μm and index of refraction $m = 1.342 - 0.0i$ (adapted from Dave, 1969).

contribute to the regular oscillations. This intuitive explanation, however, has its limitations.

The intensity pattern at small scattering angles, as computed by Dave (1969), is shown for two different sized particles in Fig. 3.27. The main curve, which covers scattering angles 0° to 10°, is for a wavelength of $\lambda = 0.40$ μm and radius of sphere $r = 6.25$ μm, and the inset in which Θ varies from 1.0° to 3.0° is for the same wavelength but for $r = 50$ μm. The size parameter $x = 2\pi r/\lambda$ is 98.2 and 785.4, respectively, for the two curves. The very strong forward peak seen for $r = 6.25$ μm is typical for large particles, but, of course, the details vary with size and index of refraction. The inset, applicable to small raindrops, does not show the forward peak, but Dave's unpublished computations for Mie scattering show four strong oscillations occurring in the first 1.0° of Θ. The maxima of the oscillations are located at approximately $\Theta = 0°$, 0.38°, 0.62°, and 0.84°, with intensities 9.76×10^{10}, 1.70×10^9, 4.09×10^8, and 1.54×10^8, respectively.

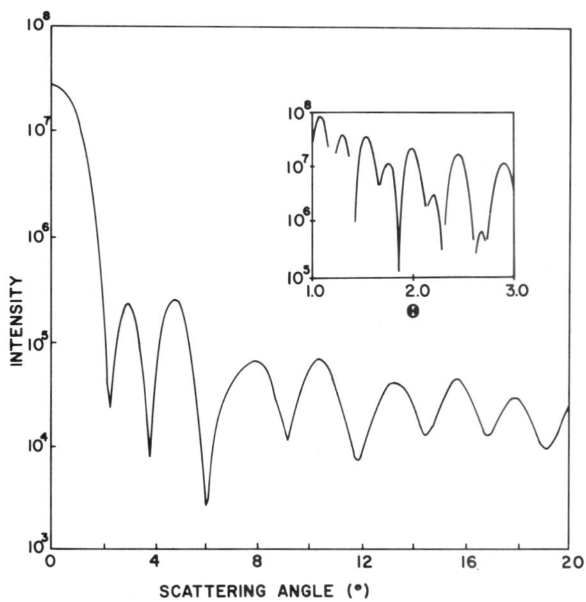

Figure 3.27 Total intensity $i_\perp + i_\parallel$ as a function of scattering angle in the region of the forward intensity peak for water spheres ($m = 1.342 - 0.0i$) of radii 6.25 μm (main curve: $x = 98.2$) and 50 μm (inset: $x = 785.4$) for wavelength $\lambda = 0.40$ μm (adapted from Dave, 1969).

Corresponding intensities computed from diffraction theory are slightly less than these, the ratios being 0.974, 0.976, 0.980, and 0.981, respectively. Since the scattering pattern is symmetric around the direction of propagation of the incident light, the maxima and minima of the scattering pattern in Fig. 3.27 would correspond to bright and dark rings around the position of the particle for an observer looking in the direction of the particle from the forward hemisphere.

As shown in Fig. 3.26, the intensity oscillations become very irregular at scattering angles $\Theta = 100°$. The intensity and degree of polarization are shown for $\Theta = 135°$ to $165°$ in Fig. 3.28. The ray of minimum deviation for the secondary rainbow would be at $\Theta = 126.4°$, an angle not included in Fig. 3.28, but Fig. 3.26 shows definite intensity maximum in that region. Supernumeraries of the primary rainbow could be expected to occur at angles somewhat beyond $\Theta = 140°$, and indeed a number of relatively pronounced maxima are located there. The original computations of Dave (1969) show these features even more clearly indicated for spheres of $r = 50.0$ μm, which are more representative of raindrops causing the rainbow.

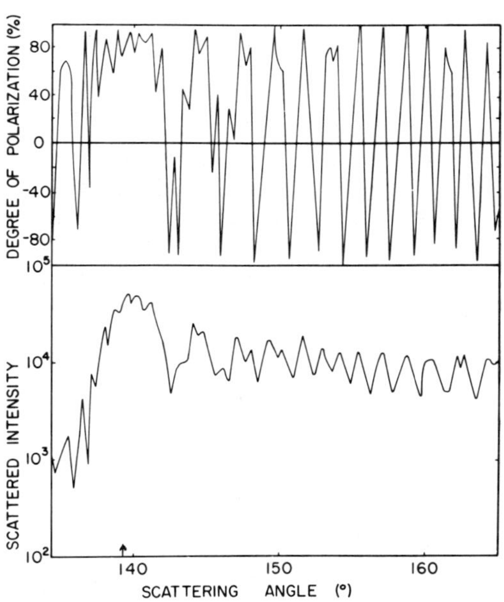

Figure 3.28 Intensity and degree of polarization as functions of scattering angle for light of wavelength $\lambda = 0.40$ μm scattered by water spheres ($m = 1.342 - 0.0i$) of radius 25.0 μm. (After Dave, 1969.)

The degree of polarization of light of $\lambda = 0.40\ \mu m$ scattered by droplets of $r = 6.25\ \mu m$ is shown for the entire range of $\Theta = 0°$ to $180°$, as computed by Dave (1969) by the relation

$$P(\Theta) = \frac{I_{\perp} - I_{\parallel}}{I_{\perp} + I_{\parallel}}$$

in Fig. 3.29. Ripples here correspond to similar ripples in the intensity field, with positive polarization generally corresponding to intensity maxima and negative polarization at intensity minima. Of particular interest in connection with later discussions of the rainbow is the high positive polarization in the region of the primary rainbow ($\Theta = 139.4°$), with strong oscillations between the primary and secondary ($\Theta = 126.4°$) rainbows and in the area of the supernumeraries of the primary rainbow.

3.4 The rainbow—geometrical optics

The rainbow, one of nature's most striking manifestations of color, has excited curiosity and wonder since antiquity. It is mentioned in the Bible as a sign of covenant between God and man, and many have tried to unravel the mysteries of its beauty. Aristotle ascribed it to reflection of sunlight by raindrops, a concept which we now know was only partly in error. A Dominican priest, Theodoric, who wrote in the early 1300s correctly attributed the primary to two refractions and one reflection of light by the raindrops. It wasn't until 1611 that the Arch-

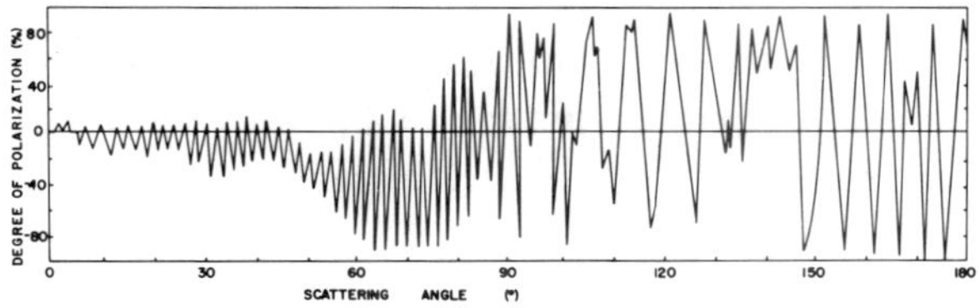

Figure 3.29 Degree of polarization as a function of scattering angle for light of wavelength 0.40 μm scattered by water spheres (m=1.342−0.0i) of radius 6.25 μm (adapted from Dave, 1969).

bishop of Spalatro in Italy properly explained the mechanism of both primary and secondary bows, but still it remained for the French physicist Rene Descartes writing in 1637 and using the newly discovered law of refraction (Snell's law), together with the tracing of numerous rays through the drop, to finally put the explanation on a firm mathematical basis. (In connection with this work, Descartes claimed first discovery of the law of refraction, although there now seems little doubt that he was preceded by W. Snellius. However, Descartes did receive at least a consolation prize of sorts when the ray which is deviated the minimum amount during its passage through a water droplet was designated as the Descartes ray.) The angles computed by Descartes were reasonably accurate, and he performed some laboratory experiments by means of a spherical flask of water to verify his results. Yet it was Newton's discovery that white light was composed of all of the colors that finally explained the characteristic color bands of the rainbow. His demonstration showed that the order of the colors in the rainbow is in perfect accord with theory (*Optics*, 1704).

Since this early work, research has continued on the phenomenon of the rainbodw. Interesting descriptions of the various aspects of rainbows, along with beautiful colored photos, were recently presented by Greenler (1980), and a more quantitative discussion applied to ray tracing techniques, also with some colored pictures, was given by Tricker (1970). Useful discussions of the geometric optics theory were provided by Humphreys (1940) and Johnson (1954), and a qualitative description of the phenomena associated with rainbows is contained in the classic book of Minnaert (1910). In his earlier work on rainbows, Van de Hulst (1957) gave the relatively complete theory of the rainbow and related scattering phenomena, and Pernter and Exner (1910) presented very detailed formalism in descriptions of various aspects of the rainbow. Of particular interest is their determination, by means of a color triangle, of the colors resulting from the superposition of eight different wavelengths on their emergence from raindrops of different sizes. As would be expected, the purity and saturation of rainbow colors are highest for the largest raindrops, and for droplets of radii of 0.5 mm and smaller the usual rainbow colors are replaced by mostly white light. This is the case for fog bows or so-called white rainbows.

3.4.1 Primary and Secondary Bows

The usual approach used to describe the rainbow by physical optics is through the technique of tracing many light rays through the drop, and thereby determining the characteristics of the emergent field. Four main effects have to be taken into account: the angular deviation experienced by each ray, attenuation of the rays by reflection at the air-liquid interfaces at the surface of the drop, the projected cross-sectional area of the drop for each incident ray (and thereby the total amount of energy contained in each ray), and the solid angle into which the energy eventually emerges from the drop. These four aspects will be introduced individually below.

The method of determining the deviation of the rays can be understood by the schematic diagram of the passage of two rays through the drop in Fig. 3.30. The rays are assumed to come from a distant source such as the sun. Ray R_1, shown as a solid line, is incident on the surface of the drop at angle τ_1, and the part which enters the drop is refracted at the air-water interface to give angle τ_1'. The relation between the angles of incidence and refraction is Snell's law, written as

$$\sin\tau = m \sin\tau' \tag{3.28}$$

where m is the index of refraction of the drop in relation to that of air. The ray is bent or rotated through angle $\tau_1 - \tau_1'$ at this first event. Reflection occurs at

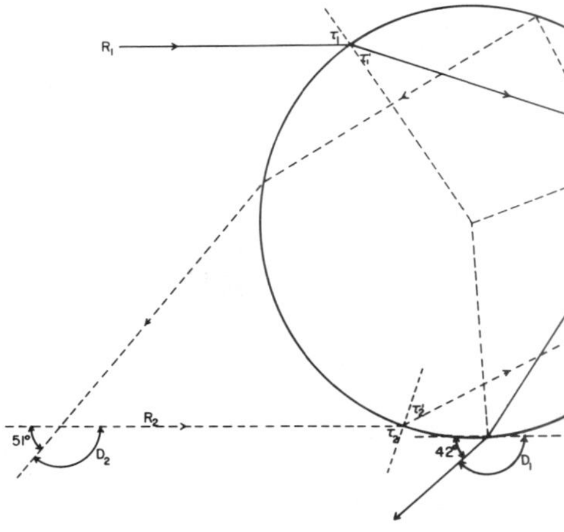

Figure 3.30 Diagram of the passage of the two rays of light through a water drop. The ray shown as a solid line undergoes a single reflection inside the drop and is responsible for the primary rainbow. The dashed line represents the ray which is reflected twice inside the drop and leads to the secondary rainbow.

the rear surface of the drop, with angle of reflection being equal to that of incidence. Since lines drawn normal to the surface at points of entry and reflection both pass through the center of the drop in Fig. 3.30, the resulting isoceles triangle makes the angles of incidence at the front surface and reflection at the back surface both equal to τ_1', and the total rotation of the ray on reflection is $180° - 2\tau_1'$. By the same reasoning it can be seen that the angle of incidence on the surface at the point of emergence from the drop is likewise τ_1', and emergence again rotates the beam $\tau_1 - \tau_1'$. All of these rotations are in the same direction, and therefore add up to a total deviation D_1 of the ray given by

$$D_1 = 2\,(\tau_1 - \tau_1') + 180° - 2\tau_1' = 2\tau_1 + 180° - 2\,(2\tau_1') \tag{3.29}$$

The dashed line in Fig. 3.30 shows a second ray, which is incident on the droplet surface at angle τ_2 and undergoes two reflections inside the drop. By reasoning entirely analogous to that for the singly reflected ray, it is easily seen that the rotation of this doubly reflected ray is

$$\begin{aligned} D_2 &= 2(\tau_2 - \tau_2') + 180° - 2\tau_2' + 180° - 2\tau_2' \\ &= 2\tau_2 + 2(180°) - 2(3)\tau_2' \end{aligned} \tag{3.30}$$

In fact a recursion relationship of this type exists for any number n internal reflections, which is

$$D_n = 2\tau_n + n(180°) - 2(n+1)\tau_n' \tag{3.31}$$

Although the effects of three, four, or more internal reflections are never seen in an actual rainbow, they exist in theory. There are three reasons for these multiply reflected rays to be invisible: (1) the higher numbers of reflections require rays to be at very near the grazing angle of the drop, thereby being subject to only a small projected area and a correspondingly small amount of energy, to enter the drop; (2) some of the small amount of energy which is available is lost at each reflection event; and (3) the deviations of the rays for three and four internal reflections are such that the emergent rays are concentrated in the bright region of the sky only 40° to 50° from the sun and their weak contribution is overwhelmed by natural brightness of the sky. Only after five internal reflections is the angle of minimum deviation sufficient to orient the bow away from the bright part of the sky, though too little energy is available to make the fifth-order bow visible. Van de Hulst (1957) found that of all the total energy scattered in the rainbow, less than 0.5% is contained in bows resulting from three or more internal reflections.

Ray-tracing techniques yield a strong enhancement of light in the directions of the primary and secondary bows, with a darker sky between the two. All of these features agree with observation. A diagram illustrating the cause for this distribution for the primary bow is shown in Fig. 3.31. The ray which strikes the surface of the drop at one particular angle of incidence is deviated less during its passage through the drop than are the rays that strike at any other angle. This unique ray, the Descartes ray, is shown as number 4, the heavy solid line, in Fig. 3.31. Both the dashed rays outside and the solid rays inside of the Descartes ray are all deviated from the original beam more than this one is, and the angle of deviation of the Descartes ray is thus termed the angle of minimum deviation D_m. There is a concentration of rays at deviation angles just smaller and just larger than D_m, which corresponds to the brightest part of the rainbow. Those

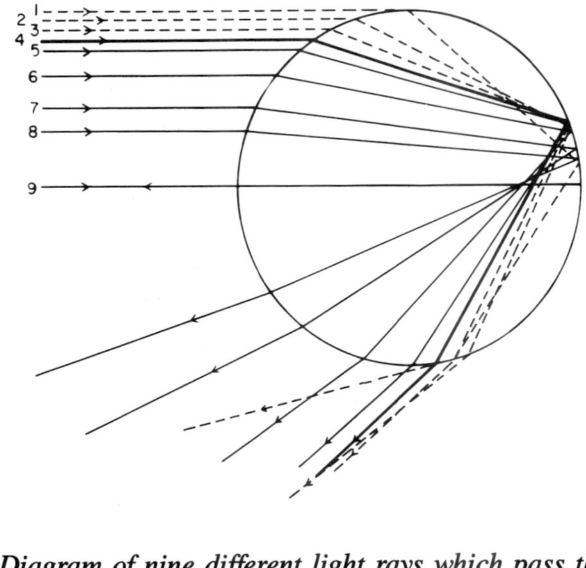

Figure 3.31 Diagram of nine different light rays which pass through a water droplet and constitute the primary rainbow. The heavy solid line (#4) represents the Descartes ray, the unique ray which undergoes the minimum deviation in the process.

at still larger angles contribute to enhanced illumination of the sky inside the bow at the expense of that outside of the bow.

In order to compute the angle of minimum deviation for n internal reflections, we differentiate Eq. (3.31) to give

$$\frac{dD_n}{d\tau} = 2 - 2\,(n + 1)\,\frac{d\tau'}{d\tau} \tag{3.32}$$

On differentiating Snell's law, Eq. (3.28), we have

$$\cos\tau = m \cos\tau' \frac{d\tau'}{d\tau} \tag{3.33}$$

and a combination of these relations gives

$$\frac{dD_n}{d\tau} = 2 - 2\,(n + 1)\,\frac{\cos\tau}{m \cos\tau'} \tag{3.34}$$

By setting $dD_n/d\tau = 0$ to get the minimum, squaring both sides, and using a trigonometric identity and Snell's law, it is readily seen that

$$\frac{1}{(n + 1)^2} = \frac{\cos^2\tau}{m^2 - 1 + \cos^2\tau} \tag{3.35}$$

or, on rearranging and simplifying,

$$\cos\tau = \left[\frac{m^2 - 1}{n(n + 2)}\right]^{1/2} \tag{3.36}$$

where τ is the incident angle of the Descartes ray on the surface of the drop. The corresponding τ' can be computed from Snell's law, and, finally, the deviation

D_n of the Descartes ray for n internal reflections can be computed from Eq. (3.31) for the value of m for water at any wavelength.

The results of such computations over the various colors of the visible spectrum are shown for the primary ($n=1$, left scale) and secondary ($n=2$, right scale) rainbows in Fig. 3.32. The refractive index of water for visible light is shown in the inset. The following features of the curves help explain the visual appearance of the various phenomena:

1. The secondary bow covers a considerably wider angle between the shortest violet and the longest red wavelengths than does the primary bow (3.19° for the secondary bow versus 1.76° for the primary). This helps explain

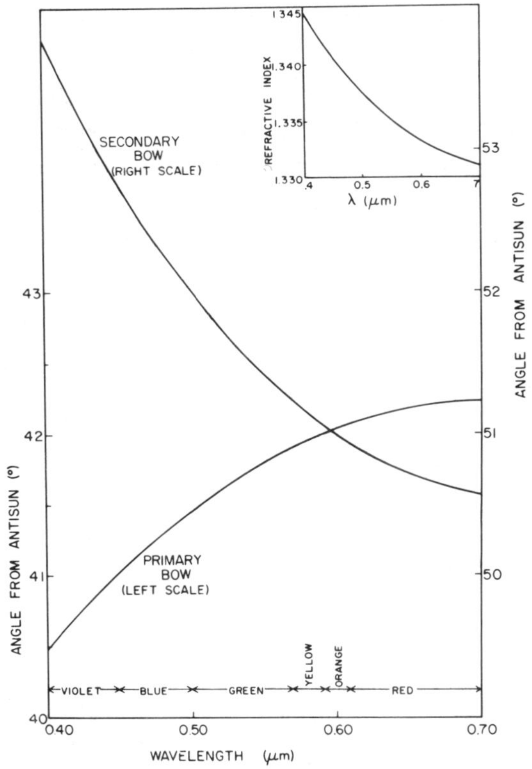

Figure 3.32 Angles from the direction of the antisun of the Descartes rays of the primary and secondary rainbows as functions of wavelength. The inset shows the spectral distribution of the index of refraction of pure water over the range of visible wavelengths.

the brightness of the secondary being less than that of the primary, but of course the extra internal reflection for the secondary, which causes additional attenuation of the ray, is also a contributing factor.

2. The longer wavelengths are spread out less than are those in the violet and blue in both primary and secondary bows. This factor tends to make the reds brighter and have higher saturation than the colors of the shorter wavelengths.

3. The different order of the colors in the two rainbows is evident, the red being on the outside and the violet on the inside of the primary bow, and the reverse for the secondary.

While the various angles shown above were computed by Snell's law, the actual intensities and degree of polarization of the emergent light are determined by the application of Fresnel's laws of reflection and refraction. For this purpose we assume the incident light to be unpolarized, and define two orthogonal components $I_\parallel$ and $I_\perp$ which are respectively in and normal to the plane defined by the directions of the incident and emergent light. For the first encounter of the light beam with the drop, the intensity of the reflected component is proportional to the coefficient

$$C_\parallel = \frac{\sin^2 (\tau - \tau')}{\sin^2 (\tau + \tau')} \tag{3.37}$$

and that for the refracted component $C_\parallel'$ is

$$C_\parallel' = 1 - \frac{\sin^2 (\tau - \tau')}{\sin^2 (\tau + \tau')} \tag{3.38}$$

Similarly, for the normal component

$$C_\perp = \frac{\tan^2 (\tau - \tau')}{\tan^2 (\tau + \tau')} \tag{3.39}$$

and

$$C_\perp' = 1 - \frac{\tan^2 (\tau - \tau')}{\tan^2 (\tau + \tau')} \tag{3.40}$$

Obviously, the light reflected at the point of entry is lost to the beam. Of that which is refracted into the drop, part is reflected inside the drop (one reflection for the primary and two for the secondary), and the remainder is lost by being refracted out of the drop. For the case of the primary rainbow, it can be seen from the diagram of Fig. 3.30 that the angles of incidence and reflection at the back surface of the drop are both τ_1' and the angle of refraction, in analogy to that of the incident beam, is τ_1. In accordance with Eqs. (3.38) and (3.39), the coefficients of reflection of the two components at the back surface of the drop are simply $C_\parallel$ and $C_\perp$. Finally, the internally reflected light is incident on the surface at the point of emergence at angle τ_1' also, and the components which actually emerge are proportional to $C_\parallel'$ and $C_\perp'$. The coefficient of transmission T for each component is the product of the three coefficients, two of refraction and one of reflection, or

$$T_\parallel = C_\parallel' C_\parallel C_\parallel' = (C_\parallel')^2 \, C_\parallel$$

$$T_\perp = C_\perp' C_\perp C_\perp' = (C_\perp')^2 \, C_\perp \tag{3.41}$$

For a unit intensity of the (unpolarized) incident ray, the intensity components of the emergent ray for any angle of incidence are

$$I_\parallel = \frac{1}{2}\left[1 - \frac{\sin^2(\tau' - \tau)}{\sin^2(\tau' + \tau)}\right]^2 \left[\frac{\sin^2(\tau' - \tau)}{\sin^2(\tau' + \tau)}\right]$$

$$I_\perp = \frac{1}{2}\left[1 - \frac{\tan^2(\tau' - \tau)}{\tan^2(\tau' + \tau)}\right]^2 \left[\frac{\tan^2(\tau' - \tau)}{\tan^2(\tau' + \tau)}\right] \tag{3.42}$$

For a secondary bow, the additional internal reflection requires another multiplication by $C_\parallel$ or $C_\perp$, resulting in the last factors of Eq. (3.42) being

$$\left[\frac{\sin^4(\tau' - \tau)}{\sin^4(\tau' + \tau)}\right]$$

and

$$\left[\frac{\tan^4(\tau' - \tau)}{\tan^4(\tau' + \tau)}\right]$$

respectively.

The results of computations of the intensity components of the emergent light for the primary and secondary rainbows are shown as functions of the angle at which the original ray is incident on the drop in Fig. 3.33. An incident intensity of unity and an index of refraction for water of 1.3320 ($\lambda = 0.65\ \mu m$) have been assumed. It is emphasized that only deviations and attenuation of the rays have been taken into account here. Component $I_{\parallel}$ consists of vibrations parallel

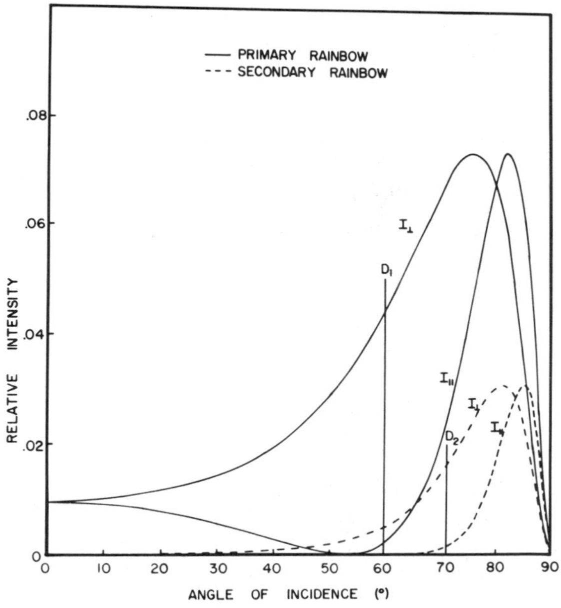

Figure 3.33 Intensity components of the primary (solid lines) and secondary (dashed lines) rainbows as a function of the angle at which the incident beam strikes the surface of the drop. Incident intensity of unity was assumed, and only deviations of the rays and attenuation through the drop are taken into account. The Descartes rays for the primary and secondary bows occur at incident angles $\tau = 59.47°$ and $71.87°$, respectively.

to the plane defined by the incident and emergent directions, and $I_\perp$ is for vibrations normal to that plane. For normal incidence $I_\parallel = I_\perp$, and from Eqs. (3.41) and (3.42) we have

$$I_\parallel = \frac{1}{2} \left[C_\parallel (C'_\parallel)^2\right] \tag{3.43}$$

and similarly for $I_\perp$. Actually, the expressions in Eq. (3.42) become indeterminate at $\tau = 0$, but Fresnel's laws give the coefficient of reflection for the sum of the components for $\tau = 0$ as

$$C = \left(\frac{m - 1}{m + 1}\right)^2 \tag{3.44}$$

which is the equivalent to

$$C_\parallel = C_\perp = \frac{1}{2}\left(\frac{m - 1}{m + 1}\right)^2 \tag{3.45}$$

For $m = 1.3320$, this expression yields 0.01013 for each coefficient for reflection. The coefficients for refraction $C'_{\parallel,\perp} = 1 - C_{\parallel,\perp}$, which for this case results in $C'_\parallel = C'_\perp = 0.098987$. Finally, for $\tau = 0$ and unit incident intensity, we have from Eq. (3.43), $I_\parallel = I_\perp = 0.009926$ for the case of a single internal reflection, as shown in Fig. 3.33. For two internal reflections and $\tau = 0$,

$$I_\parallel = I_\perp = \left[(C_\parallel)^2\ (C'_\parallel)^2\right] = 0.0001006 \tag{3.46}$$

a quantity which is too small to plot usefully in Fig. 3.33.

As the angle of incidence increases, the curves for the intensity components diverge, indicating the existence of polarization of the light. In fact, $I_\parallel$ for the primary bow is zero at approximately $\tau = 53°$, the polarizing (Brewster) angle for this index of refraction. In the secondary bow, both components are smaller than for the primary bow, the difference being due to the loss of energy on the second reflection event.

From Eq. (3.36) we can compute τ_D for the Descartes ray to be 59.47° for the primary bow and 71.87° for the secondary. At these angles, which are indicated by lines D_1 and D_2 in Fig. 3.33, components $I_\perp$ are much greater than those of $I_\parallel$ for both primary and secondary bows. This means that the light of the Descartes rays is highly polarized with the maximum of the electric vector being in the direction tangential to the rainbow at any point.

While the positions of the Descartes rays are given with respect to the angle of incidence of the light on the drop in Fig. 3.33, the visual appearance of the rainbow is more closely related to the angles of emergence of the light from the drop. By combining Eqs. (3.28), (3.31), and (3.36) we can compute the angle of emergence θ of the light with respect to the direction $\tau=0°$ to be $\theta=42.22°$ for the Descartes ray of the primary and $\theta=50.63°$ for that of the secondary for the index of refraction $m=1.3320$. The intensity components $I_\parallel$ and $I_\perp$ are shown as a function of angle of emergence for light which has entered the drop at angles less than those of the Descartes rays ($\tau<\tau_D$) and at angles greater than those of the Descartes rays ($\tau>\tau_D$) in Fig. 3.34. Included also are curves of total intensity. The positions of the Descartes rays are indicated by the arrows. It should be emphasized that only transmission by the drop has been taken into account at this state, and that the geometrical area of the drop applicable to each ray will modify this picture significantly, as shown later in Fig. 3.36. The degree of polarization, however, can be determined from the intensity components of Fig. 3.34, as geometry does not affect their relative magnitude.

The maxima of both $I_\parallel$ and $I_\perp$ are due mainly to light which strikes the drop at angles greater than those of the Descartes rays ($\tau>\tau_D$). The maxima are displaced from each other by about 8° for the primary and 5° for the secondary. They are of comparable magnitude for each rainbow, but both components are approximately 2.5 times as bright in the primary as in the secondary.

The degree of polarization for this case is the so-called Rubenson polarization, given previously as

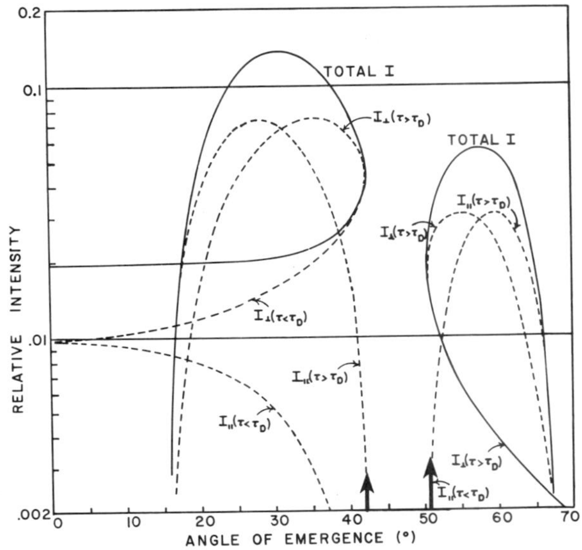

Figure 3.34 Total intensity and intensities of the orthogonal components of light in the primary and secondary rainbows as a function of angle of emergence for the case in which only transmission through the drop is considered. Geometric area of the drop applicable to the various rays changes the distribution to that of Fig. 3.33. The positions of the Descartes rays are indicated by the arrows.

$$P = \frac{I_\perp - I_\parallel}{I_\perp + I_\parallel} \tag{3.47}$$

A glance at Fig. 3.33 shows that for most angles of incidence $I_\perp > I_\parallel$, which means that the polarization of light in the rainbow is mainly positive. The maximum of the vibration of the electric vector is thus mainly tangential to the bow at any point along the bow.

As mentioned above, the model can be made more realistic by including the effects of the cross-sectional area of the drop on which the rays are incident and the solid angle into which the rays finally emerge from the drop. The geometry is shown in Fig. 3.35. The projected area onto which a given ray of unit cross section is incident is proportional to the length.

$$a - b = r[\sin(\tau + \Delta\tau) - \sin\tau] = r\cos\tau\Delta\tau \tag{3.48}$$

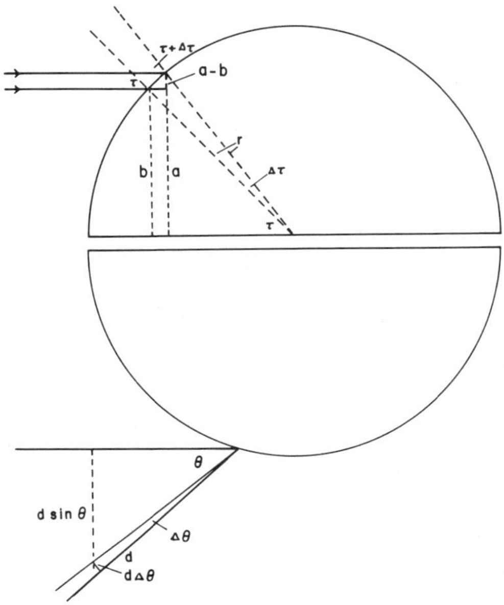

Figure 3.35 Diagram showing the projected area of a portion of a drop for an incident ray and the solid angle into which the ray eventually emerges from the drop.

The radiation is, of course, incident on the whole drop. The length of the annulus around the drop at distance b from the axis is $2\pi b = 2\pi r \sin\tau$, and the area of the annulus is

$$2\pi r^2 \sin\tau \cos\tau\Delta\tau = \pi r^2 \sin 2\tau\Delta\tau \tag{3.49}$$

For a unit incident intensity this represents the amount of energy which would be available for entry into the drop, but, of course, a portion is lost at the first surface and during its traverse through the drop.

The part of the energy which traverses the drop eventually emerges into a solid angle shown at the bottom of Fig. 3.35. This solid angle has also the form of an annulus at angle θ around the incident direction. At a distance d from the drop the area of the annulus is $2\pi d^2 \sin\theta\Delta\theta$, and by definition the solid angle is $2\pi \sin\theta\Delta\theta$. On combining this with Eq. (3.49), we have the intensity at angle θ as

$$I = TI_o \frac{r^2 \sin 2\tau}{2\sin\theta} \frac{\Delta\tau}{\Delta\theta} \tag{3.50}$$

where T represents the fractional transmission through the drop. Expressions for T were given for the parallel and normal components of the incident light in Eq. (3.41).

Since, as shown in Fig. 3.30, the magnitude of θ is $\theta = 180° - D_n$ and its sign has no relevance, we can invert Eq. (3.34) to obtain

$$\frac{\Delta\tau}{\Delta\theta} = \left[\frac{m \cos \tau'}{2(m \cos\tau') - 2(n + 1) \cos\tau} \right] \tag{3.51}$$

and on combining this with Eq. (3.50), we can write

$$\frac{I}{I_o} = T \left[\frac{r^2 \sin 2\tau}{2\sin\theta} \right] \left[\frac{m \cos\tau'}{2(m \cos\tau') - 2(n + 1) \cos\tau} \right] \tag{3.52}$$

For the primary bow $n=1$ and for the secondary $n=2$. Thus we have the intensity for the primary rainbow relative to that of the incident light as

$$\frac{I}{I_o} = \frac{T}{4} \left[\frac{r^2 \sin 2\tau}{\sin\theta_1} \right] \left[\frac{m \cos\tau'}{m \cos\tau' - \cos\tau} \right] \tag{3.53}$$

and for the secondary

$$\frac{I}{I_o} = \frac{T}{4} \left[\frac{r^2 \sin 2\tau}{\sin\theta_2} \right] \left[\frac{m \cos\tau'}{m \cos\tau' - 3\cos\tau} \right] \tag{3.54}$$

These relations are applicable to either the parallel or perpendicular component of the light, the difference being accounted for in the transmission factor T according to Eq. (3.41).

The final intensity distribution, in which both transmission by the drop and geometry of the drop are taken into account, is shown in Fig. 3.36. The components arriving at incident angles greater than $(\tau > \tau_D)$ and less than $(\tau < \tau_D)$ those of the Descartes rays are shown separately, and their sum is given as the total intensity. The intensity reaches a theoretically infinite value just at the Descartes rays of both primary and secondary bows. Away from these singular angles, however, the intensity, both individual components and total, decreases rapidly, as the appearance of a rainbow indicates it should. Within about the first 15° from the Descartes directions, the intensity is composed largely of light striking the drop at angles greater than those of the Descartes rays. This seems

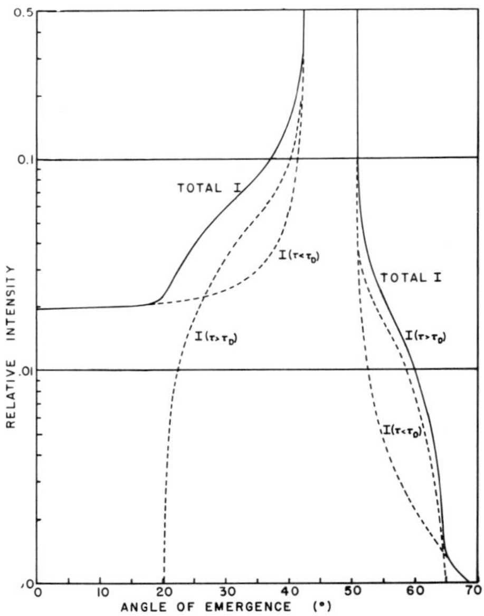

Figure 3.36 Total intensity and intensities of the orthogonal components of light in the primary and secondary rainbows as functions of angle from the direction of the antisun. Both transmission and geometry of the drop have been taken into account.

somewhat surprising in view of the small projected area of the drop represented by those large angles of incidence. This component rapidly decreases with angle from the Descartes rays, however, particularly for the primary bow, and at angles beyond 15° to 20° from the Descartes rays the intensity is due mostly to light incident on the drop at angles less than those of the Descartes rays. A strong change of gradient is indicated at an angle of about 22° from the Descartes ray for the primary, but this area would fall in the region of the normal skylight and thus not be evident in the natural phenomenon.

The Rubenson degree of polarization [see Eq. (3.47)] for the light of the primary and secondary rainbows for components incident on the drop at angles less than ($\tau < \tau_D$) and greater than ($\tau > \tau_D$) the angles of incidence for the Descartes rays, and for the sums of the components, is shown as a function of angle θ of the emergent rays in Fig. 3.37. The polarization is essentially complete in the regions of the Descartes rays, particularly for the primary rainbow, for both components and for the total. It falls off rapidly with angle away from the Descartes rays and follows relatively closely the component $\tau > \tau_D$, as the

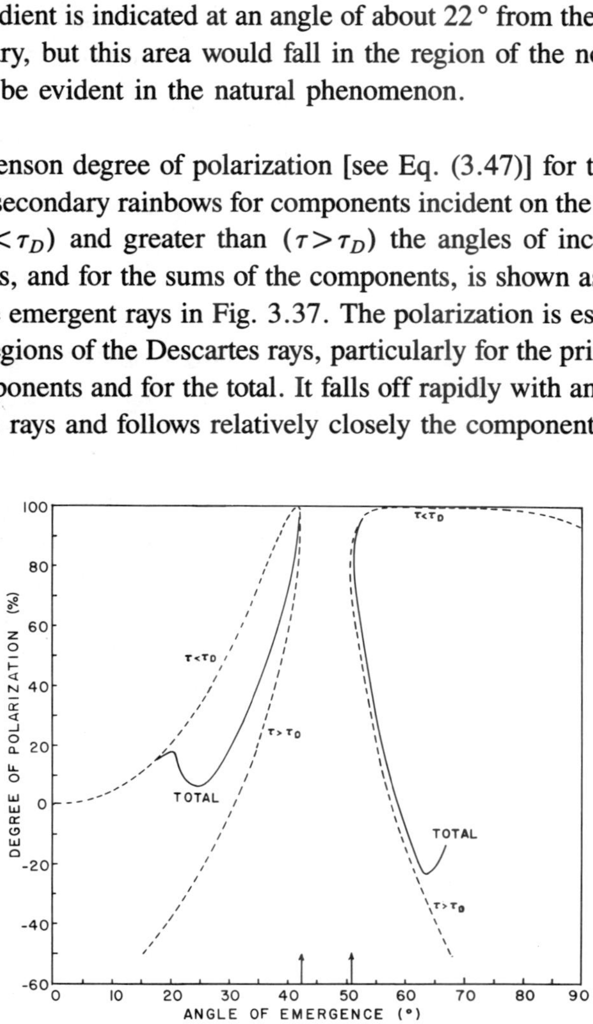

Figure 3.37 Degree of polarization as a function of emergent angle for light in primary and secondary rainbows. The dashed curves are for components striking the drop at angles less than ($\tau < \tau_D$) and greater than ($\tau > \tau_D$) those of the Descartes rays, and the solid curves are for the total.

relative magnitudes of intensity in Fig. 3.36 indicates that it should. At angles from the Descartes rays of approximately 20° for the primary and 13° for the secondary, however, the intensity of component $\tau > \tau_D$ has become so low that component $\tau < \tau_D$ becomes dominant, and the degree of polarization for the total rapidly approaches that for the stronger component. The polarization of the total light is positive at almost all angles for both rainbows, the only exception being a small region of negative polarization of about 10° width centered at $\theta = 64°$ in the secondary rainbow. This means that except for this small region, the vibration of the electric vector in both rainbows is tangential to the bow at any point along the bow.

3.4.2 Supernumerary bows

The faint-colored bands which are visible fairly frequently inside the primary rainbow, and infrequently outside the secondary, are generally termed supernumerary bows, although terms such as auxiliary rays or spurious rays are sometimes applied. They arise from interference of light waves of different phases emerging from raindrops. The different phases are produced by different pathlengths of the rays inside the drop. Referring again to Fig. 3.31, it can be seen that the rays (dashed in the diagram) which are incident on the drop at angles greater than that of the Descartes ray ($\tau > \tau_D$) generally have a shorter pathlength through the drop than do those (solid) incident at angles less than τ_D. Since some of the dashed rays are in the same direction as some of the solid rays on emergence from the drop, the differences in phase result in interference phenomena manifested as supernumerary bows. Since phases and relative intensities of the emergent rays are dependent on drop sizes, rainbows resulting from certain sizes or size distributions of drops are not accompanied by supernumeraries. This will be mentioned further below.

The physical explanation of supernumeraries was first developed by the English astronomer Sir George Airy in 1838 (Airy, 1839). In the interests of brevity we take a more intuitive approach with less mathematical manipulation than that given by either Airy or by Pernter and Exner (1910). Following Tricker (1970), we shall develop a simplified form of the Airy integral and show some of its ramifications. The wave front of constant phase of the light emerging from a drop in the immediate neighborhood of the Descartes ray is shown, with angles greatly exaggerated, in Fig. 3.38. The dashed lines are parallel to the Descartes ray, and the wave front is everywhere normal to the direction of propagation of the rays (solid lines). If x is distance along the wave front through 0, and y is displacement of the wave front from the dashed line, then the wave front is well approximated by the cubic relation

$$y = k'x^3 \tag{3.55}$$

It is to be expected that the linear dimensions of the curve will be proportional to the radius r of the drop. The constant k' of Eq. (3.55) must have the dimensions of $(L)^2$, but we can make a dimensionless constant k by writing $k' = k/r^2$. Then we have for the curve

$$y = \frac{k}{r^2} x^3 \tag{3.56}$$

If we expand a portion of Fig. 3.38 to yield Fig. 3.39, we can see that the path difference between the rays at O and M is

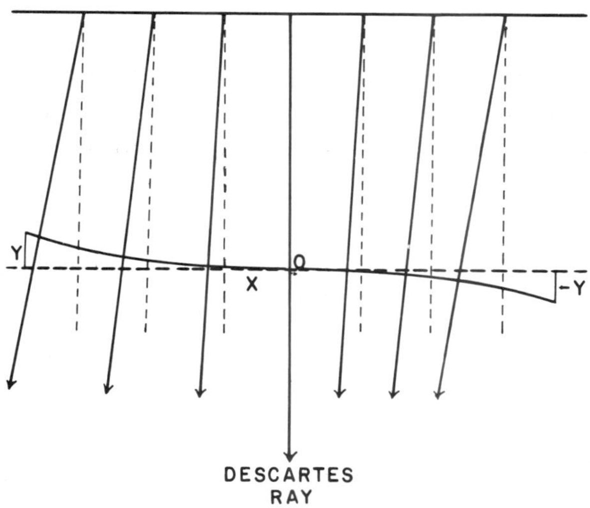

Figure 3.38 Diagram of a wave front of constant phase for light emerging from a drop in the immediate neighborhood of the Descartes ray.

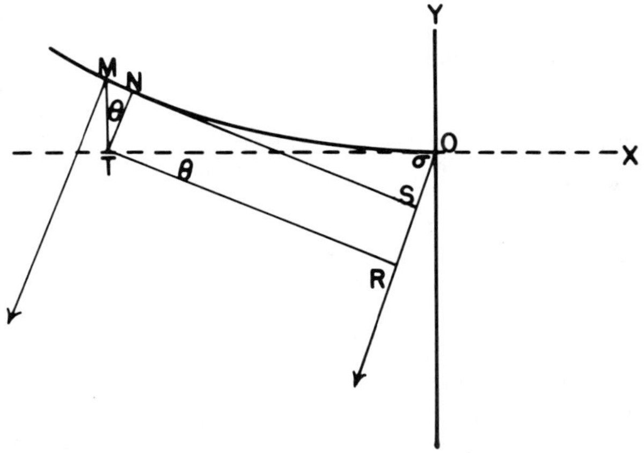

Figure 3.39 Expanded portion of Fig. 3.38 in the neighborhood of the Descartes ray.

$$\sigma = OS = OR - RS = x \sin\theta - y \cos\theta \tag{3.57}$$

or

$$\sigma = x \sin\theta = \frac{k}{r^2} x^3 \cos\theta \tag{3.58}$$

From diffraction theory first developed by Fresnel, we can represent the disturbance of the wave front from element dx at 0 by sin wt, where t is time and w is a circular frequency, and at point M the disturbance is

$$\sin(wt + \delta) = \sin(wt) \cos\delta + \cos(wt) \sin\delta \tag{3.59}$$

The phase difference δ is, by definition

$$\delta = \frac{2\pi\sigma}{\lambda}$$

where λ is wavelength. For the whole wave front the disturbance is

$$\sin(wt) \int \cos \frac{2\pi\sigma}{\lambda}\, dx - \cos(wt) \int \sin \frac{2\pi\sigma}{\lambda}\, dx \tag{3.60}$$

For an integration over the whole wave front, in which σ is both + and −,

$$\int_{-\infty}^{\infty} \sin \frac{2\pi\sigma}{\lambda}\, dx = 0$$

and the resulting amplitude of the oscillation is simply

$$A = \int_{-\infty}^{\infty} \cos \frac{2\pi\sigma}{\lambda}\, dx$$

or

$$A = \int_{-\infty}^{\infty} \frac{2\pi}{\lambda} \left[x \sin\theta - \frac{kx^3}{r^2} \cos\theta\right] dx \tag{3.61}$$

This is one form of the Airy rainbow integral. It can be cast in a more useful form by changing variables by the relations (Tricker, 1970)

$$x^3 = \frac{\pi r^2}{4k \cos\theta}\, u^3$$

$$x = \frac{\lambda z u}{4 \sin\theta} \tag{3.62}$$

$$dx = \left(\frac{\lambda r^2}{4k \cos\theta}\right)^{1/3} du$$

from which we have the final form of the Airy integral as

$$A = \left(\frac{\lambda r^2}{4k \cos\theta}\right)^{1/3} \int \cos \frac{\pi}{2}\, (u^3 - zu)\, du \tag{3.63}$$

Several different authors have evaluated this relation by various means. The behavior of the Airy curve resembles that of a diffraction pattern in the shadow formed as a light beam passes a straight edge. An inspection of Eq. (3.63) shows that the amplitude will vary with both droplet size and wavelength of the light, being proportional to $(\lambda)^{1/3}$ and $(r)^{2/3}$.

Minnaert (1940) has given the following method of estimating mean diameter of the drops forming rainbows:

Drop Diameter, mm	Appearance of Rainbow
1–2	Very bright violet and vivid green; pure red, scarcely any blue. Numerous supernumeraries, violet-pink alternating with green
0.50	Red considerably weaker, fewer supernumeraries.
0.20–0.30	No more red, for other colors bow broad and well developed. Supernumeraries more yellow. Gap between supernumeraries indicates drop diameters 0.20 mm; gap between primary and first supernumerary indicates diameters less than 0.20 mm.
0.08–0.10	Bow broader and paler, only violet vivid. First numerary has white tints and is well separated from primary.
0.06	Primary has distinct white stripe.
0.05	Mist bow. White

The appearance of the interference pattern formed by rays emerging from a raindrop was shown very graphically by Fraser (1983) by a moire pattern of light and dark bands in the emergent field. The pattern shows closely spaced bands for large drops and widely spaced bands for small drops. Fraser concluded that a primary bow is formed only by drops of radii between 0.10 and 0.2 mm, and that the appearance of a rainbow requires the existence of this size drops in the size distribution. Similarly, the first two supernumeraries require drops of radii 0.2 to 0.3 mm, and other sizes make no contribution, according to Fraser. This, of course, is at variance with the estimations of Minnaert mentioned above. Fraser goes on to predict that the minima of intensity in supernumerary bows will be separated by approximately 0.75°, regardless of the size distribution of the raindrops forming them. He has substantiated the statement by having two unbiased analysts make measurements of rainbow photographs by which they determined average separations of 0.68° and 0.70°.

In a recent analysis, Konnen and de Boer (1979) distinguished between the primary rainbows formed by the two orthogonal intensity components $I_{\parallel}$ and $I_{\perp}$ and found them to be different. They termed the rainbow from $I_{\perp}$ the common rainbow, and that from $I_{\parallel}$ the polarized rainbow. This seems an unfortunate choice of terms, but it is taken from the fact that $I_{\parallel}$ vanishes completely at the Brewster angle whereas no such mechanism operates on $I_{\perp}$. At any rate, the authors developed an extension to the Airy theory for taking account of the 180° phase shift that occurs at the Brewster angle for the reflection of $I_{\perp}$ inside the drop for the primary rainbow. Since the phase of the emergent rays is critical in formation of the supernumeraries, the positions of the maxima and minima of the supernumeraries are different for the common and polarized rainbows in the authors' terminology, as are their magnitudes. The two sets of maxima and minima are shown as a function of angle from the Descartes ray in Fig. 3.40

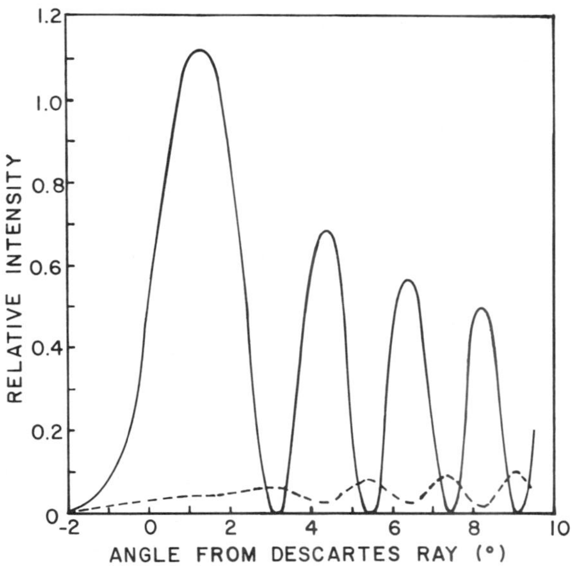

Figure 3.40 Intensities of orthogonal intensity components $I_{\perp}$ (solid curve) and $I_{\parallel}$ (dashed curve) as functions of angle from the Descartes ray for supernumeraries of the primary rainbow, as computed by Konnen and de Boer (1979) for the case of $2\pi r = 1500\lambda$ (droplet radius approximately 0.14 mm for wavelength 0.60 μm).

for the case of $2\pi r = 1500\lambda$. This corresponds to a droplet radius of approximately 0.14 mm for $\lambda = 0.60\ \mu m$. Beyond the first maximum for $I_\perp$, the supernumeraries are 180° out of phase, which means that the maxima of each set falls in the gaps of the other set. However, the magnitude of the maxima for $I_\parallel$ are so much smaller than those for $I_\perp$ that it is doubtful if the field would be visibly changed by them. Konnen and de Boer found the out-of-phase character applied for drop sizes $r < 0.6$ mm, whereas for $r > 0.6$ mm the oscillations are in phase. Thus if the sizes given by Fraser are realistic, the two types of oscillations would always be out of phase in the natural rainbow.

3.5 Ice crystal halos

Since cirrus, or ice crystal, clouds are much more common in nature than are rain showers, the halo is a more commonly observed phenomenon than the rainbow. Because of the many circles, arcs, and spots possible in a well-developed halo complex, the phenomenon can be strikingly beautiful, and although the colors visible in the halo are not usually as intense as those of the rainbow, the parhelia (sundogs) and some of the arcs often show brilliant displays of color. In addition, the relatively modest amount of color in the halo as a whole is more than compensated for by the intricate geometric patterns of light in various parts of the sky which are seen in a well-developed halo complex.

A number of famous displays of the various halo phenomena have been described in the literature. Needham (1959) gave the halo drawing from the manuscript by the Chinese emperor Chu Kao-Chih from the year 1425 which is reproduced here as Fig. 3.41. A translation of the characters below the drawing gives the title as "A solar halo with four earrings, four embracements, and four thumb-rings." The appearance of the halo signified conspiracy on the part of the ministers. The gates were to be shut and emergency orders were to be issued through the land. If within 3 days there was rain, the orders were to be cancelled. Nyberg (1984) has shown a photograph of a painting of a halo complex seen over the town of Stockholm, Sweden, in 1535. The painting now hangs in the Stockholm Cathedral and is named *Vadersolstavla* (the weather sun picture). According to Nyberg, the picture has been criticized as having a distorted view of the halo, but, be that as it may, the painting is an interesting old record of a spectacular display of the halo. Another interesting old drawing was made

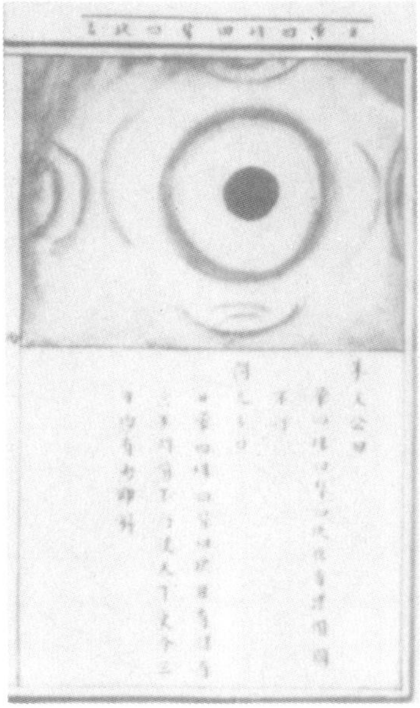

Figure 3.41 Photograph of a halo complex drawn by the Chinese Emperor Chu Kao-Chih in the year 1425. A translation of the characters below the halo gives the title as "A solar halo with four earrings, four embracements, and four thumb rings."

by the Danish astronomer Hevelius of a display seen at Gdansk in 1662 (see Greenler, 1980). The unique feature of this drawing is the portrayal for the first time of two arcs crossing the 90° halo, arcs which are now termed Hevel's halo.

Perhaps the most famous early drawing of a halo complex is that made by Tobias Lowitz in St. Petersburg, Russia, in 1790. A tracing of the drawing, as taken from the book of Pernter and Exner (1910), is shown below as Fig. 3.42. Although this carefully detailed drawing suffers some geometrical distortions arising from the effort to map a three-dimensional sky on a flat plane, it is evidence of a keen and careful observation which Lowitz left for posterity. As a result of this drawing, he has been immortalized by having his name given to some of the arcs of the halo (Muller et al., 1979). Similarly, the arcs of Parry were named for Sir William Edward Parry, English arctic explorer, as a result of a diagram he made in the Arctic during his search for the Northwest Passage in

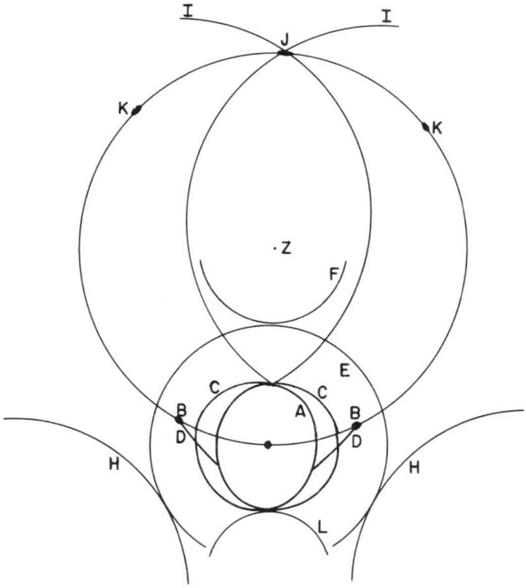

Figure 3.42 Tracing of the outlines of the halo complex observed by Tobias Lowitz in St. Petersburg in 1790. The tracing was taken from a copy of the original in Pernter and Exner (1910). The various features are the following:

A: 22° halo
B: Sundogs of 22° halo
C: Circumscribed halo
D: Lowitz arcs
E: 46° halo
F: Circumzenithal arc
G: Parhelic circle
H: Infralateral arcs
I: Anthelic arcs
J: Anthelion
K: 120° Parhelia
L: Tangential arc of 22° halo
Z: Zenith

1821. The halo complex was also a favorite subject for German woodcuts in the 16th century. For instance, Strauss (1975) shows reproductions of woodcuts of no fewer than 22 halo phenomena, each being termed ''Apparitions in the Sky,'' made by various artists from 1550–1600.

It is surprising that so few actual photographs of the halo complex have been made since the development of the photographic process in 1838. One reason, no doubt, is the requirement for special photographic equipment for covering the wide angles encompassed by the entire complex. A camera with the normal type of lens covers only a 35° angle, so it will not cover all of even the 22° halo. Use of the usual wide angle lens will yield photos of the 22° halo, but not the 46° halo and many of the arcs. One method of getting photos covering the entire dome of the sky is to use a fish-eye camera lens, but most people do not have such specialized equipment, and even those who do have it usually do not carry it with them on the off chance that they will get an opportunity to photograph a halo. Visitors to research stations in the Antarctic, however, have good assurance of seeing well-developed halo phenomena, and often equip themselves with the requisite camera equipment. By using fisheye lenses, Robert Greenler and Austin Hogan both took beautiful photos of halo phenomena at the South Pole in 1977 (Plates 4-1 and 4-2 of Greenler, 1980). Although some distortion results from fisheye lenses, most of the circles and arcs of the complex are well shown in the photos. In addition, Hogan's picture shows considerable color in sundogs and upper tangent arcs.

The principal requirement for a halo to develop is that the atmosphere in the vicinity of the observer contain many ice crystals, with little cloudiness otherwise. In low to middle latitudes the ice crystals are normally manifested as cirrus clouds which form in the high altitude and cold regions of the atmosphere. Beginning students in meteorology are normally taught that the cirrostratus clouds, which are most conducive to halo formation, form as moist air ascends over the surface of a warm front, and are thus often precursors of bad weather. While there is a grain of truth in this concept, the majority of cases of cirrus are not related to warm front activity, and halos are thus poor forecasting tools for future weather events. Although ice crystals form at high altitudes in the lower latitude areas of the earth, the low temperatures of polar regions often create ice crystals in the surface layers of the atmosphere. For instance, at the research station at the geographic south pole, the author has seen long periods of time when the air appeared to be filled with glittering diamonds as ice crystals fell slowly through the lower few hundred meters of an otherwise perfectly clear atmosphere. In this case a moist layer of air had moved in at higher altitudes and provided conditions conducive to the formation of the myriad of ice crystals which slowly sifted down to the surface. The halo phenomena which result from these perfectly formed crystals are truly spectacular. But residents of the less exotic areas of the world do not need to journey to the polar regions to observe the beauties of the halo—they need only to watch for them.

Explanations of the various features of the halo are surprisingly involved, in spite of the generally simple forms assumed by the ice crystals responsible for halos. A complete discussion of the formulae by which the phenomena are characterized is beyond the scope of this book, but the subject has been pursued in great detail in the books of Pernter and Exner (1910), Humphreys (1940), and Tricker (1970), and the individual features of the halo complex have been discussed in numerous scientific publications over the last several years (Greenler and Mallmann, 1972; Tricker, 1973, 1979; Lynch, 1979; Lynch and Schwartz, 1979; Mallmann and Greenler, 1979; McDowell, 1974, 1979; Tape, 1979; Greenler et al., 1977; Lynch, 1978). The most successful method of simulation of the features of the halo is that of Greenler (1980) and his colleagues, who used the Monte Carlo technique for numerically following numerous rays passing at various angles through the crystals and determining thereby the relative intensity distributions resulting from refraction and reflection by the crystals. Greenler did not consider polarization effects in halos, but Konnen (1983) has recently discussed polarization resulting from both refraction by the ice crystals and the birefringence of ice. These subjects will be mentioned further below. Fraser (1979) has discussed the types and sizes of ice crystals causing halos, and Tape (1983) contributes a photograph of ice crystals collected in the Antarctic at the time of a halo.

The sequence to be followed here is to first describe the ice crystal forms thought to be responsible for the halo, together with the basic relations by which the light rays are deviated by the crystals, and then to discuss briefly the orientations of the crystals and the paths of the rays by which each of the halo features is produced. It should be mentioned at the outset that not all of the features are readily explained, and that according to the simulations of Greenler (1980) some features which have never been actually observed may in fact be a part of the halo complex.

3.5.1 Forms and properties of ice crystals

Ice crystallizes in the hexagonal system, which means there are three axes which are parallel to the 60° faces of the crystal and one principal axis at right angles to the other three. The crystal faces may be developed to different extents, thereby giving many different crystalline shapes. The ends of the crystal may be developed into pyramidal shapes, resulting in numerous flat faces from which light may be reflected or through which it may be refracted. However, most of the halo features may be explained on the basis of the two simple forms shown in Fig. 3.43. The rays of light may enter the crystal through one of the 60° faces and exit through another 60° face, enter through a 60° face and exit through an end face, or enter an end face and exit through a 60° face. The first of these paths is responsible for the 22° halo, and the other two for the 46° halo. Other paths, even including reflection from the faces, are possible, as will be seen in the discussion of the various features of the complex.

3.5.2 Deviation of light rays by ice crystals

As with the rainbow discussed above, light is deviated from its original direction by its passage into and out of an ice crystal. We follow the method of Humphreys (1940) in computing this deviation. If a light ray is incident on the face of the crystal of Fig. 3.44 at angle τ_1 with respect to the surface normal, it is bent at the surface and proceeds inside the crystal at angle τ_1'. Thus the

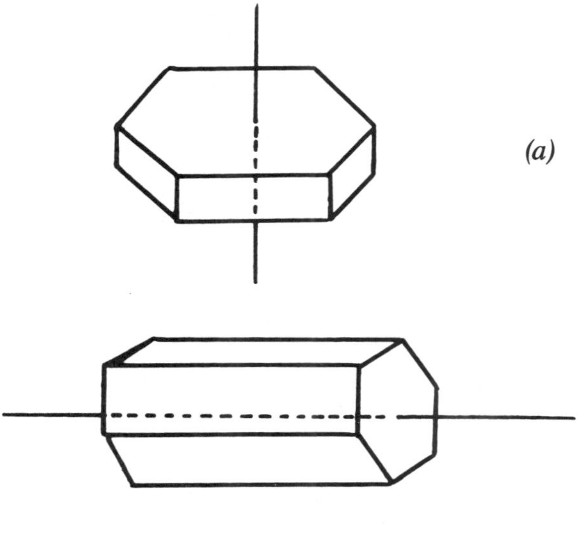

(a)

(b)

Figure 3.43 Two forms of ice crystals responsible for various features of the halo complex. (a) Tabular-type ice crystal, (b) Columnar-type ice crystal.

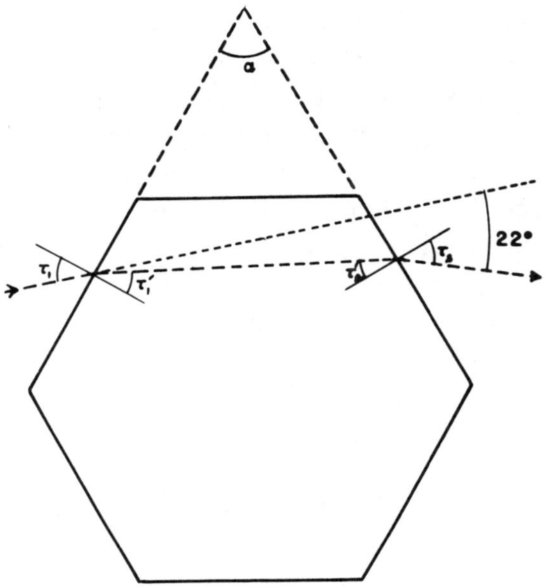

Figure 3.44 Diagram of the passage of light ray through a hexagonal ice crystal, with deviations on entry and exit adding up to a total deviation of approximately 22° and leading to the 22° halo.

deviation at this first encounter with the crystal is $\tau_1 - \tau_1'$. Similarly, at emergence, the angle of incidence on the second face is τ_2' and the angle of emergence is τ_2, so the total deviation is

$$D = \tau_1 - \tau_1' + \tau_2 - \tau_2' = \tau_1 + \tau_2 - (\tau_1' + \tau_2') \tag{3.64}$$

From the diagram we have the angles

$$a = 90° - \tau_1';\ b = 90° - \tau_2';\ a + b + \alpha = 180°$$

or

$$\tau_1' + \tau_2' = \alpha \tag{3.65}$$

and

$$D = \tau_1 + \tau_2 - \alpha \tag{3.66}$$

It is not necessary that the crystal be a hexagon for these relations to hold.

The concentration of rays which we see as a halo appears at the angle of minimum deviation D_m (the equivalent of the Descartes ray in the rainbow). On differentiating Eq. (3.66) with respect to incident angle, we have for minimum deviation

$$\frac{dD}{d\tau_1} = 1 + \frac{d\tau_2}{d\tau_1} = 0 \tag{3.67}$$

From this relation we see that for minimum deviation

$$\begin{aligned} d\tau_1 &= -d\tau_2 \\ d\tau_1' &= -d\tau_2' \end{aligned} \tag{3.68}$$

In order to compute the amount of deviation of the ray we use Snell's law of refraction for both entrance into and emergence from the crystal. Thus we have

$$\begin{aligned} \sin\tau_1 &= m \sin\tau_1' \\ \sin\tau_2 &= m \sin\tau_2' \end{aligned} \tag{3.69}$$

or

$$\sin\tau_1 \sin\tau_2' = \sin\tau_2 \sin\tau_1' \tag{3.70}$$

Here m is the index of refraction of ice with respect to that of air. By differentiating the relations (3.69), combining with Eq. (3.70), and simplifying, we can write

$$\cos(\tau_1 - \tau_2') = \cos(\tau_1' - \tau_2) \tag{3.71}$$

or

$$\tau_1 - \tau_2' = \tau_1' - \tau_2 \tag{3.72}$$

Similarly

$$\cos(\tau_1 + \tau_2') = \cos(\tau_1' + \tau_2)$$

or

$$\tau_1 + \tau_2' = \tau_1' + \tau_2 \tag{3.73}$$

On adding and subtracting Eqs. (3.72) and (3.73), we have for minimum deviation of the ray

$$\tau_1 = \tau_2;\ \tau_1' = \tau_2' = \frac{\alpha}{2} \tag{3.74}$$

This means that the ray of minimum deviation is transmitted symmetrically with respect to the crystal faces.

The discussion thus far is for the case of the incident ray being normal to the principal axis of the crystal. In general, however, the orientations of ice crystals in the atmosphere are not so constrained, and it is necessary to consider the refraction of rays skewed with respect to the principal axis. In the interest of brevity, we refer to the discussion of Tricker (1970) for the somewhat involved geometry of skewed rays and write the result directly. If h is the angle of incidence of the skewed ray with respect to the principal axis, Tricker has shown that the deviation of the ray is given by Snell's law of Eq. (3.69) if the index of refraction m is replaced by m' written as

$$m' = (m^2 - \sin^2 h)^{1/2}/\cos h \tag{3.75}$$

It is of interest to note that the minimum deviation of skewed rays is greater than that of rays incident normal to the principal axis, with the result that the skewed rays tend to make the halo more diffuse than it would be if all of the crystals were oriented normal to the incident ray.

We turn now to a discussion of the geometry and crystal orientations necessary for producing the various features of a halo complex. Extensive reference will be made to the simulations of Greenler (1980), as well as to a number of recent articles devoted to specific features. Since the halo complex observed and drawn by Tobias Lowitz at St. Petersburg, Russia, in 1790 has been so frequently reproduced, we take that as a starting point. A tracing of the Lowitz drawing was given above as Fig. 3.42. The letters have been introduced for convenience of discussion. It should be mentioned that although the various features are indicated by lines in the diagram, they are really bands of light in nature.

3.5.3 The 22° halo

The simplest and most frequently observed feature of the halo complex is the 22° halo. If myriads of crystals of the type shown in Fig. 3.43b are randomly oriented in the atmosphere, then some will be in the appropriate position and orientation at every point in the sky at an angle of about 22° from the sun to deviate some of the incident solar rays into the direction of an observer, according to the diagram of Fig. 3.44. This configuration is seen as a circle of increased brightness encircling the sun at an angle of approximately 22°. The brightest part of the 22° halo corresponds to the angle of minimum deviation for the rays striking the crystal normal to the principal axis, as mentioned above. Since this is a well-defined angle, the halo has a relatively sharp boundary on the side nearest the sun. Light from skew rays and light striking the crystal at angles other than that of minimum deviation is spread over wider angles in the direction away from the sun.

The 22° halo often shows a reddish color on the boundary nearest the sun, but appears white otherwise. This comes about because of the change of the index of refraction m for ice with wavelength of the light. For red light $m=1.307$ and for violet light $m=1.317$, with values between these two for other visible wavelengths. The angles of minimum deviation for rays incident normal to the crystal axis are 21.60° for red and 22.37° for violet light. Skew rays have larger deviations than these. Thus red light incident normal to the crystal axis is deviated least of all, with rays of other wavelengths and different degrees of skewness spread out all together at angles greater than that for red light. The superposition of different wavelengths, with even the red light at skew angles, effectively washes out any color at most angles, leaving the only visible color as a narrow band of red at an angle near 21.60°, the side nearest the sun. The total intensity of the skylight is enhanced in the region outside of the halo at the expense of that in the region toward the sun, an effect which is clearly visible in a well-developed halo.

3.5.4 Sundogs or parhelia

The 22° halo is produced by the elongated prism-type crystals, such as shown above in Fig. 3.43b which have random orientations as they fall through the air. Not all crystals assume random orientations, however, and the preferred orientations produce many of the arcs and spots of the halo complex. This is the case with the bright and often highly colored spots at angles of approximately 22° on either side of the sun. Here we designate them by the popular term sundogs, although terms such as parhelia and mock suns are often applied. Flat plate-type crystals such as shown in Fig. 3.43a fall through the air in a manner similar to that of a falling leaf, with the largest surface area, the end faces, oriented horizontally. With this orientation, only light which is proceeding in an approximately horizontal direction strikes the 60° faces of the crystals in such a manner as to undergo the minimum deviation, and thus those crystals located horizontally from the sun are in the right location to refract incident sunlight to the eye of the observer. Furthermore, the statistically preferred orientation results in very minor contributions by skew rays. The wavelengths are thus more effectively separated in sundogs than in the normal 22° halo, with the result that the colors are more vivid and spectrally pure. In fact, it is often possible to see beautiful reds, yellows, greens, and even blues very distinctly in sundogs. The overlapping colors make a long white tail which extends several degrees away from the colored part of the sundog.

For completely monodirectional radiation incident on the sides of these horizontal plate crystals, the sundogs should be a thin colored line, but as actually observed they have a significant vertical dimension. There are two reasons for this. First, the disk of the sun subtends an angle of approximately 0.5°, so the sundog must have at least this vertical dimension. Second, the falling plates are likely to vibrate a certain amount around horizontal axes of all different azimuthal orientations, thereby giving rise to skew rays which cause a general spreading of the bright spot and some overlapping of the colors.

The description thus far applies to the case of the sun on the horizon, so that the rays refracted into the observer's eye strike the vertical faces of the crystals from a normal direction and undergo the minimum deviation. Thus for low sun elevations the sundogs coincide with the 22° halo. For higher sun elevations, however, the rays seen by the observer are only skew rays, which have undergone more than the minimum deviation and are thus more than 22° from the sun. For this reason sundogs seen at the higher sun elevations lie somewhat outside of the normal 22° halo. This separation is clearly visible in a well developed halo display. In fact, the separation increases with increasing sun elevation, becoming 10° at sun elevation 50° and over 50° at the highest sun elevation at which sundogs are formed (Humphreys, 1940). If the sun elevation is greater than approximately 60.75°, the incident light is totally reflected inside the 60° prism and does not emerge as refracted light. Thus sundogs do not form at sun elevations higher than this. Because of the perceived curvature of a line of constant elevation for an observer on the ground, sundogs generally appear to be above the center of the sun. Only for the sun on the horizon do they appear at the same elevation as the sun.

3.5.5 Upper and lower tangent arcs of the 22° halo, and the circumscribed halo

The upper and lower tangent arcs of the 22° halo and the halo circumscribed around the 22° halo all arise because of the preferential orientation of columnar ice crystals such as shown in Fig. 3.43b. This type of elongated body tends to fall with the long axis horizontal. Refraction of the light entering one of the 60° faces and emerging from another face one removed gives a minimum deviation of approximately 22°. In fact, the random orientation of such crystals produce the 22° halo, and a preferred orientation of platelet crystals yields the sundogs discussed above. Horizontal orientation of the long axis of columnar crystals would tend to give enhanced light intensity (the equivalent of sundogs) at the top and bottom of the 22° halo due to rays undergoing minimum deviation, and skew rays would cause light streaks on each side of the top and bottom spots to form arcs tangent to the 22° halo. For the sun on the horizon, the upper and lower tangent arcs would be equivalent, but of course the lower tangent arc would be 22° below the horizon and thus not visible except from an elevated vantage point.

As the sun rises above the horizon, the tangential arcs undergo a pronounced modification, and eventually they bend around and form a complete elliptically shaped ring surrounding the 22° halo—the circumscribed halo. A simulation of the evolution of the arcs has been developed by Greenler and Mallmann (1972).

The geometry of the process, which is too tedious to be reproduced here, is given in detail by Pernter and Exner (1910) and Tricker (1970), among others. The configurations of the arcs, as given by Tricker, are shown for different sun elevations in Fig. 3.45a, and Fig. 3.45b shows the circumscribed halos after the arcs have been extended and joined on each side of the sun at the higher sun elevations. With the exception of the case with the sun on the horizon, the two arcs behave quite differently with increasing sun elevation. For instance, the lower arc forms a cusp and a loop at sun elevations 10° and 20°, respectively, whereas the upper tangent arc undergoes a more regular evolution from arc to circumscribed halo. The different behavior arises from the different geometry, the lower arc being about 44° below the upper arc and thus subject to quite different angles with respect to the horizontal orientation of the relevant ice crystals.

The upper tangent arc is not a particularly rare phenomenon, but the lower tangent arc and complete circumscribed halo are seen much less frequently. Pernter and Exner (1910) and Tricker (1970) reproduced a very distinctive photograph of the lower tangent arc at the time it formed a sharp cusp, the photo-

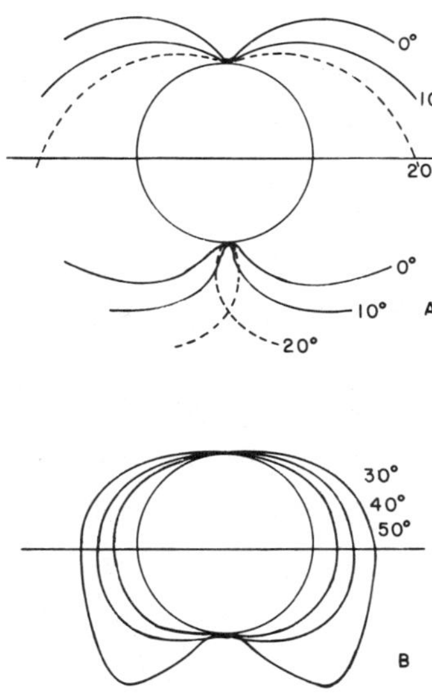

Figure 3.45 Configurations assumed by tangential arcs (a) of the 22° halo at the sun elevations indicated, and their evolution into the circumscribed halo (b) at the higher sun elevations.

graph having been taken from a mountain top and showing the cusp below the horizon. Greenler and Mallmann (1972) and Greenler (1980) used a numerical simulation technique to describe in detail the evolution of the arcs and circumscribed halo as the sun rises from the horizon to a 70° elevation. An attractive feature of the numerical simulation is that it yields information on the relative intensities, as well as on the shapes of the various features.

3.5.6 The Parry arcs

These arcs, named as a result of a sketch by Sir William E. Parry during his quest for the Northwest Passage in 1819 to 1829, may be classified as a subset of the tangential arcs of the 22° halo. Since they are similarly located with respect to the 22° halo, are separated by only 1.5° to 2°, and the region between them is often filled with a general diffuse illumination, it is sometimes difficult to distinguish between the two types. In a well-delineated case, however, the Parry arc forms a distinct band above the upper tangential arc or below the lower tangential arc.

There is evidence from Greenler's (1980) simulations and geometrical considerations that the Parry arcs are formed by refraction in columnar-shaped crystals with their long axes horizontal, as are the tangent arcs of the 22° halo, but with an additional constraint that two of the hexagonal faces are approximately horizontal also. Although the forces maintaining such a unique orientation of the crystals must be very small, it appears that this is a stable orientation for the fall of hexagonal crystals. In addition, it is to be expected that a certain amount of oscillation around the stable position would occur.

Rotation of the crystals around the long axis produces a certain amount of diffusion in the Parry arcs. Greenler et al. (1977) have obtained realistic simulations of the intensity distribution by assuming rotation angles of ±3° from the horizontal. Rotations greater than this act to fill in the space between the Parry arcs and the tangent arcs of the 22° halo. According to Greenler (1980), rotation angles of ±30° turn the Parry arcs into tangent arcs of the 22° halo.

3.5.7 The arcs of Lowitz

These arcs, named for Tobias Lowitz, the observer of the 1794 halo display at St. Petersburg (see Fig. 3.42), are also closely allied with the 22° halo. The arcs, which are labeled D in Fig. 3.42, are short streaks of light extending diagonally downward from the sundogs to intersect the 22° halo at some distance below and to either side of the sun. Mueller et al. (1979) have simulated the Lowitz arcs by considering hexagonal platelet crystals to be spinning around a major diagonal of the crystal (i.e., corner to corner in a direction parallel to the basal face) while maintaining this axis in a horizontal orientation with arbitrary azimuth. Three different paths of rays through the crystal would result from the spinning crystal, one for each of the 60° faces illuminated by the sun. It is assumed that there are crystals in all directions from the observer so that some would be in the correct position to deviate light to the eye of the observer at appropriate angles of rotation. The simulation of Mueller et al. (1979) give not only the arcs of Lowitz in their observed positions but also indicate the probable existence of similar arcs extending diagonally above the sundogs to become tangent to the 22° halo above the parhelic circle. A third feature indicated by the simulations is an arc roughly concentric with the 22° halo below the sun that extends upward to the sundogs on either side of the sun. The arcs of Lowitz are seen relatively rarely, and apparently the other two features thought to occur simultaneously with the Lowitz arcs have not yet been observed. Since all three of these arcs extend from the 22° halo to the sundogs, they can be seen as separate arcs only for sun elevations sufficiently high for the sundogs to be perceptibly separated from the 22° halo.

3.5.8 The 46° halo

The 46° halo is reproduced by light which enters the ice crystal through a 60° face, which is parallel to the optical axis, and exits through the end face oriented 90° to the optical axis, or vice versa. The refracting angle is thus 90° for the 46° halo, analogous to 60° for the 22° halo. The angle of minimum deviation for $m = 1.31$ and 90° refracting angle is 45.73°. For the formation of a complete circle at this angle from the sun, some of the innumerable ice crystals such as shown in Fig. 3.43b have to be oriented properly at every point on the circle to refract the incident sunlight in the direction of the observer. As with the 22° halo, the only color normally visible in the 46° halo is a faint band of red on the side nearest the sun, but other colors may be visible in an extraordinary display. The dispersion produced by the 90° refracting angle is greater than that for the 60° angle, so the 46° halo is wider, the colors are more widely separated, and the light is less intense than that at 22°. Consequently, the 46° halo is seen much less frequently than its smaller cousin.

3.5.9 The 46° sundogs

Just as the sundogs of the 22° halo require a preferred orientation of the ice crystals, so also do those of the 46° halo although the orientations are different for the two. The 46° sundogs require flat end faces to be vertical, such as would be the case for columnar ice crystals (Fig. 3.43b) with the long axis horizontal. In fact, these are thought to be the type of crystals primarily responsible for both the 46° halo and its attendant sundogs. The resistance of air on a falling crystal acts to orient the crystal for maximum resistance, which would make the long axis of columnar crystals horizontally oriented but the flat face of tabular crystals horizontal. This means that the requirement for vertical faces in the 46° sundogs indicates a primary contribution from columnar-type crystals. Again the 90° refraction angle spreads the colors more in the 46° sundogs than in those at 22°. This tends to make the 46° colors purer but less intense, and spreads out the white tail over a wider angle. The low intensities of both the 46° halo and its sundogs cause them to be relatively rarely seen in nature.

3.5.10 The circumzenithal arc

Another rarely seen arc, but occasionally a very spectacular one, is the circumzenithal arc. As the name implies, it is centered on the zenith. The arc length is not more than 90° and usually less. The tabular-type crystals producing it have the end faces horizontal, and thus the optical axis vertical. As mentioned previously, tabular-type crystals are stable in this orientation but columnar crystals are not, so at least some crystals of the tabular-type must exist at the time of a circumzenithal arc. Light enters the horizontal end face from above and emerges from a 60 ° side face. Minimum deviation occurs for a path through the crystal which is symmetric with respect to the 90 ° refraction angle, which is 45.73 ° for $m = 1.31$. This is the angle of the 46 ° halo. For horizontal faces, the path is symmetric and deviation a minimum for a sun elevation of 22.1 °, in which case the circumzenithal arc is tangent to the top of the 46 ° halo. This often gives rise to the erroneous designation of the circumzenithal arc as the upper tangent arc of the 46 ° halo. For sun elevations less than or greater than 22.1 °, the angle between arc and sun is larger than 45.73 °, but it is within 1° of this value for sun elevations between 16 ° and 27 ° (Humphreys, 1940). For the sun just above the horizon the circumzenithal arc is 12.07 ° above the 46 ° halo, and it converges on the zenith and disappears as the sun approaches 32.2 ° elevation. The configuration of an arc is produced by the position of the image being altered by rotation of the crystal about its principal axis or by the 60 ° faces of different crystals being in different orientations while the axis is still vertical. Since the arc is due to refraction of the light without reflection, there is no possibility of the circumzenithal arc extending to the area beyond the zenith opposite the sun.

3.5.11 The circumhorizontal arc

This is a band of light stretching around the sky at a constant elevation above the horizon in a manner similar to the circumzenithal arc. It is produced by the same tabular-type crystals as those producing the circumzenithal arc, and it is also due to refraction. The difference is that, whereas in the circumzenithal arc light enters the flat top face of the crystal from above, in the circumhorizontal arc light enters a 60 ° side face of the crystal and emerges in a downward direction from the flat bottom face of the crystal. The angle of deviation is the same for all such horizontally oriented crystals, which makes the circle of light appear at a constant elevation. Refraction across the 90 ° refraction angle separates the different colors over a relatively wide angle, as in the 46 ° halo and circumzenithal arc, which results in the circumhorizontal arc showing spectacular colors of high purity. Because of the geometry of the refraction, the circumhorizontal arc is possible only for sun elevations above 58 °.

3.5.12 The parhelic circle

The parhelic circle is a white band of light encircling the sky parallel to the horizon at the elevation of the sun. It is due to reflection of light by the ice crystals, and not by refraction as with most other features of halos. Reflection, of course, does not separate the wavelengths, so parhelic circles show no color. The faces of the crystals from which reflection occurs must have a vertical orientation, and all possible azimuthal orientations for the complete circle to appear. Either of the crystal forms shown in Fig. 3.43 may meet these criteria. The long prismatic crystal falls with the crystal axis horizontal, so the normal end faces would be vertical, and the flat crystal, falling with the crystal axis vertical, would have the 60° faces oriented vertically. Thus either or both types may generate a parhelic circle.

At first glance it may not be obvious why reflection from vertical faces of crystals would produce a circle at the same elevation as the sun, instead of at half this elevation. The problem is easily resolved, however, by realizing that for a sun elevation of say 30°, light reflected from vertical surfaces is directed downward also at 30°. Thus an observer would see the light coming from a 30° elevation angle, which of course is the assumed elevation of the sun. Random azimuthal orientations of the vertical faces would send light to the observer from all azimuths, which, in combination with the perceived 30° elevation, would be seen as a band of light encircling the sky parallel to the horizon at the height of the sun. This is the parhelic circle.

3.5.13 Supralateral and infralateral arcs

Crystals responsible for the formation of arcs tangent to the 46° halo in a position at about 45° downward and upward from the sun, the infralateral and supralateral arcs, are the same type of columnar crystals which form the 46° halo itself and the circumzenithal arc. The difference lies in the fact that for the 46° halo the crystal axes have a random orientation and for the circumzenithal arc both the long axis and two of the 60° faces are horizontal. For the lateral arcs, however, the long axis is horizontal but the 60° faces have random orientation or are spinning around the horizontal axis.

Although complete observations of the arcs are lacking, the simulations of Greenler et al. (1979) indicate that the arcs should undergo a definite evolution with changing sun elevation. For the sun on the horizon both arcs should be concentrated at the sides of the 46° halo. As the sun rises above the horizon, they should separate in complex configurations into a long arc capping the 46° halo, with more conventional tangent arcs below the parhelic circle. These are the arcs most often seen. The supralateral members disappear for sun elevations above 32°, as in that case the angle of exit from the crystal is such as to cause total internal reflection within the crystal. Infralateral arcs, however, may be very distinct at moderate sun elevations of 20° to 60°. At still higher sun elevations the numerical simulations indicate the infralateral arcs to combine into a complete band underneath the 46° halo, but as far as is known this configuration has not been observed.

A whole family of additional tangent arcs to the 46° halo are indicated as possible by the simulations of Greenler et al. (1979). A complex of ten additional arcs is shown for 0° sun elevation, with a complicated evolution of the family with increasing sun elevation. The rays giving rise to these arcs would be refracted across 90° angle between the end face and the 60° faces of tabular crystals which themselves undergo a spin around a horizontal axis between opposite corners joining the 60° faces. At the present stage the arcs resulting from this mechanism must be considered purely hypothetical, but the simulation might indicate an avenue for future observations.

3.5.14 Phenomena in the region of the anthelion

Many observations show the presence of various arcs, spots, and pillars in the vicinity of the anthelion. The anthelion is located on the parhelic circle at an azimuth directly opposite the sun. This point should not be confused with the antisolar point or antisun, which is a point below the horizon for positive sun elevations. Light reaching the observer from such a large angle from the sun as the anthelion must undergo one or more reflections from the faces of ice crystals. The reflections are mainly internal, but the parhelic circle was seen to be produced by external reflections, either internal or external, from vertically oriented crystal faces.

A large number of possibilities may be conceived from reflections of light inside ice crystals. For instance, Tricker (1970) has suggested a method whereby hexagonal columnar crystals may produce five mock suns. By this mechanism, light enters the end face of a columnar crystal, is totally internally reflected from any of the 60° side faces, and emerges from the opposite end face. Refraction on exit is exactly the opposite of that on entry, so the effect is the same as if the incident beam had been reflected directly from the side faces. The five spots of light will lie on a small circle whose axis is the long axis of the crystal and which passes through the sun. Spinning of the crystal will cause three of the spots to be elongated into arcs, whereas the other two will remain fixed at 120° from the sun and form the parhelia of 120°. Further, light reflected from the second end face, after three reflections from side faces, and emerging from the face where it entered would appear from the direction of the anthelion, and rotation or random orientation of the crystals would cause arcs to pass through the anthelion—the anthelic arcs. This latter mechanism is responsible for Tricker (1973) likening the columnar ice crystal to a kaleidoscope.

Another possible explanation of the anthelic arcs, in which a single internal reflection is required, has been advanced by Hastings (1920). Light paths for these and other phenomena involving reflection are diagrammed by Greenler (1980).

3.5.15 Light pillars

The traditional explanation of the vertical column of light extending usually, above, but sometimes below, the sun is that the light is reflected from the horizontal faces of tabular crystals. If this were the whole explanation, instead of being in the form of vertical pillars of light, there would be only light spots. In order to form pillars the flat faces of the crystals must have a statistical deviation from exact horizontal orientation. The light pillar is in some respects the equivalent of sun glint on the roughened sea surface, in which case reflection is from facets of waves with different angles of slope.

Greenler (1980) has shown at least semi-realistic light pillars resulting from crystal orientations deviating from 0° to 6° from the horizontal. The reflection will be from both faces of a flat crystal, the second reflection being due to light refracted into the crystal, undergoing an internal reflection, and being refracted back out with no change from the original direction. The quasi-horizontal orientation results in light pillars being brightest and most spectacular at low sun elevations, and indeed beautiful displays of light pillars are often seen extending above the sun after the sun has set. Light pillars are sometimes produced from street lights or other local sources when the weather is cold enough to produce ice crystals in the air near the ground.

3.5.16 The sub-sun

The sub-sun is seen only from an elevated point, such as in an airplane or on a mountain. It is an uncolored bright spot of light at some angular distance directly below the sun. The mode of production is the same as that of light pillars, but the small and concentrated nature of the spot indicates very small deviations of the ice crystals from a true horizontal orientation.

3.5.17 Polarization of the halo

The transmission and refraction of light as it enters and leaves an ice crystal produces a certain amount of polarization, but the values are low. The application of Fresnel's laws to the two components of light on the 60° faces of the hexagonal prism yields a degree of polarization of 3.7% in the region of minimum deviation of the 22° halo and arcs and spots of the 22° halo group. For the 46° halo group, in which radiation either enters or exits through the end plane, the polarization reaches about 16%. Ice crystals, however, exhibit birefringence (double refraction) resulting in complete polarization of the ordinary and extraordinary rays. Minimum deviations of these two rays are slightly different due to the different indices of refraction along the two axes. According to Konnen (1983), the angular difference between them is 0.11° for the 22° halo group and 0.15° for the 46° halo group. Because the index of refraction is slightly less for the ordinary ray than for the extraordinary ray, the inside edge of the 22° halo is due to the ordinary ray, which yields a vibration of the electric vector in the plane of scattering. This type of polarization would cover a band of (red) light less than 0.1° to 0.2° wide, which should be detectable by eye. The halo from the extraordinary ray, however, is displaced away from the edge by the angles mentioned above, and is thus not visible due to its being overwhelmed by the general light field arising by refraction of the ordinary rays of the different wavelengths.

Since reflection of light may produce high values of polarization, features such as the parhelic circle or sun pillars may be highly polarized. For instance, Lynch (1979) computed the degree of polarization of light in the parhelic circle for an index of refraction of ice of $m=1.31$ from the Fresnel laws of refraction and reflection. The results of the computations for three different sun elevations are shown in Fig. 3.46. The Brewster angle, at which 100% polarization occurs, is 52.6° for $m=1.31$. For the sun on the horizon (sun elevation 0.25°), the Brewster angle for ice crystals of the parhelic circle is at azimuth 74.7°. The azimuth of complete polarization increases with increasing sun elevations, and reaches 180° when the sun elevation is equal to the Brewster angle. At sun elevations higher than this, the polarization of the parhelic circle is a maximum at azimuth 180°, but it does not reach 100% at the higher sun elevations.

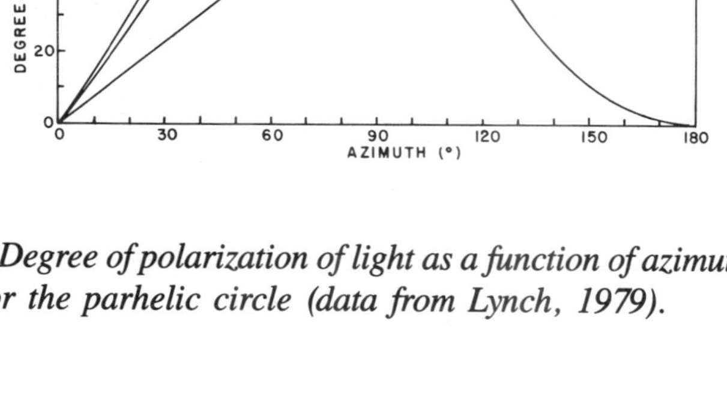

Figure 3.46 Degree of polarization of light as a function of azimuth with respect to the sun for the parhelic circle (data from Lynch, 1979).

References

Ahlberg, M.S., and I. Fangmark (1983): Characterization of coarse particles from and oil-fired power plant collected for use in biological tests. *J. Aerosol, Sci.*, *15*, 329.

Airy, G.B. (1838): On the intensity of light in the neighbourhood of a caustic, *Trans. Cambridge Phil. Soc., VI,* 397–403.

Asano, S. (1979): Light scattering properties of spheroidal particles, *Appl. Opt., 18,* 712–723.

Asano, S., and M. Sato (1980): Light scattering by randomly oriented spheroidal particles, *Appl. Opt., 19,* 962–974.

Asano, S., and G. Yamamoto (1975): Light scattering by a spheroidal particle, *Appl. Opt., 14,* 29–49.

Ashkin, A., and J.M. Dziedzic (1977): *Phys. Rev. Lett. 38,* 1351.

Barber, P.W. (1977): Resonance electromagnetic absorption of non-spherical dielectric onjects, *IEEE Trans. on Microwave Theory and Techniques, Vol. MTT–25, No. 5,* 373–381.

Barber, P.W., D-S. Y. Wang, and M.B. Long (1981): Scattering calculations using a microcomputer, *Appl. Opt., 20,* 1121–1123.

Barber, P., and C. Yeh (1975): Scattering of electromagnetic waves by arbitrarily shaped dielectric bodies, *Appl. Opt., 14,* 2864–2872.

Bullrich, K. (1964): Scattered radiation in the atmosphere and the natural aerosol, *Advances in Geophys., 10,* Academic Press, New York.

Cadle, R.D. (1966): *Particles in the Atmosphere and Space,* Van Nostrand Reinhold, New York.

Chippett, S., and W.A. Gray (1978): The size and optical properties of soot particles. *Combustion and Flame, 31,* 149–159.

Chu, C.M., and S.W. Churchill (1955): Representation of the angular distribution of radiation scattered by a spherical particle, *J. Opt. Soc. Am., 45,* 958–962.

Chylek, P. (1976): Partial wave resonances and the ripple structure in the Mie normalized extinction cross section, *J. Opt. Soc. Am., 66,* 285–287.

Chylek, P. (1977): Extinction cross sections of arbitrarily shaped randomly oriented nonspherical particles, *J. Opt. Soc. Am., 67,* 1348–1350.

Chylek, P., G.W. Grams, and R.G. Pinnick (1976): Light scattering by irregular randomly oriented particles, *Science, 193,* 480–482.

Chylek, P., J.T. Kiehl and M.K.W. Ko (1978): Optical levitation and partial-wave resonances, *Phys. Rev. A, 18,* 2229–2233.

Chylek, P., J.T. Kiehl, and M.K.W. Ko (1980): Surface wave in light scattering by spherical and non-spherical particles, in D.W. Schuerman (ed), *Light Scattering by Irregular Particles,* Plenum Press, New York, pp. 153–164.

Dave, J.V. (1969): Scattering of visible light by large water spheres, *Appl. Opt.*, *8,* 155–164.

Deirmendjian, D. (1969): *Electromagnetic Scattering on Spherical Polydispersions,* American Elsevier Pub. Co., New York.

Deirmendjian D., and R. Classen (1962): *Light Scattering on Partially Homogeneous Spheres of Finite Size,* The RAND Corporation, P—393—PR.

Del Monte, M.G., G.M.B. Marcazzan, C. Sabbioni, and A. Venture (1983): Morphological, physical and chemical characterization of particles emitted by coal-fired power plant, *J. Aerosol Sci.*, *15,* 323–327.

Fahlen, T.S., and H.C. Bryant (1968): Optical back scattering from single water drops, *J. Opt. Soc. Am.*, *58,* 304–310.

Fraser, A.B. (1979): What size of ice crystals causes the halo? *J. Opt. Soc. Am.*, *69,* 1112–1118.

Fraser, A.B. (1983): Why can supernumerary bows be seen in a rain shower? *J. Opt. Soc. Am.*, *73,* 1626–1628.

Goldie, E.C.W. (1971): A graphical guide to halos, *Weather 26,* 391–396.

Greenberg, J.M. (1972): Absorption and emission of radiation by nonspherical particles, *J. Colloid Interface Sci.*, *39,* 513–519.

Greenler, R. (1980): *Rainbows, Halos, and Glories,* Cambridge Univ. Press, Cambridge.

Greenler, R.G., and A.J. Mallmann (1972): Circumscribed halos, *Science, 176,* 128–131.

Greenler, R.G., A.J. Mallmann, J.R. Mueller, and R. Romito (1977): Form and origin of the Parry arcs, *Science, 195,* 360–367.

Greenler, R.G., J.R. Mueller, W. Hahn, and A.J. Mallmann (1979): The 46° halo and its arcs, *Science, 106,* 643–649.

Gumprecht, R.O. and C.M. Sliepcevich (1951): *Tables of Scattering Functions for Spherical Particles,* Engineering Research Inst., University of Michigan Press, Ann Arbor.

Hansen, J.E., and L.O. Travis (1974): Light scattering in planetary atmospheres, *Space Sci. Rev.*, *16,* 527–596.

Hastings, C.S. (1920): A general theory of halos, *Mo. Wea. Rev.*, *48*, 322–330.

Hidy, G.M. (ed) (1972): *Aerosols and Atmospheric Chemistry,* Academic Press, New York.

Holland, A.C., and G. Gagne (1970): The scattering of polarized light by polydisperse systems of irregular particles, *Appl. Opt.*, *9,* 1113–1121.

Humphreys, W.J. (1940): *Physics of the Air,* 3rd ed., McGraw-Hill, New York.

Irvine, W.M. (1965): Light scattering by spherical particles: Radiation pressure, asymmetry factor, and extinction cross section, *J. Opt. Soc. Am.*, *55,* 16–21.

Jennings, S.G., and R.G. Pinnick (1980): Relationships between visible extinction, absorption, and mass concentration of carbonaceous smokes. *Atmos. Environ.*, *14,* 1123–1129.

Johnson, J.C. (1954): *Physical Meteorology,* John Wiley, New York.

Junge, C.E. (1963): *Air Chemistry and Radioactivity,* Academic Press, New York.

Kattawar, G.W., and G.N. Plass (1967): Resonance scattering from absorbing spheres, *Appl. Opt. 6,* 1549–1554.

Kerker, M. (1969): *The Scattering of Light and Other Electro-magnetic Radiation,* Academic Press, New York.

Konnen, G.P. (1983): Polarization and intensity distributions of refraction halos, *J. Opt. Soc. Am.*, *73,* 1629–1640.

Konnen, G.P., and J.H. de Boer (1979): Polarized rainbow, *Appl. Opt.*, *18,* 1961–1965.

Latimer, P. (1984a): Light scattering by a structured particle: The homogeneous sphere with holes, *Appl. Opt.*, *23,* 1844–1847.

Latimer, P. (1984b): Light scattering by a homogeneous sphere with radial projections, *Appl. Opt. 23,* 442.

Latimer, P., and P. Barber (1978): Scattering by ellipsoids of revolution: A comparison of theoretical methods, *J. Colloid and Interface Sci.*, *63,* 310–316.

Lowan, A.N. (1949): *Tables of Scattering Functions for Spherical Particles,* Appl. Math., Ser. 4, Nat. Bur. Stds., Washington, D.C.

Lynch, D.K. (1978): Atmospheric halos, *Sci. Am.*, *238,* 144–152.

Lynch, D.K. (1979): Polarization models of halo phenomena. I. The parhelic circle, *J. Opt. Soc. Am.*, *69,* 1100–1103.

Lynch, D., and Pt. Schwartz (1979): Origin of the anthelion, *J. Opt. Soc. Am.*, *69*, 383–386.

Mallmann, A.J., and R.G. Greenler (1979): Origins of anthelic arcs, the anthelic pillar, and the anthelion, *J. Opt. Soc. Am.*, *69*, 1107–1112.

McCartney, E.J. (1976): *Optics of the Atmosphere*, Wiley & Sons, New York.

McDowell, R.S. (1974): The formation of parhelia at higher sun elevations, *J. Atmos. Sci.*, *31*, 1876–1884.

McDowell, R.S. (1979): Frequency analysis of the circumzenithal arc: Evidence for the oscillation of ice-crystal plates in the upper atmosphere, *J. Opt. Soc. Am.*, *69*, 1119–1122.

Minnaert, M. (1940): *Light and Color in the Open Air*, G. Bell and Sons, London.

Mueller, J.R., R.G. Greenler, and A.J. Mallmann (1979): The arcs of Lowitz, *J. Opt. Soc. Am.*, *69*, 1103–1106.

Needham, J. (1959): *Science and Civilization in China*, Vol. 3, London.

Nevitt, T.J., and C.F. Bohren (1984): Infrared backscattering by irregularly shaped particles: A statistical approach, *J. Clim. Appl. Met.*, *23*, 1342–1349.

Newton, Sir Isaac (1704): *Opticks or a Treatise on the Reflections, Refractions, Inflections and Colours of Light*, Dover Publications, New York, 1952.

Nyberg, A. (1984): An old painting of a halo phenomenon in Stockholm Cathedral, *Weather*, *39*, 84–87.

Penndorf, R. (1957): New tables of total Mie scattering coefficients for spherical particles of real refractive index ($1.33 \leq n \leq 1.50$) *J. Opt. Soc. Am.*, *47*, 1010–1015.

Pernter, J.M., and F.M. Exner (1910): *Meteorologische Optik*, Wilhelm Braumuller, Vienna and Leipzig.

Pollack, J.B., and J.N. Cuzzi (1980): Scattering by nonspherical particles of size comparable to a wavelength; A new semi-empirical theory and its application to tropospheric aerosols, *J. Atm. Sci.*, *37*, 868–881.

Roessler, D.M., D-S. Y. Wang, and M. Kerker (1983): Optical absorption by randomly oriented particles, *Appl. Opt.*, *22*, 3648–3651.

Rozenberg, G.V. (1966): *Twilight. A Study in Atmospheric Optics*, Plenum Press, New York.

Saxon, D.S. (1955): Lectures on the Scattering of Light, Sci. Rept. No. 9, Contr. AF 19 (122)—239, Dept. of Meteorology, Univ. of California, Los Angeles.

Schuerman, D.W. (ed) (1980): *Light Scattering by Irregularly Shaped Particles,* Plenum Press, New York.

Shifrin, K.S. and J.L. Zelmanovich (1966,1968): *Tables of Light Scattering*; Vol. I—*Angular Functions*; Vol. II—*Elements of Scattering Matrices and Properties of the Scattered Field,* Hydrometeorological Publishing House, Leningrad, USSR.

Shipley, S.T., and J.A. Weinman (1978): A numerical study of scattering by large dielectric spheres, *J. Opt. Soc. Am., 68,* 130–134.

Sobolev, V.V. (1972): *Scattering of Light by Planetary Atmospheres,* Nauka Pub. House, Moscow (in Russian).

Stratton, J.A. (1941): *Electromagnetic Theory,* McGraw-Hill, New York.

Strauss, W.L. (1975): *The German Single-Leaf Woodcut 1550–1600,* Vol. 1-3, Abaris Books, Inc., New York.

Tape, W. (1979): Geometry of halo formation, *J. Opt. Soc. Am., 69*, 1122–1132.

Tape, W. (1983): Some ice crystals that made halos, *J. Opt. Soc. Am., 73,* 1641–1645.

Tricker, R.A.R. (1970): *Introduction to Meteorological Optics,* Elsevier Pub. Co., New York.

Tricker, R.A.R. (1973): A simple theory of certain helical and anthelic halo arcs. The long hexagonal ice prism as a kaleidoscope, *Roy. Met. Soc. Quart. J., 99,* 649–656.

Tricker, R.A.R. (1979): Arcs associated with halos of unusual radii, *J. Opt. Soc. Am., 69,* 1093–1100.

Van de Hulst, H.C. (1957): *Light Scattering by Small Particles,* Wiley and Sons, New York.

Waterman, P.C. (1971): Symmetry, unitarity, and geometry in electromagnetic scattering, *Phys. Rev., D.3,* 825–839.

Welch, R.M., and S.K. Cox (1978): Nonspherical extinction and absorption efficiencies, *Appl. Opt. 17,* 3159–3168.

Zelmanovich, J.L. and K.S. Shifrin (1968): *Tables of Light Scattering Coefficients of Attenuation, Scattering, and Light Pressure,* Vol. III, Hydrometeorological Publishing House, Leningrad, USSR.

Zelmanovich, I.L. and K.S. Shifrin (1971): *Tables of Light Scattering: Vol. IV. Scattering of polydisperse Systems,* Hydromet. Pub. House, Leningrad, USSR.

Zerull, R.H. (1976): Scattering measurements of dielectric and absorbing nonspherical particles, *Beitr. Physik Atm., 49,* 168–188.

Zerull, R.H., and R.H. Giese (1974): Microwave and analogue studies, in T. Gehrels (ed.), *Planets, Stars, and Nebulae Studied With Photopolarimetry.* Univ. of Arizona Press, pp. 901–915.

CHAPTER 4

THE RAYLEIGH ATMOSPHERE

4.1 Historical notes on Rayleigh scattering

Attempts to explain the color of the sky extend back at least as far as Alhazen in the 11th century, and in the early 16th century Leonardo da Vinci observed the blue color of very fine mists of water droplets when seen against a black background. In the early 18th century, the concept of Newton and his contemporaries was that skylight resulted from reflection by water droplets or small platelets suspended in the air, a concept which was still held by Brewster over a century later. A fundamental difficulty with the water droplet or platelet hypothesis was exposed by the discovery of skylight polarization by Arago in 1809. There was no way reflection from water as droplets or platelets could give the observed high degrees of polarization in right-angle scattering. The classic experiments by the British physicist John Tyndall (1820–1893) on the scattering of light by vapors and hydrosols showed the probable explanation of skylight to be scattering by small particles, which were thought by Tyndall to be small water droplets in the upper atmosphere. As an indication of the importance he ascribed to the scattering problem, he commented that the intensity and polarization of skylight constituted "the two great standing enigmas of meteorology."

The enigmatic character of the problem was dispelled in 1871 by the first publication on the subject by the British physicist J.W. Strutt, later Lord Rayleigh (1884–1919). Although later papers of Rayleigh (1884, 1899) cast his ideas in terms of proper electromagnetic theory, his first explanation was based entirely on dimensional analysis (Tricker, 1970). His basic assumptions were that the particles were spherical, isotropic, much smaller than the wavelength of light, and denser than the surrounding medium. It was apparent that the amplitude A of scattered radiation would be proportional to the amplitude A' of the incident radiation. Thus the ratio A/A' would be dimensionless, but would conceivably be a function of the following quantities: volume V of the particle, distance r from particle to point of observation, wavelength λ, speed c of light, densities D and D' of the particle and its surroundings. Since A/A' has no dimension of mass, if density enters it must be as D/D', which is dimensionless and may be omitted except for a possible constant. Similarly, the only quantity involving time is c, and since A/A' has no dimension of time, c is not involved. From other considerations it was thought that A/A' would be directly proportional to V and inversely proportional to r, which means that dimensionally, A/A' must have another length squared in the denominator. Thus $A/A' \sim (V/r\lambda^2)$, and the ratio of intensities, being the squares of the amplitudes, would be

$$\frac{I}{I'} = \left(\frac{A}{A'}\right)^2 \sim \frac{V^2}{r^2\lambda^4}$$

This led to the important law that the intensity of radiation scattered by very small particles varies as the inverse fourth power of the wavelength.

In order to explain the polarization, Rayleigh considered two types of polarization of the incident beam. The first was for the incident electric vector of amplitude $A'_\perp$ vibrating in a direction normal to the scattering plane (plane defined by incident and scattered radiation). The electric vector of the scattered radiation will also be normal to the scattering plane, and its amplitude will be

$$A_\perp = K\frac{V}{r\lambda^2}A'_\perp \tag{4.1}$$

Here K is a constant incorporating material characteristics of the medium. Because of symmetry, $A_\perp$ will be independent of scattering angle Θ. The second case is for the electric vector of the incident light vibrating in a direction parallel to the scattering plane with amplitude $A'_\parallel$. From simple geometry, the amplitude $A_\parallel$ of the scattered light will vary as $\cos\Theta$, giving

$$A_\parallel = K\frac{V}{r\lambda^2}A'_\parallel \cos\Theta \tag{4.2}$$

For unpolarized incident light, the total intensity of the scattered light is $I = A_\perp^2 + A_\parallel^2$, or

$$I = K^2 \frac{V^2}{\lambda^4 r^2} (1 + \cos^2\Theta) I' \quad (4.3)$$

This is the well-known equation for Rayleigh scattering. Its consequences will be discussed in detail in following sections.

Through this relatively simple reasoning, Rayleigh was able to explain the blue color and polarization of skylight and of light due to scattering by small particles in the experiments of Tyndall and others. While later work of Rayleigh and many other scientists have refined the ideas and introduced correction for anisotropy of particles, multiple scattering among particles, and other effects, the concepts Rayleigh applied to the problem are still valid today.

Rayleigh scattering formalism on the basis of modern electromagnetic theory is outlined in section 4.3. In preparation for its use in a realistic model of the Rayleigh atmosphere, however, one should consider the Rayleigh model applied to the atmosphere of the earth.

4.2 Density profiles of the Rayleigh atmosphere

The permanent gases of the earth's atmosphere consist of approximately 21% oxygen and 78% nitrogen, with a large number of trace gases to make up the other 1%. Of the variable gases, the most abundant is water vapor, which makes up to 4% of the total under very moist conditions. Also included in the gases are ozone, carbon dioxide, and a number of others, but their total mass is negligible compared to that of the permanent gases. All of these gaseous constituents make up the Rayleigh atmosphere, but the ever-present aerosols, clouds, and other nongaseous particles are not included in the Rayleigh model. Furthermore, as normally used, the Rayleigh model takes account of only the permanent gases, although water vapor is not entirely a negligible component.

Using the equation of state for gases and the hydrostatic equation, it is a simple matter to compute the density of the Rayleigh atmosphere as a function of height if the temperature profile is known. The equation of state for a perfect gas is written as

$$P = \rho \frac{R^*}{M} T \tag{4.4}$$

where the symbols are P: pressure; ρ: mass density; R^*: gas constant; M: molecular weight; T: absolute temperature. For a mixture of gases such as that of the atmosphere, M is an effective molecular weight for n gases given by

$$M = M_1 \frac{\rho_1}{\rho} + M_2 \frac{\rho_2}{\rho} + \ldots + M_n \frac{\rho_n}{\rho} \tag{4.5}$$

For dry air the effective M is 28.9 in the C.G.S. system of units, and $R = 8.315 \times 10^7$ ergs (°C)$^{-1}$. It is convenient for some purposes to define a specific gas constant for dry air as $R_d = R^*/M$, in which case the equation of state of dry air is written as

$$P_d = \rho_d R_d T \tag{4.6}$$

It is possible to take into account the effect of water vapor on air density by the relation

$$P_m = R_d \rho_d T \left(1 + \frac{R_v}{R_d} r_v\right) \tag{4.7}$$

where R_v is the specific gas constant for water vapor and r_v is the mixing ratio (mass of water vapor per unit mass of dry air). In most cases, however, the variability of water vapor content and lack of adequate measurements make it impractical to account for water vapor effects in Rayleigh models of the atmosphere.

The hydrostatic equation representing an atmosphere in dynamic equilibrium takes the form

$$dP = -\rho g dz \tag{4.8}$$

where g is the acceleration of gravity and z is height. We have suppressed the subscript d for convenience. On combining Eqs. (4.4) and (4.8), we obtain the height dependence of pressure as

$$\frac{dP}{P} = d(ln\ P) = \frac{gM}{R^*T}\,dz \tag{4.9}$$

or

$$d(ln\ P) = \frac{gM}{R^*(T_o + \gamma z)}\,dz \tag{4.10}$$

where T_o is temperature at the surface or some other reference level and γ is the lapse rate of temperature ($\gamma = -dT/dz$). Any change of gravity or atmospheric composition with height is neglected in Eq. (4.10). If we assume a constant lapse rate of temperature, not an unrealistic assumption for some cases, then we can integrate Eq. (4.10) from a zero reference level to any height z to give

$$\frac{P}{P_o} = \left(\frac{T}{T_o}\right)^{-(gM/R^*\gamma)} \tag{4.11}$$

The equivalent expression for density is

$$\frac{\rho}{\rho_o} = \frac{P\,T_o}{P_o\,T} = \left(\frac{T}{T_o}\right)^{-[(gM/R^*\gamma)+1]} \tag{4.12}$$

For the singular case of an isothermal atmosphere ($\gamma = 0$), Eq. (4.10) reduces to

$$\frac{P}{P_o} = \frac{\rho}{\rho_o} = \exp\left(-\frac{gM}{R^*T_o}\right)\,(z - z_o) \tag{4.13}$$

It may be mentioned that the isothermal case is unrealistic for the atmosphere as a whole, although approximately isothermal conditions are not uncommon for relatively thick layers of the atmosphere.

The number density of molecules is defined by the relation

$$N = \rho\left(\frac{n_a}{M}\right) = \rho\left(\frac{6.0225 \times 10^{23}}{28.964}\right) = 2.0793 \times 10^{22}\rho \tag{4.14}$$

where n_a is Avagadro's number. Since M is assumed constant (M=28.964 kg/kilo-mole), we have the useful relation giving the ratio of N at any level to its sea level value N_o as

$$\frac{N}{N_o} = \frac{\rho}{\rho_o} = \left(\frac{T}{T_o}\right)^{-[(gM/R^*\gamma)+1]} \tag{4.15}$$

A number of models of the earth's atmosphere have been developed by various authors and agencies. A typical one is the U.S. Standard Atmosphere, 1976 (N.O.A.A., 1976), in which the number density of gaseous molecules is tabulated as a function of altitude. A curve of this quantity is shown in Fig. 4.1. Other atmospheric models, termed supplemental models by Valley (1965), applicable at different latitudes (Arctic, sub-Arctic, mid-latitude, subtropical, tropical), are

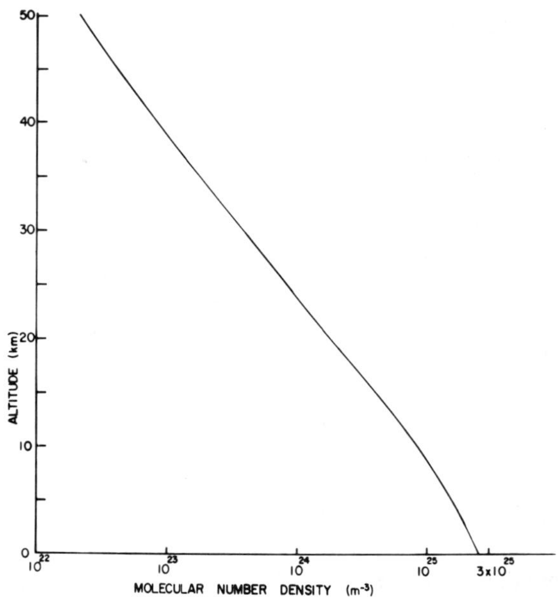

Figure 4.1 Molecular number density as a function of altitude for the U.S. Standard Atmosphere, 1976.

available, each of them except the tropical being divided into summer and winter sections. Those models give height profiles of mass density, but not molecular number density as does the standard 1976 model. However, molecular number density is easily computed from mass density by use of Eq. (4.14). Values of N at 5-km intervals from the surface to 50 km are given for a number of the auxiliary models, as well as for the standard 1976 model, in Table 4.1. (Less than 0.001 of the mass of the atmosphere is above 50 km.) The latitudes for which the various models are designed are Arctic: 75°, sub-Arctic: 60°, mid-latitude: 45°, subtropical: 30°, and tropical: 15°. The tabulation shows that the number densities for the auxiliary models do not differ from those of the standard by more than ±10% throughout most of the troposphere, although the low temperatures of the Arctic models cause high densities near the surface. Also, the high and cold tropopause for tropical models causes a deviation of about 16% from the standard at 15 km, a fact which has considerable practical importance in the choice of a calibration level for lidar observations. The largest deviations from the standard model are produced by the extreme temperatures that occur in sub-Arctic models in the 40- to 50-km regions. For instance, the deviation from the standard for the sub-Arctic winter model at 50 km is about −25%, whereas for the sub-Arctic summer model the deviation is +21%. The reversal is caused by a 50-km temperature change from 259.3 K in winter to 277.1 K in summer. The corresponding temperature for the standard model is 270.65 K.

4.3 Scattering coefficients for the Rayleigh atmosphere

4.3.1 Radiation by a molecular dipole

Consider an individual molecule consisting of a positively charged nucleus at the center surrounded by orbital electrons. The molecular mass resides almost entirely in the nucleus, which we consider to constitute a single particle. The balancing negative charge is carried by electrons of small but finite mass. Such a model is appropriate for the oxygen, nitrogen, and other gaseous molecules of the atmosphere, and, as will be seen, this type of molecule reacts to passing electromagnetic waves as a dipole antenna for extracting energy from the incident wave and re-radiating it in all directions. We speak of this process as the scattering of radiation, with Rayleigh scattering being scattering by particles of dimensions which satisfy certain criteria listed below.

TABLE 4.1 Molecular number density (m^{-3}) at 5-km altitude intervals for the U.S. Standard, 1976 model atmosphere (N.O.A.A., 1976) and for selected auxiliary models (Valley, 1965) for various latitude regions of the earth. Numbers in parentheses are the exponents of 10 by which the entries are to be multiplied, and the percentages indicate the deviations of the auxiliary models from the U.S. Standard.

Altitude (km)	U.S. Standard, 1976	Arctic winter	Arctic summer	Sub-Arctic winter	Sub-Arctic summer
0	2.5470(25) 0%	2.9458(25) 15.7%	2.6294(25) 10.9%	2.8534(25) 12.0%	2.5364(25) -0.4%
5	1.5312(25) 0%	1.5633(25) 2.1%	1.5215(25) -0.6%	1.5505(25) 1.3%	1.5063(25) -1.6%
10	8.5976(24) 0%	7.9054(24) -8.1%	8.3924(24) -2.4%	8.0656(24) -6.2%	8.6119(24) 0.2%
15	4.0493(24) 0%	3.6069(24) -10.9%	3.9244(24) -3.1%	3.6804(24) -9.1%	4.0402(24) -0.2%
20	1.8486(24) 0%	1.6212(24) -12.3%	1.8727(24) 1.3%	1.6957(24) -8.3%	1.8980(24) 2.7%
25	8.3341(23) 0%	7.1500(23) -9.4%	8.8871(23) 6.6%	7.7385(23) -7.1%	8.8302(23) 6.0%
30	2.8278(23) 0%	3.1574(23) -17.5%	4.1063(23) 9.9%	3.4196(23) -10.7%	4.1142(23) 7.5%
35	1.7597(23) 0%	— —	— —	1.5321(23) -12.9%	1.9379(23) 10.1%
40	8.3077(22) 0%	— —	— —	6.9239(22) -16.7%	9.4104(22) 13.3%
45	4.0882(22) 0%	— —	— —	3.2633(22) -20.2%	4.8117(22) 17.7%
50	2.1351(22) 0%	— —	— —	1.5974(22) -25.2	2.5792(22) 20.8

Altitude (km)	Mid-latitude winter	Mid-latitude summer	Subtropical summer	Tropical
0	2.7051(25) 6.2%	2.4784(25) -2.7%	2.3951(25) -6.0%	2.4256(25) -4.8%
5	1.5410(25) 0.6%	1.4994(25) -2.1%	1.4885(25) -2.8%	1.4968(25) -2.2%
10	8.4661(24) -1.5%	8.6484(24) 0.6%	8.7300(24) 1.5%	8.7386(24) 1.6%
15	3.9305(24) -2.9%	4.3753(24) 8.1%	4.7313(24) 16.8%	4.6985(24) 16.0%
20	1.8069(24) -2.3%	1.9655(24) 6.3%	1.9999(24) 8.2%	1.9785(24) 7.0%
25	8.2136(23) -1.4%	8.9170(23) 7.0%	8.7398(23) 4.9%	8.4107(23) 0.9%
30	3.7080(23) -3.1%	4.0989(23) 7.1%	3.9694(23) 3.7%	3.8063(23) -0.6%
35	1.6477(23) -6.4%	1.9262(23) 9.5%	1.8587(23) 5.6%	1.7822(23) 1.3%
40	7.5364(22) -9.3%	9.3663(22) 12.7%	8.9962(22) 8.3%	8.6937(22) 4.6%
45	3.6201(22) -11.4%	4.7161(22) 25.4%	4.5051(22) 10.2%	4.3597(22) 6.6%
50	1.8617(22) -12.8%	2.4997(22) 17.1%	2.3711(22) 11.1%	2.2887(22) 7.2%

Preliminary to a discussion of Rayleigh scattering in the atmosphere, we should realize that only a tiny amount of energy is scattered by a single molecule. However, as was seen in Table 4.1, there are huge numbers of molecules in a unit volume of the atmosphere (of the order of 10^{25} m^{-3} at sea level), so very significant and measurable amounts of energy may be scattered by a finite volume of atmospheric air. In such an assemblage of particles, the distances between the particles, though small by normal concepts, is large enough, taken on a statistical basis, so that the particles scatter independently. This is to say that the oscillations of the electrons of a given particle in response to the passing wave is not interfered with by its neighboring particles to a statistically significant extent.

The simple law of Rayleigh scattering is postulated on the following five assumptions:

1. The particle is much smaller in physical dimensions than the wavelength of light ($x = (2\pi r/\lambda) \ll 1$). This criterion is well satisfied for visible light by gaseous molecules of the air, for which the diameter is from one to a few Angstrom units (10^{-10} $\mu m = 10^{-4}$ μm). Generally particles of radius $r \leq 0.03\lambda$ are considered to satisfy the small-particle criterion, which for visible light includes agglomerations of many molecules and many of the Aitken nuclei. It does not include the bulk of dust, haze, and other atmospheric aerosols, for which other types of scattering laws apply (see Chapter 3).

2. The particles are nonionized. Most particles in the troposphere and stratosphere meet this criterion. Relatively large numbers of ionized particles constitute the ionosphere.

3. The index of refraction of the particles is not much different from unity. The physical significance of this requirement is that the phase of the wave inside the particle is approximately the same as that in the surrounding space. The condition is well met by ordinary air, m for air of $T = 15$ °C and $P = 1013$ mb being 1.000278 for $\lambda = 0.55$ μm. Its satisfaction is not so evident, however, for small water droplets ($m = 1.33$), sulfuric acid droplets ($m \approx 1.5$), or mineral particles such as quartz ($m \approx 1.55$). For the equal phase approximation to hold in this case, we must have the

additional criterion that $|m|x \ll 1$. This auxiliary to the small particle requirement listed above shows that particles with m appreciably greater than unity must be smaller than those of $m \approx 1$ to exhibit Rayleigh-type scattering.

4. The simplest case for Rayleigh scattering is for isotropic particles. An isotropic particle exhibits no preferred direction of oscillation of the electrons in response to the impressed wave. The diatomic nature of the nitrogen and oxygen molecules which make up 99% of the permanent gases of the earth's atmosphere, however, causes a violation of this assumption, and partially explains the discrepancy between the polarization properties observed in the actual atmosphere and those predicted on the basis of pure Rayleigh scattering. A correction for molecular anisotropy was derived by Cabannes (1929), and the law applied for scattering by the real atmospheric gases is usually termed Rayleigh-Cabannes scattering. Molecular anisotropy will be covered in section 4.3.4 and its effects in a Rayleigh atmosphere in section 4.5.7.

5. The frequency of the light is not near a resonant frequency for the particle. The mechanical system of an electron bound to a nucleus by an elastic force has a natural resonant frequency. For cases in which the frequency of the incident wave is near the resonant frequency of the molecule, the scattering amplitude may become very large. This is not a problem for near-visible or longer wavelengths in the atmosphere, as resonant frequencies of the major atmospheric gases correspond to wavelengths of 0.2 μm or less. For some of the trace gases in the atmosphere, however, resonant frequencies occur at longer wavelengths. For instance, radiation in the sodium D resonance lines at 0.5890 and 0.5896 μm, which is observed in right angle scattering, has less than 50% polarization, in contrast to the approximately 100% polarization it should have for pure Rayleigh scattering. Scattering by such trace gases is of negligible magnitude in comparison with that of the abundant gases of nitrogen and oxygen, although resonant phenomena may be of interest in themselves. The abundance and height distribution of sodium vapor in the atmosphere can be determined from its emission by resonant scattering during twilight (Chamberlain, 1978), and resonant scattering by hydrogen in the upper atmosphere is responsible for the planetary coronas that have been observed on Mars and Venus.

If a dipole such as a nitrogen or oxygen molecule is subjected to an incident electromagnetic wave of amplitude $\mathbf{A}'$ and angular frequency ω, an oscillation of the negatively charged electrons with respect to the positively charged and more massive nucleus is set up by the oscillating electric field of the wave. Oscillation of the charges induces a secondary or scattered wave that emanates from the particle and proceeds outward in all directions. The dipole moment $\mathbf{p}$ of the system is, in general, a vector of magnitude given by the product of the charge and its displacement distance, with a direction related to the polarizability α of the particle. The polarizability is a property of the particle itself, and in the most general case it has three components α_1, α_2 and α_3 oriented along three mutually orthogonal axes. Following Van de Hulst (1957), we designate these axes by the unit vectors $\mathbf{n}_1$, $\mathbf{n}_2$, and $\mathbf{n}_3$. If the vector amplitude $\mathbf{A}'$ is parallel to axis $\mathbf{n}_i$, then the dipole moment along that axis is

$$\mathbf{p}_i = \alpha_i \mathbf{A}' \tag{4.16}$$

and zero along the other axes. In the general case, however,

$$\mathbf{A}' = A_1'\mathbf{n}_1 + A_2'\mathbf{n}_2 + A_3'\mathbf{n}_3 \tag{4.17}$$

and the induced dipole moment is

$$\mathbf{p} = \alpha_1 A_1'\mathbf{n}_1 + \alpha_2 A_2'\mathbf{n}_2 + \alpha_3 A_3'\mathbf{n}_3 \tag{4.18}$$

Since A' is an oscillating wave of the form $\mathbf{A}'e^{iwt}$, the dipole moment is of the same form, $\mathbf{p}e^{iwt}$, and there is no change of frequency in the scattering process. The polarizability, a property of the material of the particle, is related to the complex index of refraction, and is itself a complex quantity of the form $\alpha_i = \alpha_{1i} - i\alpha_{2i}$, α_1, and α_2 being the real and complex parts respectively.

4.3.2 Polarizability of particles

The formalism resulting from the tensor form of the polarizability has been discussed by Van de Hulst (1957, section 6.5), and does not need to be repeated here. The results will be written, however, and their application to the problems of Rayleigh scattering in the atmosphere will be outlined in some detail.

We assume an assemblage of randomly oriented particles, each of which has a complex polarizability α with components in the $\mathbf{n}_1$, $\mathbf{n}_2$, and $\mathbf{n}_3$ directions. The averages of the polarizabilities for the assemblage are given in terms of two real quantities (Van de Hulst, 1957, p. 79) A and B written as

$$A = \frac{1}{15}\left(\alpha_1\alpha_1^* + \alpha_2\alpha_2^* + \alpha_3\alpha_3^*\right)$$
$$B = \frac{1}{30}\left(\alpha_1\alpha_2^* + \alpha_2\alpha_3^* + \alpha_3\alpha_1^* + \alpha_2\alpha_1^* + \alpha_3\alpha_2^* + \alpha_1\alpha_3^*\right) \tag{4.19}$$

(We have retained the notation of Van de Hulst here, in spite of the use of A also as an amplitude. No confusion should result.)

As shown in the next section, the polarizability affects the angular distribution of scattered light (i.e. its phase function or matrix), as well as its polarization and total intensity.

4.3.3 Angular distribution of scattered radiation

As previously, we designate a scattering plane by the directions of propagation of incident and scattered radiation and consider the properties of radiation scattered at angle Θ from the incident direction. Symmetry around the incident direction is assumed, so the distribution in the entire space surrounding the particle, the phase function or matrix, can be expressed in terms of the one angle Θ. The incident radiation is characterized by the components ($I_\parallel'$, $I_\perp'$, U', V') and the scattered radiation by ($I_\parallel$, $I_\perp$, U, V). Any state of polarization may be represented by these parameters. The intensity of radiation at distance r and propagated at angle Θ from a single particle or a typical particle of a small assemblage is given by the product of the scattering and incident matrices written as

$$\begin{vmatrix} I_\parallel \\ I_\perp \\ U \\ V \end{vmatrix} = \frac{k^4}{r^2} \begin{vmatrix} 2A+3B\cos^2\Theta+A-B & A-B & 0 & 0 \\ A-B & 3A+2B & 0 & 0 \\ 0 & 0 & (2A+3B)\cos\Theta & 0 \\ 0 & 0 & 0 & 5B\cos\Theta \end{vmatrix} \begin{vmatrix} I_\parallel' \\ I_\perp' \\ U' \\ V' \end{vmatrix} \tag{4.20}$$

This is a completely general relationship which gives the intensity and state of polarization of the scattered radiation for an arbitrary intensity and polarization of the incident radiation. When the multiplication in Eq. (4.20) is carried out, we have

$$\begin{vmatrix} I_{\parallel} \\ I_{\perp} \\ U \\ V \end{vmatrix} = \frac{k^4}{r^2} \begin{vmatrix} [(2A+3B)\ \cos^2\Theta + A - B]I'_{\parallel} + (A-B)I'_{\perp} \\ (A-B)I'_{\parallel} + (3A+2B)I'_{\perp} \\ [(2A+3B)\ \cos\Theta]U' \\ (5B\cos\Theta)V' \end{vmatrix} \tag{4.21}$$

Some special cases for this general relationship are of interest:

a. Case of right angle scattering ($\cos\Theta = 0$)
   The Stokes parameters are

$$\frac{k^4}{r^2}\left\{[(A-B)\,I'_{\parallel} + I'_{\perp}],\ [(A-B)\,I'_{\parallel} + (3A+2B)\,I'_{\perp}],\ 0,\ 0\right\}$$

Thus there is no possibility of the light scattered at 90° having any elliptical polarization (V=0) regardless of the ellipticity of the incident radiation, and the plane of polarization is either in or normal to the scattering plane (U=0) depending on the relative magnitude of the first two components.

b. Case of unpolarized incident light ($I'_{\parallel} = I'_{\perp} = \frac{1}{2}I'$ and $U' = V' = 0$).

   The parameter set for the scattered light is

$$\frac{k^4}{2r^2}\,I'\left\{[(2A+3B)\cos^2\Theta + 2(A-B)],\ (4A+B),\ 0,\ 0\right\}$$

Here also there is no possibility of elliptical polarization in the light scattered at any angle, and the scattered light is partially plane polarized with the plane of polarization in or normal to the scattering plane.

c. Case of incident light plane polarized normal to the scattering plane ($I' = I'_\perp$ and $I'_\parallel = U' = V' = 0$). The parameters of the scattered light are

$$\frac{k^4}{r^2}\, I' \{ (A - B),\ (3A + 2B),\ 0,\ 0 \}$$

giving partial plane polarization which is independent of scattering angle, and no elliptic polarization. (It should be noted that the scattering of plane polarized light by Mie-type particles produces some elliptic polarization, in contrast to this Rayleigh case.

d. Case of isotropic polarizability

In the case of isotropic particles, the polarizability is likewise isotropic with the value $\alpha = \frac{1}{3}(\alpha_1 + \alpha_2 + \alpha_3)$. In this case, the dipole moment of Eq. (4.15) for any direction is directly proportional to the impressed amplitude $\mathbf{A}'$ in that direction, the constant of proportionality being α. Thus we have

$$p_i = \alpha \mathbf{A}'_i \tag{4.22}$$

An intuitive concept of angular scattering can best be obtained from the scalar phase function $P(\Theta)$ as opposed to the complete phase matrix of Eq.(4.20). For this case of Rayleigh scattering, the distribution of (scalar) intensity is given by

$$P(\Theta) = \tfrac{3}{4}(1 + \cos^2\Theta) \tag{4.23}$$

The factor ¾ is necessary to normalize to unity the phase function integrated over the entire solid angle. Since there is no absorption, the albedo of single scattering is unity (see section 2.8). For the case of unpolarized incident light, the angular distribution of the total intensity surrounding a molecule scattering according to this relation is shown by the solid curve of Fig. 4.2. The direction of propagation of the incident radiation is shown by the arrows, and scattering angle Θ is labeled at 30° increments from the forward scattering ($\Theta = 0°$) to backward scattering ($\Theta = 180°$) directions. The distribution has rotational symmetry around both the $\Theta = 0°$–$180°$ direction and the $\Theta = 90°$ direction, and the magnitude at $\Theta = 90°$ is exactly half that in the forward and backward directions.

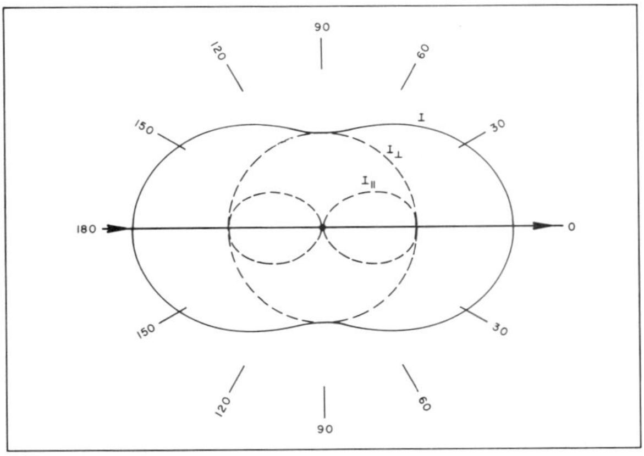

Figure 4.2 Scattering pattern of a Rayleigh particle for unpolarized incident radiation. $I_\parallel$ and $I_\perp$ are the components parallel and normal, respectively, to the scattering plane. The pattern has axial symmetry around the direction of propagation of the incident radiation.

The vibrations of the electric vector resolved into orthogonal components parallel ($I_\parallel$) and normal ($I_\perp$) to the scattering plane are shown by the dashed curves of Fig. 4.2. Component $I_\perp$ is independent of Θ and $I_\parallel$ is proportional to $\cos^2\Theta$. This gives $I_\parallel = I_\perp$ in forward and backward directions, and $I_\parallel = 0$ at $\Theta = 90°$. The degree of polarization of the scattered light is

$$P = \frac{I_\perp - I_\parallel}{I_\perp + I_\parallel} \tag{4.24}$$

and its variation with Θ is

$$P(\Theta) = \frac{1 - \cos^2\Theta}{1 + \cos^2\Theta} = \frac{\sin^2\Theta}{1 + \cos^2\Theta} \tag{4.25}$$

The angular distribution of $P(\Theta)$ for primary Rayleigh scattering is shown in Fig. 4.3. As with the intensity, the polarization field also shows rotational symmetry around $\Theta = 0°$ to $180°$ and $\Theta = 90°$ directions. Some points of interest are that $P = 0$ in forward and backward directions, radiation scattered at $\Theta = 90°$

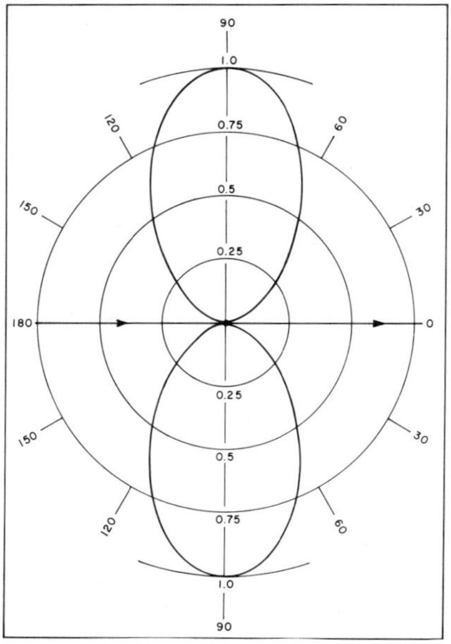

Figure 4.3 Angular distribution of the degree of polarization of radiation scattered by a Rayleigh particle for the case of unpolarized incident radiation. The pattern has axial symmetry around the direction of propagation of the incident radiation.

is completely polarized, and in all other directions there is partial polarization of the scattered radiation. In this case, of course, P is only linear polarization, as Rayleigh scattering does not produce elliptic or circular types of polarization.

4.3.4 Molecular anisotropy and depolarization factor

It has long been observed that, contrary to the predictions of the simple scattering theory, right-angle scattering in the atmosphere does not produce complete polarization of the scattered light. There are several reasons for this, perhaps the most important being scattering by the natural aerosols and the effects of multiple scattering. However, the fact that the polarizability of the gaseous molecules of the atmosphere is not isotropic has a definite and observable depolarizing effect for molecular scattering. It is the accepted practice to denote the ratio $I_{\|}/I_{\perp}$ for right-angle scattering as the depolarization factor Δ. From

Eq.(4.20), it can be seen that Δ varies with the state of polarization of the incident light. For instance, for the case of unpolarized incident light, such as direct sunlight incident on the Rayleigh atmosphere, Eq.(4.21) gives for $\Theta=90°$

$$\Delta = \frac{I_{\parallel}}{I_{\perp}} = \frac{2A - 2B}{4A + B} \tag{4.26}$$

and for plane-polarized incident light with the electric vector normal to the scattering plane ($I'_{\parallel}=0$),

$$\Delta = \frac{A - B}{3A + 2B} \tag{4.27}$$

On combining Eq.(4.26) with the Rubenson definition of the degree of polarization

$$P = \frac{I_{\perp} - I_{\parallel}}{I_{\perp} + I_{\parallel}}$$

we see that for right-angle Rayleigh scattering

$$P = \frac{1 - \Delta}{1 + \Delta} \tag{4.28}$$

If the incident light is unpolarized, right-angle scattering gives

$$P = \frac{2A + 3B}{6A - B}$$

and for plane-polarized incident light with $I'_{\parallel}=0$

$$P = \frac{2A + 3B}{4A + B}$$

The value of Δ depends on the physical characteristics of the scattering molecules, and thus is different for different gases. Several recent authors have analyzed laboratory data for determining Δ, but have obtained results which are significantly divergent. On the basis of the values of Δ developed, they have obtained similarly divergent values of the volume scattering coefficient and normal optical thickness of the Rayleigh atmosphere (see sections 4.3.6 and 4.3.7). Values of Δ for atmospheric gases at the indicated wavelengths as given by various authors are listed in Table 4.2.

Much of the difficulty in determining Δ is in the method of handling the effects of Raman scattering. If gas molecules are subjected to a beam of very nearly monochromatic light at wavelength λ_o, the scattered light has a spectrum consisting of mainly light at λ_o, but with some faint neighboring lines due to small variations around λ_o arising from vibrations and rotations of the anisotropic molecules. Although these neighboring lines are faint, they are highly depolarized, and thus may have a significant effect on the measured depolarization factor. Most laboratory measurements do not have sufficient spectral resolution to discriminate between the central wavelength and those of Raman scattering, in which case relatively high values of Δ are obtained. If, however, very high spectral resolution, as with lasers, is used or values are determined theoretically without taking into account the Raman effect, lower values of Δ result. The value of Δ selected for computations affects not only the degree of polarization of the scattered light, but it may change the volume scattering coefficient and normal optical thickness of the Rayleigh atmosphere by as much as several percent.

Returning to Eq.(4.28) we see that the different values of Δ in Table 4.2 yield very significant differences in the degree of polarization for right angle scattering. For instance, the two latest values, those of Frohlich and Shaw of

Table 4.2 Values of the depolarization factor $\Delta(\times 10^2)$ for atmospheric gases, as given by various authors.

Gas	Rowell et al. (1971)	Alms et al. (1975)	Hoyt (1977)	Frohlich & Shaw (1980)	Young (1980)
Nitrogen	2.14	2.10±0.04	—	1.17±0.06	2.10
Oxygen	5.66	—	—	1.4±0.1	5.8
CO_2	7.3	7.97±0.10	—	—	7.8
Dry air	—	—	1.39±0.06	—	2.79
Moist air	—	—	—	0.95±0.4	—

0.0095 and Young of 0.0139 give $P(90°) = 0.973$ and 0.981, respectively. The value of 0.0303, derived from some laboratory measurements for dry air by Gucker et al. (1969), and used for the computations of Bahethi and Fraser (1980), gives $P(90°) = 0.941$. Gucker's measurements included the Raman scattering effect.

A comparison of laboratory values of Δ with those for the free atmosphere is made difficult by the lack of agreement in the laboratory data, non-Rayleigh effects produced by the ever-present aerosols in the atmosphere, multiple scattering, and reflection from the underlying surface. The effects of aerosols and of multiple scattering can be minimized by making measurements at high altitudes, as was done by Tousey and Hulbert (1947) and by Packer and Lock (1951). Unfortunately, the measurements of Δ in both cases were for white light, and were thus subject to considerable multiple-scattering effects, even at high altitudes. The lowest value of Δ observed by Tousey and Hulbert from an altitude of 3.28 km (10,000 ft) was 0.10, which by Eq.(4.24) gives the highest $P = 0.818$. Although the measurements of Packer and Lock ranged in altitude up to 12.47 km (38,000 ft) their lowest value of Δ was, surprisingly, only 0.14, yielding a maximum $P = 0.610$. This, of course, is lower than polarization values frequently observed at lower altitudes, and must have been strongly influenced by the haze layers they observed to be present at high levels in the atmosphere. Measurements of P by Coulson (1980) from an altitude of 3400 m in Hawaii gave maximum values of P near sunrise of about 0.855 for the near infrared spectral region, which corresponds to $\Delta = 0.078$. While this is considerably above the values of Table 4.2 for the effects of molecular anisotropy for dry air, it indicates the possibility of a relatively small combined effect of multiple scattering, atmospheric aerosols, and surface reflection under certain circumstances.

Molecular anisotropy also has an effect on the intensity of scattering. For real polarizabilities along the three axes, we can define a mean polarizability

$$\bar{\alpha} = \frac{1}{3}(\alpha_1 + \alpha_2 + \alpha_3) \tag{4.29}$$

and an anisotropy factor

$$\gamma = \alpha_1^2 + \alpha_2^2 + \alpha_3^2 - \alpha_1\alpha_2 - \alpha_2\alpha_3 - \alpha_3\alpha_1 \tag{4.30}$$

Kerker (1957) has shown that the intensity of light scattered at 90° is enhanced by the factor

$$\frac{13\,\gamma^2}{45(\bar{\alpha})^2}$$

for unpolarized incident light, and by

$$\frac{7\gamma^2}{45(\bar{\alpha})^2}$$

for plane-polarized incident light. These values are due to the original work of Cabannes (1929). The effects of molecular anisotropy in a Rayleigh model of the atmosphere are the subject of section 4.5.7.

4.3.5 Scattering cross sections

The large numbers of molecules which are contained in even an elemental volume of atmospheric air make it desirable to deal with the macroscopic refractive index of the medium, instead of with single particles. The polarizability is related to the bulk refractive index m by the Lorentz-Lorenz law written as (Van de Hulst, 1957)

$$\alpha = \frac{3}{4\pi N}\left(\frac{m^2 - 1}{m^2 + 2}\right) \tag{4.31}$$

This law is based on the assumption of $r \ll \lambda$ and that there are many particles in a spherical volume with radius λ. Otherwise the restrictions are not stringent: m may be considerably different from unity, and particle size and shape are immaterial. The intensity and polarization of the scattered light are, of course, functions of the number density of the particles.

There are two types of scattering cross sections of relevance in discussions of light scattering by particles—total and angular. The total scattering cross section σ is the area taken normal to the direction of propagation of the incident wave encompassing the power which is equivalent to the total power scattered in all directions by the particle. The angular scattering cross section $\sigma(\Theta)$ is the analogous area encompassing the power abstracted from the original beam and scattered at angle Θ per unit solid angle. The two types of cross section are related by the relation

$$\sigma = \int_0^{4\pi} \sigma(\Theta)\, dw \tag{4.32}$$

where w is solid angle and the integration is taken over the entire 4π steradians surrounding the particle.

For a small isotropic particle of radius r, it can be shown (Stratton, 1941) that σ is given by the relation

$$\sigma = \frac{128\pi^5 r^6}{3\lambda^4} \left(\frac{m^2 - 1}{m^2 + 2} \right)^2 \tag{4.33}$$

where m is the real refractive index of the particle relative to that of its surroundings. The equivalent expression for angular scattering cross section is

$$\sigma(\Theta) = \frac{16\pi^4 r^6}{\lambda^4} \left(\frac{m^2 - 1}{m^2 + 2} \right) \sin^2\Theta \tag{4.34}$$

Two aspects of these relations are of particular interest for Rayleigh scattering. First, the scattering cross section is inversely proportional to the fourth power of the wavelength, a relation which, as mentioned previously, was first deduced by Rayleigh and which is invoked to explain the blue color of the sunlit sky. Second, the scattering cross section is critically dependent on the size of the particle, being proportional to the sixth power of the radius. This dependence will be discussed further in connection with the general Mie type scattering, but it must be realized that the Aitken nuclei and agglomerations of molecules in the atmosphere which meet the Rayleigh criterion of $r/\lambda \leq 0.03$ are much more effective in the Rayleigh regime than their relatively smaller numbers would indicate. As an example, let us take an Aitken nucleus with $r = 0.01$ μm and visible light of $\lambda = 0.5$ μm, giving the ratio $r/\lambda = 0.02$ to meet the Rayleigh criterion. If we now compare this to a molecule for which $r = 0.001$ μm, the ratio $(0.01/0.001)^6 = 10^6$, which shows that one such Aitken particle scatters as much visible light as a million molecules. Of course, the number density of molecules in the atmosphere at sea level is of the order 10^{25} m^{-3}, which means that for Aitken nuclei to scatter as much as do the gaseous molecules, the number density of Aitken nuclei would have to be the unrealistically high value of 10^{19} m^{-3}. In clear atmospheric conditions, a typical value of Aitken nuclei is 5×10^8 m^{-3}, and in polluted conditions, perhaps 10^{10} m^{-3}. Thus Rayleigh scattering is strongly dominated by the atmospheric gases. The effect of scattering by particles much larger than Aitken nuclei, which is very important in the real atmosphere, is the subject of another chapter.

The scattering efficiency Q defined by the ratio of the scattering to geometrical cross sections, is obtained by dividing Eq.(4.32) by r to give

$$Q = \frac{128\pi^4 r^4}{3\lambda^4}\left(\frac{m^2 - 1}{m^2 + 2}\right)^2 \tag{4.35}$$

On substituting some typical values of the various parameters into this relation, it can be seen that very small particles are very inefficient scatterers. For example, if $r = 0.01$ μm, $\lambda = 0.50$ μm, and $m = 1.40$, then $Q \doteq 4\times10^{-5}$, which is the fraction of the energy passing through an area of πr^2 which is scattered by the particle.

4.3.6 Volume scattering coefficient for air

Although the amount of energy scattered by a single molecule is extremely small, we can take advantage of the large number density of molecules in the atmosphere to define a scattering coefficient of moderate magnitude for a unit volume of air. As shown by Fig. 4.1 above, N is on the order of 10^{25} m^{-3} at sea level, and 10^{22} m^{-3} at 50-km altitude. In defining the volume scattering coefficient β, we assume the particles to be isotropic, to be randomly spaced and in thermal motion so that the scattering from a given molecule is independent of the existence of its neighbors. We further assume no constant phase relationships among the streams of radiation scattered by the individual molecules. In other words, the scattering is independent and incoherent, conditions which are well satisfied in the atmosphere. In this case the scattering from a volume is simply the sum of scattering by all particles in the volume. Thus, in accord with Eqs. (4.33) and (4.34), β for total scattering is

$$\beta = N\sigma = \frac{24\pi^3}{N\lambda^4}\left(\frac{m^2 - 1}{m^2 + 2}\right)^2 \tag{4.36}$$

and for angular scattering

$$\beta(\Theta) = N\sigma(\Theta) = \frac{9\pi^2}{N\lambda^4}\left(\frac{m^2 - 1}{m^2 + 2}\right)^2 \sin^2\Theta \tag{4.37}$$

Since the dimensions of σ are $[L^2]$ and of N are $[L^{-3}]$, β has the dimension $[L^{-1}]$, and similarly for $\beta(\Theta)$. In the latter case, however, the pure number sr^{-1} must be understood to apply.

These expressions show the volume scattering coefficient to be proportional to λ^{-4} if the other parameters are independent of λ. The strict proportionality is broken down, however, by m being slightly dependent on λ, as the data in Table 4.3 show. The net result is that β is proportional to approximately $\lambda^{-4.08}$ through the visible range, but becomes approximately $\sim\lambda^{-4}$ in the near infrared.

By knowing the atmospheric density as a function of altitude (see section 4.2), we can utilize Eq. (4.36) to compute the altitude profile of the volume scattering coefficient for the Rayleigh atmosphere at different wavelengths. The

TABLE 4.3 The index of refraction of dry air at $T = 15°$ C and $P = 1013$ mb as a function of wavelength, as computed by the formula of Edlen (1953)

λ (μm)	m	λ (μm)	m
0.30	1.00029156	0.60	1.00027697
0.32	1.00028902	0.70	1.00027578
0.34	1.00028699	0.80	1.00027503
0.36	1.00028531	0.90	1.00027451
0.38	1.00028392	1.00	1.00027415
0.40	1.00028275	1.50	1.00027363
0.45	1.00028052	2.00	1.00027298
0.50	1.00027896	3.00	1.00027277
0.55	1.00027782	4.00	1.00027269

calculations have been carried out by many authors, with somewhat divergent results, as mentioned in connection with the depolarization factor in section 4.4. Hoyt (1977) concluded that the computations made by Penndorf (1955) and Elterman (1968) gave values of β that were too high by about 3.5%. This was increased to 4.5% by the work of Frohlich and Shaw (1980). However, Young (1980) indicated these large discrepancies were unrealistic for conditions in the real atmosphere. In the absence of more definitive values, we accept the results of Elterman (1968) as being adequate for practical problems, and give the data of Table 4.4 to represent the vertical profiles of volume scattering coefficient for the standard atmosphere.

TABLE 4.4 Volume scattering coefficient β^R and the normal optical thickness of the Rayleigh atmosphere as functions of altitude z for selected wavelengths (data from Elterman, 1968). $N(z)$ is for the U.S. Standard Atmosphere. (The negative digits appended to β^R entries indicate the power of 10 by which the entry is multiplied.)

λ(μm)	0.30		0.34		0.40		0.50		0.60	
z(km)	β^R(km^{-1})	τ^R	β^R(km^{-1})	τ^R	β^R(km^{-1})	τ^R	β^R(km^{-1})	τ^R	β^R(km^{-1})	τ^R
0	1.446 − 1	1.222	8.492 − 2	0.717	4.303 − 2	0.354	1.716 − 2	0.145	8.156 − 3	0.054
2	1.188 − 1	0.959	6.978 − 2	0.563	3.536 − 2	0.285	1.410 − 2	0.114	6.701 − 3	0.042
4	9.672 − 2	0.744	5.680 − 2	0.437	2.878 − 2	0.221	1.148 − 2	0.088	5.455 − 3	0.032
6	7.792 − 2	0.570	4.576 − 2	0.335	2.319 − 2	0.170	9.245 − 3	0.068	4.395 − 3	0.024
8	6.207 − 2	0.431	3.645 − 2	0.253	1.847 − 2	0.128	7.364 − 3	0.051	3.501 − 3	0.018
10	4.881 − 2	0.320	2.867 − 2	0.188	1.453 − 2	0.095	5.791 − 3	0.038	2.753 − 3	0.008
15	2.299 − 2	0.147	1.350 − 2	0.086	6.841 − 3	0.044	2.728 − 3	0.017	1.297 − 3	0.004
20	1.050 − 2	0.067	6.164 − 3	0.039	3.123 − 3	0.020	1.245 − 3	0.008	5.919 − 4	0.002
25	4.732 − 3	0.031	2.779 − 3	0.018	1.408 − 3	0.009	5.614 − 4	0.004	2.669 − 4	0.001
30	2.173 − 3	0.015	1.276 − 3	0.009	6.467 − 4	0.004	2.578 − 4	0.002	1.226 − 4	0
40	4.080 − 4	0.003	2.770 − 4	0.002	1.404 − 4	0.001	5.596 − 5	0	2.660 − 5	0
50	1.212 − 4	0.001	7.119 − 5	0.001	3.607 − 5	0	1.438 − 5	0	6.837 − 6	0
	0.70		0.80		0.90		1.06		1.67	
0	4.365 − 3	0.037	2.544 − 3	0.021	1.583 − 3	0.013	8.192 − 4	0.007	1.322 − 4	0.001
2	3.586 − 3	0.029	2.091 − 3	0.017	1.300 − 3	0.010	6.731 − 4	0.005	1.086 − 4	0.001
4	2.919 − 3	0.022	1.702 − 3	0.013	1.059 − 3	0.008	5.480 − 4	0.004	8.841 − 5	0.001
6	2.352 − 3	0.017	1.371 − 3	0.010	8.528 − 4	0.006	4.415 − 4	0.003	7.122 − 5	0.001
8	1.873 − 3	0.013	1.092 − 3	0.008	6.793 − 4	0.006	3.516 − 4	0.002	5.673 − 5	0
10	1.473 − 4	0.010	8.589 − 4	0.006	5.342 − 4	0.005	2.765 − 4	0.002	4.462 − 5	0
15	6.939 − 4	0.004	4.045 − 4	0.003	2.516 − 4	0.002	1.302 − 4	0.001	2.101 − 5	0
20	3.168 − 4	0.002	1.847 − 4	0.001	1.149 − 4	0.001	5.946 − 5	0	9.593 − 6	0
25	1.423 − 4	0.001	8.326 − 5	0.001	5.178 − 5	0	2.681 − 5	0	4.325 − 6	0
30	6.560 − 5	0	3.824 − 5	0	2.378 − 5	0	1.231 − 5	0	1.986 − 6	0
40	1.424 − 5	0	8.300 − 6	0	5.162 − 6	0	2.672 − 6	0	4.311 − 7	0
50	3.659 − 6	0	2.133 − 6	0	1.327 − 6	0	6.867 − 7	0	1.108 − 7	0

4.3.7 Normal optical thickness of Rayleigh atmosphere

Included also in Table 4.4 are values of the normal optical thickness τ^R of the Rayleigh atmosphere computed from the relation

$$\tau^R(\lambda,z) = \int_z^\infty \beta^R(\lambda,z)\, dz \tag{4.38}$$

Here the integration is taken from altitude z to the top of the atmosphere. More complete data for the wavelength range of most interest can be obtained from Fig. 4.4, in which $\tau^R(\lambda,z)$ is shown as a function of wavelengths for 16 different altitudes in the Rayleigh atmosphere.

4.4 Computational methods for atmospheric scattering

A relatively large number of different methods have been developed for computing the radiative fields resulting from scattering and absorption in planetary atmospheres. The methods are applicable to the Rayleigh model, but are generally

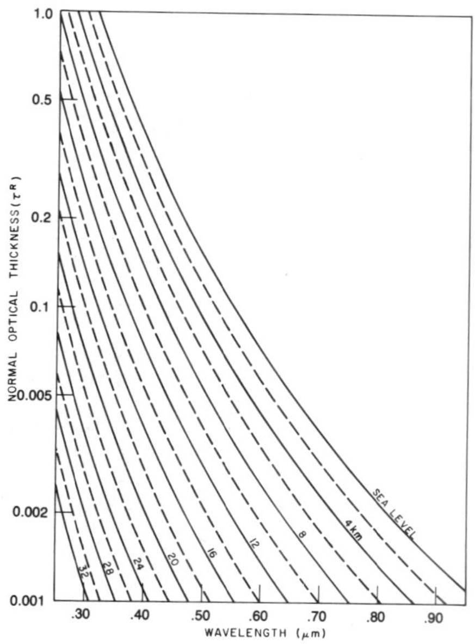

Figure 4.4 Normal optical thickness of the Rayleigh atmosphere as a function of wavelength for 17 different altitudes above sea level.

applicable to more realistic cases of the turbid atmosphere as well. A critique of the following methods has been included in a study sponsored by the Radiation Commission of the International Association of Meteorology and Atmospheric Physics (I.U.G.G.). (Lenoble, 1974):

A. Exact analytical methods
   1. Singular Eigenfunctions method
   2. Wiener-Hopf technique
   3. Reduction to *H*-function and *X*- and *Y*- functions

B. Approximate methods
   1. Similarity relations
   2. Eddington approximation
   3. Two-stream approximation
   4. Exponential kernel approximation
   5. Perturbation method

C. Computational methods
   1. Monte Carlo method
   2. Method of spherical harmonics
   3. Method of discrete ordinates
   4. Method of Gauss-Seidel iteration
   5. Method of successive orders of scattering
   6. Matrix operator method
   7. Doubling or adding method
   8. Principles of invariance
   9. Invariant imbedding

A summary of all of the various methods is beyond the scope of the present discussion, but because of its sound physical basis and its relatively widespread use, particularly for the Rayleigh atmosphere, we will outline the method of discrete ordinates in terms of the X and Y functions of Chandrasekhar (1950). The formulas applicable to a plane-parallel model of the Rayleigh atmosphere are taken without essential change from Chandrasekhar's treatise. We assume a parallel beam of radiation of net flux $\pi \mathbf{F}$ to be incident from direction $(-\mu_o, \phi_o)$ on top of a plane-parallel Rayleigh atmosphere of total optical thickness τ_1. The problem is to find the intensity and state of polarization of the skylight emerging from the bottom $(\tau = \tau_1)$ and earthlight emerging from the top $(\tau = 0)$ of this atmospheric model. The method takes into account all orders of scattering by the molecules of the atmosphere and, in principle, permits the inclusion of effects of reflection from the planetary surface. Monochromatic radiation of wavelength λ is assumed throughout.

The basic equation of radiative transfer was given for the scalar case in Chapter 2, Eq. (2.10). Here we adopt the vector form, which includes the effects of polarization. The incident flux $\pi\mathbf{F}(-\mu_o, \phi_o)$ is expressed in terms of the Stokes parameters as

$$\pi\mathbf{F}(-\mu_o,\phi_o) = \pi[F_e,F_r,F_u,F_v] \tag{4.39}$$

F_e and F_r are, respectively, the components in and normal to the plane of the sun's vertical. (For uniformity of notation we choose subscript e, instead of Chandrasekhar's ℓ, for the first Stokes parameter and have used Ω in place of (μ,ϕ). The vector equation of transfer is (Chandrasekhar, 1950, p. 250)

$$\mu \frac{d\mathbf{I}}{d\tau}(\Omega,\Omega') = \mathbf{I}(\tau;\Omega) - \frac{1}{4\pi}\int_{-1}^{1}\int_{0}^{2\pi} \mathbf{P}(\Omega,\Omega')\,\mathbf{I}(\tau;\Omega')\,d\mu'\,d\phi'$$

$$- \frac{1}{4} e^{-\tau/\mu_o}\,\mathbf{P}(\Omega,\Omega')\,\mathbf{F}(\Omega_o) \tag{4.40}$$

Here $\mathbf{P}(\Omega;\Omega')$ is in general a (4×4) phase matrix for scattering from direction $\Omega' = (\mu',\phi')$ into direction $\Omega = (\mu,\phi)$. However, for the unpolarized radiation direct from the sun $V=0$ and the phase matrix is reducible with respect to Stokes parameter V (Chandrasekhar, p. 249). In this case $\mathbf{P}$ becomes a (3×3) matrix for computing the Stokes parameters I_e, I_r, and U of the scattered radiation field. It can be represented as the product of a diagonal matrix $\mathbf{Q}$ and three other matrices involving the geometry. Thus we have

$$\mathbf{P}(\Omega,\Omega') = \mathbf{Q}[\mathbf{P}^{(0)}(\mu,\mu') + (1-\mu^2)^{1/2}(1-\mu'^2)^{1/2}\,\mathbf{P}^{(1)}(\Omega,\Omega')$$

$$+ \mathbf{P}^{(2)}(\Omega,\Omega)] \tag{4.41}$$

Here

$$\mathbf{Q} = \begin{vmatrix} 1 & 0 & 0 \\ 0 & 1 & 0 \\ 0 & 0 & 2 \end{vmatrix} \tag{4.42}$$

$$\mathbf{P}^{(0)}(\mu,\mu') = \frac{3}{4}\begin{vmatrix} 2(1-\mu^2)(1-\mu'^2)+\mu^2\mu'^2 & \mu^2 & 0 \\ \mu'^2 & 1 & 0 \\ 0 & 0 & 0 \end{vmatrix} \tag{4.43}$$

$$\mathbf{P}^{(1)}(\Omega,\Omega') = \frac{3}{4}\begin{vmatrix} 4\mu\mu'\cos(\phi'-\phi) & 0 & 2\mu\sin(\phi'-\phi) \\ 0 & 0 & 0 \\ -2\mu'\sin(\phi'-\phi) & 0 & \cos(\phi'-\phi) \end{vmatrix} \tag{4.44}$$

$$\mathbf{P}^{(2)}(\Omega,\Omega') = \frac{3}{4}\begin{vmatrix} \mu^2\mu'^2\cos2(\phi'-\phi) & -\mu^2\cos2(\phi'-\phi) & \mu^2\mu'\sin2(\phi'-\phi) \\ -\mu'^2\cos2(\phi'-\phi) & \cos2(\phi'-\phi) & -\mu'\sin2(\phi'-\phi) \\ -\mu\mu'^2\sin2(\phi'-\phi) & \mu\sin2(\phi'-\phi) & \mu\mu'\cos2(\phi'-\phi) \end{vmatrix} \tag{4.45}$$

The objective is to use these expressions to compute the vector intensity of the skylight and earthlight emerging from the bottom and top, respectively, of a plane parallel Rayleigh atmosphere.

The vector intensities of these two radiation fields are given in terms of a transmission matrix $\mathbf{T}(\tau_1;\Omega,\Omega_o)$ and reflection matrix $\mathbf{S}(\tau_1;\Omega,\Omega_o)$ by the relations

$$\mathbf{I}(\tau_1;-\mu,\phi) = \frac{1}{4\mu}\,\mathbf{T}(\tau_1;\Omega,\Omega_o)\,\mathbf{F}(\Omega_o) \tag{4.46}$$

$$\mathbf{I}(0;+\mu,\phi) = \frac{1}{4\mu}\,\mathbf{S}(\tau_1;\Omega,\Omega_o)\,\mathbf{F}(\Omega_o) \tag{4.47}$$

The negative and positive signs of μ indicate radiation directed downward and upward, respectively. Matrices $\mathbf{T}$ and $\mathbf{S}$ take the expanded form

$$\mathbf{T}(\tau_1;\Omega,\Omega_o) = \mathbf{Q}\left[\frac{3}{4}\,\mathbf{T}^{(0)}(\tau_1;\mu,\mu_o) + (1-\mu^2)^{1/2}(1-\mu_o^2)^{1/2}\mathbf{T}^{(1)}(\tau_1;\Omega,\Omega_o) + \mathbf{T}^{(2)}(\tau_1;\Omega,\Omega_o)\right] \tag{4.48}$$

and

$$\mathbf{S}(\tau_1;\Omega,\Omega_o) = \mathbf{Q}\left[\frac{3}{4}\,\mathbf{S}^{(0)}(\tau_1;\mu,\mu_o) + (1-\mu^2)^{1/2}(1-\mu_o^2)^{1/2}\mathbf{S}^{(1)}(\tau_1;\Omega,\Omega_o) + \mathbf{S}^{(2)}(\tau_1;\Omega,\Omega_o)\right] \tag{4.49}$$

On combining these with the phase relations (4.43) to (4.45) we have for the azimuth dependent terms of skylight $\mathbf{T}^{(i)}\,(i=1,2)$

$$\mathbf{T}^{(i)} = \frac{\mu\mu_o}{\mu-\mu_o}\,P^{(i)}(-\mu,\phi;-\mu_o,\phi_o)\left[y^{(i)}_{(\mu)}\,x^{(i)}_{(\mu_o)} - x^{(i)}_{(\mu)}\,y^{(i)}_{(\mu_o)}\right] \tag{4.50}$$

and for earthlight $\mathbf{S}^{(i)}\,(i=1,2)$

$$\mathbf{S}^{(i)} = \frac{\mu\mu_o}{\mu+\mu_o}\,P^{(i)}(\mu,\phi;-\mu_o,\phi_o)\left[x^{(i)}_{(\mu)}\,x^{(i)}_{(\mu_o)} - y^{(i)}_{(\mu)}\,y^{(i)}_{(\mu_o)}\right] \tag{4.51}$$

The quantities in brackets are the fundamental X and Y functions which were developed by Chandrasekhar for this problem. They are further specified below. The azimuth independent matrices $\mathbf{T}^{(0)}(\tau_1;\mu,\mu')$ and $\mathbf{S}^{(0)}(\tau_1;\mu,\mu')$ are expressed as

$$\mathbf{T}^{(0)}(\mu,\mu') = \frac{\mu\mu'}{\mu-\mu'}\left\{\begin{bmatrix}\xi(\mu) & \sqrt{2}\eta(\mu)\\ \sigma(\mu) & \sqrt{2}\theta(\mu)\end{bmatrix}\begin{bmatrix}\psi(\mu') & \chi(\mu')\\ \sqrt{2}\phi(\mu') & \sqrt{2}\zeta(\mu')\end{bmatrix} - \begin{bmatrix}\psi(\mu) & \sqrt{2}\phi(\mu)\\ \chi(\mu) & \sqrt{2}\zeta(\mu)\end{bmatrix}\begin{bmatrix}\xi(\mu') & \sigma(\mu')\\ \sqrt{2}\eta(\mu') & \sqrt{2}\theta(\mu')\end{bmatrix}\right\} \tag{4.52}$$

and

$$\mathbf{S}^{(0)}(\mu,\mu') = \frac{\mu\mu'}{\mu+\mu'}\left\{\begin{bmatrix}\psi(\mu) & \sqrt{2}\phi(\mu)\\ \chi(\mu) & \sqrt{2}\zeta(\mu)\end{bmatrix}\begin{bmatrix}\psi(\mu') & \chi(\mu')\\ \sqrt{2}\phi(\mu') & \sqrt{2}\zeta(\mu')\end{bmatrix} - \begin{bmatrix}\xi(\mu) & \sqrt{2}\eta(\mu)\\ \sigma(\mu) & \sqrt{2}\theta(\mu)\end{bmatrix}\begin{bmatrix}\xi(\mu') & \sigma(\mu')\\ \sqrt{2}\eta(\mu') & \sqrt{2}\theta(\mu')\end{bmatrix}\right\} \tag{4.53}$$

The eight functions ψ, ϕ, χ, ζ, ξ, η, σ, θ, termed the scattering functions, are linear combinations of the functions X_e, Y_e, X_r, Y_r satisfying the two pairs of nonlinear integral equations of the type

$$x_i(\mu) = 1 + \mu\int_0^1 \Psi_i(\mu')\left[x_i(\mu)x_i(\mu') - y_i(\mu)y_i(\mu')\right]\frac{d\mu'}{\mu+\mu'} \tag{4.54}$$

$$y_i(\mu) = e^{-\tau/\mu} + \mu\int_0^1 \Psi_i(\mu')\left[y_i(\mu)x_i(\mu') - x_i(\mu)y_i(\mu')\right]\frac{d\mu'}{\mu+\mu'} \tag{4.55}$$

with $\Psi_e(\mu) = \frac{3}{4}(1-\mu^2)$, $\Psi_r(\mu) = \frac{3}{8}(1-\mu^2)$ and $i=e,r$. Equations (4.54) and (4.55) are the definitions of the X and Y functions. The characteristic function $\Psi(\mu)$ is an even polynomial in μ which has the constraint

$$\int_0^1 \Psi(\mu)\, d\mu \leq \frac{1}{2} \tag{4.56}$$

For the purpose of computations, Coulson et al. (1960) changed slightly the forms of the original relations given by Chandrasekhar (1950). The intensity components of the scattered light may be expressed as

$$I_{e,r}(\tau;\Omega) = I_{e,r}^{(0)}(\mu,\mu_o) + I_{e,r}^{(1)}(\tau;\Omega)\cos\phi + I_{e,r}^{(2)}(\tau;\Omega)\cos 2\phi \tag{4.57}$$

where azimuth is taken with respect to that of the sun. The azimuth independent terms of Eq. (4.57) assume the following forms for the diffuse skylight from the bottom of the atmosphere:

$$I_e^{(0)}(\tau;-\mu,-\mu_o) = \frac{C}{\mu - \mu_o}\left[K\xi(\mu) + L\eta(\mu) - M\psi(\mu) - N\phi(\mu)\right] \tag{4.58}$$

$$I_r^{(0)}(\tau;-\mu,-\mu_o) = \frac{C}{\mu - \mu_o}\left[K\sigma(\mu) + L\theta(\mu) - M\chi(\mu) - N\zeta(\mu)\right] \tag{4.59}$$

Here $C = (3/32)(\mu_o F_o)$ and

$$\begin{aligned} K &= \psi(\mu_o) + \chi(\mu_o) & L &= 2[\phi(\mu_o) + \zeta(\mu_o)] \\ M &= \xi(\mu_o) + \sigma(\mu_o) & N &= 2[\theta(\mu_o) + \eta(\mu_o)] \end{aligned} \tag{4.60}$$

For diffuse earthlight from the top of the atmosphere, the equivalent relations for the azimuth independent components are

$$I_e^{(0)}(o;+\mu,-\mu_o) = \frac{C}{\mu + \mu_o}\left[K\psi(\mu) + L\phi(\mu) - M\xi(\mu) - N\eta(\mu)\right] \tag{4.61}$$

$$I_r^{(0)}(o;+\mu,-\mu_o) = \frac{C}{\mu + \mu_o}\left[K\chi(\mu) + L\zeta(\mu) - M\sigma(\mu) - N\theta(\mu)\right] \tag{4.62}$$

The azimuth dependent terms in Eq. (4.57) contain as factors the functions

$$
\begin{aligned}
I_e^{(1)}(\tau;-\mu,-\mu_o) &= \frac{4C}{\mu-\mu_o}\,\mu\mu_o(1-\mu^2)^{1/2}(1-\mu_o^2)^{1/2}W^{(1)}(\mu,\mu_o) \\
I_r^{(1)}(\tau;-\mu,-\mu_o) &= 0 \\
I_e^{(2)}(\tau;-\mu,-\mu_o) &= -\frac{C}{\mu-\mu_o}\,\mu^2(1-\mu_o^2)W^{(2)}(\mu,\mu_o) \\
I_r^{(2)}(\tau;-\mu,-\mu_o) &= \frac{C}{\mu-\mu_o}\,(1-\mu_o^2)W^{(2)}(\mu,\mu_o)
\end{aligned}
\tag{4.63}
$$

and

$$
\begin{aligned}
I_e^{(1)}(o;+\mu,-\mu_o) &= -\frac{4C}{\mu+\mu_o}\,(1-\mu^2)^{1/2}(1-\mu_o^2)^{1/2}M^{(1)}(\mu,\mu_o) \\
I_r^{(1)}(o;+\mu,-\mu_o) &= 0 \\
I_e^{(2)}(o;+\mu,-\mu_o) &= -\frac{C}{\mu+\mu_o}\,\mu^2(1-\mu_o^2)M^{(2)}(\mu,\mu_o) \\
I_r^{(2)}(o;+\mu,-\mu_o) &= \frac{C}{\mu+\mu_o}\,(1-\mu_o^2)M^{(2)}(\mu,\mu_o)
\end{aligned}
\tag{4.64}
$$

The relations (4.63) are for skylight and those of (4.64) are for earthlight. Factors $W^{(j)}$ and $M^{(j)}$, $j=1,2$ are given in terms of the X and Y functions by

$$
\begin{aligned}
W^{(j)}(\mu,\mu_o) &= X^{(j)}(\mu_o)Y^{(j)}(\mu) - Y^{(j)}(\mu_o)x^{(j)}(\mu) \\
M^{(j)}(\mu,\mu_o) &= X^{(j)}(\mu_o)X^{(j)}(\mu) - Y^{(j)}(\mu_o)Y^{(j)}(\mu)
\end{aligned}
\tag{4.65}
$$

$X^{(j)}$ and $Y^{(j)}$ satisfy Eqs. (4.54) and (4.55) if we set

$$
\begin{aligned}
\Psi^{(1)}(\mu) &= \frac{3}{8}(1-\mu^2)(1+2\mu^2) \\
\Psi^{(2)}(\mu) &= \frac{3}{16}(1+\mu^2)^2
\end{aligned}
\tag{4.66}
$$

On combining the azimuth independent and dependent terms for the intensity components, we have for skylight the final expressions

$$I_e(\tau; -\mu, -\mu_o) = \frac{C}{\mu - \mu_o} \Big\{ [K\xi(\mu) + L\eta(\mu) - M\psi(\mu) - N\phi(\mu)]$$

$$+ 4\mu\mu_o(1 - \mu^2)^{1/2}(1 - \mu_o^2)^{1/2} W^{(1)}(\mu,\mu_o) \cos\phi$$

$$- \mu^2(1 - \mu_o^2) W^{(2)}(\mu,\mu_o) \cos 2\phi \Big\} \tag{4.67}$$

$$I_r(\tau; -\mu, -\mu_o) = \frac{C}{\mu - \mu_o} \Big\{ [K\sigma(\mu) + L\theta(\mu) + M\chi(\mu) + N\zeta(\mu)]$$

$$+ (1 - \mu_o^2) W^{(2)}(\mu,\mu_o) \cos 2\phi) \Big\} \tag{4.68}$$

and for earthlight the equivalent expressions

$$I_e(o; +\mu, -\mu_o) = \frac{C}{\mu + \mu_o} \Big\{ [K\psi(\mu) + L\phi(\mu) - M\xi(\mu) - N\eta(\mu)]$$

$$- 4\mu\mu_o(1 - \mu^2)^{1/2}(1 - \mu_o^2)^{1/2} M^{(1)}(\mu,\mu_o) \cos\phi$$

$$- \mu^2(1 - \mu_o^2) M^{(2)}(\mu,\mu_o) \cos 2\phi \Big\} \tag{4.69}$$

$$I_r(o; +\mu, -\mu_o) = \frac{C}{\mu + \mu_o} \Big\{ [K\chi(\mu) + L\zeta(\mu) - M\sigma(\mu) - N\theta(\mu)]$$

$$+ (1 - \mu_o^2) M^{(2)}(\mu,\mu_o) \cos 2\phi \Big\} \tag{4.70}$$

The Stokes parameter U for skylight and earthlight is given by the expressions

$$U(\tau; -\mu, -\mu_o) = \frac{2C}{\mu - \mu_o} \left[2\mu_o(1 - \mu_o^2)^{1/2}(1 - \mu^2)^{1/2} W^{(1)}(\mu,\mu_o) \sin\phi \right.$$

$$\left. - \mu(1 - \mu^2) W^{(2)}(\mu,\mu_o) \sin 2\phi \right] \tag{4.71}$$

$$U(o; +\mu, -\mu_o) = \frac{2C}{\mu + \mu_o} \left[2\mu_o(1 - \mu_o^2)^{1/2}(1 - \mu^2)^{1/2} M^{(1)}(\mu,\mu_o) \sin\phi \right.$$

$$\left. + \mu(1 - \mu_o^2) M^{(2)}(\mu,\mu_o) \sin 2\phi \right] \tag{4.72}$$

For unpolarized incident light, as is the case for radiation from the sun, parameter $V=0$.

This completes the expressions for scattered radiation emerging from the bottom and top of the Rayleigh atmosphere according to the discrete ordinate method of Chandrasekhar (1950). The X and Y functions play a crucial role in the theory. Numerical values of the X and Y functions and the scattering functions of Eqs. (4.52) and (4.53) have been tabulated by Sekera and Blanche (1952) for optical thickness $\tau=0.15$, 0.25, and 1.0 and by Sekera and Ashburn (1953) for $\tau=0.02$, 0.05, and 0.10. Unfortunately both of these sets of tables had only limited distribution and are not generally available. Similar tables, however, were published by Chandrasekhar and Elbert (1954) for $\tau=0.05$, 0.10, 0.15, 0.20, 0.25, 0.50, and 1.0. The three sets of tables were used by Coulson, Dave, and Sekera (1960) for extensive computations of light emerging from the bottom and top of the Rayleigh atmosphere.

Extension of the X and Y functions to values of $\tau^R > 1.0$ were of great interest in the 1960s for possible application to the atmospheres of other planets. It was found, however, that the series would not converge for the larger values of τ^R and that strong oscillations developed in the angular distributions of the functions. As a result of these difficulties other methods were developed for application to the larger optical thicknesses, as well as to generally turbid atmospheres. Among these methods are the doubling method, method of successive orders of scattering, matrix operator method, spherical harmonics method, Monte Carlo method, and certain other methods discussed by Lenoble (1974). As a result, the radiation fields in a Rayleigh atmosphere have been computed for optical thicknesses up to 16 as shown in section 4.5.3 below.

4.5 Polarization of Skylight in a Rayleigh Atmosphere

4.5.1 Background

Although scattering of light in the real atmosphere, which always contains dust, haze, and other non-Rayleigh type particles, may seem to be mainly of academic interest, such is not the case. In fact, results obtained from exact solution for Rayleigh scattering are an important first step for the proper interpretation of polarization as observed in the real atmosphere. The number of molecules in a unit volume of atmospheric gases is so large compared to the number of aerosol particles in the volume that molecular scattering is responsible for the major part of the observed polarization. A contributing factor in the strong influence of gaseous particles on the observed polarization is that small particles are much more efficient polarizers at most scattering angles than are the larger aerosol particles. Thus the effect of the aerosols in the atmosphere can be considered a first-order correction to the values computed for the pure Rayleigh atmosphere. When theoretical deviations of the polarization computed for Rayleigh and turbid atmosphere models with different aerosol distributions are compared with the observed deviations, a good estimate of the actual aerosol characteristics can be determined. Furthermore, in some very clear atmospheric conditions the Rayleigh model itself is not a bad approximation of reality.

For purposes of discussion we divide the Rayleigh polarization fields into two components, that of the radiation directed downward to the surface from the sunlit sky and that directed to space from the top of the sunlit atmosphere. To distinguish between the two, the first of these will be termed skylight and the second, earthlight. This section is for skylight.

The polarization of skylight for the Rayleigh atmosphere was studied extensively following Rayleigh's explanation of small particle scattering in 1871 (see section 4.1), but the computational problem for taking multiple scattering effects into account was formidable. Early in this century, Ahlgrimm (1915) was successful in computing the effects of secondary scattering, and two decades later Blickhan (1934) combined the results for secondary scattering with measurements of atmospheric turbidity to show the close relationship of skylight polarization with aerosols in the atmosphere.

Beginning around 1950 three simultaneous developments made it possible for the first time to combine theory and observation in a most complete and logical manner. First, Chandrasekhar (1950) was able to utilize the representation of polarized light that had been defined by Stokes (1852) a century earlier to develop an elegant method of solving the problem of radiative transfer in all its complexity for a plane-parallel model atmosphere overlying a planetary surface which is either completely absorbing or which has reflection which follows the Lambert law of reflection. Although later solutions of radiative transfer have been developed by several authors, Chandrasekhar's solution has been widely applied to computing polarization and intensity in the Rayleigh atmosphere.

A second mid-century development was that of fast electronic computers. Without such powerful computers, the theoretical developments would have been of marginal use, but the possibility of using numerical techniques in combination with large memory capacity of modern computers has permitted an abundance of radiative transfer computations based on various techniques by many authors. As a result, the computation of radiative transfer in a Rayleigh atmosphere has become commonplace. Typical results of such computations are shown below.

Finally, the development since World War II of very sensitive photomultiplier tubes and other photosensitive devices has opened up new avenues for the measurement of light, and particularly its state of polarization. An early electronic photopolarimeter by Sekera and Collaborators (1955) was used for extensive sets of polarization measurements under different conditions, and later polarizing radiometers based on even more powerful techniques have been utilized by several authors. A description of these instruments is reserved for the section (Chapter 11) on measurements.

Exploitation of the new-found computing possibilities for the radiative transfer problem was not long in coming. Chandrasekhar (1950) and Chandrasekhar and Elbert (1954) computed the polarization of skylight for a Rayleigh atmosphere, as did Sekera and Collaborators (1955) and Sekera (1956). Coulson (1959) studied the polarization of earthlight emerging from the top of a Rayleigh atmosphere and Coulson et al. (1960) developed an extensive set of tables of polarization and intensity of both skylight and earthlight for the Rayleigh atmosphere. These tables are the source of much of the data shown below. Later computations of skylight polarization for the Rayleigh atmosphere have been carried out by many authors, some of whom used different methods which confirm the results of Coulson et al. (1960), and some of whom extended the computations to large optical thicknesses or to anisotropic scattering. For instance, Dave and Furukawa (1966) used a modified discrete ordinate method to perform the computations for a Rayleigh optical thickness of 10.0 and then extended them to optical thickness infinity by tables and formulas given by Chandrasekhar (1950). Plass and Kattawar (1970) used a Monte Carlo technique for optical thicknesses up to 73.25. Unfortunately, there are two difficulties with the Plass and Kattawar results. First, the Monte Carlo technique, being statistical in nature, is always subject to statistical scatter of the data, and is thus not well adapted to showing specific details of the fields. A more fundamental problem in the results of Plass and Kattawar, however, is that in order to improve the statistics they integrated the intensity components over large solid angles and then used the simple relation $P=(I_\perp - I_\parallel)/(I_\perp + I_\parallel)$, the so-called Rubenson polarization, to compute the degree of polarization for one given direction. This procedure gives erroneous results because of the change of the plane of polarization with direction in the hemisphere. For precise results in this case the entire Stokes vector must be used, for which $P=(Q^2+U^2+V^2)^{1/2}/I$. Thus the polarization data given by Plass and Kattawar are reliable only in a qualitative sense.

Herman and Browning (1965) used an iterative technique to extend the computations to $\tau=4.0$. By use of the matrix method Howell and Jacobowitz (1970) computed the atmospheric polarization, but only for $\tau=0.25$. Polarization computations for Rayleigh optical thicknesses up to $\tau=16$ have been given by Plass and Kattawar. Other studies of the polarization for a Rayleigh atmosphere have been carried out by deBary and Bullrich (1964), and Bahethi and Fraser (1980), among others. Representative results of the various studies will be discussed below.

4.5.2 Polarization in sun's vertical for small and moderate optical thicknesses

The large number of computations of skylight polarization for small and moderate optical thicknesses ($\tau \leq 1.0$) for a plane-parallel model of the atmosphere all give results that are consistent with those tabulated by Coulson et al. (1960).

The distribution of the degree of polarization through the sun's vertical, as given by Coulson et al., is shown for five different wavelengths for a solar zenith distance of 53.1° in Fig. 4.5. Included also is a curve representing only single scattering, the polarization for which is given by the angular relation

$$P = \frac{\sin^2\Theta}{1 + \cos^2\Theta} \tag{4.73}$$

where molecular anisotropy is neglected. The sea level optical thickness for the five wavelengths shown are, in increasing order, 1.0, 0.50, 0.25, 0.10, and 0.02. The position of the sun is indicated by the arrow. The general features of this type of distribution, including the negative values of polarization ($I_\parallel > I_\perp$) in the vicinity of the sun, strong positive polarization ($I_\perp > I_\parallel$) throughout most of the plane, and a well-defined maximum at an angle of about 90° from the

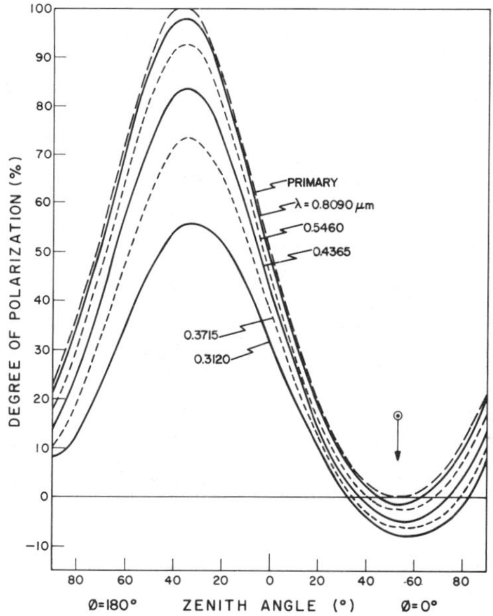

Figure 4.5 Degree of polarization as a function of angle in the plane of the sun's vertical for a plane-parallel Rayleigh atmosphere at five different wavelengths for a solar zenith angle of 53.1° (μ_0=0.60, A=0, τ=0.02, 0.10, 0.25, 0.50, 1.00).

sun, are typical of both theoretical and measured features of skylight polarization. The details, however, vary with sun elevation, surface reflection, and atmospheric constituents, as well as with wavelength.

The variation of the polarization field with wavelength shown by Fig. 4.5 is due entirely to the effects of multiple scattering in the atmosphere. In the absence of multiple scattering the degree of polarization would be as given by the top curve of the family, with $P=0$ at scattering of $\Theta=0°$ and $P=100\%$ at $\Theta=90°$, independent of wavelength. The only neutral point (point of zero polarization) for primary scattering is at $\Theta=0°$, P being >0 at all other scattering angles. Multiple scattering, however, is generally a depolarizing mechanism, causing a decrease of the magnitude of P at the larger scattering angles and, at the same time, introducing negative polarization at small scattering angles. This comes about by changing the relative magnitudes of $I_\perp$ and $I_\parallel$ in the relation $P=(I_\perp - I_\parallel)/(I_\perp + I_\parallel)$ given above as Eq. (4.24). The shift of P in the negative direction shows that multiple scattering increases $I_\parallel$ at the expense of $I_\perp$ throughout the entire plane, and at the smaller scattering angles $I_\parallel > I_\perp$. This creates a region of negative polarization in the vicinity of the sun and causes the appearance of neutral points in the polarization distribution. A discussion of the neutral points is included as section 4.7 below.

Since the scattering efficiency of molecular size particles is critically dependent on wavelength, in fact being proportional to λ^{-4}, multiple scattering effects in the Rayleigh atmosphere are very pronounced at ultraviolet wavelengths, but are relatively minor in the near infrared. This dependence results in the ultraviolet having a low polarization maximum in the positive branch of the curve with a large region of negative polarization surrounding the sun. In contrast, the longer wavelengths are highly polarized in the positive branch, and have relatively negligible areas of negative polarization. A practical consequence of this difference is that the effects of non-Rayleigh scattering by dust, haze, and other aerosols in the actual atmosphere are masked by multiple scattering effects to a lesser extent in the long wavelength region than in short visible or ultraviolet wavelengths. Thus remote sensing of atmospheric aerosols is most effective in the longer wavelength region of the spectrum, whereas observations of features, such as sky intensity, which are dependent on a long optical pathlength through the atmosphere, should emphasize the shorter wavelengths.

The shift of the polarization field across the sky with changing position of the sun can be seen by the curves of Fig. 4.6, in which the degree of polarization throughout the sun's vertical is shown for seven different positions of the sun. Solar zenith angles are given at the curves, and solar positions are indicated by the arrows along the abscissa. These data are for a Rayleigh optical thickness of $\tau=0.15$, which corresponds to a wavelength of 0.495 μm at sea level. When the sun is low on the horizon ($\theta_o=84.3°$), the maximum of approximately $P=90\%$ is located about 5.7° from the zenith, for a scattering angle of approximately 90°. In this case a region of negative polarization, reaching as much as $P=-7.5\%$, exists near both the solar and antisolar horizons ($\phi=0°$ and 180°). Neutral points appear in the sky above the sun (Babinet point) and above the antisun (Arago point). The distribution is approximately symmetric around the position of the maximum, as could be anticipated from the symmetric pattern for Rayleigh scattering. No data for still larger solar zenith angles are shown, as the assumption of a plane parallel atmospheric model is not a valid approximation for the sun very near the horizon.

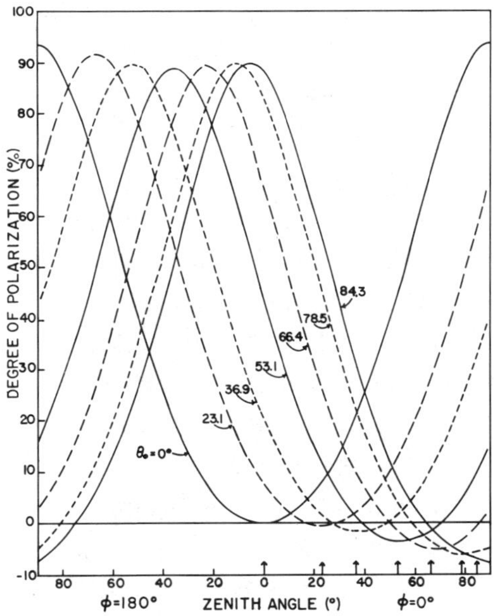

Figure 4.6 Degree of polarization as a function of angle in the plane of the sun's vertical for seven different solar zenith angles for a wavelength of 0.495 μm. ($\tau=0.15$, $A=0, \mu_0=0.10$, 0.20, 0.40, 0.60, 0.80, 0.92, 1.00).

As the sun rises in the sky, the pattern shifts across the sky in such manner as to always maintain the maximum at a scattering angle of approximately 90°. At the same time the region of negative polarization shrinks, entirely vanishing at the azimuth of the antisun and becoming smaller in both polarization magnitude and angular extent in the region of the sun. At the time the Arago point vanishes at the antisolar azimuth, another neutral point, the Brewster point, makes its appearance in the region of the sky below the sun. The shrinking of the magnitude and extent of negative polarization in the vicinity of the sun continues until the negative polarization vanishes completely just as the sun reaches the local zenith. At this instant, the Babinet and Brewster points merge into a single point of zero polarization in the direction of the zenith. Also at this time, the polarization field is symmetric around the zenith, with strong polarization in the region of the horizon at all azimuths.

The dependence of the polarization field on position of the sun shown in Fig. 4.6 is generally representative of that at other wavelengths as well, although details vary. The features due to multiple scattering, including the decrease of the positive polarization and enhancement of the negative branch of the polarization curve, become more pronounced with decreasing wavelength (increasing optical thickness) at all sun elevations. It is interesting, however, that the extent and magnitude of negative polarization reaches its fullest development at $\tau \approx 1.0$, a feature which appears, from the limited number of computations available for $\tau > 1.0$, to be characteristic of all solar zenith angles.

4.5.2.1 Magnitude of polarization maximum

The magnitude of the polarization maximum, $P_{\max}$, is shown by Figs. 4.5 and 4.6 to vary not only with wavelength, but also shows a slight variation with solar position as well. This magnitude, as determined by graphical interpolation from the data of Coulson et al. (1960), is listed for seven solar zenith angles and seven wavelengths in Table 4.5. The data are probably accurate to better than 0.5% in polarization magnitude. Although the change of $P_{\max}$ with θ_o is only of the order of 2% to 3%, it shows a definite pattern, with low values of $P_{\max}$ occurring in the mid-range of solar zenith angles and generally higher values with the sun near either the horizon or zenith. The reason for this pattern might be thought to lie in the combined optical path encountered by incident solar radiation at zenith angle θ. The geometrical path for this sum would be $\sec\theta_o + \sec\theta$. If, further, the amount of multiple scattering, and hence depolarization, is directly related to the combined pathlength in the atmosphere, then it would seem reasonable to expect the largest P_{max} to occur for the shortest geometrical pathlength.

The fact that such simple intuitive reasoning is inadequate to explain reality is obvious when the geometrical pathlength $\sec\theta_o + \sec\theta$ shown in Table 4.6 is compared with the P_{max} tabulations of Table 4.5. The effect is just the opposite of that anticipated: short pathlengths are generally accompanied by low polarization and long pathlengths with high polarization.

A clue toward explaining the discrepancy might be obtained from the position of P_{max} with respect to that of the sun. Fraser (Appendix C of Sekera and Collaborators, 1955) has computed the position of P_{max} by differentiation of P

TABLE 4.5 Magnitude of the polarization maximum (P_{max} in %) for seven zenith angles (θ_o) for the seven wavelengths (λ) included in the tabulations of Coulson et al. (1960). Entries were obtained by graphical interpolation from tabulated values, and are thought to be accurate to $\pm 0.5\%$.

$\theta_o(°)\backslash\lambda(\mu m)$	0.3120	0.3715	0.4365	0.4950	0.5460	0.6440	0.8090
84.3	56.9	74.5	85.1	89.8	92.6	95.5	98.0
78.5	56.0	73.6	84.6	89.7	92.7	95.8	98.8
66.4	55.8	73.0	83.9	89.1	92.5	95.5	98.9
53.1	55.5	73.0	83.9	88.5	92.5	95.2	98.5
36.9	55.5	73.6	84.5	90.0	93.3	95.8	98.0
23.1	55.6	75.1	86.6	91.5	94.5	97.0	98.3
0	57.9	77.3	88.7	93.4	95.7	97.9	99.1

TABLE 4.6 Geometrical pathlength represented by the sum $\sec\theta_o + \sec\theta$, where θ_o is solar zenith angle and θ is zenith angle of the direction of P_{max}, as a function of θ_o. It is assumed that $\theta_o + \theta = 90°$.

$\theta_o(°)$	$\sec\theta_o + \sec\theta$	$\theta_o(°)$	$\sec\theta_o + \sec\theta$
84.3	11.073	36.9	1.915
78.5	6.036	23.1	3.636
66.4	3.589	0.0	53.09 (for $\theta_o = 88.9°$)
53.1	1.915		

versus θ and setting the derivative to zero. If we differentiate the relation $P=(I_\perp - I_\parallel)/(I_\perp + I_\parallel)$ with respect to $\mu=\cos\theta$ and set it to zero, we have, after some manipulation

$$\frac{dI_\perp}{I_\perp d\mu} - \frac{dI_\parallel}{I_\parallel d\mu} = 0 \tag{4.74}$$

or

$$\frac{d(\ln I_\perp)}{d\mu} = \frac{d(\ln I_\parallel)}{d\mu} \tag{4.75}$$

at the position of the maximum. Fraser obtained finite differences of the logarithms of $I_\perp$ and $I_\parallel$ from tabulations of those quantities and plotted curves of the right- and left-hand sides of Eq. (4.75) as functions of μ. The intersection of the curves then gave the position of P_{max}. The results of this procedure, given in Fig. 4.7 for four of the seven wavelengths discussed above, show that P_{max} is closest to the sun in the mid-range of θ_o, with the distances approaching 90° at both small and large solar zenith angles. This means that for the plane-parallel Rayleigh atmosphere P_{max} has its lowest values at the same time its distance from the sun is a minimum, and both of these minima correspond to the minimum geometrical pathlength of the radiation through the atmosphere.

The explanation for these apparently contradictory phenomena probably lies in the intensity distribution of the skylight. The minimum intensity for Rayleigh scattering tends to be located at an angle of 90° from the sun. When the sun

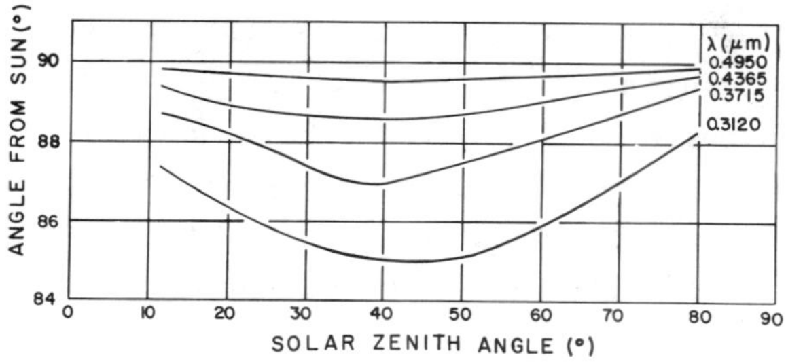

Figure 4.7 Position of the polarization maximum with respect to that of the sun as a function of solar zenith angle for four different wavelengths in a plane-parallel Rayleigh atmosphere. (Adapted from Fraser, Appendix C, Sekera and Collaborators, 1955.)

is in its mid-range of zenith angles, the horizon under the polarization maximum and intensity minimum is very bright and with little polarization. The molecules seen by an observer viewing in the mid-range of θ are illuminated by this strong and poorly polarized light from near the horizon, and the degree of polarization in the viewed direction is thereby decreased. Furthermore, since the contribution from near the horizon decreases with increasing elevation angle, the position of P_{max} is shifted toward the zenith. As the sun moves toward the zenith the horizon intensity decreases and its polarization increases, in which case the magnitude of P_{max} is not so strongly decreased and its position is permitted to return nearer to the 90° scattering angle it would have in the absence of multiple scattering. When the sun is near the horizon and P_{max} is near the zenith, there are no strong intensity gradients near P_{max} to cause an appreciable shift of its position, and no strong sources of poorly polarized light in the vicinity to cause much of a decrease of the magnitude of P_{max}. Only in the mid-ranges of θ_o and θ are both intensity and polarization distributions appropriate for causing the maximum decrease of P_{max} and the maximum shift of its position. An additional point, which tends to bolster the case for this explanation, is the fact that any shift of position of P_{max} that occurs is always in the direction away from the horizon, and never in the direction of the bright and poorly polarized light source in the region of the horizon.

4.5.2.2 Polarization at the zenith

Due to the simplicity of making polarization measurements in a fixed direction independent of sun elevation and azimuth, measurements of the degree of polarization in the direction of the zenith have been made by a number of authors beginning with Jensen (1899). The polarization maximum occurs near the zenith when the sun is on the horizon, and the zenith polarization rapidly decreases as the sun elevation increases and the point of maximum descends toward the antisolar horizon. At those lower latitude locations at which the position of the sun reaches the zenith, the zenith polarization decreases to zero, but most measurements have been made at the higher latitude locations at which the sun does not reach the zenith.

Early observations of zenith polarization have been correlated with atmospheric turbidity by Jensen (1899), Dorno (1919), Gockel (1920), and Tichanowski (1926). While the popularity of this measurement has decreased in favor of that of the polarization maximum since more versatile and flexible polarimeters have come into general use, zenith polarization is still a useful indicator of atmospheric turbidity, and it is thought likely that this type of measurement may be the subject of rekindled interest as more workers become involved with polarization investigations in the future. In fact, some recent measurements of polarization made during periods of twilight have been used to indicate turbidity conditions in the higher levels of the atmosphere (Steinhorst, 1977; Coulson, 1983), but the twilight problem is left to a separate section of this book. For the interpretation of zenith measurements throughout the day, however, data on zenith polarization in the Rayleigh atmosphere will be useful as a basis for comparison with the observations.

The degree of polarization at the zenith in a plane-parallel Rayleigh atmosphere is shown as a function of solar zenith angle for seven different wavelengths (at sea level) and no surface reflection in Fig. 4.8. The data for the diagram are from Coulson et al. (1960). The curves extend only to $\theta_o = 84.3°$, as the plane-parallel assumption on which the computations are based invalidates results for the sun very near the horizon. The steep slope of the curves in the $\theta_o = 35°$ to $80°$ range occurs as the strong polarization gradients such as those of Fig. 4.6

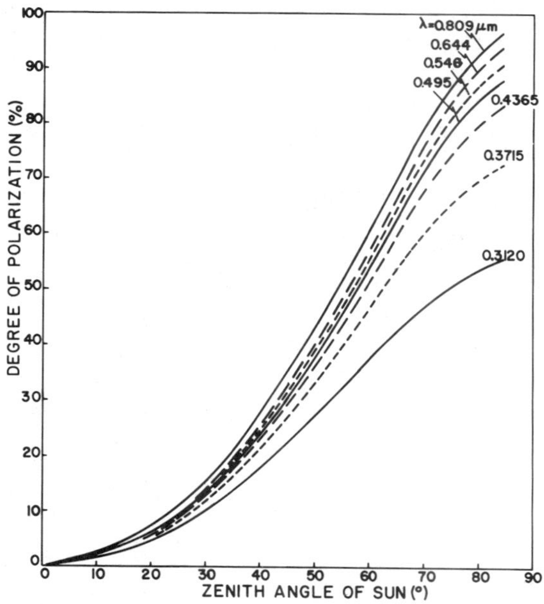

Figure 4.8 Degree of polarization at the zenith as a function of solar zenith angle at seven different wavelengths in a plane-parallel Rayleigh atmosphere.

shift across the zenith sky with changing position of the sun. The only neutral point which ever reaches the zenith for this atmospheric model is that which coincides with the sun when the sun is at the zenith. As will be seen in later sections, however, one or more neutral points have been observed at the zenith under certain conditions of atmospheric turbidity.

The data of Fig. 4.8, for which the surface reflectance $A=0$ are tabulated, along with those for $A=0.25$ and 0.80, in Table 4.7. A surface reflectance of $A=0.25$ corresponds roughly to that of a desert surface for visible wavelengths, and a new snow surface is typified by $A=0.80$. A Lambert law of reflection was assumed in the computations. The polarization values at $\theta_o=0°$ have all been set to zero, as they must be from geometry, even though roundoff errors

TABLE 4.7 Degree of polarization in the direction of the zenith at various wavelengths (λ) and zenith angles (θ_o) of the sun for three different values of Lambert surface reflectance $A=0$, 0.25, and 0.80 (Data from Coulson et al., 1960).

		Wavelength (μm)						
A	θ_o(°)	0.3120	0.3715	0.4365	0.4950	0.5460	0.6440	0.8090
0.0	84.3	0.558	0.727	0.833	0.882	0.910	0.940	0.966
	78.5	0.530	0.690	0.787	0.832	0.859	0.886	0.915
	66.4	0.441	0.556	0.625	0.657	0.678	0.697	0.729
	53.1	0.306	0.374	0.412	0.431	0.445	0.454	0.484
	36.9	0.153	0.180	0.195	0.203	0.210	0.213	0.233
	23.1	0.060	0.070	0.075	0.077	0.081	0.082	0.091
	0.0	0.000	0.000	0.000	0.000	0.000	0.000	0.000
0.25	84.3	0.498	0.666	0.778	0.829	0.857	0.885	0.906
	78.5	0.470	0.621	0.710	0.749	0.771	0.792	0.811
	66.4	0.382	0.478	0.529	0.550	0.564	0.575	0.593
	53.1	0.260	0.310	0.335	0.345	0.353	0.357	0.374
	36.9	0.128	0.147	0.155	0.159	0.163	0.164	0.174
	23.1	0.050	0.056	0.059	0.060	0.063	0.063	0.066
	0.0	0.000	0.000	0.000	0.000	0.000	0.000	0.000
0.80	84.3	0.364	0.536	0.666	0.724	0.754	0.781	0.797
	78.5	0.339	0.481	0.568	0.603	0.622	0.637	0.648
	66.4	0.263	0.339	0.379	0.395	0.404	0.410	0.417
	53.1	0.171	0.207	0.225	0.232	0.238	0.240	0.246
	36.9	0.082	0.095	0.101	0.104	0.107	0.107	0.110
	23.1	0.032	0.036	0.038	0.039	0.042	0.041	0.039
	0.0	0.000	0.000	0.000	0.000	0.000	0.000	0.000

in the original computations cause a small non-zero value to appear in the original tabulations.

4.5.3 Polarization in sun's vertical for large optical thickness

Computations of polarization are much less numerous for optical thicknesses of $\tau > 1.0$ than for $\tau \leq 1.0$. There are likely two reasons for this. First, the computer requirements increase rapidly as τ increases above 1.0, and, second, the Rayleigh atmosphere of the earth is well represented by optical thicknesses of $\tau \leq 1.0$. At wavelengths much below $\lambda = 0.312$ μm (which corresponds to $\tau = 1.0$ at sea level), absorption by ozone, oxygen, and other gases is the main determinant of the radiative regime of the earth's atmosphere, so computations for the larger optical thicknesses are of little practical concern for most atmospheric problems of the earth. The larger optical thicknesses are applicable, however, to the atmospheres of some other planets, and, of course, to other media such as lakes and oceans.

On the basis of the changes of the polarization field with optical thickness shown above, in which case $\tau \leq 1.0$, it would be anticipated that the positive branch of the polarization curve would decrease in magnitude with increasing optical thickness, that the negative branch would increase, that the neutral points would continue to move away from the sun or antisun, and that the general configuration of the fields would remain about the same otherwise. Computations for the larger values of τ show, however, that these expectations are not fulfilled in general, and some new features of the polarization field for which there were no indications at the smaller optical thicknesses, appear for optical thicknesses much above $\tau = 1.0$. In fact, the negative branch of the polarization curve reaches its maximum development at about $\tau = 1.0$, and it rapidly diminishes and eventually disappears entirely as τ increases up to about 2.5. The precise value of τ at which negative polarization vanishes is somewhat dependent on position of the sun and surface reflection.

Computations of polarization in the sun's vertical for optical thicknesses $\tau > 1.0$ have been carried out by Dave and Furukawa (1966) for various values of τ between 1.0 and 10 for skylight and up to $\tau = \infty$ for earthlight, Herman and Browning (1965) for $\tau \leq 4.0$, and by Kattawar et al. (1976) for $\tau \leq 16$. The results of Dave and Furukawa will be shown here as typical of all such computations for thick Rayleigh atmospheres.

Figure 4.9 shows the degree of polarization in the sun's vertical for $1.0 \leq \tau \leq 10$, according to the computations of Dave and Furukawa (1966) for a solar zenith angle of 53.1°. The magnitude of P_{max} decreases rapidly with increasing τ, but it remains on the antisolar side of the zenith up to $\tau = 4.0$. However, the horizons have polarization values nearly as high as P_{max} at $\tau = 4.0$. At $\tau = 10$, the normal configuration of P_{max} has been supplanted by very low polarization values throughout a wide range of angles surrounding the zenith. Any directionality due to position of the sun has been lost at $\tau = 10$, which means that the polarization field should be independent of azimuth, with a value of $P \approx 0.10$ near the horizons, for all $\tau > 10$. This is borne out by the computations of Kattawar et al. (1976) for $\tau = 16$.

As mentioned above, the negative branch of the polarization curves of Fig. 4.9 reach their maximum development at about $\tau = 1.0$, beyond which the magnitude of negative polarization and the angular separation of the neutral

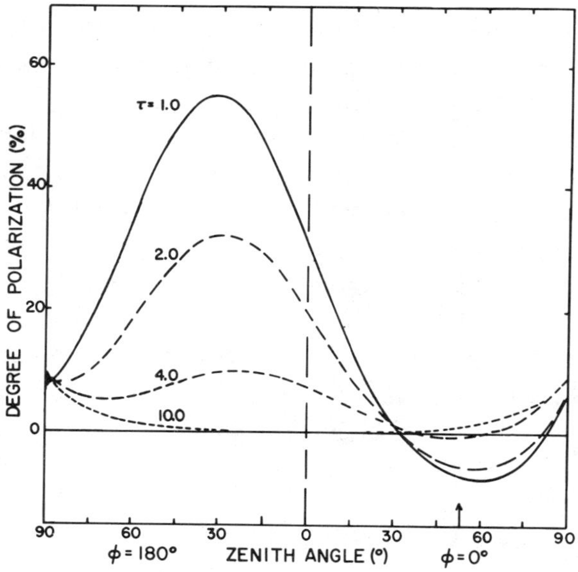

Figure 4.9 Degree of polarization versus zenith angle in the sun's vertical for the case of large optical thicknesses of the Rayleigh atmosphere for a solar zenith distance of 53.1°. The position of the sun is indicated by the arrow. (Adapted from Dave and Furukawa, 1966.)

points both decrease. In fact the negative branch in this case had almost disappeared at $\tau=4.0$, beyond which negative polarization and neutral points would both be nonexistent and the entire plane would have positive polarization.

The same type of behavior is shown for a solar zenith angle of 84.3° in Fig. 4.10. In this case, however, negative branches of the curves, with their attendant neutral points, occur both near the sun and the antisolar horizon. Here also, the maximum negative values are for $\tau=1.0$. An anomalous behavior of the neutral points is shown for $\tau=1.4$ and 2.0 in this diagram. For $\tau=1.4$, a double Arago point occurs in the antisolar region of the plane and the Brewster point has moved to about the position of the sun. For $\tau=2.0$ the Brewster point has migrated to the region above the sun and in essence constitutes one part of a double Babinet point, with the only negative polarization being located a considerable distance above the sun. The characteristics of the neutral points for all optical thicknesses will be discussed further in a later section.

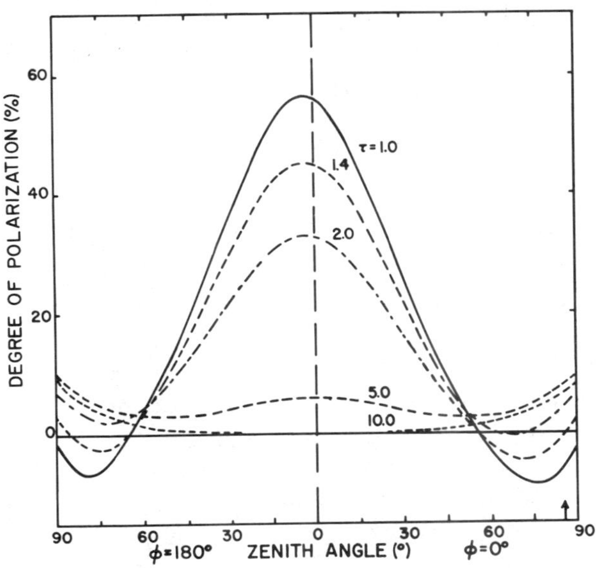

Figure 4.10 Degree of polarization versus zenith angle in the sun's vertical for the case of large optical thicknesses of the Rayleigh atmosphere for a solar zenith distance of 84.3°. The position of the sun is indicated by the arrow. (Adapted from Dave and Furukawa, 1966.)

The degree of polarization for the sun at the zenith, as computed by Dave and Furukawa (1966), is given for various optical thicknesses and zenith angles of observation in Table 4.8. The polarization field, which is independent of azimuth, has a neutral point at the zenith. The polarization increases monotonically from zenith to horizon, but there is a large area of the sky surrounding the zenith in which $P<1.0\%$. The region of the horizons, however, has more than 6% positive polarization, even for $\tau=10.0$.

4.5.4 Distribution of polarization over the hemisphere of the sky

The discussions above have concentrated on the distribution of the degree of polarization in the plane of the sun's vertical for various wavelengths or optical thicknesses for a plane-parallel Rayleigh atmosphere and no reflection from the underlying ground surface. Indeed, the distribution in the sun's vertical contains more information than that in any other meridian plane across the sky. The neutral points are normally located in that plane, as is the maximum polarization for most parameter combinations in the Rayleigh atmosphere. In some cases P_{max} migrates out of the sun's vertical, but in those cases there are differences of only a few percent at most between the real P_{max} and that in the sun's vertical. It is useful, however, in studies of the polarization in the solar almucantar and its variation with atmospheric turbidity, to know the hemispheric distribution that would exist in the pure Rayleigh atmosphere.

This hemispheric distribution is shown for sea level wavelengths of 0.644 and 0.312 μm ($\tau=0.05$ and 1.0), zero surface reflection, and a solar zenith angle of 78.5° in Figs. 4.11 and 4.12, respectively. The data for constructing the

TABLE 4.8 Degree of polarization of skylight for the sun at the zenith in a plane parallel Rayleigh atmosphere for various optical thicknesses and zenith angles of observation, according to the computations of Dave and Furukawa (1966).

$\tau \backslash \theta_o$ (°)	0	60.0	72.5	84.3
1.0	0.0	0.374	0.487	0.550
2.0	0.0	0.209	0.258	0.301
3.0	0.0	0.107	0.132	0.169
5.0	0.0	0.033	0.050	0.085
10.0	0.0	0.015	0.032	0.065

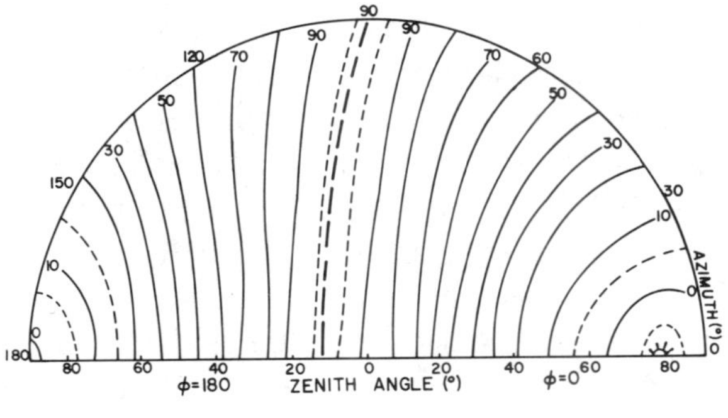

Figure 4.11 Distribution of the degree of polarization over the hemisphere of the sky for a wavelength of 0.644 μm (τ=0.05) and solar zenith angle of 78.5°. Only one-half of the hemisphere is given here and in subsequent diagrams, the other half hemisphere being a mirror image of this one. The plane of the sun's vertical is represented by the straight bottom line, with 0 being the zenith and the star being the position of the sun. The outside semicircle is the horizon.

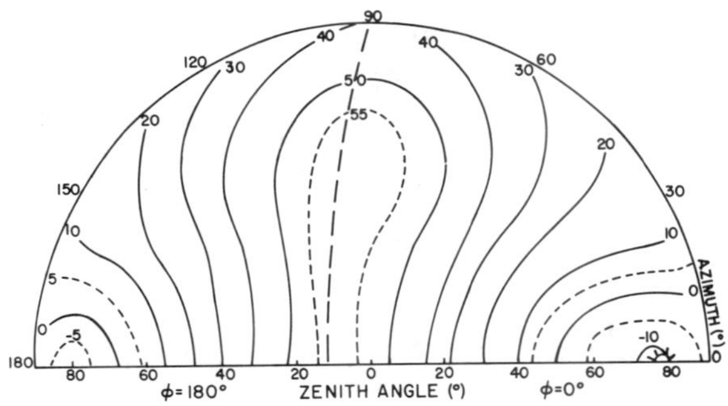

Figure 4.12 Distribution of the degree of polarization over the hemisphere of the sky for a wavelength of 0.312 μm (τ=1.0) and solar zenith angle of 78.5°. For an explanation of the geometry, see caption of Fig. 4.11.

diagrams are from the tabulations of Coulson et al. (1960). The isopleths have been drawn in by eye and labeled in percent. Only one-half of the hemisphere is shown, the other half being a mirror image of this one. The plane of the sun's vertical is represented by the straight bottom line, with 0 being the zenith and the star indicating the position of the sun. The outside semicircle is the horizon, with azimuth angle being taken with respect to the azimuth of the sun.

For the case of $\lambda = 0.644$ μm shown in Fig. 4.11, the maximum polarization of more than 95% stretches in an unbroken band across the sky from a point just on the antisolar side of the zenith in the sun's vertical to the horizon at an azimuth about 90° from that of the sun. The total variation of P in this band is only about 0.2%, with the highest P being near the horizon. The controlling parameter is scattering angle, as shown by the configuration of the curves around the sun and near the antisolar horizon. A line of 90° scattering angle, as computed for this geometry from the law of cosines for a spherical triangle, is shown by the heavy dashed line in the region of the maximum. Multiple-scattering effects are relatively small for this red wavelength, so the maximum is at, or just slightly less than, a scattering angle of 90°. Lines of zero polarization occur in the region surrounding the sun and near the antisolar horizon, with the Babinet and Arago points located at the points these zero lines intersect the sun's vertical. We cannot consider the regions inside these lines as negatively polarized, since the concept of negative polarization is merely a convenience arising from the definition of the Rubenson degree of polarization (Eq. 4.23) applicable only to the sun's vertical or the plane of scattering. Outside of this plane the state of linear polarization is given in terms of the three Stokes parameters $I_{\parallel}$, $I_{\perp}$, U by the relations

$$P = \frac{[(I_{\perp} - I_{\parallel})^2 + U^2]^{1/2}}{(I_{\perp} + I_{\parallel})}$$

$$U = (I_{\perp} - I_{\parallel}) \tan 2\chi$$

where χ is the angle of the plane of polarization taken with respect to the vertical direction. The degree of polarization P is considered to be a positive quantity from the square root relation, so only in the case $U=0$ does a negative sign of P have meaning. In general $U=0$ primarily in the sun's vertical, although, as discussed below, $U=0$ at certain other lines over the hemisphere whose positions depend on τ and θ_o.

A somewhat different distribution of polarization over the sky occurs at ultraviolet wavelengths, as shown for $\lambda = 0.312$ μm ($\tau = 1.0$) in Fig. 4.12. Multiple scattering keeps the maximum at less than 58%, causes P_{max} to be located outside of the sun's vertical about midway between the sun's vertical and horizon, and causes the highest polarization to occur at scattering angles of less than 90°. Although scattering angle is obviously important here also, it does not control the field to the same extent as it does at longer wavelengths. Multiple scattering at the shorter wavelengths causes an increase in size of the regions enclosed by lines of $P=0$, with an attendant movement of the Babinet and Arago points in the directions away from the sun and antisolar point, respectively. Fig. 4.12 shows the horizons to have lower polarization values than the region 20° to 30° above the horizon at virtually all azimuths at this solar angle, in contrast to the increased horizon polarization in the solar half of the hemisphere at longer wavelengths.

Figure 4.13 shows the hemispheric distribution of polarization for $\lambda=0.495$ μm ($\tau=0.15$) and zenith angle of the sun of 66.4°. The pattern here also shows the dominant influence of scattering angle, with a broad band of P_{max} of slightly greater than 85% stretching across the sky at a scattering angle of approximately 90°. In this case the sun elevation is such that the Arago point is no longer in evidence, but it has been replaced by the Brewster point just above the solar horizon. The pattern for $\lambda=0.495$ μm but for the sun relatively near the zenith ($\theta_o=23.1$°) is shown in Fig. 4.14. Except for a shift across the sky, the distribution is similar to that of the previous diagram. The band of P_{max} of about 90% is relatively near the horizon, as is the line of 90° scattering angle, and the region of the horizon is highly polarized at all azimuths.

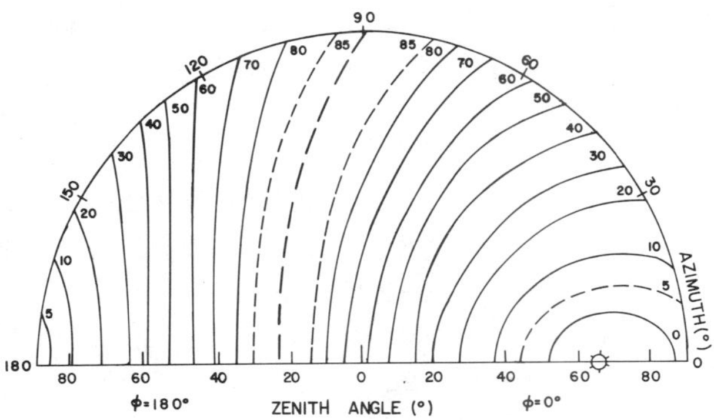

Figure 4.13 Distribution of the degree of polarization over the hemisphere of the sky for a wavelength of 0.495 μm ($\tau=0.15$) and solar zenith angle of 66.4°. For an explanation, see the caption of Fig. 4.11.

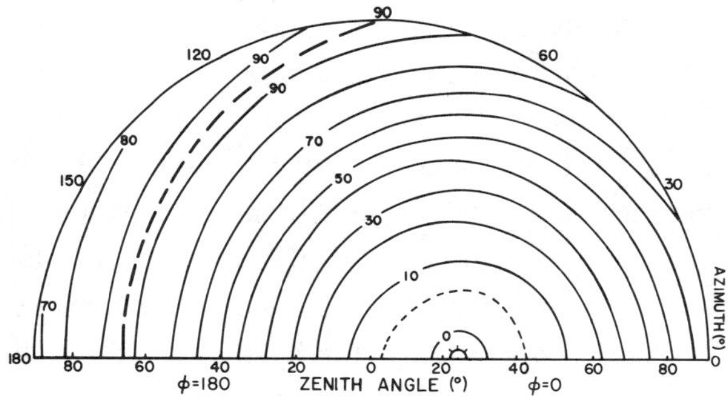

Figure 4.14 Distribution of the degree of polarization over the hemisphere of the sky for a wavelength of 0.495 μm ($\tau=0.15$) and solar zenith angle of 23.1°. For an explanation of the geometry, see the caption of Fig. 4.11.

If the sun were at the zenith, of course, P_{max} would coincide with the horizon at all azimuths, the pattern would be symmetric around the zenith direction, and the single neutral point would coincide with the sun at the zenith. For the $\theta_o=23.1°$ shown here, the Babinet and Brewster points are still separate, although they are both relatively close to the sun. Data for the sun at the zenith, for which case the field is independent of azimuth, are given for seven wavelengths (optical thicknesses) and 13 zenith angles of observation in Table 4.9. The entries are taken from the tabulations of Coulson et al. (1960), and rounded off to the fourth decimal place. Roundoff errors in the computations themselves result in significant deviations from zero at the zenith, particularly for the smallest optical thickness.

The above hemispheric distributions of polarization are considered as typical for a plane-parallel Rayleigh atmosphere with no reflection from the underlying surface. Data for a total of 49 such distributions are included in those tabulations, the parameters for which are seven solar zenith angles and seven optical thicknesses. The assumptions of plane parallelism for the atmosphere and no surface reflection are reasonably restrictive, of course, to say nothing of the neglect of dust, haze, and other particulate matter in the atmosphere through the Rayleigh scattering assumption. The effects of these assumptions will be considered further in later sections.

TABLE 4.9 Degree of polarization as a function of zenith angle of observation for seven different wavelengths (optical thicknesses at sea level) for the case of the sun at the zenith. The polarization distribution for the complete hemisphere is represented, as the field is independent of azimuth in this case. (Data from Coulson et al., 1960.)

	$\lambda(\tau)$ (μm)			
θ_o(°)	0.312(1.0)	0.372(0.50)	0.436(0.25)	0.495(0.15)
88.9	0.5787	0.7733	0.8871	0.9338
84.3	0.5532	0.7545	0.8708	0.9170
78.5	0.5265	0.7181	0.8237	0.8656
71.3	0.4808	0.6438	0.7314	0.7664
66.4	0.4393	0.5796	0.6540	0.6835
58.7	0.3633	0.4688	0.5231	0.5449
50.2	0.2763	0.3490	0.3850	0.3994
43.9	0.2150	0.2679	0.2933	0.3035
32.9	0.1219	0.1491	0.1609	0.1659
23.1	0.0605	0.0741	0.0783	0.0796
16.3	0.0302	0.0380	0.0380	0.0410
11.5	0.0150	0.0210	0.0186	0.0197
0.0	0.0003	0.0007	0.0035	0.0075

	0.546(0.10)	0.644(0.05)	0.809(0.02)
88.9	0.9573	0.9790	0.9914
84.3	0.9403	0.9608	0.9724
78.5	0.8874	0.9056	0.9157
71.3	0.7856	0.8001	0.8077
66.4	0.7007	0.7124	0.7176
58.7	0.5583	0.5662	0.5678
50.2	0.4095	0.4142	0.4123
43.9	0.3117	0.3140	0.3115
32.9	0.1716	0.1718	0.1664
23.1	0.0838	0.0837	0.0799
16.3	0.0429	0.0405	0.0429
11.5	0.0238	0.0280	0.0366
0.0	0.0044	0.0047	0.0274

4.5.5 Polarization in the almucantar of the sun

Several authors (e.g., Pyaskovskaya-Fesenkova, 1960) have measured the polarization distribution in the almucantar of the sun (small circle oriented horizontally through center of sun) as an indicator of atmospheric turbidity. The attractive feature of this procedure is that the pathlength through the atmosphere as seen by an observer at the surface is invariant in the solar almucantar, and thus interpretation of the data is facilitated. In the sun's vertical, on the other hand, or in any other plane in which zenith angle is variable, the change of the atmospheric pathlength with angle introduces complications in data analysis which are avoided in the solar almucantar. As far as is known, however, no computations for the Rayleigh atmosphere have been specifically for the solar almucantar, and polarization values in that case must be interpolated from hemispheric distributions such as those given in Figs. 4.11 to 4.14 above. This procedure degrades the precision of the final data, as a double graphical interpolation is required; one to develop the hemispheric polarization field and a second to obtain values from the field. The final values have a probable error of perhaps ±2% in most cases.

The degree of polarization in the solar almucantar at $\lambda=0.546\ \mu m$ ($\tau=0.10$) and six different solar zenith angles, as derived by graphical interpolation from the tables of Coulson et al. (1960), is plotted as a function of azimuth angle

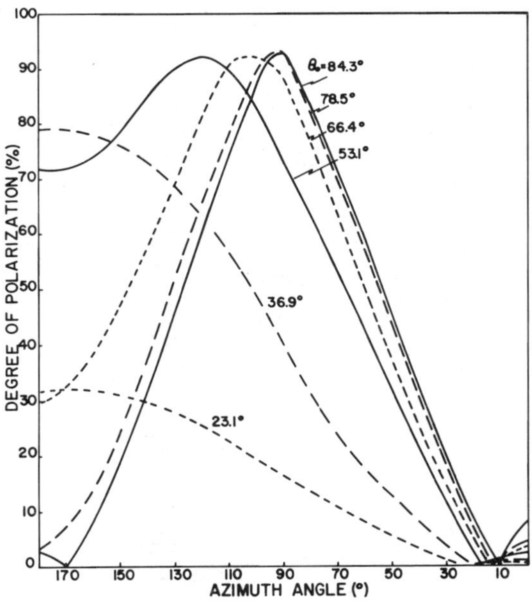

Figure 4.15 Degree of polarization as a function of azimuth in the solar almucantar for a wavelength of 0.546 μm (τ=0.10) for six different zenith angles of the sun.

taken with respect to the azimuth of the sun in Fig. 4.15. The four curves for $\theta_o \geq 53.1$ all reach definite maxima of about the same magnitude as those found in the sun's vertical. This could have been anticipated from the characteristic pattern of polarization over the hemisphere shown in Figs. 4.11 to 4.14. The geometry is such that a 90° scattering angle is achieved somewhere in the solar almucantar for all $\theta_o \geq 45°$. The dominant influence of scattering angle thus gives P_{max} for the solar almucantar of approximately the same magnitude as that for the sun's vertical, and in fact the polarization field in Fig. 4.12 gives P_{max} one to two percent higher in the solar almucantar than in the sun's vertical for the mid-range of zenith angles of the sun. For $\theta_o < 45°$, the scattering angle never reaches 90° anywhere in the solar almucantar, a fact which is responsible for the low polarization values shown for $\theta_o = 36.9°$ and 23.1° in Fig. 4.15. This means that measurements of polarization in the solar almucantar would be most useful for $\theta_o < 45°$, and indeed such measurements have been carried out at the relatively high-latitude regions of Europe and the Soviet Union where large solar zenith angles are prevalent throughout the day. Such measurements will be discussed in a later chapter.

The degree of polarization in the solar almucantar for three different wavelengths for a solar zenith angle of 78.5° is shown in Fig. 4.16. The strong dependence of P_{max} on wavelength, particularly in the ultraviolet region where multiple scattering is prevalent, is shown by the curves, as is the dominance of scattering angle in defining the field. These effects were seen in previous diagrams, and no unexpected features are evident here.

4.5.6 Dispersion of polarization

Polarization measurements made in the real atmosphere have shown the dispersion of polarization (i.e., the change of polarization with wavelength) to be a relatively sensitive indicator of atmospheric turbidity (Sekera, 1956). To interpret measurements of this quantity in natural atmospheric conditions, however, it is useful to determine polarization dispersion in a Rayleigh atmosphere and consider deviations from that as a perturbation due to other parameters such as atmospheric aerosols or surface reflection. The measurements of polarization dispersion have generally been concentrated on the polarization maximum, as are the modeling results discussed here. Discussions of perturbations in polarization dispersion arising from surface reflection and aerosol scattering are reserved for later sections.

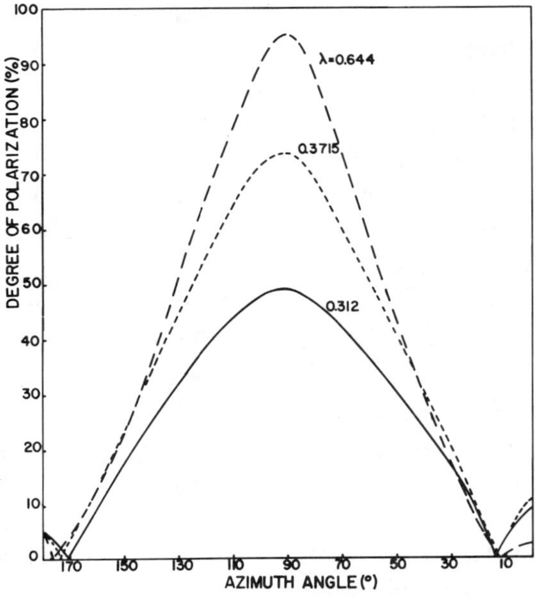

Figure 4.16 Degree of polarization as a function of azimuth in the solar almucantar for three different wavelengths and a solar zenith angle of 78.5°.

The dispersion of P_{max} for the plane-parallel Rayleigh atmosphere, as obtained by graphical interpolation in the tables of Coulson et al. (1960), is shown for three solar zenith angles θ_o in Fig. 4.17. The data are for the sun's vertical, and do not reflect the slightly higher P_{max} which occurs outside this plane in some cases. The change of P_{max} with θ_o, already listed explicitly in Table 4.5 above, is evident in these curves, but the major feature of Fig. 4.17 is the very strong polarization dispersion which occurs in the short wavelength region, with only modest dispersion at wavelengths above $\lambda = 0.50$ μm.

During the period of optical measurements before the time of electronic-type polarimeters, Jensen (1928) and other observers studied the dispersion of polarization by means of red and blue optical filters. While the effective wavelengths of such broadband filters are difficult to assess, the peak wavelength for the blue filter might be taken as 0.48 μm and that of the red filter as 0.65 μm. The curves of Fig. 4.17 show a change of P_{max} from approximately 88% to 96% over this wavelength range for $\theta_o = 53.1°$, and a comparable change for

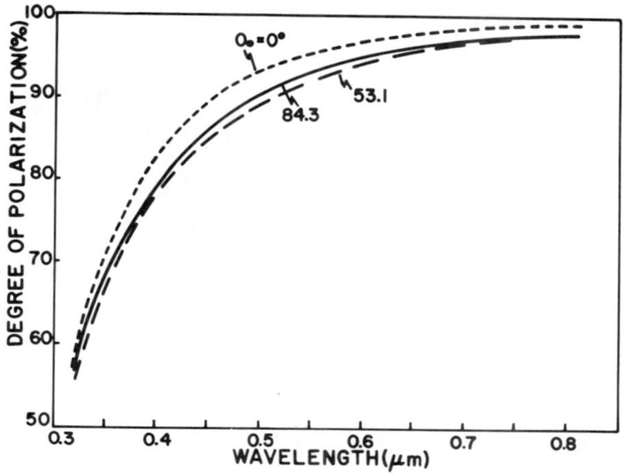

Figure 4.17 Dispersion of the polarization maximum in a plane-parallel Rayleigh atmosphere for three different solar zenith angles.

θ_o=84.3°. A larger change, and presumably more sensitive indicator of atmospheric conditions, would result from the use of shorter wavelengths, particularly for the blue, but then problems with sensitivity of the eye might interfere with the measurement. A greater total change of P_{max} could be achieved by taking advantage of the increasing slope of the curves with decreasing wavelength by the use of modern electronic polarimeters. In fact, Sekera (1956) took measurements at λ=0.46 and 0.625 μm as representative of blue and red, respectively, but even a farther shift of the blue toward short wavelengths would probably enhance the sensitivity of the polarization dispersion as an indicator of atmospheric conditions.

4.5.7 Effects of molecular anisotropy

The general theory of molecular anisotropy was covered in section 4.3.4. Here we discuss its effects in the plane-parallel Rayleigh atmosphere. Its effects have generally been considered small in relation to other depolarizing effects in the actual atmosphere, and have often been neglected in light-scattering computations. However, Dave (1964) emphasized the need to include molecular anisotropy in radiative transfer computations, and Hansen and Hovenier (1974) included the effect in an analysis of the atmosphere of Venus. In a later study,

Bahethi and Fraser (1980) computed the effects of molecular anisotropy at wavelength $\lambda = 0.55$ μm in a clear Rayleigh atmosphere and a Rayleigh atmosphere with scattering by haze and absorption by ozone.They took into account both the change of optical thickness and the modifiction of the scattering phase matrix due to molecular anisotropy. The four models, all of which were plane parallel and overlying a Lambert surface of albedo A, are those given in Table 4.10. The optical thickness due to molecules (τ^R), aerosols (τ^A), and ozone (τ^O) are included in the table. The overall effect of anisotropy was to increase τ for the Rayleigh atmosphere by 5.3% and for the turbid atmosphere by 1.2%.

In conjunction with an increase of optical thickness by molecular anisotropy, there was also a change of the angular distribution of the intensity and polarization of the scattered radiation. A summary of the changes of intensity produced by molecular anisotropy in the plane of the sun's vertical for radiation scattered upward from the top and downward from the bottom of the atmospheric models is given in Table 4.11. Three main points are evident in the data. First, as would

TABLE 4.10 Model atmospheres used by Bahethi and Fraser (1980) for computations of the effects of molecular anisotropy on intensity and polarization of scattered radiation.

Model		Anisotropy	τ^R	τ^A	τ^O	Total	% increase
Rayleigh	*A*	without	0.0924	0	0	0.0924	—
	B	with	0.0973	0	0	0.0973	5.3
Turbid	*C*	without	0.0924	0.2979	0.0269	0.4172	—
	D	with	0.0973	0.2979	0.0269	0.4221	1.2

TABLE 4.11 Percentage change of the intensity of radiation upward from the top and downward from the bottom of the atmospheric models of Table 4.9 computed by Bahethi and Fraser with and without molecular anisotropy. Lambert reflectivity of the underlying surface is given by A.

			$\theta = (°)$								
			$\phi = 0°$					$\phi = 180°$			
Radiation direction	A	Models	80	60	40	20	0	20	40	60	80
Up	0	*A* to *B*	3.5	5.4	6.8	7.3	5.5	4.4	3.5	3.3	2.7
Up	0	*C* to *D*	-1.0	0.6	2.1	3.2	3.2	2.8	2.2	1.4	1.3
Down	0	*A* to *B*	2.5	3.3	3.8	4.5	6.0	6.8	6.8	5.4	3.2
Down	0	*C* to *D*	-0.9	-1.0	-0.4	0.2	1.3	2.6	3.2	2.7	1.4
Up	0.50	*C* to *D*	-0.9	-0.1	0.1	0.1	0.1	0.2	0.2	0.1	0.5
Down	0.50	*A* to *B*	2.8	3.4	3.8	4.5	5.6	6.3	5.9	5.1	3.4
Down	0.50	*C* to *D*	-0.8	-1.0	-0.3	0.5	1.6	2.5	2.8	1.8	1.1

be expected, molecular anisotropy is more effective in changing the intensity in the Rayleigh atmosphere than in the turbid atmosphere at essentially all angles of scattering. For instance, the maximum change of 7.3% in the first case is to be compared with 3.2% in the latter. Only for the case of upward radiation near the nadir direction over a highly reflecting surface is there an indication of greater effect in the turbid atmosphere, but the maximum difference of only 0.2% may be within computational error and not significant. Secondly, the maximum effect of molecular anisotropy occurs in the direction near right-angle scattering in both upward and downward streams of radiation. Finally, a highly reflecting surface almost eliminates the effect of molecular anisotropy in the upward radiation at the top of the atmosphere. In this case a large part of the upward stream is due to radiation reflected from the surface and transmitted directly on its outward traverse of the atmosphere, and, since this component has not been subject to atmospheric scattering, its addition to the total reduces the fractional effect due to molecular anisotropy. As this direct component is not included in the downward stream at the bottom of the atmosphere, however, the effect of molecular anisotropy in the downward stream is not so strongly influenced by a highly reflecting surface.

The effects of molecular anisotropy on the degree of polarization are shown for radiation scattered upward from the top and downward from the bottom of the model atmospheres in Figs. 4.18 and 4.19, respectively. The maximum depolarization occurs for approximately right-angle scattering for both upward and downward streams and for both Rayleigh and turbid atmospheres. The magnitude of the maximum depolarization is 6.9% for both cases of the Rayleigh atmosphere and 3.5% for the turbid atmosphere. The effect becomes negligible as the scattering angles approach 0° or 180°, and the positions of the neutral points are essentially unaffected by molecular anisotropy.

Although molecular anisotropy must be considered for accurate radiative transfer calculations, its effects on both intensity and polarization of scattered light are relatively modest in comparison to those of scattering by atmospheric aerosols and reflection from the underlying surface. The strong depolarizing effects of aerosols is made evident by comparing curves *A* and *C* in Figs. 4.18 and 4.19, in which case aerosol scattering decreases the polarization by approximately 46%. This is to be compared with the 3.5% and 6.9% for molecular anisotropy. Furthermore, aerosol scattering causes a significant shift of the posi-

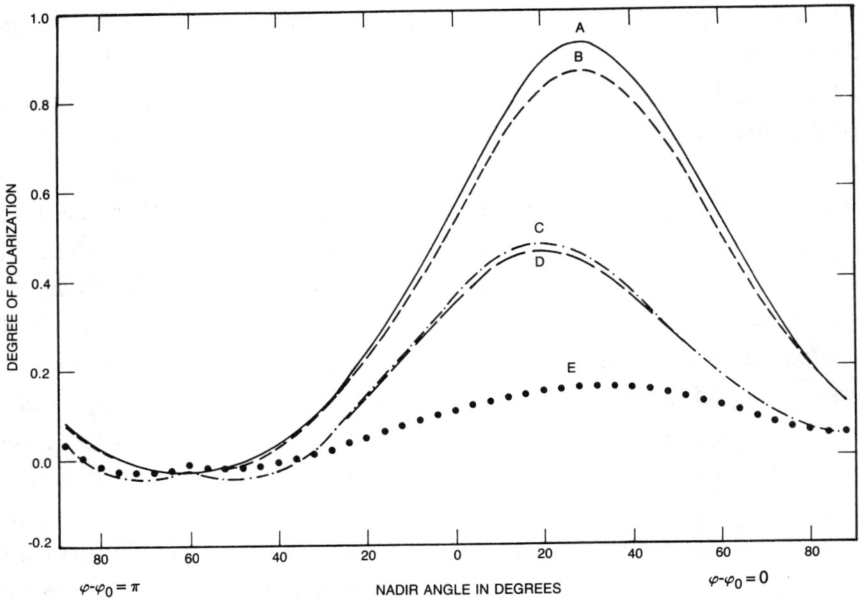

Figure 4.18 Degree of polarization of earthlight (radiation directed to space) from the top of various atmospheric models with (models B, D, and E) and without (models A and C) molecular anisotropy. Models A and B are for the plane-parallel Rayleigh atmosphere, and models C, D, and E are for a turbid atmosphere. Surface reflectivity A=0 for models A, B, C, and D, but A=0.25 for model E. (After Bahethi and Fraser, 1975.)

tions of neutral points. An additional decrease of 10% for the downward radiation, and more than 30% for upward radiation, is contributed by reflection from a surface of a moderate 25% albedo. This is not to say that the effects of molecular anisotropy are negligible, but they do have to be put in perspective.

4.6 Effects of surface reflection

4.6.1 Lambert surface reflection

A portion of the solar radiation incident at the top of the atmosphere is transmitted by the atmosphere to the surface, where a further portion is reflected from the surface and enters the solar regime of the atmosphere directed in the upward direction from the bottom. A part of this surface-reflected radiation

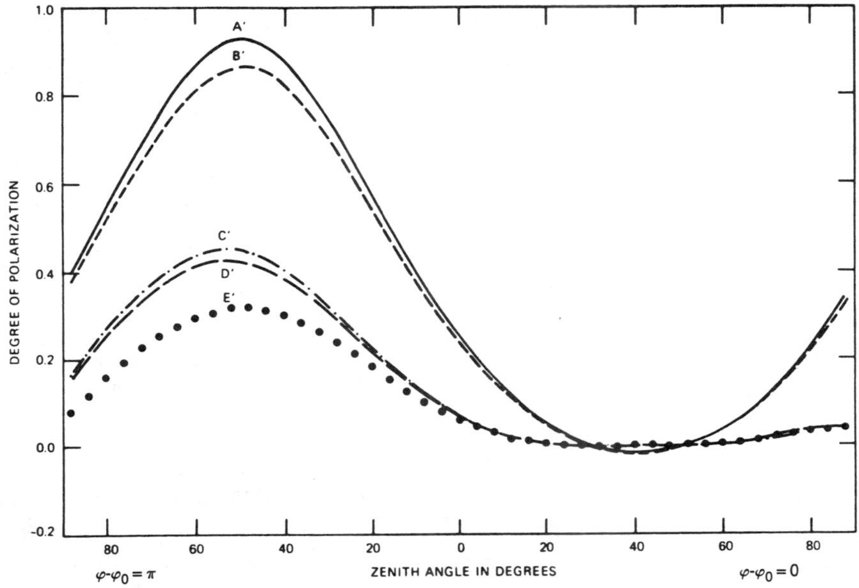

Figure 4.19 Same as for Fig. 4.18 except this case is for skylight (radiation directed downward at the bottom) from the model atmospheres. (After Bahethi and Fraser, 1975.)

emerges to space from the top of the atmosphere, after being transmitted either directly or diffusely in the outward traverse, and is of great importance in problems in remote sensing of surface properties from satellites. This aspect will be discussed in Chapter 5. Another part of the surface-reflected radiation is returned to the surface by scattering within the atmosphere, and thus adds to surface irradiation a component of surface-reflected radiation. This component is likewise subject to backscatter by the atmosphere, and in principle some of the radiation undergoes an infinite number of reflection-backscatter cycles between surface and atmosphere.

The geometry of the surface reflection problem is shown in Fig. 4.20. Parallel neutral radiation of net flux πF_o per unit time, unit frequency interval, and unit area normal to its direction of propagation is incident from zenith angle θ_o on the top of a plane parallel Rayleigh atmosphere of total optical thickness τ_1. The radiation that emerges from the bottom of the model atmosphere can be conveniently divided into four components, each component having its own history

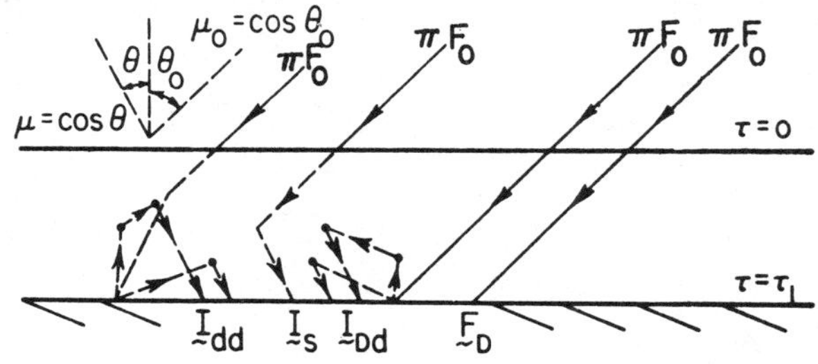

Figure 4.20 Schematic diagram of the four different components of solar radiation which are incident at the ground.

of transmission, reflection, and scattering by the atmosphere-surface system. The four components are listed as follows, where subscript D denotes direct transmission and subscript d either diffuse transmission or diffuse backscatter of surface-reflected radiation. In cases of double subscripts, the first refers to the initial traverse of the atmosphere, and the second to a subsequent process.

COMPONENT NO. 1: Radiation that is transmitted directly through the atmosphere, the original parallel radiation having suffered only attenuation in the process.

This radiation is given by $\pi\mu_o\mathbf{F}_o(-\mu_o,\phi_o)e^{-\tau_1/\mu_o}$ where $\mathbf{F}_o\ (-\mu_o,\phi_o)$ is a column matrix of the Stokes parameters (F_e,F_r,F_u,F_v). Here $\theta_o=\cos^{-1}\mu_o$ is the zenith angle, ϕ_o the azimuth of the direction of propagation, and e and r refer to the directions in the plane normal to the direction of propagation of the radiation which are parallel and perpendicular, respectively, to the vertical plane through the direction of propagation. Since the radiation is unpolarized, $F_e=F_r=½F_o$ and $F_u=F_v=0$. It is assumed throughout that no elliptic polarization is introduced, in which case $F_v=V=0$ and no further discussion is required.

The direct extra-atmosphere incident flux of $\pi\mu_oF$ units is attenuated during its downward atmospheric traverse and reaches the surface as

$$\mathbf{F}_D(\tau_1;-\mu_o,\phi_o) = \pi\mu_o\mathbf{F}_o(-\mu_o,\phi_o)\ e^{-\tau_1/\mu_o} \tag{4.76}$$

where τ is the wavelength-dependent normal optical thickness of the atmosphere.

COMPONENT NO. 2: Radiation that is transmitted directly through the atmosphere, reflected from the surface, and backscattered again to the surface by the atmosphere.

The outward intensity $I_{Dg}(\tau;+\mu,\phi)$, at the nadir angle $\theta=\cos^{-1}\mu$ and azimuth ϕ, at the ground due to the surface reflected part of F_D is

$$\mathbf{I}_{Dg}(\tau;\mu,\phi) = \frac{1}{2}\mathbf{F}_D\rho(\mu,\phi;\mu_o,\phi_o)\ \mathbf{N}(\mu,\phi;-\mu_o,\phi_o) \tag{4.77}$$

where $\mathbf{N}(\mu,\phi;\mu_o,\phi_o)$ is a column matrix with elements $(1+P\cos 2\chi$, $1-P\cos 2\chi$, $2P\sin 2\chi)$. The degree of polarization P, the angle χ between the vertical plane and the plane of polarization, and the bidirectional reflectance ρ are all obtained from measurements of the reflection properties of a given surface. The part of $\mathbf{I}_{Dg}$ that is backscattered by the atmosphere and is again incident at the surface, but this time as diffuse radiation, is

$$\mathbf{I}'_{Dg}(\tau_1;-\mu,\phi) = \frac{1}{4\pi\mu}\int_o^{2\pi}\int_o^{1}\mathbf{S}(\mu,\phi,\mu',\phi')\ \mathbf{I}_{Dg}(\mu',\phi')\ d\mu' d\phi' \tag{4.78}$$

The scattering matrix $\mathbf{S}$, first defined by Chandrasekhar (1950) and discussed in section 4.4 above, requires an involved but basically not difficult computational procedure. Because of the complicated nature of $\mathbf{S}$ and the necessity of introducing data on $\mathbf{I}'_{Dg}(\mu',\theta')$ at discrete directions in the outward hemisphere, numerical integration techniques are normally required for determining $\mathbf{I}'_{Dg}(\tau,-\mu,\phi)$.

COMPONENT NO. 3: Radiation that is scattered one or more times on its downward traverse of the atmosphere and reaches the surface in the form of diffuse skylight.

The intensity $\mathbf{I}_s(\tau_1;-\mu,\phi)$ of this diffusely transmitted radiation is (cf. Chandrasekhar, 1950)

$$\mathbf{I}_s(\tau_1;-\mu,\phi) = \frac{1}{4\mu}\ \mathbf{T}(\mu,\phi;\mu_o,\phi_o)\ \mathbf{F}_o(-\mu_o,\theta_o) \tag{4.79}$$

The atmospheric transmission matrix $\mathbf{T}(\mu,\phi;\mu_o,\phi_o)$ was also discussed in section 4.4.

COMPONENT NO. 4: Radiation that is diffusely transmitted on its downward traverse of the atmosphere, reflected from the surface, and backscattered again to the surface.

We also include in this component the diffusely transmitted radiation that has undergone two or more reflection-backscatter cycles. The upward intensity at the ground of the surface-reflected part of this diffuse radiation $I_d(-\mu,\phi)$ is

$$\mathbf{I}_{dg}(\tau_1;+\mu,\phi) = \frac{1}{2}\int_o^{2\pi}\int_o^1 \rho(\mu,\phi;\mu',\phi')\mathbf{I}_d(-\mu',\phi')\mathbf{N}(\mu,\phi;\mu',\phi')\mu' d\mu' d\phi' \tag{4.80}$$

The part of $I_{dg}(\tau_1;+\mu,\phi)$ that is returned to the surface by backscattering in the atmosphere is

$$\mathbf{I}_{dd}(\tau_1;-\mu,\phi) = \frac{1}{4\pi\mu}\int_o^{2\pi}\int_o^1 \mathbf{S}(\mu,\phi;\mu',\phi')\mathbf{I}_{dg}(\mu',\phi')d\mu' d\phi' \tag{4.81}$$

There are two difficulties in evaluating $\mathbf{I}_{dd}(\tau_1;+\mu,\phi)$. First, the combination of (4.48) with (4.49) constitutes a quadruple integration of an integro-differential equation, a process which is very expensive in computer time. The second and more fundamental difficulty is that the available data on the reflection properties of surfaces are generally insufficient for determining $\rho(\mu,\phi;\mu',\phi')$ and $\mathbf{N}(\mu,\phi;\mu',\phi')$ for all of the required combinations of incidence and reflection angles.

Fortunately, there is a very simple method of determining the scalar form $I_{dd}(\tau_1;-\mu)$ of this diffuse component. If we assume for this one component Lambert-type reflection from the ground, then we can write (cf. Chandrasekhar, 1950)

$$I_{dd}(\tau_1;-\mu) = \frac{A\mu_o F_o}{1 - A\bar{s}}\left[\frac{\gamma_e(\mu_o) + \gamma_r(\mu_o)}{2} - e^{-\tau_1/\mu_o}\right]\left[1 - \frac{\gamma_e(\mu) + \gamma_r(\mu)}{2}\right] \tag{4.82}$$

where A is the total reflectance of the surface, given by

$$A = \int_o^{2\pi}\int_o^1 \rho\mu d\mu d\phi \tag{4.83}$$

In Eq. (4.82) $\bar{s}$ is a function of only optical thickness and $\gamma_e(\mu_o)$, $\gamma_r(\mu_o)$, $\gamma_e(\mu)$, and $\gamma_r(\mu)$ are functions of both optical thickness and wavelength. Values of $\bar{s}$ and the γ's are tabulated by Chandrasekhar and Elbert (1954), but they may also be readily computed from the original expressions.

The only flux attributable to the original directly transmitted radiation which is not included is either $\mathbf{F}_o(\tau_1;-\mu_o,\phi_o)$ or $\mathbf{I}_{Dg}(\tau_1;-\mu,\phi)$ is a small flux $F'_{Dd'}(\tau_1;-\mu,\phi)$ resulting from reflection-backscatter cycles of orders higher than the first, i.e., all orders of reflection and backscatter of $\mathbf{I}_{Dd}(\tau_1;-\mu,\phi)$. The flux $F'_{Dd}(\tau_1;-\mu)$ of the original diffuse component is

$$F_{Dd}(\tau_1;-\mu) = \int_o^{2\pi}\int_o^1 I_{Dd}(\tau_1;-\mu,\phi)\ \mu d\mu d\phi \tag{4.84}$$

and its reflection from a Lambert surface gives the isotropic intensity in the outward direction

$$(I_{dd})_g(\tau_1;+\mu) = \frac{A}{\pi}F_{Dd}(\tau_1;-\mu) \tag{4.85}$$

The downward flux due to the backscatter of $(I_{Dd})_g$ for the first cycle is

$$F_{Dd}^{(1)}(\tau_1;-\mu) = \pi\bar{s}(I_{Dd})_g\ (\tau_1;+\mu) = A\bar{s}F_{Dd} \tag{4.86}$$

and equivalent expression for the nth cycle is

$$F_{Dd}^{(n)} = (A\bar{s})^n F_{Dd} \tag{4.87}$$

Since $A \leq 1$ and $\bar{s} < 0.5$, this represents a convergent series, and as $n \rightarrow \infty$, the total contribution due to reflection and backscatter of F_{Dd} is obtained from the sum of terms in a geometrical progression as

$$F_{Dd'}(\tau_1;-\mu) = F_{Dd}\left(\frac{A\bar{s}}{1 - A\bar{s}}\right) \tag{4.88}$$

In general, $F_{Dd'}$ is a very small part of the total flux incident on the surface. For instance, for A=0.25 and $\bar{s}$=0.296 (for τ_1=0.50), the computations show that $F_{Dd'}$ is only about 0.006 F_D for θ_o=53°. For smaller optical thicknesses the ratio would be even less.

The whole problem of the effects on skylight of radiation reflected from the surface is greatly simplified if it is assumed that the surface follows Lambert's law of reflection. This means that the surface reflects a fixed fraction of A of the incident radiation and the reflected radiation is unpolarized and isotropic in the outward hemisphere. Because of its convenience in computational schemes, the Lambert approximation has found wide application by many authors, in spite of the fact that natural surfaces on the earth, and presumably other planets, are known to deviate strongly from the Lambertian character in some cases. The approximation is worse for the earthlight (i.e., radiation directed to space from the top of the atmosphere) than it is for skylight. Other types of reflection, such as Fresnel reflection or reflection determined by laboratory measurements, have been considered by various authors and will be discussed further below. However, it is useful to see the effects introduced in the skylight regime of a Rayleigh atmosphere by Lambert surface reflection.

For the case of Lambert reflection we can define two orthogonal intensity components of skylight $I_e^*(\tau;\mu,\mu_o)$ and $I_r^*(\tau;\mu,\mu_o)$ which are due entirely to surface reflections, and which augment the corresponding components of skylight in the absence of surface reflection. For these additional components we have the scalar forms

$$I_e^*(\tau;-\mu,\phi) = \frac{A\mu_o F_o}{4(1 - A\bar{s})}\left[\gamma_e(\mu_o) + \gamma_r(\mu_o)\right]\left[1 - \gamma_e(\mu)\right]$$

$$I_r^*(\tau;-\mu,\phi) = \frac{A\mu_o F_o}{4(1 - A\bar{s})}\left[\gamma_e(\mu_o) + \gamma_r(\mu_o)\right]\left[1 - \gamma_r(\mu)\right] \tag{4.89}$$

where the functions γ_e, γ_r and $\bar{s}$ are the same as in Eq. (4.82). Then the orthogonal components of skylight intensity are $I_e+I_e^*$ and $I_r+I_r^*$; the total intensity is

$$I(\tau;\mu,\phi;\mu_o\phi_o) = I_r + I_r^* + I_e + I_e^* \tag{4.90}$$

Stokes parameter Q is

$$Q = (I_r + I_r^*) - (I_e + I_e^*) \tag{4.91}$$

and the degree of polarization

$$P = \frac{(Q^2 + U^2 + V^2)^{1/2}}{I} \tag{4.92}$$

These last three quantities have been tabulated in Coulson et al. (1960), which is the source of some of the data for the figures which follow. Also used are some unpublished computations kindly made available by Dr. Bruce W. Fitch (1977).

The degree of polarization of skylight at $\lambda=0.650$ μm ($\tau=0.047$) is shown as a function of zenith angle in the sun's vertical of a plane-parallel Rayleigh atmosphere for four different values of Lambert surface reflectance and two zenith angles of the sun in Fig. 4.21. For the sun near the horizon ($\theta_o=78.5°$ — solid curves) the effect of surface reflection is relatively modest, causing a decrease of P_{max} of approximately 10°, 18°, and 26° for $A=0.25$, 0.50. and 0.75, respectively, as the unpolarized radiation from the surface is backscattered by the overlying atmosphere. Only very minor depolarization effects occur throughout much of the plane, and the addition of unpolarized light to the already unpolarized light in the direction of the neutral points has essentially no influence on the positions of the neutral points.

A considerably greater effect of surface reflection occurs for the case of the sun nearer the zenith, as shown for $\theta_o=23.1°$ by the dashed curves of Fig. 4.21. Three factors combine to increase the influence of surface reflection for high sun elevations over that which occurs for the sun nearer the horizon. First, even in the absence of an atmosphere, the surface irradiance, being proportional to $\cos\theta_o$, is stronger for high than for low sun elevations. For the case shown, the ratio is cos 23.1°/cos 78.5°= 4.61. Second, the ratio of transmis-

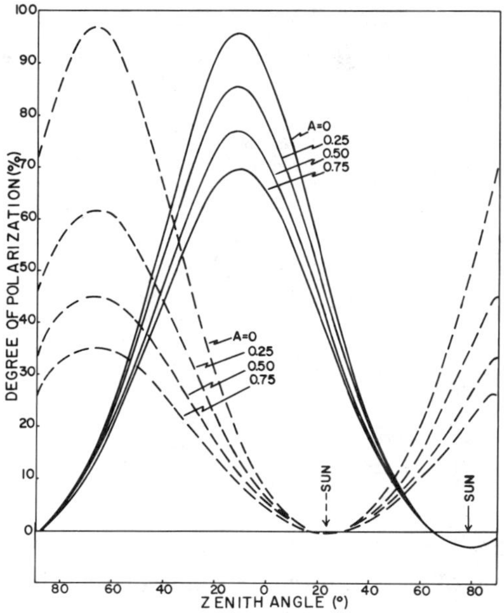

Figure 4.21 Degree of polarization as a function of angle in the sun's vertical in a plane-parallel Rayleigh atmosphere at a wavelength of 0.65 μm for two different solar zenith angles (θ_o=78.5°—solid curves; θ_o=23.1°—dashed curves) and four different values of Lambert surface reflectance. (Data from Fitch, 1977.)

sions of the direct solar beam, being proportional to $e^{(\tau/\cos\theta_o)}$, is 0.950/0.790=1.20 for the two angles considered. This means that the direct beam irradiance of the surface, by far the dominant source for this small optical thickness, is 5.53 times as strong for $\theta_o=23.1°$ as for $\theta_o=78.5°$. A third factor is that the fraction of the incident beam which eventually shows up as scattered (and highly polarized) skylight is less for $\theta_o=23.1°$ than $\theta_o=78.5°$. The combination of all these factors produces very strong depolarization of skylight by Lambert surface reflection for high sun elevations and low optical thicknesses, as shown by the dashed curves of Fig. 4.21. Here also, however, the positions of the neutral points are essentially unchanged by Lambert surface reflection. The small increase in depolarization shown by the dashed curves near the solar horizon is probably an artifact due to the plane-parallel assumption.

The equivalent case for a wavelength of 0.411 μm (τ=0.333) is shown in Fig. 4.22. The more pronounced multiple scattering of this shorter wavelength decreases the polarization maximum and shifts the neutral points from their

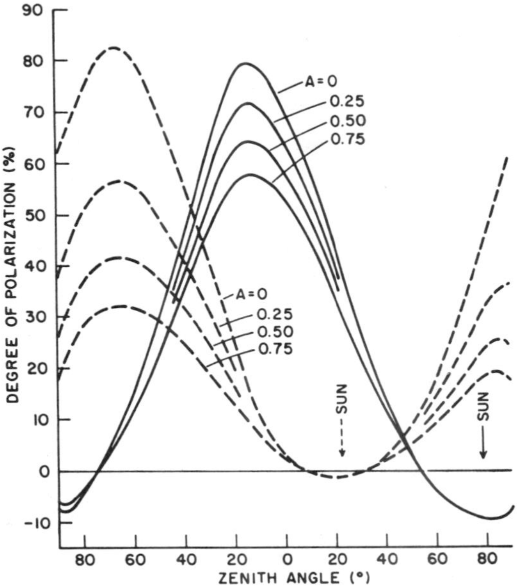

Figure 4.22 Degree of polarization as a function of angle in the sun's vertical in a plane-parallel Rayleigh atmosphere at a wavelength of 0.411 μm for two different solar zenith angles (θ_o=78.5°—solid curves; θ_o=23.1°—dashed curves) and four different values of Lambert surface reflectance. (Data from Fitch, 1977.)

positions for λ=0.65 μm, but the effects of surface reflection are qualitatively the same for the two wavelengths. Quantitatively, the depolarization effects of surface reflection are somewhat suppressed by stronger molecular scattering of the incident solar beam at the shorter wavelength here than in the previous case, but the overall configurations are not significantly different.

The dispersion of the polarization maximum is shown for the four different surface reflectances at two solar zenith angles in Fig. 4.23. The long atmospheric pathlength of the incident beam for ϕ_o=78.5° results in a strong dispersion of P_{max} because of the negative fourth-power wavelength dependence of Rayleigh scattering and the fact that relatively little of the scattered radiation arises from the surface reflection, as outlined above. For the sun near the zenith, however, the atmospheric pathlength of the solar beam is short, the surface is strongly irradiated, and a dominant portion of the scattered light consists of

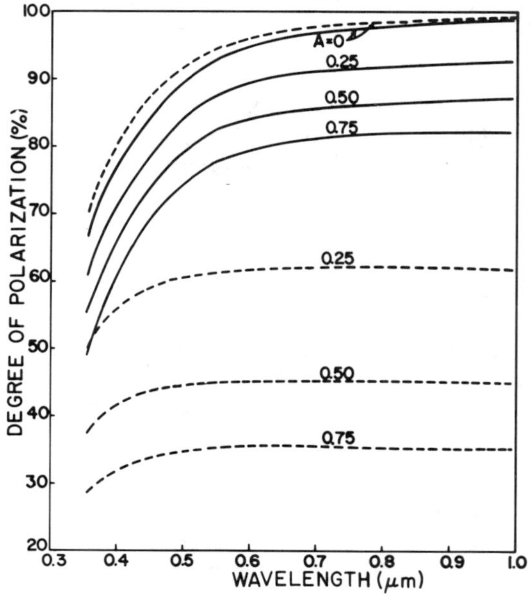

Figure 4.23 Dispersion of the polarization maximum for four different surface reflectances for two solar zenith angles (θ_o=78.5°—solid curves; θ_o=23.1°—dashed curves) in a plane-parallel Rayleigh atmosphere.

the diffuse and unpolarized contribution from the Lambert reflecting surface. Since the surface reflectance is assumed to be independent of wavelength, little dispersion of P_{max} results from scattering of this surface contribution. Only at the shortest wavelengths, where scattering of the original incident radiation is very strong, is there appreciable polarization dispersion for high sun elevations, even for the modest surface reflectance of $A=0.25$. It should be realized, of course, that most real surfaces do not correspond closely to the Lambert model and that appreciable polarization and directionality in the reflected radiation would modify the picture from that shown here. The Lambert approximation is useful, however, for indicating the influence of surface reflection in the overall skylight regime.

Studies of the radiative effects of Lambert surface reflection have generally been based on the assumption that the surface is of infinite horizontal extent with the reflected radiation isotropically distributed throughout the outward hemisphere. The problem of discontinuous surface properties has not been solved in any detail. However, a step in this direction was taken in a study by Coulson (1968), in which the angular reflectance pattern of a Lambert surface was taken as discontinuous, with the same discontinuous pattern applying to every position on the infinite plane. Each small element of the surface was assumed to have a Lambert reflectance of A=0.80 through azimuths 0°–90° and 270°–360°, while the reflectance through azimuths 90°–270° was 0.05. An observer standing at an arbitrary point on the surface would see one-half of the infinite plane as it might look with a cover of new snow, while the opposite half plane would appear as dark as a black loam soil. The dividing line between the two types of surface would always be in the 90°–270° plane through the observer's position, regardless of where it was located on the surface. The problem was to determine the characteristics of the portion of the skylight due to atmospheric backscattering of the surface-reflected radiation for this model. A unit normal flux of unpolarized incident radiation was assumed, the angle of incidence being immaterial as the Lambert assumption obviates any dependence on incident angle.

The method employed for the computations was that outlined in connection with Eqs. (4.76) to (4.88) above. Both intensity and polarization of the skylight component due to surface reflection were computed. Values of the intensity would be required in order to combine this surface-reflected component with the skylight resulting from atmospheric scattering in the absence of a surface to get the total, but here we concentrate only on the polarization of the component resulting from surface reflection.

The distribution of the degree of polarization in the vertical plane at 0°–180° azimuths is shown for three different wavelengths in Fig. 4.24. Included also is a schematic diagram of the assumed surface reflectance. The effect of scattering angle is evident in the polarization distribution, the maximum polarization occurring at about θ=45° in both hemispheres. This is the approximate region at which the scattering angle for surface-reflected light would reach closest to right-angle scattering on a statistical basis, and would thus correspond closest to the polarizing angle for primary Rayleigh scattering. Low polarization values near the zenith and at the horizons is evidence of mainly backward and forward scattering, respectively, although multiple scattering effects are significant near the horizontal directions and are responsible for the negative values at θ=0. The generally lower polarization values at λ=0.32 μm than at the longer wavelengths is a result of greater multiple scattering of short-wavelength radiation.

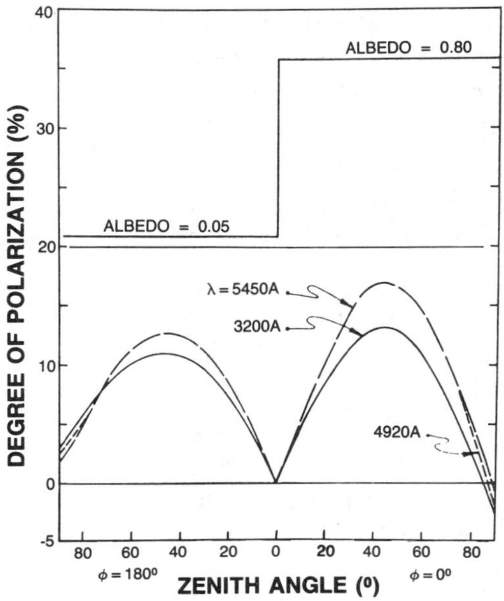

Figure 4.24 Degree of polarization of surface-reflected radiation of various wavelengths which is scattered into the downward hemisphere by an overlying Rayleigh atmosphere. Half of the hemisphere is assumed to have a Lambert reflectance of 0.80 and the other half a Lambert reflectance of 0.05.

The physical explanation of the higher polarization over the high reflectance half-plane than over that with low reflectance is inherent in the polarization pattern of primary Rayleigh scattering, and not in multiple scattering effects. The dominant contribution for scattering in both hemispheres of the sky is that leaving the surface from the high-reflectance half of the scattering pattern. The portion of that which is scattered back down is scattered mainly at intermediate scattering angles ($\Theta \approx 90°$) over the high-reflectance portion, but over the low-reflectance portion the scattering angles for light from the $A = 0.80$ portion are generally much greater ($\Theta \approx 180°$).

Since the Rayleigh scattering pattern gives high polarization at $\Theta \approx 90°$ and low polarization at $\Theta \approx 180°$, the skylight due to surface reflection has higher polarization over the section of $A = 0.80$ than over the low-reflectance section.

As far as is known, no computations equivalent to these have been made for more realistic and complicated discontinuities for Lambert surface reflection, but the effects on skylight of reflection from real surfaces such as soils, sands, and vegetation have been studied by several authors. These will be discussed in a later section. It is worth mentioning, however, that the data presented here and in previous diagrams show quite conclusively that surface reflection, even the unpolarized radiation from a Lambert surface, plays a very important role in determining skylight characteristics, and the partially polarized light reflected from bright natural surfaces has the potential for a still more important part in the skylight regime.

4.6.2 Fresnel surface reflection

A second theoretical law of reflection studied by various authors in connection with its effect on skylight is that of Fresnel (mirror-type) reflection, such as that of a surface of still water or smooth clear ice. As explained in any physics text which deals with Fresnel's laws, the reflectance of such a smooth, dielectric surface and the state of polarization of the reflected radiation are functions of the index of refraction m of the reflecting material, as well as angle of incidence and state of polarization of the incident radiation. For a Fresnel surface in a horizontal orientation which is illuminated by a pencil of parallel radiation from zenith angle θ_o and azimuth ϕ_o, the incident and reflected pencils are in the same vertical plane, and the reflected pencil, also parallel, is reflected into direction θ,ϕ in such manner that $\theta=\theta_o$, $\phi=\phi_o+180°$. The matrices of incident light $\mathbf{I}(-\mu_o)$ and reflected light $\mathbf{I}_g(\mu,m)$ are related by the reflection matrix $\mathbf{R}(\mu,m)$ through the equation

$$\mathbf{I}_g(\mu,m) = \mathbf{R}(\mu,m)\,\mathbf{I}(\mu_o) \tag{4.93}$$

where $\mathbf{R}(\mu,m)$ for linear polarization may be written as (Fraser, 1964)

$$\mathbf{R}(\mu,m) = R(\mu,m)\begin{vmatrix} q^2(\mu,m) & 0 & 0 \\ 0 & 1 & 0 \\ 0 & 0 & q(\mu,m) \end{vmatrix} \tag{4.94}$$

In this relation we have

$$R(\mu,m) = \left[\frac{\mu - (m^2 - 1 + \mu^2)^{1/2}}{\mu + (m^2 - 1 + \mu^2)^{1/2}}\right]^2 \tag{4.95}$$

$$q(\mu,m) = \frac{1 - \mu^2 - \mu(m^2 - 1 + \mu^2)^{1/2}}{1 - \mu^2 + \mu(m^2 - 1 + \mu^2)^{1/2}} \tag{4.96}$$

The intensity components, total intensity, and degree of polarization of the radiation $\mathbf{I}_g\,(\mu,m)$ reflected by a Fresnel surface such as still water, are shown as a function of angle of incidence $\theta_o = \cos^{-1}\mu_o$ in Fig. 4.25. It has been assumed that the incident light is unpolarized and of unit intensity, and that $m=1.346$. The well-known low reflectance of water for radiation at nearly normal incidence is shown by the diagram. For instance, the total reflectance is only 2%–3% for incidence angles of up to 40°, and it is only at angles beyond 65° that the reflectance reaches 10%. At still larger angles the total reflectance increases rapidly, and complete reflection is reached at grazing incidence $\theta_o=90°$.

The two orthogonal components I_{gr} and I_{ge}, which are due to vibrations, respectively, parallel and normal to the reflecting surface, undergo their own variations. While I_{gr} is a monotonically increasing function of incident angle,

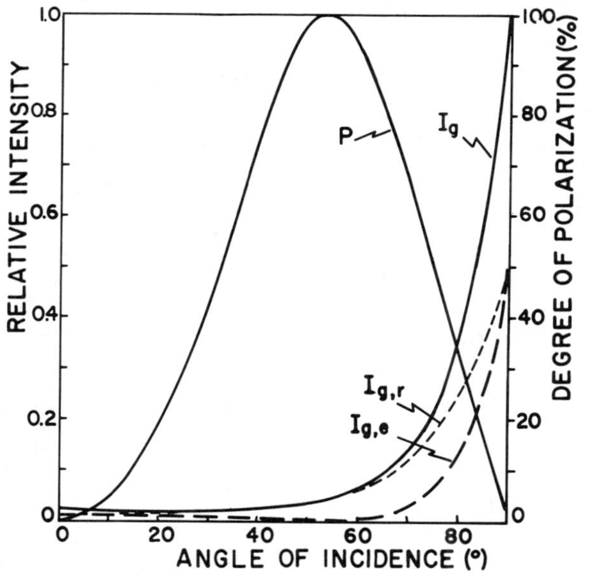

Figure 4.25 Total intensity, intensity components, and degree of polarization of light reflected from a smooth water surface as functions of angle of incidence of the light. The index of refraction of water of 1.346 gives the angle of incidence for complete polarization (the Brewster angle) of 53.4°.

I_{ge} first decreases with increasing θ_o, reaches zero at Brewster's angle (53.4° for $M=1.346$), and then increases to equal I_{gr} at $\theta_o=90°$. Since for this principal plane we have $U_g=0$ for the reflected light, the degree of polarization is given by the Rubenson relation, or

$$P_g = \frac{I_{gr} - I_{ge}}{I_{gr} + I_{ge}} \qquad (4.97)$$

This means that $P_g=0$ at normal incidence and at grazing incidence, where in both cases $I_{gr}=I_{ge}$, and $P=1.0$ at Brewster's angle, where $I_{ge}=0$. In fact this is the definition of Brewster's angle. At all other angles $0<P<1.0$, as shown by the curve.

There are several consequences of this reflection pattern for the polarization of skylight in a Rayleigh atmosphere overlying such a Fresnel surface. First, the total reflectance is so low at all incidence angles out to beyond the Brewster angle that very little radiation is provided by such a surface for subsequent scattering by the overlying atmosphere. Second, even though the reflected radiation may be highly polarized, its intensity is so low that its addition has little influence on the overall skylight polarization field. Finally, at large angles of incidence the reflectance of the Fresnel surface becomes high, but for the atmosphere-surface model under consideration, the total amount of energy reaching the surface at large θ_0 is strongly restricted by the geometric $\cos\theta_o$ factor and low atmospheric transmission mentioned previously in connection with Lambert surfaces. Thus the skylight effects of Fresnel reflection are generally minimal, as pointed out by Sekera (1961). The one case in which Fresnel reflection has an important effect is its influence on the positions of the neutral points. This and other aspects of neutral point behavior will be discussed in section 4.7.

Other types of surface reflection have been considered for specific problems. Takashima et al. (1976) computed the intensity and polarization of the radiation directed to space (i.e., the earthlight) from the top of an atmosphere overlying a surface of hybrid nature consisting of part-Lambert and part-Fresnel reflection elements, but the skylight was not included in the study. A surface with specular reflection from statistically oriented facets, similar to what has been observed for a rough sea surface, was assumed by Takashima (1983) for computing the intensity of the earthlight, but polarization was not included. Fraser

(1981) studied the effects on the neutral points of skylight of the reflection of radiation at $\lambda = 0.70\ \mu m$ from a wind-roughened sea surface, the reflection matrix for which was given by Mullamaa (1964). He found that this type of reflection caused the neutral points to move out of the sun's vertical as did most other cases of specular reflection, but the total change of polarization magnitude was very small. Fitch (1981) used laboratory measurements of the reflection properties of soil and sand to develop reflection matrices suitable for computing the intensity and polarization of earthlight for three different wavelengths and three solar zenith angles. The skylight was not included in his computations.

4.7 The neutral points of skylight polarization

It has been found by many researchers in atmospheric optics (e.g., Jensen, 1942; Dorno, 1919) that the positions of the neutral points (points at which the light is unpolarized or neutral) are sensitive indicators of the amount and type of atmospheric turbidity. Dramatic changes of the positions of these points were observed after the large volcanic eruptions of Krakatoa in 1883 and Mount Katmai in 1912, but less spectacular, yet consistent, changes have been found in conditions of low-level turbidity such as that of populated areas in Europe and elsewhere (Neuberger, 1940). More attention has been devoted to the measurement of neutral point positions than to any other feature of skylight polarization. Their demonstrated sensitivity to turbidity is a primary reason, but a very important secondary reason is the ease with which the neutral points can be observed. The Savart polariscope, a very simple device which was normally mounted on a protractor or quadrant with a plumb line for measuring angles, was used extensively over a period of more than a century for neutral point observations. Such measurements began before the middle of the 19th century (Babinet, 1840; Brewster, 1842, 1847, 1861) and continued up to the time the more versatile electronic polarimeters came into existence after World War II. In fact, the Savart polariscope is still sometimes employed as a convenient device for neutral point measurements (Volz, 1983), but being a visual-type instrument, the measurements are usually only in white light.

The positions of the three neutral points are shown schematically in Fig. 4.26. Under normal conditions they are located only in the sun's vertical. The Arago point (AR), named for its 1809 discovery by the French astronomer D.F.J. Arago, is located 20° to 25° above the antisolar point (AS), and its location in the portion

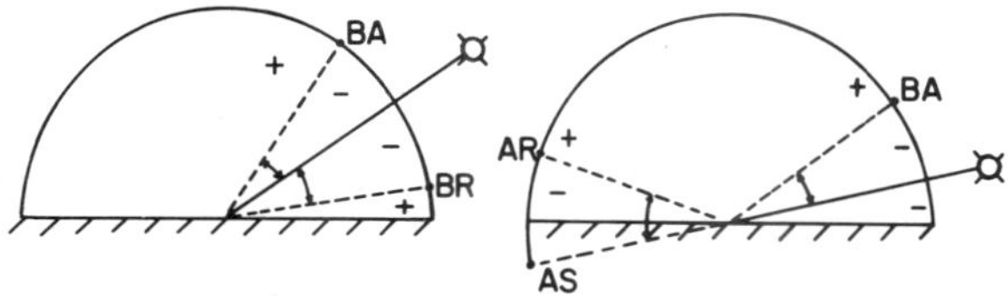

Figure 4.26 Schematic diagram showing the normal positions of the neutral points of Babinet (BA), Brewster (BR), and Arago (AR). The plane of the paper represents the sun's vertical, and the antisolar point is AS.

of the sky opposite the sun makes it the easiest of the three to observe. The other two points, first observed by the French meteorologist J. Babinet in 1840 and by the Scottish physicist Sir David Brewster in 1842, are located above and below the sun in the positions indicated. While the Babinet point, located 20° to 25° above the sun, is also readily observed by visual means, the Brewster point is embedded in the brightest and least polarized region of the sky, and is therefore difficult to observe visually. There is, however, no problem in observing any of the three by the electronic polarimeters in current use. Under normal atmospheric conditions, only two of the three are visible at a given time, as the Arago point sets below the horizon at the same time the Brewster point appears above the horizon, and vice versa. The Babinet point is, of course, visible from before sunrise until after sunset unless the sun is very close to the local zenith, when both Babinet and Brewster points disappear.

The behavior of the neutral points, as observed under different conditions of atmospheric turbidity is postponed for a later section; here we restrict the discussion to the neutral points in a plane parallel Rayleigh atmosphere. The positions determined for this theoretical model have been found useful as a reference against which anomalies observed in the real atmosphere may be evaluated.

The point at which the degree of polarization is zero, a neutral point, is characterized by the double requirement

$$Q = I_r - I_e = 0$$

$$U = 0 \tag{4.98}$$

While either of these conditions may be met at certain points outside of the sun's vertical (see section 4.4.9 on neutral lines), the two requirements are met simultaneously only at points in the sun's vertical for the Rayleigh atmosphere with either no surface reflection or Lambert-type surface reflection. It is seen below that Fresnel-type reflection sometimes causes neutral points to occur outside the sun's vertical even in the Rayleigh model. As shown in Fig. 4.26, negative polarization ($I_r < I_e$) occurs between the Arago point and either the antisolar point or the horizon, between the Babinet point and the sun, and between the sun and either the Brewster point or the horizon. Light from all other parts of the sun's vertical is positively polarized ($I_r > I_e$). At the neutral points the negatively polarized component just balances that with positive polarization, making the composite neutral.

Various explanations have been offered to account for the existence of negative polarization. Brewster (1847) thought it was due to either "reflection," "refraction," or a combination of the two in the plane normal to the sun's vertical, and Jensen (1942) reported observations of the reflection from water which introduced such negative polarization into the skylight. Later work by Chandrasekhar (1950) and others, however, has shown that such negative polarization can be introduced purely by multiple scattering in the Rayleigh atmosphere, and does not depend on surface reflection, although, as mentioned above, reflection by Fresnel's laws has the potential for modifying the negative/positive ratio of skylight. Of course in the real atmosphere, scattering by dust, haze, and other aerosol particles may introduce polarization of either sign, depending on particle characteristics and incident radiation, which is the reason neutral point positions are such sensitive indicators of atmospheric turbidity.

The most convenient method of representing neutral point positions is in terms of their angular distance from the sun for the Babinet and Brewster points, and the antisun for the Arago point. These distances of the three neutral points are shown as a function of sun elevation for three different wavelengths in Fig. 4.27. The data for the positions were obtained by graphical interpolation from the tables of Coulson et al. (1960). When the sun is on the horizon the Arago point is the same distance above the antisolar horizon as the Babinet point is above the sun. As the sun rises in the sky both distances increase, particularly at the shortest wavelengths, but the increase is not enough to prevent the Arago point from sinking toward the antisolar horizon. At some sun elevation the Arago

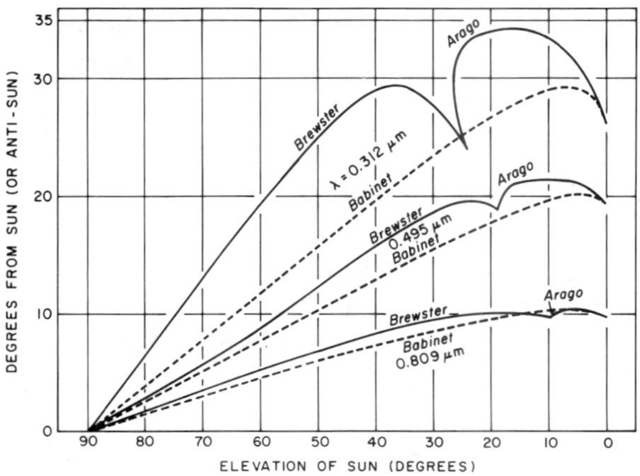

Figure 4.27 Angular distance of the neutral points from the sun for the Babinet and Brewster points and the antisun for the Arago point as a function of sun elevation for three different wavelengths in a plane-parallel Rayleigh atmosphere.

point sinks rapidly, reaching the antisolar horizon at just the time the Brewster point appears on the horizon underneath the sun. The Arago point curve for $\lambda = 0.312$ μm shows the theoretical possibility of a double Arago point developing at a sun elevation of about 24°, at which time a secondary Arago point appears on the horizon underneath the primary. As the sun climbs above 24° the primary point descends very rapidly, and the secondary rises above the horizon, equally rapidly, to meet the primary. For a sun elevation of 26° the primary has descended to 5° above the horizon and the secondary has moved up to about 2° above the horizon. At this instant four neutral points appear in the sky; two Arago points, the Babinet point, and the Brewster point which has made its appearance on the solar horizon. At sun elevation of 26.2° the two Arago points merge at about 3.5° above the horizon and then vanish, leaving the entire region of the sky above the antisolar horizon with positive polarization. The doubling of the Arago point would be very difficult to observe because of the very small magnitude of the polarization in the vicinity of a neutral point, but at least there should be a relatively wide zone of essentially neutral light in ultraviolet wavelengths just above the antisolar horizon at a sun elevation of about 25°. It is not known whether this feature has been actually observed. The only cases in which double Arago points have been reported from visual observations appear to have been associated with reflection from a water surface or scattering by a strong haze layer. According to the data presented here the double Arago point should exist only in the ultraviolet, and would thus not be visible to the eye. In any case it is of minor significance and would probably be overridden by aerosol effects in the real atmosphere. Ozone absorption, by reducing multiple scattering, also has the potential for modifying neutral point positions.

As shown in Fig. 4.27, the Brewster and Arago points are farther from the sun and antisun, respectively, than the Babinet point is from the sun, the differences decreasing with increasing wavelength. The Babinet and Brewster points move closer to the sun as the sun rises higher in the sky, the two merging into a single neutral point coincident with the sun when the sun reaches the zenith. The cusps at the juncture of the Brewster and Arago curves mark the positions at which the points are coincident with the horizons at the solar and antisolar azimuths, respectively.

From curves such as those of Fig. 4.27, Coulson (1954) constructed a table of the distances of the neutral points from the sun or antisun for a number of wavelengths to use as references for evaluation of neutral point behavior as observed under different atmospheric conditions. The most useful portions of that table are reproduced here as Table 4.12 for six selected wavelengths spaced from the ultraviolet, through the visible, and into the near-infrared wavelengths for sun elevations from 0° to 50° or 60°. By interpolations among the entries, it is possible to obtain data with which to compare observations taken in the field.

The data of Table 4.12 is for the case of zero surface reflectance, but reflection from a Lambert surface causes only a relatively minor shift of neutral point positions. The maximum effect is at the shortest wavelengths for the Brewster and Arago points. For the extreme case of surface reflectance $A=0.80$, the top curve of Fig. 4.27 would be shifted upward by 3° to 4° for sun elevations in the 30° to 40° range, the sense of the shift being always to increase the neutral point distances. At other wavelengths and other solar elevations, the effect of such high surface reflectance is a neutral point shift of not more than 1° to 2° at the most, and generally less than that. For the case of more normal reflectance values such as apply to soils or vegetation, Lambert surface reflection has a completely negligible effect on neutral point positions.

The behavior of the neutral points for optical thicknesses of $\tau > 1.0$ has been given by Dave and Furukawa (1966) for sun elevations of 36.9° and 5.7°

TABLE 4.12 Angular distance in degrees of the Babinet and Brewster points from the sun, and of the Arago point from the antisun as a function of sun elevation for six different wavelengths in a plane-parallel Rayleigh atmosphere and zero surface reflectance. (After Coulson, 1954.)

Sun elev. (°)	$\lambda=0.0809\ \mu m,\ \tau=0.02$			$\lambda=0.618\ \mu m,\ \tau=0.06$			$\lambda=0.546\ \mu m,\ \tau=0.10$		
	Ba	Br	Ar	Ba	Br	Ar	Ba	Br	Ar
0	9.8	[9.7	9.8	14.5	[14.2	14.5	17.0	[16.8	17.0
2	10.1	at sun	10.2	15.0	at sun	15.3	17.5	at sun	17.9
4	10.3	elev.	10.4	15.2	elev.	15.5	17.8	elev.	18.2
6	10.4	9.7]	10.4	15.1	14.2]	15.5	17.6	16.8]	18.3
8	10.4		10.2	15.0		15.4	17.4		18.4
10	10.3	10.0		14.8		15.3	17.1		18.3
14	10.1	10.1		14.4		14.4	16.6		17.8
18	9.8	10.1	[9.7	13.8	15.1		15.9	17.1	17.2
22	9.4	10.1	at	13.1	14.5	[14.2	15.2	17.3	
26	9.0	10.4	sun	12.3	14.2	at sun	14.4	16.8	[16.8
30	8.5	10.6	elev.	11.5	13.1	elev.	13.6	16.0	at sun
35	8.0	9.4	9.7]	10.5	12.1	14.2]	12.4	14.8	elev.
40	7.4	8.3		9.6	10.1		11.3	13.5	16.8]
45	6.8	7.6		8.6	9.8		10.0	12.1	
50	6.1	6.8		7.6	8.6		8.8	10.5	

Sun elev. (°)	$\lambda=0.495\ \mu m;\ \tau=0.15$			$\lambda=0.436\ \mu m;\ \tau=0.25$			$\lambda=0.312\ \mu m;\ \tau=1.0$		
	Ba	Br	Ar	Ba	Br	Ar	Ba	Br	Ar
0	19.3	[18.9	19.3	22.0	[21.6	22.0	26.1	[24.0	26.1
2	20.0	at	20.4	23.2	at	23.4	28.2	at	28.9
4	20.2	sun	21.0	23.5	sun	24.2	28.9	sun	30.6
6	20.1	elev.	21.3	23.4	elev.	24.7	29.1	elev.	31.8
8	19.9	18.9]	21.4	23.1	21.6]	25.1	29.1	24.0]	32.7
10	19.6		21.4	22.7		25.2	28.9		33.3
14	18.9		21.3	21.8		25.1	28.2		34.0
18	18.2		20.0	20.9		24.6	27.2		34.1
22	17.3	19.6		19.9	21.8		26.0		33.6
26	16.4	19.3	[18.9	18.8	23.1	[21.6	24.7	25.4	30.9
30	15.4	18.6	at	17.7	22.5	at	23.3	27.8	
35	14.2	17.4	sun	16.3	20.8	sun	21.5	29.3	[24.0
40	12.9	15.8	elev.	14.9	19.0	elev.	19.6	29.0	at
45	11.7	14.1	18.9[	13.4	17.1	21.6]	17.7	27.3	sun
50	10.4	12.4		12.0	16.2		15.8	25.0	elev.
55	9.1	10.6		10.6	13.3		13.8	22.2	24.0]
60	7.8	8.9		9.1	11.4		11.8	19.2	

and no surface reflection. These computations, as well as those of Kattawar et al. (1976), show that the neutral points reach their largest angular distances from the sun or antisun in the range of optical thicknesses $\tau = 1.0$ to 2.0. Beyond this they move rapidly toward their respective references until they coincide with the reference directions at $\tau \approx 4.0$, and vanish from the sky, leaving only positive polarization at larger values of τ.

The computations of both Dave and Furukawa (1966) and of Kattawar et al. (1976) show that the Babinet and Brewster points behave in the normal manner, aside from their disappearance at $\tau > 4.0$, at the higher sun elevations. However, at low sun elevations the Babinet and Arago points assume a double configuration for optical thicknesses in the $\tau = 1.0$ to 2.0 range. For instance, at sun elevation 5.7°, Dave and Furukawa showed a double Babinet point from about $\tau = 1.0$ to slightly beyond $\tau = 2.0$, and a double Arago point over the range of approximately $\tau = 1.1$ to 1.6. Kattawar et al. confirmed the existence of this phenomenon for sun elevations 10.8° and 2.2° with no surface reflection, but they also showed that the doubling feature was eliminated for the very high Lambert surface reflectance of $A = 1.0$.

The neutral point behavior is shown as a function of optical thickness, as given by Kattawar et al. (1976), for a sun elevation of 32.5° in Fig. 4.28. The Babinet and Brewster points diverge rapidly from the solar direction with increasing τ, reaching their maximum distances of 22° and 28° at $\tau = 1.6$ and 0.9, respectively. Beyond this they converge rapidly to the sun, and disappear at $\tau \approx 4.0$. For no surface reflection, the Arago point does not exist at this solar elevation. However, these computations show a very large effect of the high surface reflectance of $A = 1.0$ on the position of the Brewster point, and only for $A = 1.0$ does the Arago point come into existence. In fact, the Arago point is shown to reach an angle of fully 44° from the anti-sun for $A = 1.0$, whereas its absence for $A = 0$ shows its distance would have to be less than 32.5° without surface reflection. Such a large influence of surface reflection has not been seen in any other computational results, and even in this series by Kattawar et al., effects of surface reflection do not appear anomalously high for the Babinet point at $A = 1.0$, or for any of the points for $A = 0.20$. Thus it would seem that such an extreme effect should be viewed with reservation until further confirmation is obtained.

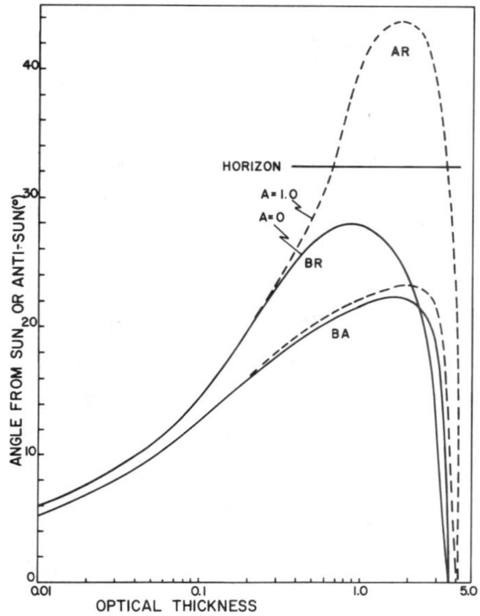

Figure 4.28 Angular distance of the Babinet and Brewster points from the sun and of the Arago point from the antisun as a function of optical thickness of a plane-parallel Rayleigh atmosphere for Lambert surface reflectance of A=0 (solid curves) and A=1.0 (dashed curves) at a sun elevation of 32.5°. (Adapted from Kattawar et al., 1976.)

In contrast to results usually obtained for Lambert-type surface reflection, reflection from surfaces such as still water which follow the Fresnel law of reflection is known from both observation and theory to have a very pronounced effect on the positions of neutral points, even though it has a minimal effect on the polarization maximum and other features of the polarization field. The addition of even a small amount of highly polarized light, such as that provided by Fresnel reflection at certain angles, to the region of essentially unpolarized light in the vicinity of a neutral point may cause a large shift of the neutral point position without much change of the field elsewhere.

In general the neutral points are confined to the plane of the sun's vertical, as discussed above, but there have been at least two sets of observations in which the neutral points were displaced in azimuth and appeared out of this plane. Cornu (1884), observing in turbid atmospheric conditions after the eruption of Krakatoa, observed four neutral points, one each to the right and left of the sun and antisun. However, the anomalous positions at that time were undoubtedly caused by the volcanic debris in the atmosphere, and were not related to the effects of surface reflection. Soret (1888), on the other hand, observed anomalies of neutral point positions which were definitely associated with reflection from surfaces of water, both lakes and the sea. Generally, neutral points appeared at angles of from 8° to 19° to the right and left of the sun at the same time that the normal Babinet and Brewster points disappeared from the sun's vertical. Fog or haze in the atmosphere appeared to augment the anomalies, but they also were seen under very clear conditions. From many observations, Soret concluded that the displacement of neutral points from the sun's vertical to positions alongside the sun is not a rare phenomenon in the vicinity of water surface.

Rare or not, no observations of anomalous neutral point positions near water seem to have been reported since Soret's time, but model computations indicate that Fresnel-type surface reflection can cause such behavior, even in a Rayleigh atmosphere. Sekera (1961) studied the problem of sea surface reflection in relation to skylight polarization, and found neutral point displacements somewhat similar to those observed by Soret. In a related effort, Fraser extended the work of Sekera and computed the positions of the neutral points for a Fresnel surface underlying a Rayleigh atmosphere (Fraser, 1968) and an atmosphere containing various amounts and types of aerosols (Fraser, 1981). The results of the first of these studies will be discussed here, with the effects of aerosols reserved for the discussion of atmospheric turbidity.

In an effort to explain Soret's (1888) visual observations of neutral points appearing outside the sun's vertical, Fraser (1968) found that the Babinet and Brewster points disappear from the sun's vertical only under certain conditions of wavelengths and solar zenith angle. The computations show that Fresnel reflection does not cause the Arago point to disappear from the sun's vertical, but only to move somewhat closer to the antisun, and no observations of Fresnel reflection causing anomalous Arago points outside of that plane have been reported. The double Arago point of Cornu (1884) mentioned previously, in which a displacement in azimuth was observed, was due to volcanic aerosols, and not surface reflection.

The positions of the Babinet and Brewster points in the sun's vertical for a Fresnel surface underlying a plane parallel Rayleigh atmosphere, as given by the computations of Fraser (1968), are shown in Fig. 4.29. The four wavelengths 0.312, 0.372, 0.436, and 0.495 μm correspond to sea level optical thicknesses of 1.0, 0.50, 0.25, and 0.15, respectively. The neutral points remain in the sun's vertical at all sun elevations below 82° for the two shortest wavelengths, and even if they moved out of this plane at the higher sun elevations, they would be so near the sun as to be very difficult to detect by observation. At $\lambda = 0.436$ and 0.495 μm, particularly at the longer of these, the neutral points disappear from the sun's vertical through relatively wide ranges of sun elevation, and the computations show them to appear at those times at various angular distances on either side of the sun's vertical and 2° to 3° below the solar almucantar. This behavior corresponds roughly to that observed by Soret (1888) near water surfaces, although the azimuths observed by Soret were somewhat larger than the computations indicate. At still longer wavelengths, such as $\lambda = 0.644$ μm, the Babinet and Brewster points do not appear in the sun's vertical at any sun elevation.

The azimuthal displacements of the neutral points when they are displaced from the sun's vertical are shown as a function of sun elevation for three wave-

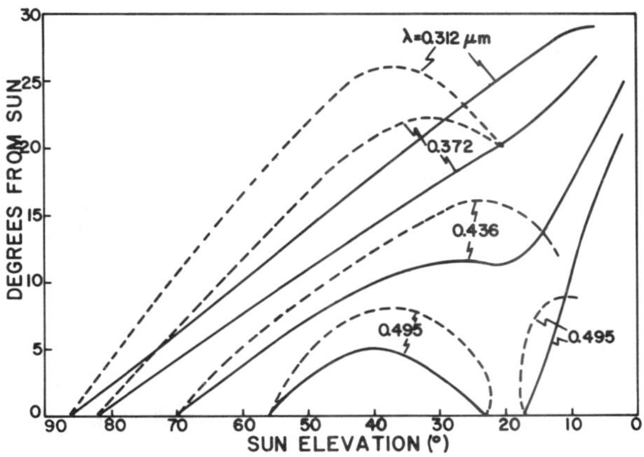

Figure 4.29 Angular distance of the Babinet (solid curves) and Brewster (dashed curves) points from the sun in the plane of the sun's vertical as a function of sun elevation at four different wavelengths for the case of a plane-parallel Rayleigh atmosphere overlying a Fresnel-type reflecting surface. (Adapted from Fraser, 1968.)

lengths in Fig. 4.30. The azimuthal displacements are observed to be relatively sensitive to both wavelength and sun elevation, as might be anticipated from the geometry of the problem. An interaction of the following processes must be responsible for the introduction of the negatively polarized light necessary to modify the neutral point positions as shown:

1. A change of reflectance and polarizing properties of the Fresnel surface with changing sun elevation.

It was shown in Fig. 4.25 that both total reflectance and the state of polarization of the light reflected from a Fresnel surface are both critically dependent on angle of incidence of the light. The curves of Fig. 4.30 show that the neutral points are displaced strongly at a sun elevation of about 20°, corresponding to an angle of incidence of 70° for the direct solar beam, and Soret's neutral point observations were in the range of sun elevations of 10° to 20°. Fig. 4.25 shows reflected light at an incidence angle of 70° to be relatively strong and to have a degree of polarization of about 70%. Backscatter of this highly polarized light entering the base of the atmosphere can strongly influence the weak polarization field previously existing near the neutral points, and thus change their positions. At higher sun elevations the total reflectance is much diminished, and even though the reflected light may still be highly polarized,

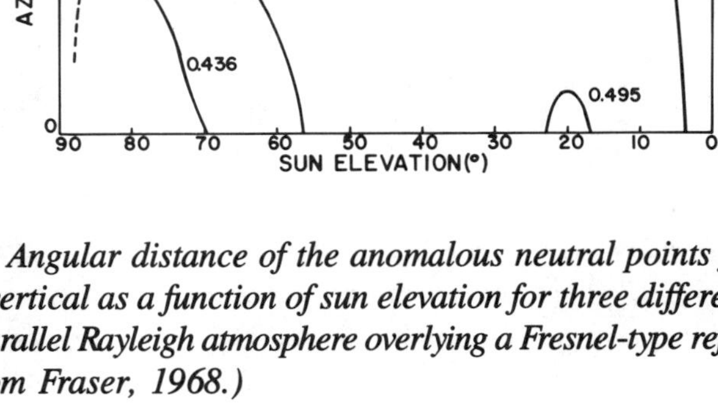

Figure 4.30 Angular distance of the anomalous neutral points from the plane of the sun's vertical as a function of sun elevation for three different wavelengths in a plane-parallel Rayleigh atmosphere overlying a Fresnel-type reflecting surface. (Adapted from Fraser, 1968.)

particularly at solar elevation angles of 35° to 40°, its contribution competes with strong scattering by the overlying atmosphere.

2. A change of the amount of radiation incident on the surface with changing sun elevation.

The total radiation incident on the surface is $\mathbf{F}_D + \mathbf{I}_{Dd} + \mathbf{I}_s + \mathbf{I}_{dd}$, where the individual components were given above in Eqs. (4.76), (4.78), (4.79), and (4.81). As indicated there, each of the components is a function of sun elevation in such manner that the radiation incident at the surface is highest at the highest sun elevations. This feature tends to offset the decrease of reflectance for the Fresnel surface with decreasing angle of incidence of the direct solar beam. Of course, the diffuse light is incident at all angles, some of which cause very high polarization of the reflected light. The addition of the composite to the field with low polarization near the neutral points causes the neutral point shift as shown.

3. A change of magnitude of the surface-reflected component, as well as its angular distribution, with changing wavelength.

For the shorter wavelengths the strong scattering by the molecular atmosphere causes both a high intensity of light which is unaffected by surface reflection and a small flux with largely diffuse character at the surface. The polarization effects due to surface reflection for this combination are thus minor, and the polarization field for the skylight is dominated by atmospheric scattering of the original incident solar beam. On the other hand, at the longer wavelengths the atmosphere attenuates the solar beam relatively little, surface reflection is strong, and the original skylight is weak. The light due to reflection from the Fresnel surface is thus dominant, and the negative polarization introduced is sufficient to cause a large shift of neutral point positions. This is the case for $\lambda = 0.644\ \mu m$ in Fig. 4.30, and it would apply to the longer wavelengths as well.

Fraser (1968) pointed out the fact that Soret's observations show the anomalous neutral points to be situated 15° to 20° from the sun's vertical at a sun elevation of 10° to 20°, whereas Fig. 4.30 shows a displacement of only about 2° for $\lambda = 0.495\ \mu m$ at 20° sun elevation. However, there is a strong dependence on wavelength, as shown by the curve for $\lambda = 0.644\ \mu m$, and Soret's observations were in broadband visible light. Thus the discrepancy between observation and theory may not be as significant as Fraser indicated. Furthermore, the atmospheric aerosols often present during Soret's observations changed the picture from that of a pure Rayleigh atmosphere, in which case any discrepancy may well be explained on the basis of scattering by atmospheric aerosols.

4.8 The neutral lines and lines of U=0

As discussed above, the neutral points of skylight polarization occur wherever the two conditions

$$Q = I_r - I_e = 0$$

$$U = Q \tan 2\chi = 0 \tag{4.99}$$

are met. Here χ is the angle between the plane of polarization and the vertical plane at the relevant azimuth. By symmetry, χ must be 0° or 90° in the sun's vertical, depending on whether I_r is less than or greater than I_e. In either case U=0. At the neutral points Q=0, which makes χ indeterminate, as it must be if P=0.

We can also have Q=0 at lines passing through the neutral points and continuing over the sky outside of the sun's vertical, the requirement for $I_r = I_e$ being that χ=45°. In fact Dorno (1919) mapped these lines of Q=0, which he termed neutral lines, over the sky in great detail. It should be made clear, however, that the term neutral line is somewhat misleading, as $P \neq 0$ at neutral lines except as they intersect the sun's vertical at the neutral points, where indeed P=0. Outside the sun's vertical P does not vanish at the neutral lines.

It is useful for discussions of the neutral lines and lines of U=0 to examine the specific expressions for I_r, I_e and U. For this purpose we adopt the notation of Coulson et al. (1960) and write the expressions for Stokes parameters for skylight as given by the discrete ordinate method of Chandrasekhar (1950). On rewriting Eq. (4.57) for convenience, we have for the intensity components

$$I_{e,r}(\mu,\phi) = I_{e,r}^{(o)}(\mu,\mu_o) + I_{e,r}^{(1)}(\mu,\mu_o)\cos\phi + I_{e,r}^{(2)}(\mu,\mu_o)\cos 2\phi$$

where, as usual, $\cos\theta = \mu$, $\cos\theta_o = \mu_o$, and azimuth ϕ is taken with respect to the azimuth of the sun. From the expressions for Stokes parameter $U(\mu,\phi)$ for skylight in Eq. (4.67), we see that if $\phi = 0°$ or $180°$, $\sin\phi = \sin 2\phi = 0$, and the plane of the sun's vertical always has $U=0$. For determining the lines of $U=0$ outside of the sun's vertical, we set Eq. (4.67) to zero and simplify to give

$$\cos\phi = \frac{\mu_o(1-\mu^2)^{1/2}[X^{(1)}(\mu_o)\,Y^{(1)}(\mu) - Y^{(1)}(\mu_o)X^{(1)}(\mu)]}{\mu(1-\mu_o^2)^{1/2}[X^{(2)}(\mu_o)\,Y^{(2)}(\mu) - Y^{(2)}(\mu_o)\,X^{(2)}(\mu)]} \tag{4.100}$$

From this equation and tabulations of the X and Y functions, we can determine the position of the neutral lines for each combination of μ and μ_o. From Eq. (4.67), however, we can see a constraint on the position of lines of $U=0$. Parameter U has dimensions of intensity, so each term of the equation also has dimensions of intensity, and the ratio of the two terms given in Eq. (4.100) must be a positive quantity. This means that $\cos\phi = \cos(-\phi) \geq 0$, or $-90° \leq \phi \leq +90°$. Thus lines of U=0 are confined to the solar half of the hemisphere, except in the sun's vertical, which has U=0 throughout.

Computations based on Eq. (4.75) were carried out by Coulson and Sekera (1955) to give the configurations of the lines of U=0 shown for two sun elevations in Fig. 4.31. Each line forms a loop, the size of which depends on sun elevation, and intersects the sun's vertical in the normal direction at the zenith and at the sun. Lines of U=0 exist only in the solar half of the hemisphere.

For the neutral lines, at which $Q=0$ or $I_r = I_e$, we set Eq. (4.67) equal to Eq. (4.68) and obtain the condition for a neutral line as

$$\cos\phi = \frac{K[\sigma(\mu) - \xi(\mu)] + L[\theta(\mu) - \eta(\mu)] - M[\chi(\mu) - \psi(\mu)] - N[\zeta(\mu) - \phi(\mu)]}{4\mu\mu_o(1 - \mu^2)^{1/2}(1 - \mu_o^2)^{1/2}\,W^{(1)}(\mu,\mu_o)} \tag{4.101}$$

We see immediately that the constraint on $\cos\phi$ that applied to lines of $U=0$ does not apply to the neutral lines, which means that neutral lines may exist in either half of the hemisphere.

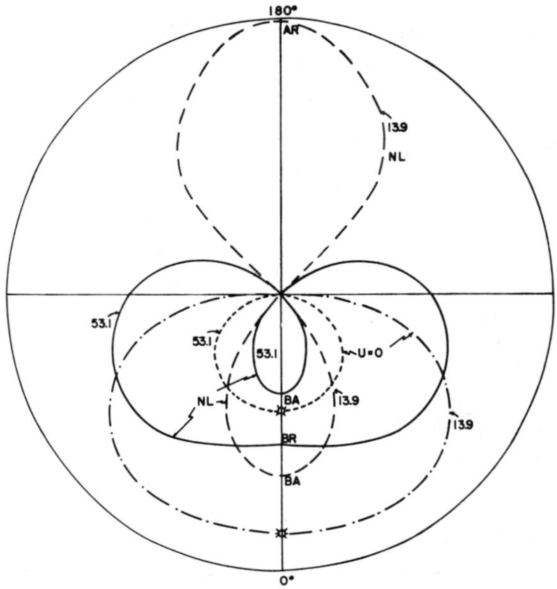

Figure 4.31 Pattern of neutral lines (NL) and lines of U=0 over the hemisphere of the sky for sun elevations 13.9° and 53.1°. The center represents the zenith, the sun is at azimuth ϕ=0, and radial distance corresponds to zenith angle.

Computations based on Eq. (4.76) give the configuration of the neutral lines for the two sun elevations 13.9° and 53.1° shown in Fig. 4.31. They form lemniscates, the configuration of which depends on sun elevation. Each branch intersects the sun's vertical at an angle of 45° at the zenith and in the normal direction at one of the neutral points. For the sun low on the horizon, one branch encloses a part of the sky above the sun and passes through the Babinet point, and the other branch encloses a larger part of the sky in the antisolar half of the hemisphere and passes through the Arago point. For a higher sun elevation at which the Arago point has set below the horizon and the Brewster point has appeared, the branch of the lemniscate above the sun has the same form as in the previous case, but has shrunk considerably in size. The other branch, however, has a different form, having opened out in the antisolar hemisphere and swung around to a large loop in the solar half of the hemisphere to pass through the Brewster point. Both branches decrease in size with increasing sun elevation, and eventually degenerate to a point as the sun reaches the zenith.

Although the distribution over the sky of the plane of polarization will be shown explicitly below, its general configuration can be seen from Fig. 4.31. It will be remembered that the angle χ between the plane of polarization and the vertical direction is, by convention, positive if rotated clockwise from the vertical, and negative if counterclockwise. Furthermore $\chi=0°$ along lines of $U=0$, and $\chi=\pm45°$ along the neutral lines. The only exception is that $\chi=90°$ where the polarization is positive in the sun's vertical, and $\chi=0$ in the region of negative polarization near the neutral points. For discussion purposes, we choose the curves for sun elevation 13.9° in Fig. 4.31, and concentrate on the right half of the diagram. The left half would be a mirror image. In the antisolar half of the hemisphere, in the sun's vertical $\chi=\pm90°$, and along the lemniscate $\chi=-45$. Thus $-45°<\chi<-90°$ inside the loop, and $0<\chi<-45$ between the loop and the line $U=0$ in the solar half hemisphere. At $U=0$, $\chi=0$, and $0°<\chi<45°$ from the line $U=0$ to the other part of the lemniscate. At the lemniscate, $\chi=\pm45°$, inside the loop $+45°<\chi<90°$, and along the sun's vertical inside the loop $\chi=90°$. Between the Babinet point and sun and between the sun and horizon, the sun's vertical has $\chi=0°$. It appears from the diagram that the zenith would be a neutral point, but this comes about because the azimuth is indeterminate there. The plane of polarization is well defined, however, since it is normal to the sun's vertical. The pattern shows that the neutral lines and lines of $U=0$ intersect only at the zenith and in the sun's vertical at the neutral points, which means that no neutral points can occur outside the sun's vertical for this model of the atmosphere. This is, of course, not true for the case of Fresnel surface reflection, as discussed above.

Chandrasekhar and Elbert (1954) have computed the positions of the neutral lines at seven sun elevations, as shown in Fig. 4.32. They used the term positive polarization for regions where $I_r>I_e$, and negative polarization for regions where $I_r<I_e$. However, it is only in the sun's vertical that these distinctions have any validity.

Although the computations for Fig. 4.32 were for $\lambda=0.495$ μm ($\tau=0.15$) and no surface reflection, the configurations shown are quite representative of those for other wavelengths and Lambert surface reflectances. The neutral lines do not depend on the polarization magnitude, but only on the plane of polarization, and the plane is relatively independent of wavelength and Lambert surface reflectance. Only in the regions of the neutral points would there be an appreciable

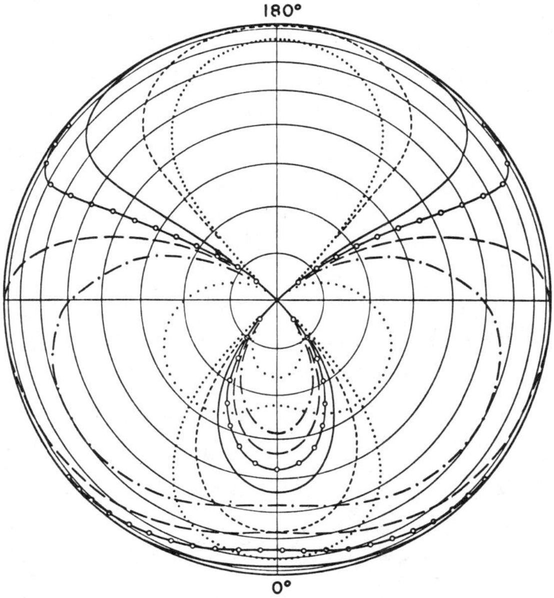

Figure 4.32 The configuration of the neutral lines over the hemisphere of the sky for a wavelength of $\lambda=0.495$ μm ($\tau=0.15$) in a plane parallel Rayleigh atmosphere with no surface reflection. (After Chandrasekhar and Elbert, 1954.) Sun elevations for the various curves are the following:

. . . . 0°	———— 31.3°	— — —46.1°	. . . . . 70.1°
- - - - 13.9°	o—o—o—o 38.8°	—·—·— 53.1°	

shift of the neutral lines, corresponding to the shift of neutral points, with wavelength.

Computations by Coulson and Sekera (1955) gave the configuration of the lines of $U=0$ as shown for eight different solar zenith angles and four optical thicknesses in Fig. 4.33. The positions of the lines are insensitive to optical thickness. Multiple scattering, although very influential in determining the degree of polarization, has little effect on the plane of polarization. Likewise, Lambert surface reflection has negligible effect on the plane of polarization. A feature of the diagram which is worthy of mention is the fact that the lines of $U=0$ do not intersect the sun's vertical at the sun, but has a deviation of 0.5° to 3.3° toward the zenith from the sun. The effect is produced by multiple scattering,

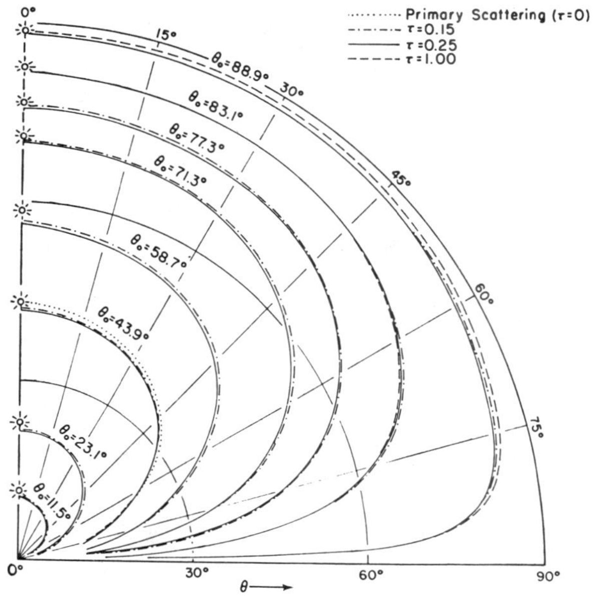

Figure 4.33 Positions of the lines of U=0 for different zenith distances of the sun and different values of optical thickness of the plane-parallel Rayleigh atmosphere. (After Coulson and Sekera, 1955.)

as for $\tau=0$ it does not occur. The precise amount otherwise depends on optical thickness and solar zenith angle. For instance, at $\theta_o=58.7°$, the deviation is 3.3° at $\tau=1.0$, but only about half that at $\tau=0.15$. For the sun near either the horizon or zenith, however, the deviations are 1° or less at all optical thicknesses. The significance, if any, of this small effect is not known.

4.9 The plane of polarization

The plane of polarization for the case of a plane-parallel atmosphere is defined by the angle χ between the maximum of the electric vector and the vertical direction at the azimuth in question. Its overall distribution across the sky was evident from Fig. 4.31, but here we show its configuration in detail. The most important determinant of its behavior is simply geometry: the plane of polarization is mainly oriented in a direction normal to the plane defined by the source, observer, and observed point in the sky. This is the so-called scattering plane, which was shown in Fig. 2.3 of Chapter 2 for the plane-parallel atmosphere. Geometry is the only determinant for χ for primary Rayleigh scattering, in which case χ assumes the configuration shown in Fig. 4.34. Only one-half of the hemisphere is shown, the other half being a mirror image of this one. The arrows are drawn to make the same angle with the bottom line of the diagram as the plane of polarization would make with the plane of the meridian at each azimuth. The line $U=0$ and the two branches of the neutral lines (dashed) are approximately as shown in Figs. 4.31–4.33, except here the single neutral point coincides with the sun.

More realistic configurations are shown for wavelength 0.495 ° μm and solar zenith angles 43.9 ° and 85.4 ° in Figs. 4.35 and 4.36, respectively. Here the arrows indicate more directly the value of χ at each azimuth. In Fig. 4.35 the neutral lines pass through the Babinet and Brewster points, and their positions are thereby modified from those for primary scattering. The lines of $U=0$, however, still pass near the sun, and the general pattern is similar to that for primary scattering. For the large solar zenith angle of Fig. 4.36, the Arago point is above the horizon, and the two branches of the lemniscate of the neutral lines are evident in the diagram.

The plane of polarization of skylight undergoes only minor changes with wavelength, so the patterns shown in Figs. 4.35 and 4.36 are generally representative of those in other spectral ranges. The only significant changes are in the

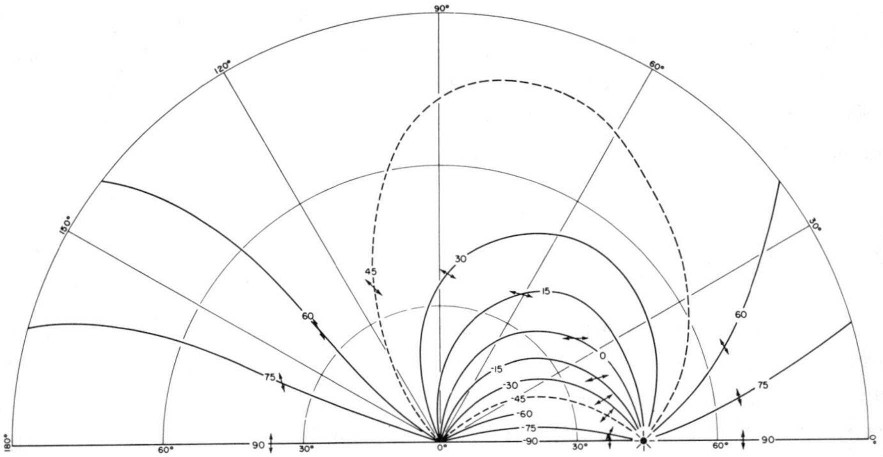

Figure 4.34 Orientation of the plane of polarization for primary scattering in a plane-parallel Rayleigh atmosphere for a solar zenith angle of 43.9° (After Coulson and Sekera, 1955.)

regions of the neutral points, at which the pattern shifts in conformity with the shift of the neutral points with wavelength. Since reflection from a Lambert-type surface adds no polarized light, it has a negligible effect on neutral point positions and a similarly negligible effect on the plane of polarization.

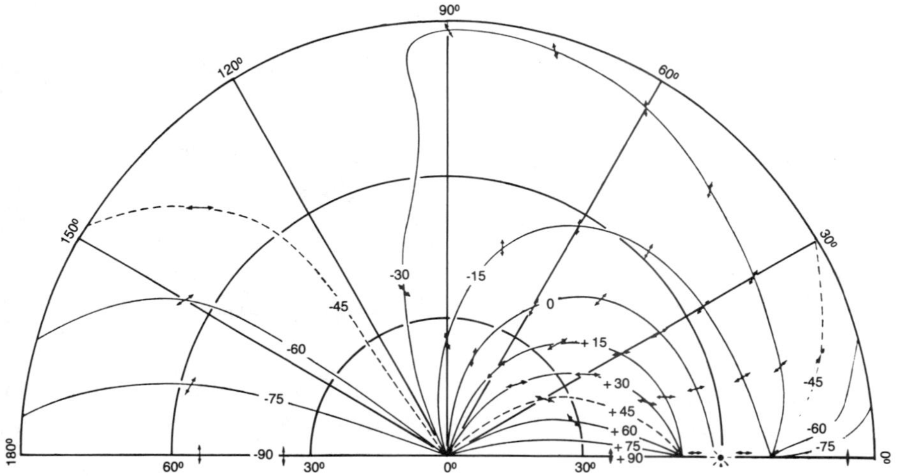

Figure 4.35 Orientation of the plane of polarization at a wavelength of 0.495 μm (τ=0.15) and solar zenith angle of 43.9° in a plane-parallel Rayleigh atmosphere. (After Coulson and Sekera, 1955.)

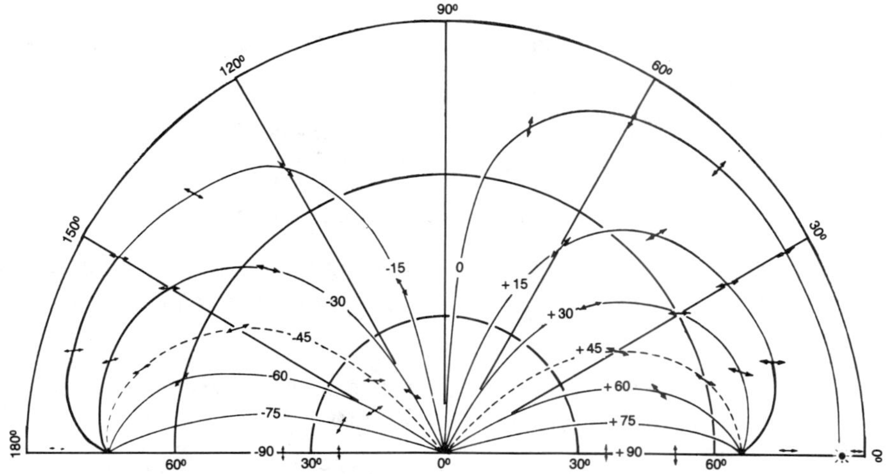

Figure 4.36 Same as Fig. 4.35 except for a solar zenith angle of 85.4°. (After Coulson and Sekera, 1955.)

References

Ahlgrimm, F. (1915): Zur Theorie der atmospharischen Polarisation, JB. *Hamburg wiss. Anst., 32,* 1–66.

Alms, G.R., A.K. Burnham, and W.H. Flygare (1975): Measurement of the dispersion in polarizability anisotropies, *J. Chem. Phys., 63,* 3321–3326.

Babinet, J. (1840): Sur un nouveau point neutre dans l'atmosphere, *Comp. Rend. Acad. Sci. Paris, 11,* 618–620.

Bahethi, O.P., and R.S. Fraser (1980): Effect of molecular anisotropy on the intensity and degree of polarization of light scattered from model atmospheres, *App. Opt., 19,* 1333–1337.

Blickhan, F. (1934): Vergleischende Messungen der Polarisation des Himmelslichtes in Frankfurt a. M. und am Taunusobservatorium, *Beitr. Geophys. 42,* 208–227.

Brewster, Sir David (1842): On the existence of a new neutral point and two secondary neutral points, Rept. British Assoc. Adv. of Sci., Pt. 2, 13.

Brewster, Sir David (1847): On the polarization of the atmosphere, *Phil. Magazine, London, XXXI,* 444–454.

Brewster, Sir David (1861): Observations on the polarization of the atmosphere, made at St. Andrews in 1841, 1842, 1843, 1844, and 1845, *Trans. Royal Soc. Edinburgh, XXIII,* 211–240.

Cabannes, J. (1929): La diffusion moleculaire de la lumiere, Les Presses Universitaires de France, Paris.

Chamberlain, J.W. (1978): *Theory of Planetary Atmospheres,* Vol. 22, Int. Geophys. Series, Academic Press, New York.

Chandrasekhar, S. (1950): *Radiative Transfer,* Clarendon Press, Oxford.

Chandrasekhar, S., and D.D. Elbert (1954): The illumination and polarization of the sunlit sky on Rayleigh scattering, *Trans. Am. Phil. Soc., B, 44,* 643–728.

Cornu, M.A. (1884): Observations relatives a la couronne visible actuellement autour du Soleil, *Comp. Rend., 99,* 488–493.

Coulson, K.L. (1954): Neutral points of skylight polarization in a Rayleigh atmosphere, Sci. Rept. No. 7, Contr. AF19(122)-239, Dept. of Meteorology, Univ. of California, Los Angeles.

Coulson, K.L. (1959): Characteristics of radiation emerging from the top of a Rayleigh atmosphere, Part I: Intensity and polarization, *Planetary and Space Sci.*, *1,* 165–284.

Coulson, K.L. (1968): Effect of surface reflection on the angular and spectral distribution of skylight, *J. Atm. Sci.*, *25,* 759–770.

Coulson, K.L. (1980): Characteristics of skylight at the zenith during twilight as indicators of atmospheric turbidity: I. Degree of polarization, *Appl. Opt.*, *19,* 3469–3480.

Coulson, K.L. (1983): Effects of the El Chichon volcanic cloud in the stratosphere on the polarization of light from the sky, *Appl. Opt.*, *22,* 1036–1050.

Coulson, K.L., J.V. Dave, and Z. Sekera (1960): *Tables Related to Radiation Emerging from a Planetary Atmosphere with Rayleigh Scattering,* University of California Press, Berkeley and Los Angeles.

Coulson, K.L. and Z. Sekera (1955): Distribution of polarization and orientation of the plane of polarization of sky radiation over the entire sky in a Rayleigh atmosphere, Appendix B, Final Report, Contr. AF19(122)-239, Dept. of Met., Univ. of California, Los Angeles.

Dave, J.V. (1964): Importance of higher order scattering in a molecular atmosphere, *J. Opt. Soc. Am.*, *54,* 307–315.

Dave, J.V., and P.M. Furukawa (1966): Intensity and polarization of the radiation emerging from an optically thick atmosphere, *J. Opt. Soc. Am.*, *56,* 394–400.

DeBary, E., and K. Bullrich (1964): Effects of higher order scattering in a molecular atmosphere, *J. Opt. Soc. Am.*, *54,* 143–1416.

Dorno, C. (1919): Himmelshelligkeit, Himmelspolarisation und Sonnenintensitat in Davos (1911 bis 1918), *Meteor. Zeit.*, *36,* 109–124, 181–192.

Edlen, B. (1953): The dispersion of standard air, *J. Opt. Soc. Am.*, *43,* 339–344.

Elterman, L. (1968): UV, Visible and IR attenuation for altitudes to 50 km, 1968, AFCRL–68–0153, Env. Res. Pap. No. 285. U.S. Air Force.

Fitch, B.W. (1977): Unpublished data, private communication.

Fitch, B.W. (1981): Effects of reflection by natural surfaces on the radiation emerging from the top of a planetary atmosphere, Doctoral Dissertation, Dept. of Land, Air, and Water Resources, University of California, Davis.

Fraser, R.S. (1968): Atmospheric neutral points over water, *Journ. Opt. Soc. Am., 58,* 1029–1031.

Fraser, R.S. (1981): Atmospheric neutral points outside of the principal plane, *Contrib. Atm. Phys., 54,* 286–297.

Frohlich, C., and G.E. Shaw (1980): New determination of Rayleigh scattering in the terrestrial atmosphere, *Appl. Opt., 19,* 1773–1775.

Gockel, A. (1920): Beitrage zur Kenntnis von Farbe und Polarisation des Himmelslichtes, *Ann. Phys., Leipzig (4), 62,* 283–292.

Gucker, F.T., S. Basu, A.A. Pulido, and G. Chin (1969): Intensity and polarization of light scattered by some permanent gases and vapors, *J. Chem. Phys., 50,* 2526–2535.

Hansen, J.E., and J.W. Hovenier (1974): Interpretation of Venus polarization, *J. Atm. Sci. 31,* 1137–1158.

Herman, B.M., and S.R. Browning (1965): A numerical solution to the equation of radiative transfer, *J. Atm. Sci., 22,* 559–566.

Howell, H.B., and H. Jacobowitz (1970): Matrix method applied to multiple scattering of polarized light, *J. Atm. Sci., 27,* 1195–1206.

Hoyt, D.V. (1977): A redetermination of the Rayleigh optical depth and its application to selected solar radiation problems, *J. Appl. Met., 16,* 432–436.

Jensen, C. (1899): Beitrage zur photometrie des Himmels, *Meteor. Zeit. 14,* 488–499.

Jensen, C. (1928): Die Himmelsstrahlung, *Handbuch der Physik,* Bd. 19, Springer, Berlin, 70–152.

Jensen, C. (1942): Die Polarisation des Himmelslichtes, *Handbuch der Geophysik,* Bd. 8, Berlin, 527–620.

Kattawar, G.W., G.N. Plass, and S.J. Hitzfelder (1976): Multiple scattered radiation emerging from Rayleigh and continental haze layers. 1: Radiance, polarization and neutral points, *Appl. Opt., 15,* 632–647.

Kerker, M. (1969): *The Scattering of Light and other Electromagnetic Radiation,* Academic Press, New York.

Lenoble, J. (ed.) (1974): *Standard Procedures to Compute Atmospheric Radiative Transfer in a Scattering Atmosphere,* Radiation Commission, Internat. Assoc. Meteor. Atm. Physics (I.U.G.G.), Vol. I.

Mullamaa, Yu. A.R. (1964): Atlas of the optical character of the disturbed sea surface, Ins. Fiziki i Astron., Akad. Nauk Estonskii SSR, Tartu (110 pp.)

N.O.A.A. (1976): *U.S. Standard Atmosphere, 1976,* Nat. Ocean. Atm. Admin., Washington, D.C.

Neuberger, H. (1940): Studies in Atmospheric turbidity in Central Pennsylvania, Penn. State Coll. Studies, No. 9, 36 pp.

Packer, D.M., and C. Lock (1951): The brightness and polarization of the daylight sky at altitudes of 18,000 to 38,000 feet above sea level, *J. Opt. Soc. Am., 41,* 473–478.

Penndorf, R. (1955): Tables of the refractive index for standard air and the Rayleigh scattering coefficients for the spectral region between 0.2 and 20 μm and their application to atmospheric optics, AFCRL–TN–206, U.S. Air Force.

Plass, G.N., and G.W. Kattawar (1970): Polarization of the radiation reflected and transmitted by the earth's atmosphere, *Appl. Opt., 9,* 1122–1130.

Plass, G.N., G.W. Kattawar, and S.J. Hitzfelder (1976): Multiple scattered radiation emerging from Rayleigh and continental haze layers. 2: Ellipticity and direction of polarization, *Appl. Opt., 15,* 1003–1011.

Pyaskovskaya-Fesenkova, E.V. (1960): Some data on the polarization of atmospheric light, Doklady Akad. Nauk, USSR, 131, (2) 297–299.

Rayleigh, Lord (1884): Presidential address, Brit. Assoc. Rept., Montreal, 1–23.

Rayleigh, Lord (1889): On the transmission of light through an atmosphere containing many small particles in suspension, and on the origin of the blue of the sky, *Phil. Mag., 47,* 375–384.

Rowell, R.L., G.M. Aval, and J.J. Barrett (1971): Rayleigh-Raman depolarization of laser light scattered by gases, *J. Chem. Phys. 54,* 1960–1964.

Sekera, Z. (1956): Recent developments in the study of the polarization of skylight, *Advances in Geohpysics,* Vol III, Academic Press, New York (pp. 43–104).

Sekera, Z. (1961): The effect of sea reflection on the sky radiation, *Union Geodesique et Geophysique Internat Monograph, 10,* 66–72.

Sekera, Z., and E.V. Ashburn (1953): Tables Relating to Raleigh Scattering of Light in the Atmosphere, *NAVORD* Rept. 2061, U.S. Naval Ordnance Test Station, Inyokern, CA.

Sekera, Z., and G. Blanche (1952): Tables Relating to Rayleigh Scattering of Light in the Atmosphere, *Sci. Rept. No. 3,* Contr. AF 19(122)–239, Dept. of Meteorology, Univ. of California, Los Angeles.

Sekera, Z., and Collaborators (1955): Investigation of the Polarization of Skylight, *Final Report,* Contr. AF19(122)–239, Dept. of Meteorology, Univ. of California, Los Angeles, 68 pp.

Soret, M.J.L. (1888): Influence des surfaces d'eau sur la polarisation atmospherique et observation de deux points neutres a droite et a gauche de Soleil, *Comp. Rend., 107,* 867–870.

Steinhorst, G. (1977): Stratospheric aerosol concentration determined by an iterative method from twilight polarization measurements, *Contrib. Atmos. Phys., 50,* 508–523.

Stokes, G.G. (1852): On the composition and resolution of streams of polarized light from different sources, *Trans. Camp. Phil. Soc., 9, 399.*

Stratton, J.A. (1941): Electromagnetic Theory, McGraw–Hill, New York.

Strutt, J.W. (Lord Rayleigh) (1871): On the light from the sky, its polarisation and colour, *Phil. Mag., 41,* 107–120, 274–279.

Takashima, T. (1983): A new approach to evaluation of the atmospheric effects on upwelling radiance from the ocean, *Papers in Meteor. and Geophys., 34,* 75–81.

Takashima, T., C.I. Taggart, and E.G. Morrissey (1976): A hybrid mode of diffuse and specular reflector for computation of the emergent radiation by the adding method, *Astrophys. and Space Sci., 40,* 157–165.

Tichanowski, J.J. (1926): Resultate der Messungen der Himmelpolarisation in verschiedenen Spektrumabschnitten, *Meteor. Zeit., 43,* 288–292.

Tousey, R., and E.O. Hulbert (1947): Brightness and polarization of the daylight sky at various altitudes above sea level, *J.O.S.A., 37,* 78–92.

Tricker, R.A.R. (1970): *Introduction to Meteorological Optics,* Elsevier Pub. Co., New York.

Valley, S.L. (ed.) (1965): *Handbook of Geophysics and Space Environments,* Air Force Cambridge Research Laboratories, Bedford, MA.

Van de Hulst, H.C. (1957): *Light Scattering by Small Particles,* Wiley, New York.

Volz, F.E. (1983): Volcanic turbidity, skylight scattering functions, and neutral points in New England 1982/1983, Fifth Conf. on Atm. Rad., Am. Meteor. Soc., 31 Oct.–4 Nov. 1983, Baltimore, MD.

Young, A.T. (1980): Revised depolarization corrections for atmospheric extinction, *Appl. Opt.*, *19*, 3427–3428.

CHAPTER 5

POLARIZATION OF THE DAYTIME SKY

5.1 Historical note

The polarization of the sky during the sunlit period has been measured by many authors, beginning with D.F.J. Arago (1786–1853) in 1809. As mentioned above, of the three neutral points which normally occur in skylight polarization, the first was discovered by Arago at the time of his original observations of 1809, while the other two were first observed by Jacques Babinet (1794–1872) and Sir David Brewster (1781–1868) in France and Scotland, respectively, in the 1840s. All three of these neutral points have been named for their discoverers. The variability of the polarization maximum was studied by Bernard, Rubenson, and Crove, also in the first half of the 19th century, in its dependence on atmospheric conditions.

Although progress on the theory of skylight polarization was significant in the first two or three decades after 1850, thanks to the work of G.G. Stokes (1819–1903), John Strutt (Lord Rayleigh, 1842–1919), and others, it was only after the eruption of Krakatao in 1883 that general interest in skylight polarization was rekindled. In fact, for several decades after this momentous event, the principal use of polarization measurements was to detect and study the effects of volcanic dust in the upper atmosphere. Polarization effects of the Krakatao eruption were observed most extensively by Cornu (1884), who also developed a new type of photopolarimeter for the purpose (Cornu, 1890). The basic Cornu design was later modified by Martens (1900) to give the most advanced visual instrument for skylight measurements, and one which continued to be used until it was supplanted by electronic polarimeters developed in the mid-20th century. A second type of polarimeter used in the 1880s was designed by Pickering (1884), but there is no indication that its development was connected with the volcanic cloud. A minor flurry of interest in skylight polarization occurred after the several eruptions of Mount Pelee in 1902–1903 (Kimball, 1903), but many more observations by Kimball (1913a,1913b), Dorno (1919), and Jensen (1928) followed the Mount Katmai eruption of 1912. Polarization effects of volcanic clouds will be discussed further.

Before the time of World War II, measurements of skylight polarization were made mainly by visual means. The Cornu-Martens polarimeter was employed mainly in measuring the degree of polarization, whereas the Savart polariscope was a popular device for measuring the positions of the neutral points. With the birth of the electronics era, new techniques and devices opened up the possibility of not only automatic measurements of light and its polarization, but measurements at wavelengths beyond the sensitivity range of the human eye were within reach for the first time. It is a historical fact, however, that observers such as Pyaskovksaya-Fesenkova (1958, 1960) and Stamov (1961) continued to acquire useful data by visual techniques well past mid-century.

With the advent of electronic techniques for observations and electronic computers for rapid computations, several centers of research developed mainly in the United States and Europe, for the studies of skylight polarization and its relation to atmospheric conditions. Perhaps the largest group, and one with which the present author was associated, was at the University of California, Los Angeles under the direction of the late Professor Z. Sekera. The UCLA center was active for over two decades, and it produced major advances in both theory and measurements. Some of the observations made there during the 1950s and early 1960s are discussed below. A smaller group was formed at the University of Arizona, where the major emphasis was, and continues to be on the theory of radiative transfer. Individuals at the National Center for Atmospheric Research, the Air Force Cambridge Research Laboratories, the National Aeronautics and Space Administration, Texas A&M University, The RAND Corporation, the IBM Corporation, and other facilities have also made significant contributions in polarization studies. Prof. S. Chandrasekhar at the University of Chicago merits special mention for his great advances in the theory.

In Japan, S. Ueno and his colleagues at the Kanazawa Institute of Technology have made significant advances in the theory of radiative transfer, particularly in its application to remote sensing, and T. Takashima at the Meteorological Research Institute of Japan was responsible for the adding method for computation in inhomogeneous atmospheres, as well as for studies of surface reflection and remote sensing of the atmosphere from space. M. Tanaka at the Geophysical Institute of Tohoku University has computed the effects of atmospheric turbidity on intensity and polarization of emergent radiation.

Both observation and theory have been pursued over at least three decades in the USSR, mainly at the Astrophysical Institute of the Kazakhstan Academy of Science, Alma Ata, and at the Institute of Atmospheric Physics, Moscow. The first of these centers was initiated well before the era of electronic polarimeters and was guided by the husband and wife team of V.G. Fesenkov and E.V. Pyaskovskaya-Fesenkova. The work of Pyaskovskaya-Fesenkova, observing through the clear atmosphere of Kazakhstan, extended over nearly three decades and was a model of dedication to a single scientific objective. Her colleagues and successors have continued the investigations with modern instrumentation and equipment. The Moscow group, set up and guided by the late Prof. G.V. Rozenberg, has made significant contributions in both theory and observations. Important theoretical advances in the theory of radiative transfer through turbid media have been achieved by Dr. K.S. Shifrin and his colleagues, most recently at the Institute of Oceanology, Leningrad.

A very active research group was assembled at the Max Planck Institute, Mainz, West Germany, by Dr. K. Bullrich and his colleagues. The Mainz group has been particularly active in studies of atmospheric aerosols, and many skylight observations have been made in different locations around the world. In France, Dr. Jacqueline Lenoble has developed and directed a research activity emphasizing radiative transfer theory and laboratory studies of radiation in various types of fluids. Prof. H.C. Van de Hulst of the Leiden Observatory, The Netherlands, has been a very prolific theoretician and author, and any library in atmospheric optics is not complete without the books of Van de Hulst.

We end this brief and admittedly incomplete historical note in recognition of the early theoretical and observational work by a group in India under the direction of the late Prof. K. Ramanathan of the Physical Research Laboratory, Ahmedabad. Of particular interest are the skylight polarization measurements made during periods of twilight and the analysis of the very complex twilight problem (Dave and Ramanathan, 1956; Dave, 1956; Shah, 1969).

The remainder of this chapter will be devoted to observations of skylight polarization taken during the main sunlit period of the day. They are thought to be representative of skylight properties at various altitudes and geographical locations under different turbidity conditions in the atmosphere. Measurements made by the present author will be emphasized, and much of the material is presented here for the first time. These results will be supplemented by data from other sources as necessary for purposes of comparison and to make the picture as complete as possible.

5.2 Skylight polarization in a clean atmosphere at high altitude

As is well known, the principal interest in measurements of skylight polarization is their sensitivity to dust, haze, and pollution in the atmosphere. The degree of polarization at the maximum is diminished by the effects of aerosol scattering of the incident light, and at the same time the neutral points of the polarization field are shifted from their normal positions. Thus, for more than a century, atmospheric scientists have utilized skylight polarization as an indicator of the existence of these larger types of particles in the atmosphere. There is at present considerable question and some controversy as to the effects of human activity on the possible buildup of atmospheric particulates. If we are injecting aerosols into the atmosphere faster than the atmosphere is cleansing itself, it is not illogical to expect some long-term climatic effects to result from the buildup of the worldwide burden of aerosols. Unfortunately, it is not clear whether the overall effect would be to cause a cooling of the planet, as some scientists are predicting, or whether the planet would undergo a warming trend as a result of the aerosols. The most acceptable analysis indicates that the trend could go either way, depending on the characteristics and location of the aerosols themselves. Neither is it known whether there is a progressive buildup of the total atmospheric burden, since evidence on this point is conflicting. In some locations, such as over the North Atlantic and in the Caucasus, there is strong evidence of a relatively rapid buildup, whereas in the South Pacific area the measurements indicate stability of the aerosol content. The longest record of reliable measurements on which assessment of aerosol loading can be made is that at the Mauna Loa Observatory in Hawaii. This is an ideal site for aerosol monitoring, located at an altitude of approximately 3400 m (11,150 ft) on the northwest flank of Mauna Loa. In this oceanic island setting far from major land masses the natural aerosol loading is minimal, and pollution from the few local sources is retained below the temperature inversion associated with the trade wind regime of the area. In fact, the 25-year record at the Mauna Loa Observatory shows the atmosphere there to be as clean as that found anywhere in the world.

In order to take advantage of this unique site for obtaining a baseline set of skylight polarization data in the clean atmosphere, a polarimeter was installed at the Observatory in January 1977 and operated on each of 96 days over a period of 6 months. Typical results extracted from the large number of measurements are shown in the following sections.

5.2.1 Method of measurement

The measurements were all made with the computer-controlled polarizing radiometer outlined briefly by Coulson and Walraven (1972) and described in more detail in Chapter 11 of this book. Briefly, it is a rotating analyzer-type polarimeter which measures the relative intensity, degree of linear polarization, and orientation of the plane of polarization in the following eight wavelengths (given in units of micrometers): 0.320, 0.365, 0.400, 0.500, 0.600, 0.700, 0.800, 0.900. Four of the interference-type filters are installed in each of the two channels of the instrument. The dual-channel design permits simultaneous measurements at two wavelengths. Several modes of operation are available in the computer program controlling the instrument. The mode most frequently used and the one most applicable to the present discussion, was a scan of the sky at 5° steps through the vertical plane of the sun (the sun's vertical). The 35 data points for each of two filters from 5° above the solar horizon, through the zenith, to 5° above the antisolar horizon were taken in approximately 6 minutes. Upon completion of this scan, the filters were changed automatically and the next scan was begun. The four scans necessary for measurements in all eight wavelengths were thus completed in approximately 24 minutes. The azimuthal orientation of the instrument was updated to coincide to within 0.1° with the azimuth of the sun before each measurement. This was accomplished by computation of the solar position by a computer program and automatic interrogation and readjustment of a shaft encoder of the instrument. A similar arrangement on the elevation drive was used to close an aperture shutter for protection of the photomultiplier tube at any time the instrument was oriented within 1° of the sun (the angular field of view). In the data plots, curves have been continued in the normal configuration across any data gap resulting from a point missed by being too close to the sun.

In order to obtain high precision of the data, the instrument was used in a photon-counting mode. The analyzer was rotated in 45° steps around the optical axis, and a photon count was taken at each step. A sine curve was fitted by a harmonic analysis routine to the resulting eight photon counts after subtraction of a measured dark count from each. The physical quantities of intensity, degree of linear polarization, and orientation of the plane of polarization were obtained from the fitted curve from the mean count rate, amplitude of the curve, and the phase of the curve, respectively. The uncertainties associated with these quantities were computed by the well-known chi-squared method, and were printed out for each measurement. The uncertainties varied between the two channels of the instrument, among the four wavelengths in each channel, and particularly with the intensity of the light. The photon count rate was controlled by a bank of four neutral density filters which were set automatically by the computer. The count rate in each channel was maintained between 2.5×10^5 and 2.0×10^6 counts per second at any time the skylight was sufficiently intense to yield the lesser of these rates. In this case, the uncertainty in each intensity measurement was generally less than 0.1% of the intensity value. Since the degree of polarization and orientation of the plane of polarization were computed from the eight intensity values, their precision also was a function of the total intensity and was generally lower for the shortwave channel than for the longwave channel. The statistical uncertainties are indicated by error bars on some of the curves plotted below, but in general the probable error of measurement was less than 0.1% for the degree of polarization. During periods of twilight, of course, the low count rates produced larger errors than these. The statistical uncertainties could be reduced by choosing longer integration times, but the decreased speed with which the measurements could be taken made it undesirable to increase the integration time much beyond 1 second.

5.2.2 Description of measurement site

The Mauna Loa Observatory, the site of the 1977 polarization measurements, is at an altitude of 3400 m (11,150 ft) on the northwest flank of Mauna Loa on the island of Hawaii (19.5° N. lat., 155.6° W. long.). The atmospheric pressure at this high altitude was about 685 mb, which is less than 68% of standard sea level pressure. This means that the normal optical thickness of the atmosphere at Mauna Loa is much less than at sea level. The Rayleigh component $\tau^R(\lambda)$ of the total optical thickness $\tau(\lambda)$ as computed for the 1959 A.R.D.C. standard atmosphere for an altitude of 3400 m is shown for the eight wavelengths of the measurements in Table 5.1. Shown also in the table are the aerosol optical thicknesses for that altitude as given by the atmospheric model of Elterman (1968). It is likely, however, that the values of τ^A, derived as an average of 119 aerosol profiles at midlatitudes, are too high, perhaps up to an order of magnitude or more, for the clear atmosphere above Mauna Loa. The very small values of total optical thickness that exist at Mauna Loa are specifically evidenced by the measurements of wavelength-integrated transmission of solar radiation at the Mauna Loa Observatory reported by Ellis and Pueschel (1971). Those data, which were selected to represent the clearest conditions measured over a 13–year period, yield a total optical thickness of 0.068. This, of course, is indicative only, as it includes the effects of absorption by water vapor and ozone, as well as attenuation by aerosols and the Rayleigh atmosphere. Later measurements covering the calendar year of 1980 at the Mauna Loa Observatory, made by a multi-wavelength sunphotometer and reported by Shaw (1982), are summarized in Table 5.2. The entries show the most frequently observed values of optical thickness $\tau^A + \tau^O$, and their approximate range $\Delta\tau$ due to the combination of aerosols and ozone in the atmosphere above Mauna Loa. The Rayleigh optical thickness τ^R has been deducted from the data. The relatively large values of $\tau^A + \tau^O$ shown for $\lambda = 0.566$ and 0.616 μm are mainly due to absorption in the Chappuis band of ozone, and will thus have little effect on the polarization data.

TABLE 5.1 Normal optical thickness of the Rayleigh atmosphere $\tau^R(\lambda)$ and of the aerosol component $\tau^A(\lambda)$, according to the atmospheric model of Elterman (1968), at the altitude of the Mauna Loa Observatory for the eight wavelengths of the skylight measurements.

$\lambda(\mu m)$	$\tau^R(\lambda)$	$\tau^A(\lambda)$	$\lambda(\mu m)$	$\tau^R(\lambda)$	$\tau^A(\lambda)$
0.320	0.625	0.096	0.600	0.047	0.058
0.365	0.333	0.091	0.700	0.025	0.052
0.400	0.246	0.077	0.800	0.015	0.048
0.500	0.098	0.064	0.900	0.009	0.045

TABLE 5.2 Most frequently observed values of optical thickness ($\tau^A + \tau^O$) due to a combination of aerosols and ozone in the atmosphere above Mauna Loa, and their approximate range $\Delta\tau$, as measured by Shaw (1982) at 10 different wavelengths over the calendar year 1980.

$\lambda(\mu m)$	$\Delta\tau^A + \tau^O$	$\Delta\tau$	$\lambda(\mu m)$	$\tau^A + \tau^O$	$\Delta\tau$
0.383	0.028	0.022–0.048	0.616	0.043	0.035–0.047
0.415	0.026	0.020–0.048	0.674	0.021	0.015–0.039
0.460	0.021	0.016–0.043	0.690	0.013	0.006–0.021
0.493	0.023	0.017–0.045	0.850	0.011	0.003–0.025
0.566	0.047	0.039–0.049	1.009	0.009	0.005–0.021

Although the wavelengths of Shaw's measurements are different from those of the polarization measurements, the data show that the atmosphere above Mauna Loa is normally very clean, but also that there can be significant perturbations of the turbidity component with increases of aerosol optical thickness by more than a factor of 2. As will be seen below, the effects of such aerosol variations are evident in the polarization data.

Another feature of the site at the Mauna Loa Observatory, important in interpreting the skylight data is the albedo of the underlying surface. A schematic diagram of the various components of reflected and scattered light entering the field of view of the polarimeter is shown in Fig. 5.1. In the absence of reflection by any surface located below the level of the Observatory, the only light entering the field of view would be I_p and I_m due to primary and multiple scattering of the incident sunlight by the atmosphere. In reality, however, radiation reflected from various surfaces below the level of the Observatory enters the atmosphere from below and a portion is scattered into the field of view of the instrument. Since the reflection process alters the polarization characteristics of such radiation, as well as its intensity and angular distribution, atmospheric scattering of the reflected radiation introduces components which have different properties than those of I_p and I_m with a consequent modification of the skylight characteristics.

The principal reflecting surfaces in the vicinity of Mauna Loa are the land surfaces, mostly dark lava and vegetation, the surface of the sea, and cumulus clouds with tops at 2000- to 2500-m altitude. These surfaces contribute components I_e, I_s, and I_c, respectively, to Fig. 5.1. The additional component due to scattering by aerosols and gases in the atmospheric layer below the Obser-

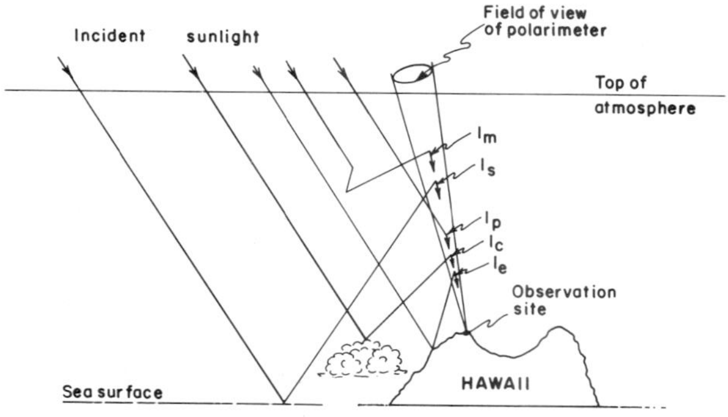

Figure 5.1 Schematic diagram showing the main components of skylight incident at the Mauna Loa Observatory. (I_p is the intensity due to primary scattering of direct sunlight, I_m is the intensity due to multiple scattering, I_c is the intensity due to reflection by clouds, I_s is the intensity due to reflection by sea surface, I_e is the intensity due to reflection by lava, vegetation, and other materials of the land).

vatory site is included in I_c. Probably the least important of the reflected components is that from the sea surface I_s. For high sun elevations the albedo of the sea is only 4% to 5%, in which case I_s is small compared to the other components. For low sun elevations the sea surface has high albedo, the reflected radiation is strongly directional, and it may be highly polarized. Fraser (1968) has shown that the positions of the neutral points are significantly affected by reflection from a still water surface, the neutral points, being in a region of very little polarization, are particularly sensitive to the addition of polarized components. Two factors tend to decrease the influence of sea reflection on the polarization field at Mauna Loa. First, the Observatory is a minimum of about 36 km, and a maximum of about 80 km from the ocean. Radiation reflected from the ocean surface would be strongly attenuated, particularly at low sun elevations, by the long pathlength through the turbid atmosphere between the ocean and the Observatory. In addition, any clouds which exist between ocean and Observatory would be very effective in interrupting the flow of this type of radiation.

Component I_e, due to reflection from the surrounding lava and vegetated surfaces must also be small. The albedo of the dark lava which covers the area near the Observatory is only 4% to 6% percent, and for wavelengths below 0.7μm the albedo of green vegetation is equally low. For wavelengths 0.7 to 0.9 μm vegetation reflects more strongly, but areas with extensive vegetation on Hawaii are at least 10 km from the Observatory and are confined mainly to the windward side of the island. Thus it is logical to assume that component I_e is not very significant in modifying the skylight over Mauna Loa.

The most important effect of reflection is largely due to clouds. The extent of cloudiness in the region of the island is extremely variable, as is the reflectivity of the clouds themselves, so component I_c will have large variations. One simplifying feature is that the radiation reflected from clouds has low polarization. Thus the addition of cloud-reflected light to the skylight field will result in a decrease in the degree of polarization at the maximum without changing the positions of the neutral points. These effects will be mentioned further below. We turn now to the skylight measurements themselves.

5.2.3 Skylight polarization in the plane of the sun's vertical as measured in Hawaii

5.2.3.1 Angular distribution in the sun's vertical

The plane through the local zenith that passes through the center of the sun has traditionally been termed the sun's vertical. This convention will be adopted here, although other authors have used terms such as the principal plane or sun's meridian. The polarization field in the sun's vertical carries more information than that in any other plane across the sky, as the neutral points are found normally only in the sun's vertical and the polarization maximum is in, or very close to, this plane. The measurements to be discussed in this section were made by this author at the Mauna Loa Observatory by keeping the polarimeter oriented in the sun's vertical and making measurements at 5° intervals in that plane from 5° above the solar horizon, through the zenith, to 5° above the antisolar horizon. Thus one such scan of the sky yielded measurements of the relative intensity, degree of linear polarization, and orientation of the plane of polarization in two wavelengths at 35 points across the sky. The operational routine for the measurements was not always the same from day to day, as there were many features to investigate and eight wavelengths for the measurements, but on several occasions the entire sunlit period was spent making repeated scans of the sky with a filter change after each scan. The four scans required for one complete cycle through the filters could be accomplished in about one-half hour, which corresponds to a change of sun elevation of approximately 7° at times of low sun elevation and less than this during the main part of the day. It was found that the principal effect of changing sun elevation was a shift of the position of the polarization field in the sky such as to maintain the location of the maximum at approximately 90° from the sun, with a small region of negative polarization surrounding the sun.

These features can be seen explicitly in the diagram of Fig. 5.2, in which the degree of polarization in the sun's vertical from the zenith to 5° above the solar and antisolar horizons is shown for six different sun elevations at a wavelength of 0.70 μm. The data were taken during the forenoon of February 19, 1977. The forenoon was favored for all of the measurements taken at the Mauna Loa Observatory, as the atmosphere was most stable and least turbid during that period, since orographic clouds, which often developed by noon or shortly

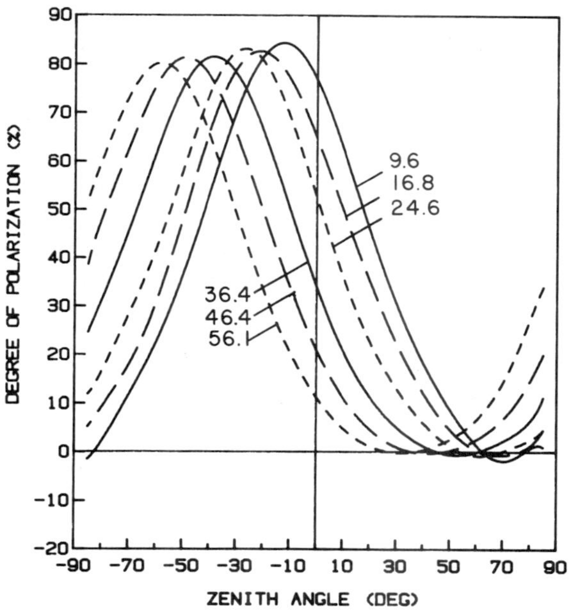

Figure 5.2 Degree of polarization of skylight at $\lambda=0.70$ μm as a function of zenith angle in the sun's vertical as measured at six selected sun elevations at the Mauna Loa Observatory in the very clean atmosphere of February 19, 1977.

thereafter as the sun heated up the dark lava surrounding the site, were at a minimum in the forenoon. In addition, the trade wind cumulus clouds, which constituted a bright and variable reflecting surface below the level of the Observatory, were normally less extensive in the morning hours than later in the day. The measurements of February 19 were chosen for Fig. 5.2 because the atmosphere was very clear on that day, and the trade wind cumulus clouds, as judged by both visual estimates and a photo from a weather satellite, covered less than one-tenth of the sky throughout a wide area surrounding the island. Thus the very minimum of surface reflectance, combined with the low turbidity of the atmosphere and the high altitude of the site, make these and other data taken on February 19 ideally suited to producing high values of skylight polarization. In fact, at no other time or place during all of the thousands of scans of the sky made by the author over a 30-year period has the skylight been more highly polarized than it was on that particular day at Mauna Loa.

Many of the features shown by computations of polarization in a Rayleigh model of the atmosphere (see Chapter 4) are evident in Fig. 5.2. Although the 85% maximum of the early morning data is somewhat lower than that for the Rayleigh case with surface albedo A=0 shown by the solid curves of Fig. 4.21, it is almost the same as that for albedo A=0.25. While it is unlikely that the albedo at Mauna Loa was as high as 0.25 on the day in question, other effects must be considered in the comparison. First, if molecular anisotropy had been considered in the computations, the maximum shown in Fig. 4.21 would have been reduced by nearly 4%. Second, the natural atmosphere, even at Mauna Loa, always contains some aerosols, which tend to depress the maximum polarization. On consideration of these two additional factors, it is not difficult to resolve the small differences between the measured polarization maximum and that predicted by the theory.

The curves of Fig. 5.2 show a gradual decrease of the polarization maximum with increasing sun elevation, the decrease amounting to approximately 5% as the sun elevation changed from 9.6° to 56.1°. This is also is in accord with theoretical expectations. The data of Table 4.5 for the Rayleigh atmosphere showed a 2% to 3% change of magnitude, with lowest values generally occurring for sun elevations in the 30° to 50° range. This change is smaller than the 5% change shown by the measurement, but Table 4.5 is for A=0. For appreciable albedo there is a much larger change of polarization maximum with sun elevation, and the albedo at Mauna Loa, while small, was certainly greater than zero. Thus the change in the maximum shown in Fig. 5.2 is explainable by a combination of the change expected for a Rayleigh atmosphere with $A=0$ plus an added effect due to a certain amount of reflection from the underlying surface. The small regions of negative polarization in the vicinity of the sun, with values of -1.0% to -1.5%, are well confirmed by theory, as are the neutral points of Fig. 5.2.

The distribution of polarization in the sun's vertical for wavelength $\lambda=0.320$ μm for the very clear atmosphere of February 19, 1977, at Mauna Loa are shown in Fig. 5.3. These data were taken with the second channel of the polarimeter simultaneously with those of the previous figure. The two most obvious differences between the measurements at the two wavelengths are the decreased magnitude of the positive maximum in the ultraviolet and the increased importance of the negative branch of the curves at the shorter wavelength. For instance, the approximately 85% maximum at $\lambda=0.70$ μm at sun elevation 9.6° is compared to about 65% at the same time for $\lambda=0.32$ μm. The 20% difference

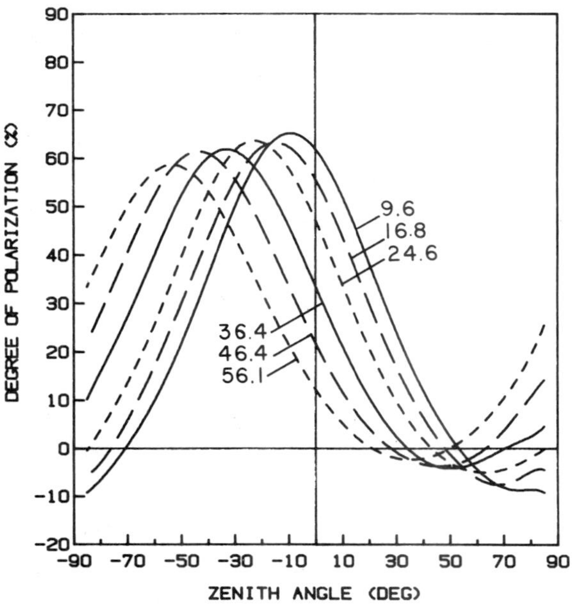

Figure 5.3 Degree of polarization of skylight at $\lambda=0.32$ μm as a function of zenith angle in the sun's vertical as measured at six selected sun elevations at the Mauna Loa Observatory in the very clean atmosphere of February 19, 1977.

persists at other sun elevations as well. Enhancement of the negative branch in the ultraviolet is very pronounced, both in magnitude of the negative polarization and the area of the sky over which the polarization is negative. At $\lambda=0.70$ μm of (Fig. 5.2) the maximum negative value was approximately -1.0%, and the area covered was confined to the immediate vicinity of the sun. In the ultraviolet, however, a negative value of nearly -10% was observed at the lowest sun elevation, and the minimum magnitude reached was -2.5% (Fig.5.3). The Babinet and Brewster points were separated by a minimum of 30° at $\lambda=0.320$ μm, whereas they essentially merged together at the higher sun elevations at $\lambda=0.70$ μm. The Arago point existed only at sun elevation 9.6° at the longer wavelength, whereas it persisted for another 15° increase of sun elevation for the ultraviolet.

5.2.3.2 Polarization 90° from sun

In one mode of operation, the computer could keep the polarimeter oriented at 90° from the sun in the sun's vertical and cause the filters to be changed in sequence to give measurements in the eight different wavelengths with a repetition rate of less than 1 minute. The 90° orientation corresponds very closely to the position of the polarization maximum P_{max}, so this mode of operation gave frequent values of P_{max} and thus permitted a study of the variation of P_{max} with time or sun elevation. Unfortunately, of course, a dedication of the instrument to P_{max} measurements prevented its use for other purposes, so such series were relatively infrequent in the measurement program. They did, however, yield some valuable information on the behavior of P_{max}.

Two selected days on which P_{max} was monitored in this manner at Mauna Loa were March 9 and 10, 1977. The atmosphere above the Observatory was very clear on both days, as indicated by the zenith polarization at sunrise being approximately 85% for $\lambda=0.80$ μm. This criterion was used throughout the period as an indicator of atmospheric clarity. There was, however, a considerable coverage of trade wind cumulus clouds below the level of the Observatory. Visual estimates from the Observatory gave 30% to 50% undercast on March 9, and 60% to 70% undercast on March 10, whereas satellite photos of a wide area around the islands gave estimates of 30% on March 9 and 80% on March 10. Measurements of P_{max} on those days at wavelengths of $\lambda=0.365$ and 0.80 μm are shown as a function of sun elevation in Fig. 5.4. The curves show a very significant decrease of polarization in the period from sunrise to a sun elevation of 20° to 25°, the decrease being most pronounced on March 10 at λ=0.80 m. Presumably the greater reflection from underlying clouds caused P to be decreased more and remain lower on March 10 than on the previous day. The effect of cloud reflection is much less pronounced at λ=0.365 μm than at 0.80 μm, a feature which is in accord with the theoretical expectations discussed previously. The onset of the sharp decrease of P_{max} in the late forenoon (sun elevation 55° to 64°) coincided approximately with a change of wind at the Observatory from predominantly downslope flow preceding the change to a flow up the mountain from below after the change. Presumably the upslope wind carried aerosol particles and water vapor from below the level of the temperature inversion as the inversion was destroyed by solar heating, thereby increasing the general atmospheric turbidity and causing the relatively rapid decrease of skylight polarization in the late forenoon.

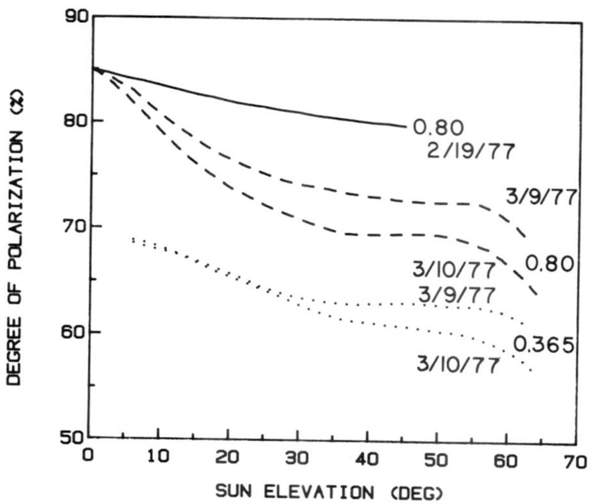

Figure 5.4 Degree of polarization at 90° from the sun in the sun's vertical at wavelengths of λ=0.365 and 0.80 μm as a function of sun elevation for March 9 and 10, 1977, at the Mauna Loa Observatory. A curve for the clear-day case of February 19, 1977, is given for comparison.

5.2.3.3 Spectral dependence of polarization

The dispersion (change with wavelength) of the polarization field in the sun's vertical is shown explicitly in Fig. 5.5 for the virtually cloudless day of February 19, 1977, at Mauna Loa. For comparison purposes the curves have all been shifted slightly along the abscissa to correspond to a sun elevation of 18°. The maximum of this angular shift was only about 3°, and the previous diagrams have shown the changes of the polarization magnitude with changing sun elevation to be a second-order effect. Thus the data of Fig. 5.5 are valid, to a close approximation, for sun elevation 18° at four representative wavelengths of the measurements. While there is some obvious noise in the data, the expected pattern of behavior is clearly shown. There is little spectral dependence of the curves at the longer wavelengths, as the clear atmosphere and high altitude of the site resulted in mainly primary scattering of the longer wavelengths by molecules and the few aerosol particles present. At wavelengths of $\lambda=0.50$ μm

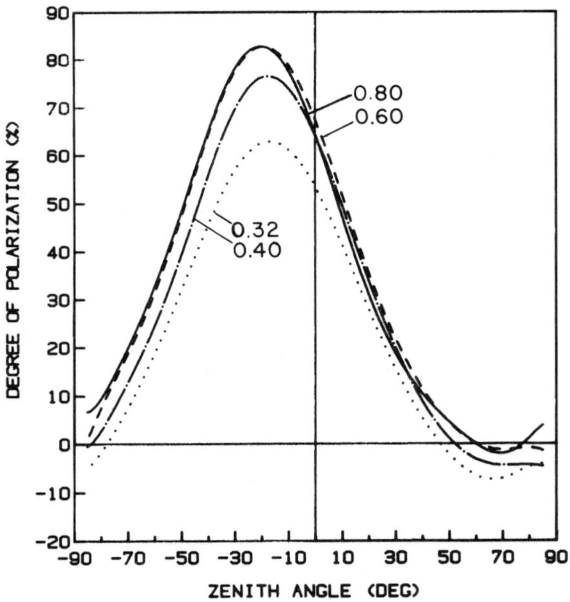

Figure 5.5 Degree of polarization of skylight as a function of zenith angle in the sun's vertical as observed in four different wavelengths at the Mauna Loa Observatory. The curves have been shifted slightly in abscissa to correspond to a sun elevation of 18°.

and below, however, multiple scattering, together with some probable aerosol influence, had a very significant effect in reducing the polarization maximum, increasing the magnitude of the negative polarization branch, and shifting the positions of neutral points. In fact, the Arago point existed at this sun elevation only at wavelengths $\lambda=0.50$ μm and below, while the Brewster point was absent at the short wavelengths but well defined at $\lambda=0.70$ μm and above. The region of negative polarization surrounding the sun was very pronounced in the ultraviolet, but of minor significance at near infrared wavelengths.

The spectral dependence of the polarization maximum P_{max} at sun elevations of 10° and 45° as determined from the measurements at Mauna Loa is shown by the solid curves in Fig. 5.6. A certain amount of graphical interpolation was necessary to obtain values at specific sun elevations. Curves (dashed) computed for the Rayleigh atmosphere corresponding to the altitude of Mauna Loa at a sun elevation of 11.5° and surface albedo A=0 and 0.25 are given for comparison. Althrough the magnitude of P_{max} shown by the measurements is different for the two sun elevations, an effect seen already in Figs. 5.2 and 5.3,

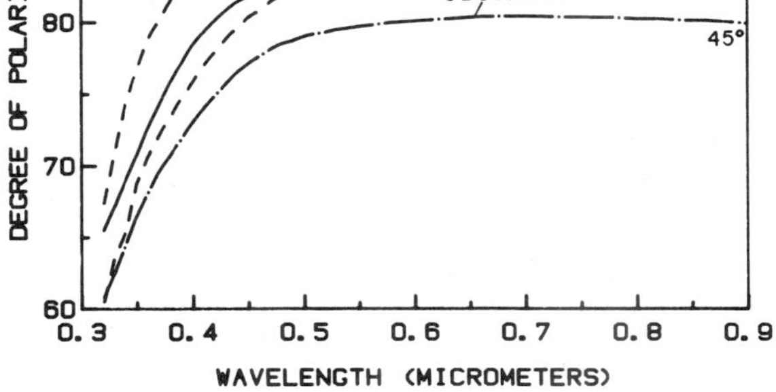

Figure 5.6 Spectral dependence of the polarization maximum as measured on February 19, 1977, at the Mauna Loa Observatory at sun elevations of 10° and 45°, and as computed for the Rayleigh atmosphere at sun elevation 11.5° with surface albedo A=0 and 0.25.

its spectral dependence is closely analogous at the two sun elevations. It is likely that the discrepancies shown are within the error of measurement. The critical dependence that occurs in the ultraviolet region can be ascribed to the similarly critical dependence of Rayleigh scattering on wavelength. This is confirmed by the similar slopes of the measured and computed curves at wavelengths below $\lambda=0.40$ μm and by the comparable magnitudes of P_{max} for the two cases. The minor changes of P_{max} between $\lambda=0.60$ and 0.90 μm for the measurements, however, indicates minor effects of Rayleigh scattering at these wavelengths in comparison to other processes. This interpretation is bolstered by the continued slope of the Rayleigh curves and their much larger values in the longer wavelength region.

The virtual absence of wavelength dependence of P_{max} in the red and near-infrared regions may be ascribed to a combination of aerosol and surface reflection effects. The depression of the magnitude of P_{max} for the measurements from that for the Rayleigh atmosphere was probably due to aerosol scattering, even though atmospheric turbidity at the time was very small. The Rayleigh optical thicknesses for Mauna Loa are only 0.047 and 0.009 at $\lambda=0.60$ and 0.90 μm, respectively. Although the aerosol optical thickness at the time is unknown, it was probably comparable to or larger than these values. If the efficiency of scattering for aerosols was inversely proportional to wavelength, as is not unusual for aerosols, then one would expect, with other things being equal, that P_{max} would be higher at $\lambda=0.90$ than at 0.60 μm. The measurements show little difference between them. Two different processes might explain the equivalence. First, if surface reflection increased with increasing wavelength, then the depolarization effects of surface reflected light would tend to counter the normal increase of P_{max} with wavelength. The known increase of reflectance with wavelength of both mineral surfaces and of plants in the near-infrared region, certainly make a contributon toward erasing the normal increase of P_{max} with wavelength. It is unlikely, however, that surface reflection could have reached values greater than 0.25, as would have been necessary according to the curves of Fig. 5.5. A second contributing factor must have been a change of the aerosol scattering pattern with wavelength, but the absence of data on the aerosols which existed at the time makes it impossible to assess the aerosol contribution.

5.2.3.4 Effects of surface reflection at Mauna Loa

In view of the effect of surface reflection on the polarization field mentioned above, it is worth looking further at the Mauna Loa data from this standpoint.

The island of Hawaii is a relatively large island, roughly circular, and about 125 km in diameter. The mountain mass of Mauna Loa dominates more than half of the island and consists of lava flows of various sizes and ages. The older lava flows are mainly covered by forests below an altitude of 2000 m. The more recent flows, regardless of altitude, still consist of bare lava, and the entire mountain above 2500 m is essentially devoid of vegetation. The lava is universally of dark color, mainly black or dark brown, and has a reflectance of not more than 0.05%. The forests, as is the case for most plants, have a low reflectance of a few percent in the visible region, with a pronounced increase at wavelengths beyond 0.75 μm. There is little seasonal effect of vegetation in this tropical region, and the only significant changes in albedo of the solid surface is in the relatively infrequent periods of surface snow cover on the upper reaches of the mountain. No snow was present during the 1977 series of measurements. The surface of the sea near the island was normally roughened by trade winds, and thus had a low reflectance of only a few percent except at times of very low sun elevation. In addition, 36 to 80 km in horizontal distance and over 3 km vertical distance between the observation site and the nearest water surface tended to decrease the effects of reflection from the sea on skylight polarization at Mauna Loa.

This means that the only significant reflection effects in the skylight observations are those of clouds that often prevail over the sea and on the flanks of the mountain surrounding the Observatory. These trade wind cumulus often formed a solid cloud mass of over a kilometer in vertical extent with tops at 1 to 1.5 km below the level of the Observatory. The sunlight reflected from the cloud tops constituted a strong source of diffuse radiation with low polarization, and had a significant depolarizing effect on the skylight at Mauna Loa. The day-to-day variations of the clouds caused corresponding variations in the skylight polarization, and tended to mask the effects of changes in the turbidity component of the atmosphere.

A further comparison of the measurements made under very clear conditions on February 19 with theoretical computations for the same sun elevation and wavelength in the Rayleigh atmosphere is shown in Fig. 5.7. The computations were made specifically for this comparison and kindly made available by Dr. Bruce Fitch for use here. The dashed curves are from computations, the top one being for Lambert surface albedo $A=0$ and the bottom on for $A=0.25$.

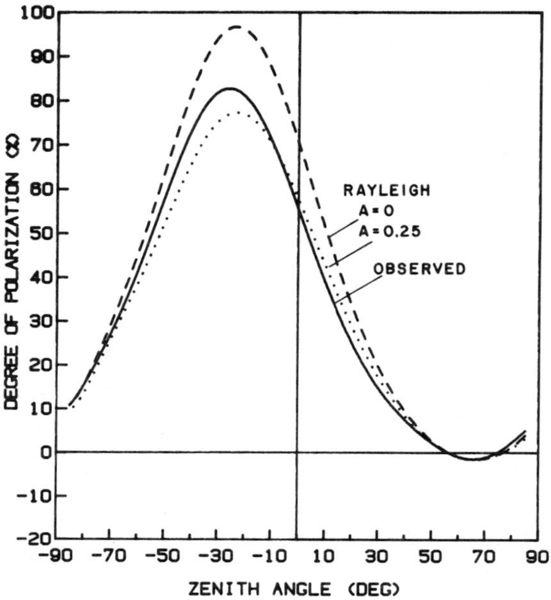

Figure 5.7 Degree of polarization as a function of zenith angle in the sun's vertical as given by computations for the Rayleigh atmosphere with surface albedo A=0 and 0.25 (broken curves) and as measured at the Mauna Loa Observatory on February 19, 1977. (Sun elevation 24.1°, wavelength 0.70 μm.)

The measurements, as shown by the solid curve correspond more closely to a surface albedo of $A=0.25$ than to $A=0$, and, in fact, through a range of angles from the Babinet point to slightly past the zenith, the measured polarization was lower than that for $A=0.25$. This discrepancy must be attributed principally to the effects of aerosols. The fact that the measured polarization was higher than that computed for $A=0.25$ in the region at and beyond the polarization maximum indicates a surface albedo considerably below $A=0.25$. As mentioned above, the small amount of cloudiness at the time of the measurements, together with the low albedo of both sea surface and dark lava surrounding the site, bolster the conclusion of small albedo effects. The predominantly forward-scattering pattern for aerosols, however, is consistent with the explanation of relatively low polarization at scattering angles of up to 70° in Fig. 5.7.

5.2.3.5 Polarization in turbid atmosphere at Mauna Loa

The atmosphere at Mauna Loa is not always as clear as might be indicated by the above discussion. On occasion, disturbed weather conditions create thunderstorms in the area which cause particles from the sea or from local sources of pollution to be carried in updrafts to high altitudes in the atmosphere. The effects of such high-level turbidity on polarization during twilight are shown in a later section. A second and more persistent type of turbidity at Mauna Loa is that due to dust transported across the western Pacific during the spring months from the large desert areas of northern China and Mongolia. At first glance, it seems unlikely that significant quantities of dust would be carried a quarter of the way around the world, but computations of atmospheric trajectories show exactly that to be the case. Further evidence of eastern Asia as the source of the dust comes from examinations of the dust particles themselves, and from the correlation between dust episodes in Hawaii with observations of dust storms in northern China a few days previously.

These episodes of high-level dust in the atmosphere are almost universally confined to the spring months of March, April, and May. Although they do not necessarily occur annually, they do occur most years, and some years see more than one dust intrusion. A dusty period during the polarization measurements of 1977 at Mauna Loa occured during the first few days of May. Distinct dust layers were visible in all directions from the Observatory, and the fact that the peak of Mount Haleakala on the island of Maui at a distance of about 140 km was obscured by the dust indicated that the dust was in the middle troposphere. Measurements made during the time showed a strong decrease of the polarization maximum, particularly at the longer wavelengths, but the general configuration of the curves was not changed much otherwise. Selected data taken on several scans of the sky during the dust episode are shown by the curves of Fig. 5.8 for a wavelength of 0.70 μm. The normal shift of position with changing sun elevation is obvious, but the maxima are depressed by 10% to 15% from their values in clear conditions. For a sun elevation of 81.0° the polarization depression reaches nearly 25%, due no doubt to the region of the maximum polarization coinciding with the dust layers near the horizon. The neutral points were mainly absent at the higher sun elevations, a feature which has been observed to occur in other cases of a turbid atmosphere (see section 7.2). In addition, a careful analysis shows that the position of the maximum, instead of being at

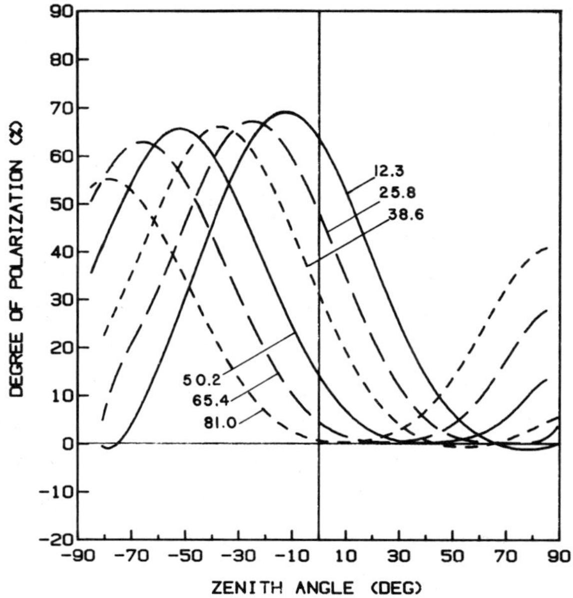

Figure 5.8 Degree of polarization of skylight at λ=0.70 μm as a function of zenith angle in the sun's vertical as measured at six selected sun elevations at the Mauna Loa Observatory during an episode of Asian dust in the atmosphere over Hawaii (May 2, 1977; λ=0.70 μm).

approximately 90° from the sun as in a clear atmosphere, has shifted by 2° to 4° away from the sun in this case. This is in accord with a similar shift observed by Sekera and Collaborators (1955) in the turbid atmosphere of southern California.

The magnitude of the polarization maximum as measured at eight wavelengths during the dust episode of May 2, 1977, is shown for four different sun elevations in Fig. 5.9. An equivalent curve taken under the clear conditions of February 19, 1977, is given for comparison. While the precision of the data for the curves is only perhaps ±1%, obtained by graphical interpolation for the specific sun elevations, it is sufficient to show a major effect of the dust at the longer wavelengths. In the short wavelength range, the dust appears to have had only minor influence, the degree of polarization at λ= 0.32 μm being essentially the same as that for the clear atmosphere. Presumably this comes about by the polarization at the short wavelengths being controlled mainly by

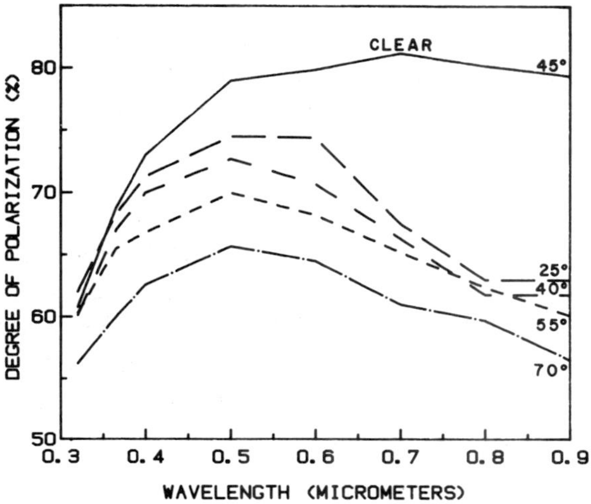

Figure 5.9 Magnitude of the polarization maximum as a function of wavelength at four sun elevations from skylight measurements during the dust episode of May 2, 1977, at the Mauna Loa Observatory. A similar curve for the clear atmosphere for a sun elevation of 45° on February 19, 1977, is shown for comparison.

multiple scattering in the lower atmosphere. The relative increase of optical thickness by the dust was relatively small in comparison to the large Rayleigh optical thickness of the clear atmosphere. At the longer wavelengths, however, optical thickness of the dust was large compared to that of the Rayleigh atmosphere, and the strong diminution of the polarization magnitude was the result. Since the strongest diminution occured in the $\lambda=0.8$ and $0.9\ \mu m$ wavelengths, it would seem to support the idea that the particles were generally in the larger submicrometer size range.

5.3 Skylight polarization at the zenith

A popular mode of measuring the polarization of the sunlit sky has been to make the measurements with the polarimeter oriented in the direction of the zenith. This is particularly convenient, as the instrument remains stationary, thereby eliminating the need for a goniometer on the mount and the continual measurement of angles. This method was used frequently in Europe in the first decades of this century, but modern technology has eased the problems of instrument orientation and control. Zenith polarization measurements are still useful, however, particularly in the case of twilight studies, and the simplicity of this method remains an attractive feature.

Under normal conditions the maximum of zenith polarization occurs when the sun is approximately on the horizon. A steady decrease occurs as the sun elevation increases and the point of maximum polarization, which is normally about 90° from the sun, descends toward the horizon. This pattern is modified by optical disturbances in the atmosphere, and those modifications are useful indicators of the disturbances. The daily variation of zenith polarization has the same general character as that of maximum polarization, but the range of variations is much larger at the zenith.

The variations in zenith polarization as a function of sun elevation, according to observations by various authors, are shown in Fig. 5.10. The two solid curves are for data from Mauna Loa, the higher of the two for a case of very clear

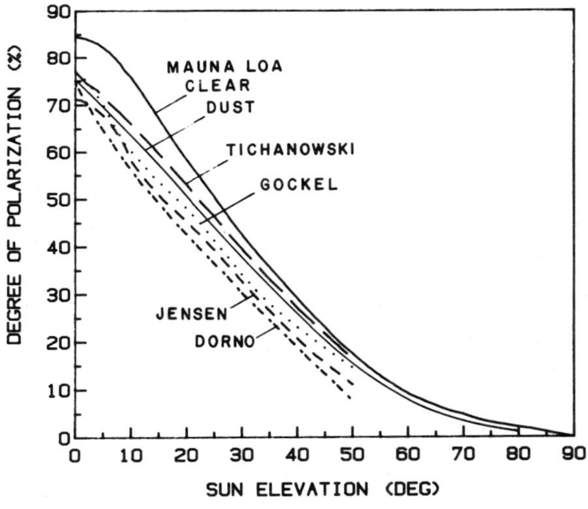

Figure 5.10 Degree of polarization at the zenith as a function of sun elevation as observed at $\lambda=0.70$ μm in clear and turbid atmospheric conditions at the Mauna Loa Observatory (solid curves) compared to that reported by various authors from measurements in Europe. Jensen (1899, 1928, 1932), Gockel (1920), Dorno (1919), Tichanowski (1926, 1927).

atmosphere and the lower one for an episode of Asian dust carried over Hawaii. The dust decreased the zenith polarization at sunrise by about 10%. The difference decreased with time as the optical path of the sunlight through the atmosphere became shorter, but there remained about a 2% difference up until solar noon. By combining data for several clear days, it was possible to extend the curve to the case of zenith sun, as the sun passes directly through the zenith twice per year at the latitude of Mauna Loa (19.5° N). At this time, of course, the zenith polarization is identically zero. The other data of Fig. 5.10 were taken in Europe, all by visual polarimeters. The most complete studies were those of Jensen (1899, 1928, 1932) in the industrialized cities of Hamburg and Arnsberg of Germany. The observations of Gockel (1920), also in Germany, show values comparable to those of Jensen. The data of Dorno (1919), taken in the clear atmosphere of relatively high altitudes in Switzerland, appear anomalously low. The fact that they were in white light does not entirely explain their depression, as observed by Jensen and Gockel. The data of Tichanowski (1926, 1927) are for the red region of the spectrum, as defined by a red absorption-type filter, and were mostly from the southern regions of Russia. There is a striking correspondence between Tichanowski's data and that of the dust episode at Mauna Loa. Although there are considerable variations among the observations, they all show the same general trend, and the discrepancies, with the possible exception of those of Dorno, can all be resolved on the basis of observational conditions.

5.4 Skylight polarization over snow surfaces

As has been stated many times, the polarization and intensity of light from the sunlit sky are sensitive indicators of atmospheric turbidity. Measurements of skylight, however, have generally been confined to midlatitude regions, in which case the atmosphere has a relatively high content of turbidity and large short-term variations. On the other hand, the atmosphere over very high-latitude regions is far removed from significant sources of pollution, so atmospheric conditions in the high latitudes should be a better index of secular trends in turbidity of the total atmosphere than are low latitude atmospheric conditions. Thus it is better to look at skylight measurements made in the very high-latitude regions of the earth. These same measurements, of course, are representative of the effects of high surface reflection, since the data available from high latitudes have been taken over extensive areas of snow cover. Because of the geometry, the data are confined to the relatively low sun elevations which prevail at high latitudes.

Bullrich et al. (1966) reported on skylight measurements made at Camp Century in North Greenland (77.2° N lat., 61.1° W long.) at an altitude of 2000 m in four different wavelengths. Typical results for wavelengths of 0.401 and 0.709 μm are shown by the curves of Fig. 5.11 for a sun elevation of 10° to 12°. The strikingly low values of the polarization, being in the 30% to 50% range at the maximum, are due to the high reflectivity of the surface and the nearly unpolarized character of the reflected light. The maximum was found to decrease with increasing sun elevation to a somewhat greater extent than at locations with low surface reflection because of the increase of surface illumination, and, hence, of surface-reflected light, as the sun rises above the horizon. The curves of Fig. 5.11, as well as other data from the measurements, show a slight increase of P_{max} with increasing wavelength through the visible region and with little spectral dependence in the near infrared. This same type of behavior was seen in the very clean atmospheric conditions in Hawaii, but it is exacerbated by reflection from snow because of the slight decrease of snow reflectivity with

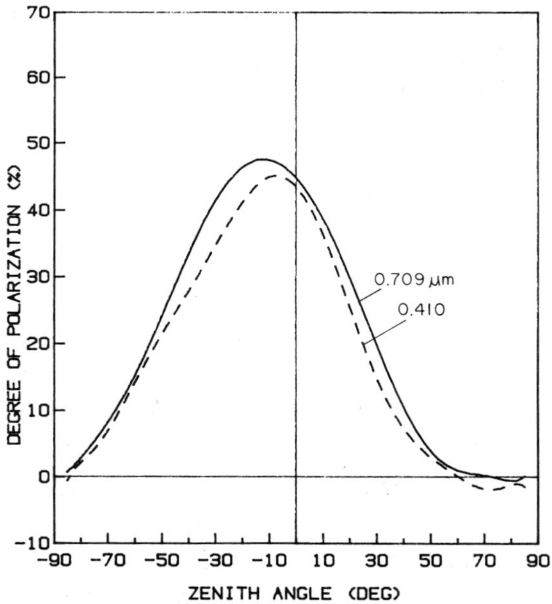

Figure 5.11 Degree of polarization of skylight as a function of zenith angle in the sun's vertical at two different wavelengths as reported by Bullrich et al. (1966) from measurements at Camp Century in North Greenland.

increasing wavelength. Incidentally, Bullrich et al. state that "when turbidity is small, and reflectivity is low, the polarization in the short wavelength range is always higher than for long wavelengths." The measurements in Hawaii show that this is not a general rule, as only under conditions of appreciable atmospheric turbidity was there a decrease of polarization with increasing wavelength at that location. Furthermore, computations for a Rayleigh atmosphere show increasing polarization with increasing wavelength. The reverse behavior, as stated by Bullrich et al. and observed at most lower altitude stations, is undoubtedly an effect of atmospheric aerosols producing depolarization preferentially at the longer wavelengths.

Computations of the polarization in a Rayleigh atmosphere, as given by Coulson et al. (1960) were extended by Bullrich et al. (1966) to the case of surface albedo $A=1.0$. A comparison of those results with the Greenland measurements show the Rayleigh atmosphere to have higher polarization than the measurements, even for a theoretical albedo of $A=1.0$. Bullrich et al. ascribe the difference to the effects of a few aerosol particles in the atmosphere, but mention that an additional factor may be the neglect of molecular anisotropy in the Rayleigh computations. For an atmospheric model with aerosol particles of 1 μm radius, computations by Bullrich et al. showed that it required only one particle per cm^3 to reduce the computed polarization to less than that given by the Greenland observations.

A short series of skylight polarization and intensity measurements was made by Coulson et al. (1974), over snow-covered terrain at McMurdo, a sea level station on the coast of Antarctica. The instrument was a rotating analyzer-type polarizing radiometer with six interference filters in the range 0.321 to 0.872 μm. Data for all six wavelengths taken under conditions of clear sky and good visibility on the morning of February 6, 1974, are shown in Fig. 5.12. Although the polarization maxima were lower in Greenland than at McMurdo, the maximum at McMurdo was still only 60%. The surface albedo was no doubt higher in Greenland than it was at McMurdo, as the temperature at McMurdo was at approximately freezing at the time of the measurements and the snow surface was considerably aged with a coarse crystalline structure. The low temperatures at Camp Century in Greenland would maintain a powder type of snow with a higher reflectivity than that at McMurdo.

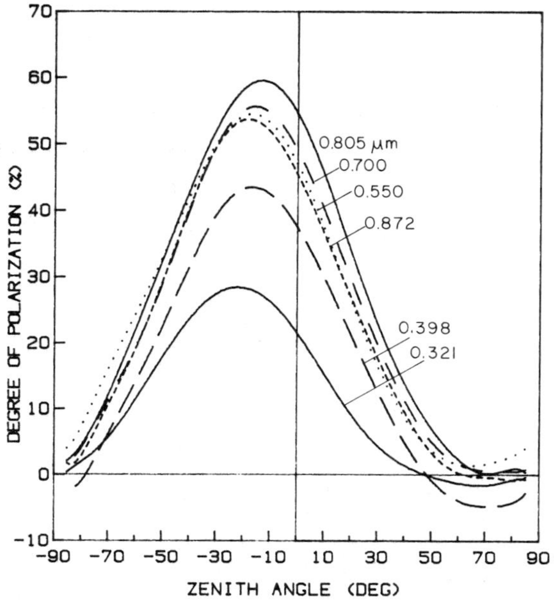

Figure 5.12 Degree of polarization of skylight as a function of zenith angle in the sun's vertical at six different wavelengths from measurements over the snow surface taken at McMurdo on the Antarctic coast (Coulson et al., 1974).

The measurements of Bullrich et al. (1966) did not show regions of negative polarization near the sun or antisun, and the authors did not mention the subject. An extrapolation of their data, as shown in Fig. 5.11, does indicate the probable existence of negative branches of the curves, though the extrapolation is open to question. The data of Fig. 5.12 for McMurdo, however, show definite regions of negative polarization at the shorter wavelengths, but not at wavelengths of λ=0.70 μm and beyond. In fact, the magnitude of negative polarization is greater than 5% at λ=0.398 μm, which is comparable to that at other locations. The measurements for the curve of λ= 0.321 μm were made at a higher sun elevation (18°) than those for other wavelengths (11°–17°), and the strong surface-reflected and largely unpolarized light for the higher sun elevations causes low polarization magnitude in both positive and negative branches of the curve without much change in the neutral points.

A long series of skylight measurements was taken at the geographic South Pole during the time Bruce Fitch spent an entire year there. From October 1974 to October 1975, he made hundreds of scans of the sky during the sunlit period. The instrument used was the same polarizing radiometer used for the measurements at McMurdo the previous year. Typical distributions of the degree of polarization in the sun's vertical are shown for all six wavelengths in Fig. 5.13. These data were taken on October 19, 1974, and since the sun elevation is essentially constant throughout a given day at the South Pole, all of the curves of Fig. 5.13 are for the same sun elevation of 14.0°. The general distribution of the polarization is similar to that observed over snow at McMurdo and Greenland, with relatively low values of the polarization maximum at all wavelengths and regions of negative polarization at the shorter wavelengths only.

Since the South Pole station is at a pressure altitude of approximately 3000 m, the normal optical thickness for the six wavelengths of the polarizing radiometer is different at the South Pole than at McMurdo, and thus the polarization, even for the case of a Rayleigh atmosphere, must be different at the two locations. The Rayleigh optical thicknesses for the six wavelengths of the

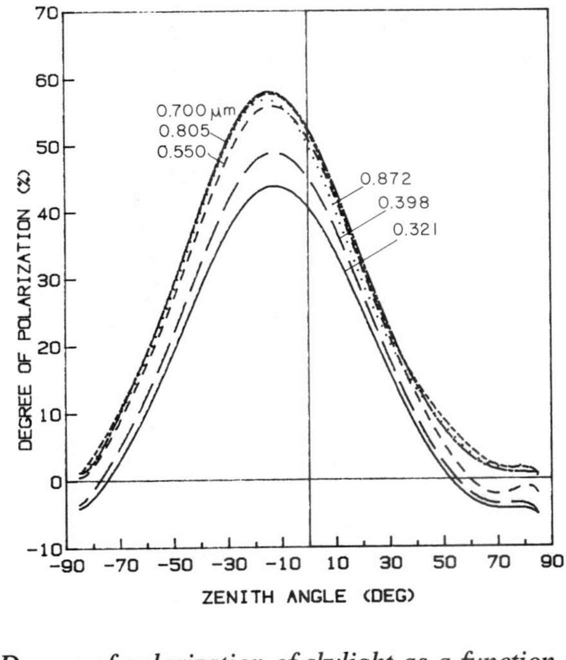

Figure 5.13 Degree of polarization of skylight as a function of zenith angle in the sun's vertical at six different wavelengths from measurements over the snow surface at the geographic South Pole (Coulson et al., 1974).

polarimeter used at McMurdo and the South Pole, as well as those applicable to the measurements of Bullrich et al. (1966) in Greenland, are listed in Table 5.3.

Measurements at the South Pole at wavelengths of 0.398 and 0.70 μm at sun elevation 14.0° are compared with computed data for a Rayleigh atmosphere with surface albedo A=0.80 in Fig. 5.14. The computed data taken from Coulson et al. (1960), are for sun elevation 11.5° and optical thicknesses 0.02 and 0.25, which corresponds to wavelengths of approximately 0.40 and 0.738 μm at the altitude of the South Pole station. The fact that the measured values are considerably lower than the computations at both wavelengths may be explained by a combination of molecular anisotropy, surface albedo that is higher than the 0.80 assumed in the computations, and possible aerosol effects. Although the atmosphere in the interior of the Antarctic continent is undoubtedly the cleanest in the world, occasional incursions of air from coastal regions do bring in aerosol particles, and the sublimation of water vapor on ice nuclei results in frequent episodes of ice crystals in the atmosphere. When this occurs, very significant changes of skylight polarization are produced, as discussed by Fitch and Coulson (1983). At the time of the measurements in Fig. 5.14, however, there were no obvious ice crystals in the atmosphere. The albedo of snow at the South Pole is on the order of 0.90 in visible and near infrared (Fitch, 1976), and approaches 0.98 in the ultraviolet, whereas a value of 0.80 was assumed in the computations. This difference is certainly a significant factor in explaining the discrepancies shown by the diagram. An additional effect not taken into account is possible

TABLE 5.3 Normal optical thickness of the Rayleigh atmosphere as a function of wavelength at the three stations with polarization measurements over snow.

	Normal optical thickness		
Wavelength (μm)	McMurdo	South Pole	Greenland
0.321	0.930	0.570	—
0.398	0.380	0.250	—
0.401	—	—	0.275
0.550	0.0972	0.066	—
0.639	—	—	0.0405
0.700	0.0363	0.025	0.0265
0.709	—	—	0.0265
0.805	0.0218	0.014	—
0.872	0.0155	0.010	—
1.010	—	—	0.0065

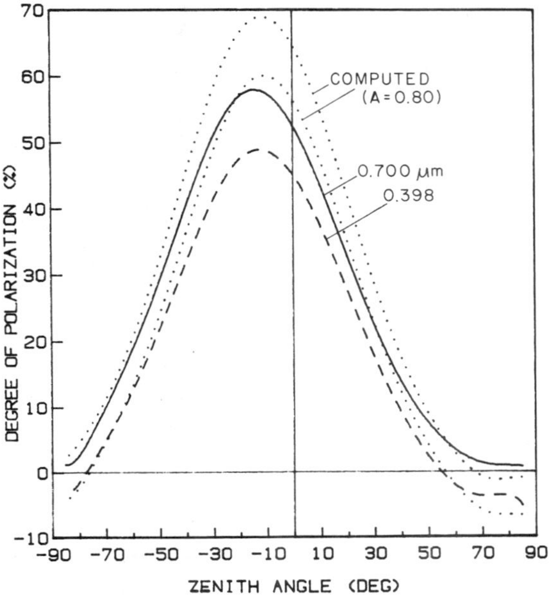

Figure 5.14 Degree of polarization as a function of zenith angle in the sun's vertical as measured at the South Pole at wavelengths 0.398 and 0.70 μm and as computed for a Rayleigh atmosphere with surface albedo A=0.80 at equivalent wavelengths 0.400 and 0.738 μm (Rayleigh optical thicknesses 0.25 and 0.02 at the altitude of the South Pole).

slight polarization in the reflected light. Lambert-type reflection was assumed in the computations. Coulson et al. (1975) ascribed the disappearance of negative polarization in the vicinity of the sun at longer wavelengths to a small amount of positive polarization introduced into the skylight regime by the reflection of the snow surface being non-Lambertian in character. It is known that at least under some circumstances snow reflection introduces a certain amount of polarization into the reflected light.

The values of P_{max} measured at the three stations with snow cover are shown as a function of wavelength in Fig. 5.15. The fact that P_{max} is generally higher at the South Pole than at McMurdo is to be expected from the higher altitude of the South Pole station, with its consequently smaller optical thickness, and cleaner atmosphere. The lower values observed by Bullrich et al. (1966) in Greenland is more difficult to explain, as the Camp Century station is at about

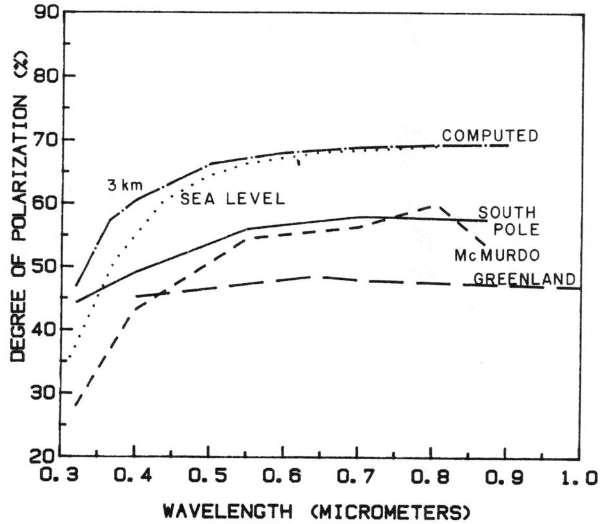

Figure 5.15 Polarization maximum in the sun's vertical as a function of wavelength as measured at three stations with snow cover and as computed for sea level and 3-km altitude for surface albedo A=0.80.

2-km altitude and far from the sea surface and from significant sources of pollution. Furthermore, the surface reflectivity of the snow there cannot be significantly higher than that at the South Pole. The only remaining parameter appropriate for explaining the discrepancy is the aerosol content of the atmosphere. Bullrich et al. were able to get a sufficient decrease of P_{max} in theoretical computations to agree with observations by assuming only one particle per cm^3 with mean radius of 1.0 μm. This is certainly a small aerosol loading, giving a horizontal visual range of 190 km and appropriate for a remote location such as Greenland. The relatively large size of these few particles is probably responsible for approximate independence of P_{max} with respect to wavelength.

Also included in Fig. 5.15 are curves of P_{max} determined from the tables of Coulson et al. (1960) for the Rayleigh atmosphere and surface albedo $A=0.80$ for sea level and an altitude of 3 km appropriate to the South Pole station. A relatively strong spectral dependence of P_{max} at short wavelengths and its virtual independence in the near infrared are similar in observations and theory, although they differ in details. The discrepancy in magnitude may be explained on the basis of an albedo of the snow surface higher than the $A=0.80$ assumed in the computations, the neglect of molecular anisotropy in the computations, possible polarization introduced by reflection from the snow, and minor effects of aerosols.

5.5 Skylight polarization measured at various locations

Although the polarization measurements made in the unique conditions at the Mauna Loa Observatory are useful as a baseline for comparison, the utility of skylight measurements will be greatest in locations with more usual atmospheric conditions. Such locations include lower altitude sites with larger optical thicknesses of the atmosphere, sites in or close to population centers, and even those in industrialized areas with a heavy burden of atmospheric aerosols. An effort will be made in this section to present skylight data taken under widely diverse conditions by a number of authors in the United States and elsewhere.

The two most useful parameters of skylight polarization in indicating atmospheric conditions are the magnitude of the polarization and the positions of the neutral points. The neutral points are the subject of Chapter 7; here we concentrate on other aspects of the polarization field as they have been observed by many authors under various atmospheric conditions in different geographic locations.

As shown above, the degree of polarization reaches its maximum P_{max} in the sun's vertical at approximately 90° from the sun. The magnitude of P_{max} is a sensitive indicator of atmospheric aerosols, and its precise position in the sky depends on wavelength, the turbidity of the atmosphere, and elevation of the sun. Aerosols affect the polarization field in two ways, both of which generally decrease the magnitude of P_{max}. First is their contribution to the total optical thickness $\tau(\lambda)$ of the atmosphere, and the consequent contribution to multiple scattering. Several terms compose $\tau(\lambda)$, which is written as:

$$\tau(\lambda) = \tau^R(\lambda) + \tau^A(\lambda) + \tau^O(\lambda) + \tau^W(\lambda) + \tau^G(\lambda)$$

The superscripts denote, respectively, the Rayleigh atmosphere, atmospheric aerosols, ozone, water vapor, and a combination of many trace gases. The effects of the last three of these is mainly in absorption, and while they may cause significant changes of skylight intensity, their polarization effects, though not identically zero, are minor compared to those of aerosols and the Rayleigh atmosphere. Since the Rayleigh optical thickness $\tau^R(\lambda)$ is due almost entirely to the oxygen and nitrogen of the atmosphere, its variability at a given location is likewise due simply to changes of the total atmospheric mass above the site, which, of course, is indicated by surface pressure. Day-to-day or seasonal changes of pressure seldom exceed 5% of the total, which is a general indication of the variability of $\tau^R(\lambda)$. In contrast, $\tau^A(\lambda)$ is extremely variable in both time and space, making changes of orders of magnitude over short periods of time or distances not at all uncommon. The existence of the aerosols promotes multiple scattering between aerosol and Rayleigh particles and among aerosol particles themselves. Thus the contribution of aerosols to $\tau(\lambda)$ has a significant effect on P_{max}.

A second effect of aerosols is due to the fact that the scattering pattern of aerosols is generally different from that of the Rayleigh atmosphere (see Chapter 3). The aerosol phase matrices for scattering are critically dependent on particle size, complex index of refraction, concentration, and even particle shape, so single scattering by aerosols not only affects the overall magnitude of P_{max} but it also promotes great variability of P_{max} with different types of aerosols.

Measurements of P_{max} taken under various atmospheric conditions by different authors are summarized in this section. Comparisons will be made with P_{max} in the Rayleigh atmosphere where feasible. Discussed in order will be data from moderate altitudes in continental locations (Cactus Peak, California, and McDonald Observatory, Texas), low-altitude continental locations (Ely, Nevada, and Davis, California) and finally low-altitude sites in industrial areas (Germany) and areas with photochemical smog (Los Angeles and Pasadena, California).

5.5.1 Cactus Peak, California

The site of polarization measurements by Sekera and Collaborators (1955) in July 1954, was an observatory at an altitude of approximately 1770 m on the summit of Cactus Peak, a small mountain in the Mojave Desert area of California. The measurements were in four narrow spectral ranges centered at wavelengths 0.365, 0.460, 0.515, and 0.625 μm. The polarimeter (Sekera, 1956) utilized a rotating retardation plate with fixed analyzer in front of the photomultiplier detector, and theoretically yielded data on all four Stokes parameters. In operation, however, the ellipticity of the light was too small to measure reliably with the system, so only relative intensity, degree of polarization, and orientation of the plane of polarization were measured. Measurements that will be discussed below taken in Nevada, Los Angeles, and Pasadena utilized the same instrument.

The area surrounding Cactus Peak is a typical desert surface, with sage brush, mesquite, and other desert plants dotting the landscape. There is much bare sand and rock between the sparse vegetation. It was estimated, from comparisons to other areas for which reflection data are available, that the average albedo of the surface was perhaps 10% in the blue and ultraviolet regions rising monotonically to 30% or more in the longer wavelengths. For comparing measurements with theory, an overall albedo of 25% will be assumed for Cactus Peak.

Data representative of the measurements of P_{max} at Cactus Peak are shown as a function of sun elevation for four different wavelengths in Fig. 5.16. These observations were made under clear atmospheric conditions on the morning of July 22, 1954. The early morning values of P_{max} are high, particularly at longer wavelengths, indicating a very clear atmosphere. The rapid decrease of P_{max} at $\lambda=0.625$ μm with increasing sun elevation is a result of high surface reflectance in the red, the reflected light rapidly increasing in intensity as the surface illumination becomes stronger. The optical thickness of the Rayleigh atmosphere at the altitude of Cactus Peak for $\lambda=0.625$ μm is approximately $\tau=0.046$. The variation of P_{max} with sun elevation for $\tau=0.05$ and surface albedo $A=0.25$ for the Rayleigh atmosphere is shown by the theoretical curve in Fig. 5.16. The data were taken from Coulson et al. (1960). Although P_{max} is higher for the Rayleigh atmosphere than that given by the observations, the decrease of P_{max} with increasing sun elevation is comparable for the two. A combination of aerosols and molecular anisotropy may well account for the differences. An increase of aerosols during the afternoon, due to increased atmospheric turbulence as the surface heated up, is responsible for slightly lower values of P_{max} in the afternoon than those in Fig. 5.16, as already mentioned by Sekera (1956). It is frequently observed in conditions of light to moderate turbidity that the highest

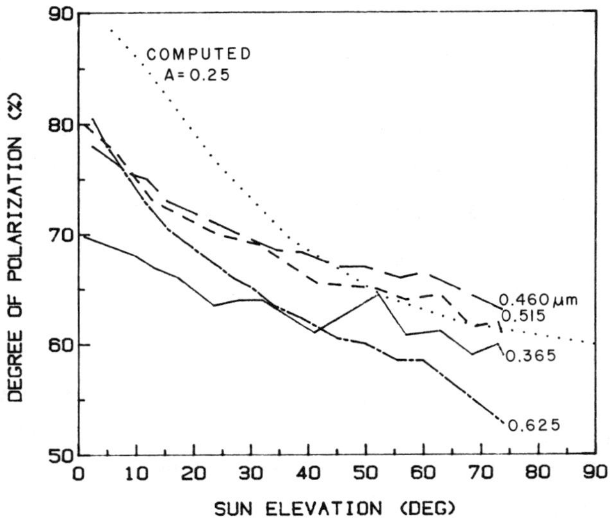

Figure 5.16 Magnitude of the polarization maximum as a function of sun elevation from measurements at four different wavelengths in clear atmospheric conditions at Cactus Peak, California. A curve for the Rayleigh atmosphere of optical thickness 0.05 with surface albedo A=0.25 is given for comparison.

values of P_{max} will fall in the mid-wavelength range, as shown in Fig. 5.16 at wavelengths of $\lambda = 0.460$ and 0.515 μm, and will yield considerably lower values at at shorter and longer wavelengths.

5.5.2 McDonald Observatory

The McDonald Observatory is located in western Texas at an altitude of approximately 2230 m. The surrounding surface has a cover of grass, which is green in winter and brown in summer, plus rocks interspersed between small deciduous and coniferous trees. The average albedo of the area is similar to, but perhaps slightly lower than that of Cactus Peak. Gehrels (1962) utilized the telescopes at the McDonald Observatory in 1959 and 1960 to measure skylight polarization at 90° from the sun in the sun's vertical at nine wavelengths from $\lambda = 0.325$ to 0.943 μm. Representative results of the McDonald measurements for clear-sky conditions at different sun elevations are shown in Fig. 5.17. Although there is considerable scatter in the data, the wavelength dependence,

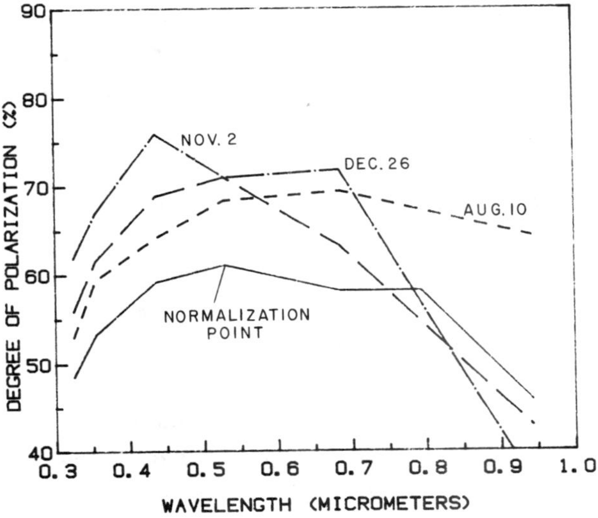

Figure 5.17 Wavelength dependence of the polarization maximum as determined by Gehrels (1962) at the McDonald Observatory. The solid curve represents an attempt by Gehrels to normalize the clear sky data to P_{max}=61.1% at λ=0.529 μm.

with a maximum $P_{\max}$ in the visible range and lower values in the ultraviolet and near infrared, is similar to that in other continental locations. For reasons that were not explained, Gehrels attempted to normalize the averages of polarization values to $P_{\max}=61.1\%$ at $\lambda=0.529$ μm and thereby obtained the solid curve of Fig. 5.17. The same general wavelength distribution is shown, but the normalization procedure appears to have minimized the overall variability of the measured values and the normalization point of 61.1% seems artificially low. The anomalously high value of 63.8% at $\lambda=0.943$ μm on August 10 is at least partially due to the low sun elevation (9.4°) at which the measurement was taken, with a consequent low intensity of surface illumination.

5.5.3 Ely, Nevada

A short series of polarization measurements was made under the supervision of the late Prof. Z. Sekera in 1954 near the town of Ely, Nevada. Although these data were not published, they are appropriate for use here to represent polarization over a low altitude surface which is covered with snow. Ely is in a semi-desert area of east central Nevada at an altitude of approximately 2170 m. The measurements were made during the afternoon of February 28, 1954, with the same six-filter polarimeter used for the measurements described below, taken in Los Angeles and Pasadena. This particular date was chosen because the ground was covered with about 12 cm of new snow which had fallen during the previous night. The strong cold front that had produced the snow was followed by very clear air. The horizontal visual range at the time of the measurements was estimated at 80 to 100 km. Continuous measurements were made from mid-forenoon until after sunset of this day.

The behavior of P_{max} as a function of sun elevation is shown for the six wavelengths of the measurements over the snow surface in Fig. 5.18. Two curves from computations of P_{max} for a Rayleigh atmosphere overlying a Lambert surface with 80% albedo are included for comparison. The observations show P_{max} to be highest in the longer wavelengths with low sun elevation, and that the polarization decreases quite sharply with increasing sun elevation. Although both

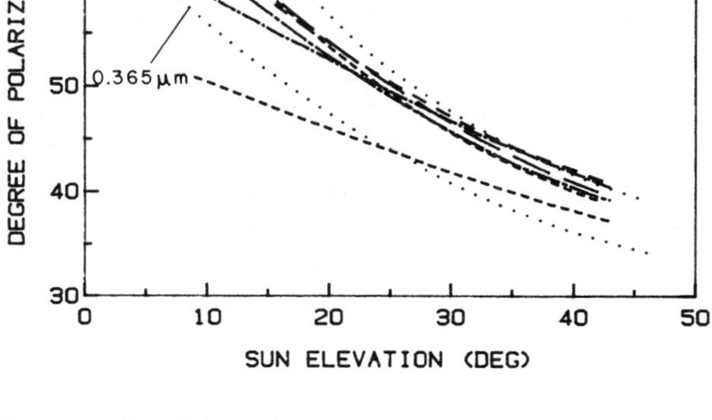

Figure 5.18 Magnitude of the polarization maximum at six different wavelengths as a function of sun elevation as observed over a new snow surface at Ely, Nevada, and as computed for a Rayleigh atmosphere overlying a Lambert surface of albedo A=0.80. (Coulson et al., 1974.)

of these features agree qualitatively with theory, neither is as pronounced as it is for the computations. At low sun elevations the observed polarization is lower than the theoretical in all wavelengths. It is unlikely that the snow albedo could have been much above 80%, in which case the discrepancy must be due to a combination of molecular anisotropy and aerosol scattering, even in the very clear atmosphere prevailing at the time. The sharper decrease of P_{max} for the computations than for the measurements is evidence that the actual albedo of the snow surface was not as high as 80%. If this were the case, depolarization effects of higher albedo for the computations would be matched at the higher sun elevations by depolarization effects due to aerosol scattering, resulting in P_{max} being about the same for the two sets of data.

In addition to the visual clarity of the atmosphere at the time of the measurements, an indication of relatively minimal aerosol scattering is given by the spectral distribution of P_{max} at three sun elevations shown in Fig. 5.19. As seen above, the normal dispersion at continental locations is for the highest values of P_{max} to be in the 0.4- to 0.5-μm range, with a significant decrease

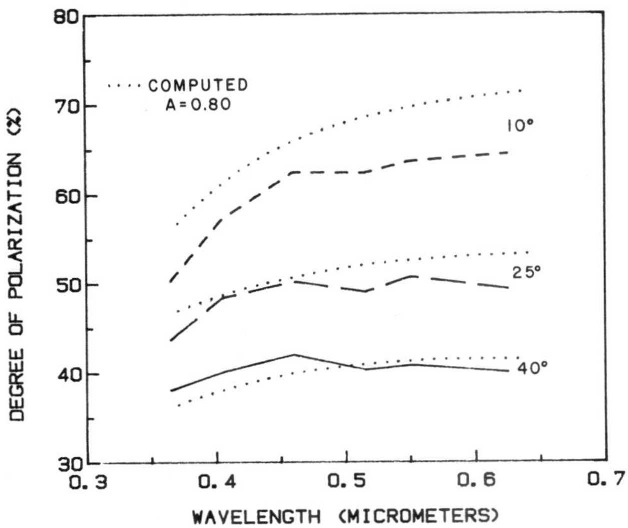

Figure 5.19 Magnitude of the polarization maximum as a function of wavelength as measured over a snow surface in Ely, Nevada, at three different sun elevations (Coulson et al., 1974) and as computed for a Rayleigh atmosphere with surface albedo A=0.80.

in the red and near-infrared regions. There is a tendency toward this feature in the measurements, particularly at sun elevation 40°, but the effect is not pronounced. The data do, however, fall below the computed magnitudes for the Rayleigh atmosphere at the longer wavelengths, indicating the existence of non-Rayleigh scattering in the observations.

Two effects not considered in the above discussion may arise from the nonideal nature of reflection by snow. First, it is known that the light reflected from snow generally is partially polarized in the sense that for the reflected light, $I_r > I_e$, which means positive polarized polarization. This positive polarization is strongest at low sun elevations. Subsequent backscatter of the partially polarized light would tend to bolster the magnitude of P_{max} at low sun elevations. The data of Figs. 5.18 and 5.19 do not appear to be affected much by this feature. The second characteristic of snow reflection is that the albedo tends to decrease with increasing sun elevation. This feature may help to explain that the slopes of the observed curves for the Ely data in Fig. 5.18 were less than those of the computed curves.

5.5.4 Davis, California

A relatively extensive observational program was carried out by the present author on the campus of the University of California, Davis, using two different polarizing radiometers developed at that institution. The dual channel instrument with eight interference filters was used for all of the polarization measurements made in Hawaii, and the single channel polarimeter with six interference filters was used for the measurements in the Antarctic. The city of Davis is located at essentially sea level in the broad and level valley of the Sacramento River. Aside from some low mountains of the Coast Range about 35 km to the west, the nearest significant land feature is the Sierra Nevada mountain range 150 km to the east. The only large source of pollution is the city of Sacramento 30 km to the east, but since the wind flow in the region is usually from west or northwest the Sacramento pollution seldom reaches Davis. On occasion some of the smog of the San Francisco–Oakland complex reaches Davis, but this is relatively infrequent. A highly localized source of smoke near Davis are fires set by farmers to dispose of rice straw and other agricultural wastes produced in the area. The effect of such smoke on skylight polarization is shown for one case in Fig. 5.21.

Since the area surrounding Davis is devoted mainly to intensive farming, the surface albedo is typical of that of green vegetation, with some fields of bare Yolo loam-type soil. The albedo of the soil, as measured by Coulson and Reynolds (1971), is wavelength dependent, being 6% to 8% in the blue and ultraviolet with a monotonic increase up to 20% in the near infrared. Agricultural crops in the growing stage have 4% to 5% albedo at wavelengths below $\lambda = 0.5$ μm and in the chlorophyll absorption bands at $\lambda = 0.6$ to 0.67 μm, albedo of about 10% in the green region, with a rapid increase up to 60% or more from $\lambda = 0.67$ to 1.0 μm and beyond.

The large complement of data for Davis shows a variety of polarization configurations for different atmospheric conditions. They are too extensive for a complete presentation here. Under very clear conditions, such as after a cold frontal passage, the polarization field is similar to that observed at Mauna Loa, but the larger optical thickness of a sea level location produces more multiple scattering and somewhat lower polarization values. We choose instead to present data for cases which are more or less unique to the Davis area.

The summer weather in Davis is typical of that of a stagnant high-pressure system, with light winds, no clouds, and a temperature inversion in the 800- to 1200-m altitude range. In such case, there is usually a blue haze over the area which restricts the visual range to 25 to 30 km. This was the situation on July 25, 1973, at which time the data of Fig. 5.20 were taken. The rapid decrease of P_{max} immediately after sunrise is produced by a relative increase of aerosol effects when low-level turbidity receives more intense incident radiation as the sun rises higher above the horizon. For the visible region, P_{max} reaches a broad and minor minimum in the 30° to 50° solar elevation range, as is normally observed elsewhere. The most anomalous feature for Fig. 5.20 is the very low magnitude of P_{max} at $\lambda = 0.80$ μm. A decrease of P_{max} at the longer wavelengths is typical of hazy conditions, but this is more extreme than in most cases. The high albedo of vegetated surfaces in the near infrared is a contributing factor, but it appears that the haze prevalent at the time was particularly rich in the larger particles. Similar results have been observed on other occasions of haze at Davis.

A second anomalous atmospheric condition occurred at Davis on October 11, 1976 as smoke from agricultural fires blanketed the area and plumes of denser

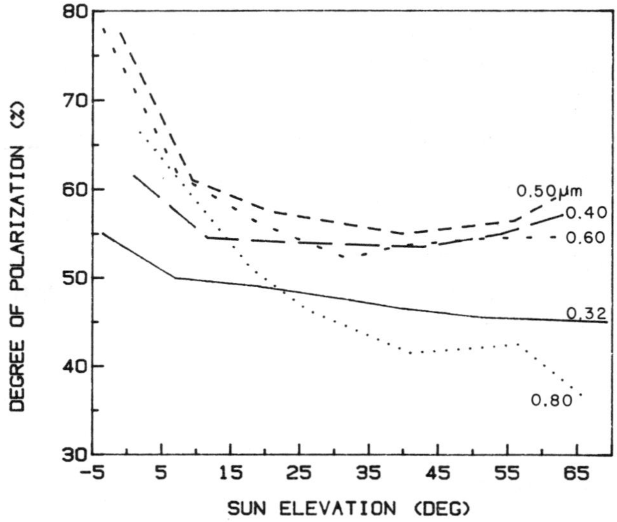

Figure 5.20 Magnitude of the polarization maximum as a function of sun elevation for five different wavelengths as measured at Davis, California, under conditions of moderate haze on July 25, 1973. (Coulson et al., 1974.)

smoke drifted across the sky. The distribution of skylight polarization in the sun's vertical for this case is shown for two different wavelengths in Fig. 5.21. Although the polarization was decreased significantly at $\lambda=0.80$ μm, a more pronounced effect occurred at $\lambda=0.40$ μm. The smoke particles from such low-temperature fires are high in carbonaceous materials, and are thus highly absorbing. The absorption minimizes multiple scattering, which itself would tend to increase the polarization. However, the heavy burden of aerosols in such conditions would be more influential in the polarization field than would Rayleigh scattering, the result being the low polarization values seen, particularly at $\lambda=0.40$ μm, in the data. The times of passage of smoke plumes across the sky are clearly evident in the diagram.

To supplement the data of anomalous conditions, it is of interest to see the distribution of polarization over the entire dome of the sky for more normal conditions at a sea level location. Fig. 5.22 shows this hemispheric distribution for $\lambda=0.365$ μm observed in Davis under clear skies and 50-km visual range on April 10, 1969. One-half of the hemisphere is shown, the other half being a mirror image of this one. The heavy dashed line shows the position of the

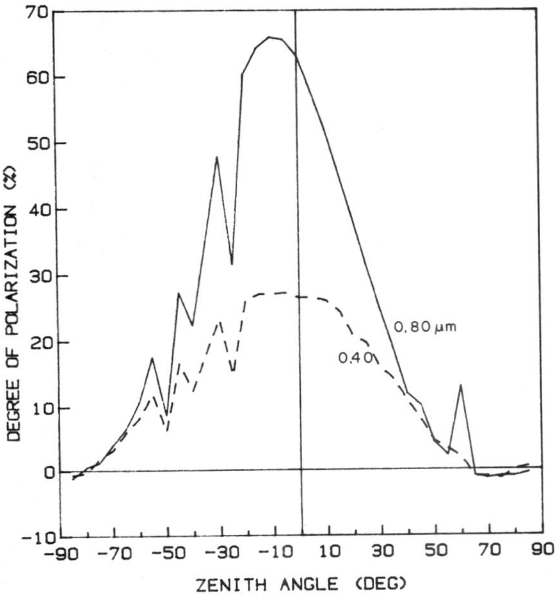

Figure 5.21 Degree of polarization of skylight as a function of zenith angle in the sun's vertical as measured in two different wavelengths at Davis, California, under conditions of smoke from agricultural fires prevalent in the low levels of the atmosphere. (Coulson et al., 1974.)

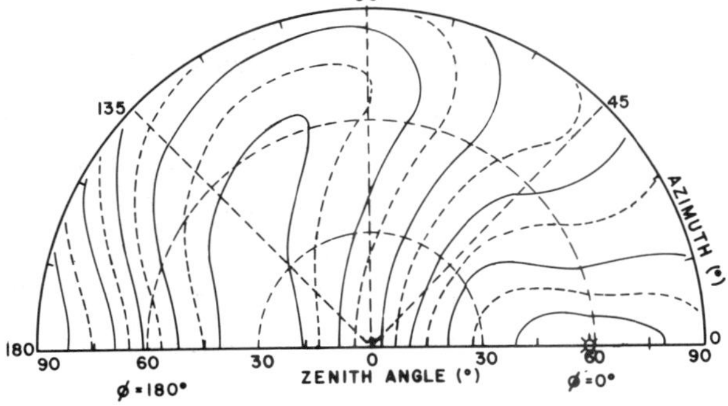

Figure 5.22 Hemispheric distribution of skylight polarization at λ=0.365 µm as observed in clear atmospheric conditions at Davis, California on April 10, 1969. The heavy dashed line shows the position of the 90° scattering angle. (Coulson et al., 1974.)

90° scattering angle. As expected from theory, the strong influence of scattering angle is shown by the data, although multiple scattering and aerosol effects even in this very clear atmosphere limited the polarization maximum to approximately 54%. The Babinet and Brewster points are clearly shown to be in their normal positions, with strong polarization gradients in the region of the zenith and near the antisolar horizon. There is a slight indication of a minor maximum outside the sun's vertical, but the data are insufficient to define it precisely.

5.5.5 Los Angeles, California

A large number of skylight measurements have been made in and near Los Angeles. Beginning with the early work of Sekera and Collaborators (1955) in the early 1950s, observational programs have been carried out by deBary (1964), Holzworth and Rao (1965), Coulson (1971), and others. The polarimeter used by Sekera and Collaborators for the early measurements was a dual-channel device with three changeable absorption-type filters in each channel. The optical system, which was basically the same for the two channels, consisted of collimator tube, filter, photomultiplier with appropriate electronics, and a rotating retardation plate with fixed analyzer. A rotation of the retardation plate in front of the analyzer produced an oscillating intensity at the photomultiplier, the characteristics of which could be interpreted in terms of the intensity and degree of polarization of the incident skylight. This instrument was used over a period of several years for measurements on the UCLA campus, in Pasadena, at Cactus Peak, and at various sites in Nevada. The precision of the measured intensity and degree of polarization suffered from electronic problems arising from the attempt to compensate for intensity changes by varying the dynode voltage on the photomultiplier. In general, measurements of the maximum degree of polarization and of relative intensity were accurate to ±1% to 2%. An additional difficulty with interpretation of the data was the wide bandpass of the absorption-type filters installed in the device.

Some of these problems were eliminated in the second polarimeter developed at UCLA and used by Holzworth and Rao (1965). This was a three-channel instrument with one interference-type filter in each channel. It was also based on a rotating retardation plate with fixed analyzer, and although intensity changes were still compensated by varying dynode voltage on the photomultiplier, improved electronics resulted in considerably improved precision of the data. In fact, Holzworth and Rao stated that measurements of the degree of polarization were accurate to ±0.1%.

The instrument used by deBary (1964) for skylight measurements in Los Angeles and briefly described by Bullrich (1964) was a single-channel device of the rotating analyzer type. It had five interference filters in the spectral range $\lambda=0.405$ to 0.783 μm. The measurements, which were confined to the sun's vertical, were used to evaluate some theoretical models of the atmospheric aerosols existing at the time, but unknown particle characteristics, temporal and spatial variations of the aerosols, and basic limitations of the models themselves resulted in only marginal success in the effort.

The effects of the well known air pollution of Los Angeles on skylight were investigated by Coulson (1971) by means of a single-channel polarimeter of the rotating analyzer type fitted with six interference filters in the $\lambda=0.320$ μm to 0.795 μm spectral range. This instrument was developed at the University of California, Davis. Because of the importance of ultraviolet radiation in the production of smog, emphasis in the measurements was on wavelengths of $\lambda=0.320$ and 0.365 μm. The intensity and polarization over the entire dome of the sky were included, as were the skylight characteristics of a solid and uniform overcast of low stratus clouds over the Los Angeles basin. Selected data from these measurements are given below.

The primary interest in the skylight data from Los Angeles and Pasadena arises from the nature of its polluted atmosphere during the summer and fall months. During this period a diurnal sea breeze regime frequently results in the air over Pasadena having higher levels of pollution than Los Angeles itself, even though it is located many kilometers northeast. The UCLA campus, where the measurements of deBary and Holzworth and Rao were made, is nearer the coast and is less subject to episodes of heavy pollution than is the downtown area.

The surface albedo of the Los Angeles region, typical of that of cities in general, is perhaps 10% to 15%, with the lower of these values being for ultraviolet and visible wavelengths and the higher for the near infrared.

An example of the skylight polarization measurements made by Sekera and Collaborators (Sekera, 1956) on the UCLA campus is shown by the polarization maximum P_{max} in Fig. 5.23 as a function of sun elevation at five wavelengths observed during the morning and afternoon of September 17, 1954. Moderate to heavy haze over the area in the forenoon limited the visual range to about 3-km, in which case P_{max} was generally less than 50% at all wavelengths. About noon the wind blew away the haze, increased the visual range to 20 km, and caused a dramatic increase of P_{max}. The dispersion was characterized by P_{max} being higher at $\lambda = 0.625$ μm than at shorter wavelengths, even during the very hazy period. In the clearer conditions P_{max} exhibited a monotonic increase from $\lambda = 0.365$ μm to 0.625 μm and reached a surprisingly high value of fully 85% at $\lambda = 0.625$ μm near sunset. This corresponds closely to the highest polarization values achieved in the very clean atmosphere at the high altitude site in Hawaii.

A reversal of this behavior of P_{max} as a function of time of day was observed in Pasadena on November 5, 1954 (Sekera, 1956), as shown by Fig. 5.24. In this case a smog layer built up and moved into the area, decreasing the visual

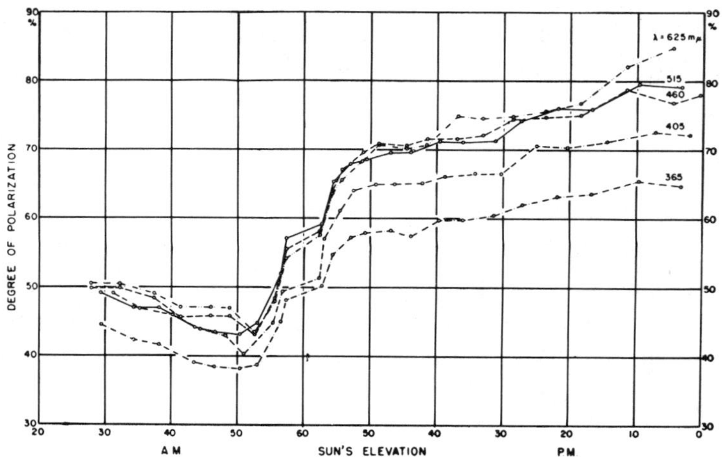

Figure 5.23 Magnitude of the polarization maximum as a function of sun elevation as measured at five different wavelengths on the UCLA campus in Los Angeles on September 17, 1954. (After Sekera, 1956.)

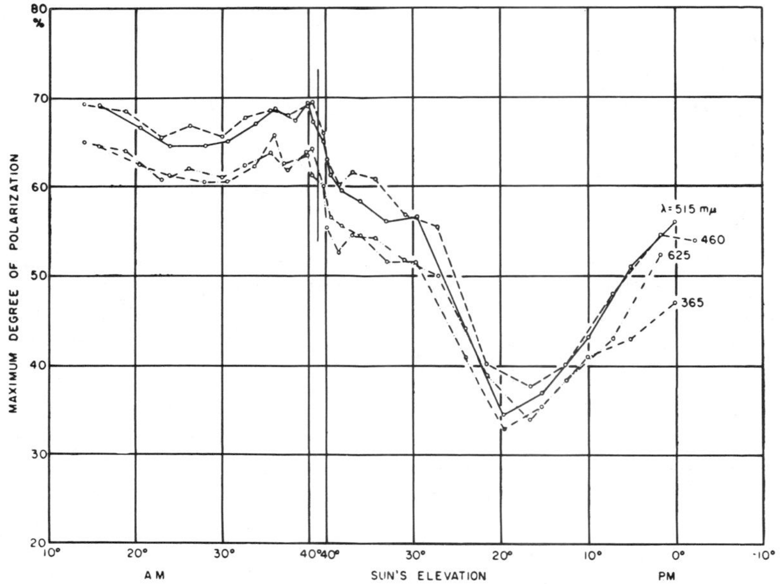

Figure 5.24 Magnitude of the polarization maximum as a function of sun elevation as measured at four different wavelengths in Pasadena, California, on November 5, 1954. (After Sekera, 1956.)

range from its forenoon value of 15 to 20 km to less than a third of this during midafternoon. The increased turbidity caused P_{max} to dip from 60% to 70% to 35% to 40% at all wavelengths. The dispersion of P_{max} was more typical of urban conditions than shown in the previous diagram, the highest values here being at $\lambda = 0.460$ and 0.515 μm, with significantly lower values at the longer and shorter wavelengths.

A comparison of the Pasadena data with that computed for a Rayleigh atmosphere with surface albedo $A = 0.10$ is shown by the plot of relative deviation of P_{max} with sun elevation in Fig. 5.25. Relative deviation is computed from the relation $(P_m - P_c)$, where P_m and P_c are, respectively, the measured and computed values of P_{max}. Relative deviations of 0% to 30% in moderately clear conditions gave way to values of -50% to -60% in very turbid conditions. One factor which may contribute to the relative deviations decreasing with increasing wavelength throughout the observations is a possible increase of surface albedo with increasing wavelength, but there is little doubt that the maximum effect is due to scattering by aerosols.

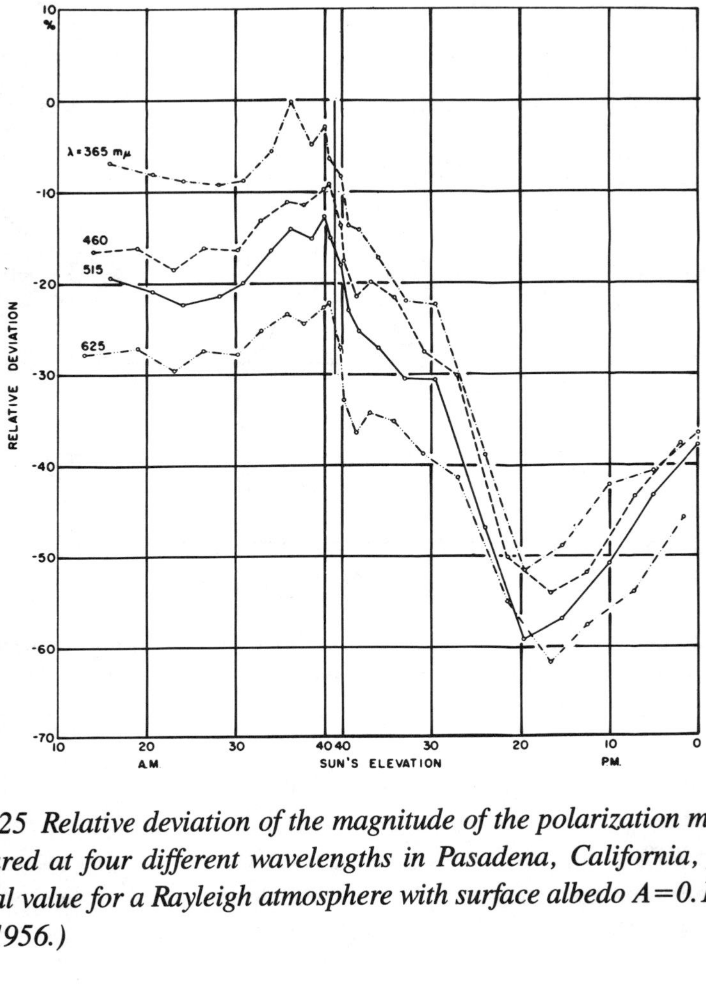

Figure 5.25 Relative deviation of the magnitude of the polarization maximum, as measured at four different wavelengths in Pasadena, California, from the theoretical value for a Rayleigh atmosphere with surface albedo $A = 0.10$. (After Sekera, 1956.)

The effect of the diurnal smog regime on skylight polarization in Los Angeles is shown in more detail by the measurements of Coulson (1971). The ultraviolet region was emphasized in this study because of its dominant role in the formation of photochemical smog. Curves of the degree of polarization at $\lambda = 0.320$ μm on a smoggy day in Los Angeles are shown as a function of zenith angle in the principal plane in Fig. 5.26. The estimated visual range at the time of these measurements decreased from 4 km in the early morning to less than 2 km in mid-forenoon. The general shape of the curves is not altered much by smog, but the magnitudes of both positive and negative regions of polarization are greatly

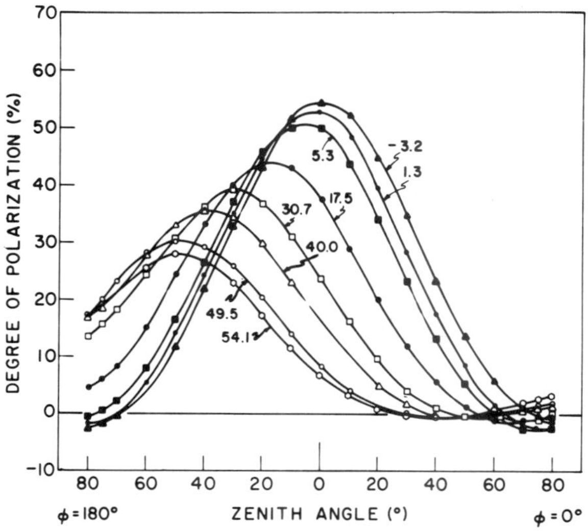

Figure 5.26 Degree of polarization of skylight as a function of zenith angle in the sun's vertical as measured for $\lambda=0.32$ μm at eight different sun elevations under conditions of moderate to heavy smog in Los Angeles, California, on September 26, 1968. (After Coulson, 1971.)

diminished by the effects of aerosols. The sharp decrease of polarization with increasing sun elevations is mainly due to the usual buildup of smog in the area during the forenoon, but the normal decrease of P_{max} with increasing sun elevation in the clear atmosphere enhances the effect.

Measurements of the degree of polarization made on a relatively clear day (September 23, 1968) and on a smoggy day (September 26, 1968) in Los Angeles are compared with computed values for the Rayleigh atmosphere with surface albedo 25% in Fig. 5.27. Some interpolation of the data was necessary to adapt the curves to a sun elevation of 30°. Both positive and negative polarization maxima were greatly decreased from the Rayleigh case, even for a relatively clear atmosphere (visual range of 25 km). For smoggy conditions the positive maximum was decreased by about a factor of 2, and the negative maximum by an even greater relative amount by the aerosols.

The very pronounced effect of smog on the polarization maximum is shown by the plot of P_{max} versus sun elevation for two wavelengths in Fig. 5.28. All

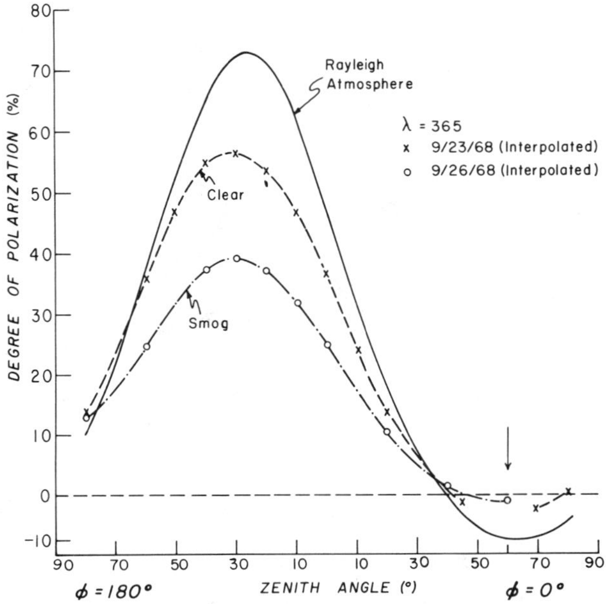

Figure 5.27 Degree of polarization of skylight as a function of zenith angle in the sun's vertical as interpolated to a sun elevation of 30° from measurements at λ= 0.365 μm for relatively clear conditions and for conditions of moderate to heavy smog in Los Angeles, California. The solid curve is for a Rayleigh atmosphere with surface albedo A=0.25.

of the measurements were taken in Los Angeles except for one clear case in Davis. Even moderate smog decreases P_{max} by 8% to 10% in the early morning, and by about 30% during the smoggiest part of the day, and another 6% to 8% decrease occurs in heavy smog. Measurements at other wavelengths (Coulson, 1971) show that the polarization effects of smog are not a strong function of wavelength, due, no doubt, to the fact that the extremely high concentrations of aerosols produce a great amount of multiple scattering at all wavelengths. The relatively few efforts at modeling the polarization effects of smog have been largely unsuccessful in obtaining agreement with measurements. DeBary (1964) achieved reasonably good representations of the intensity field for cases of light to moderate turbidity, but there was little relation between her computations and actual measured polarization. Kano (1968) achieved reasonable agreement at

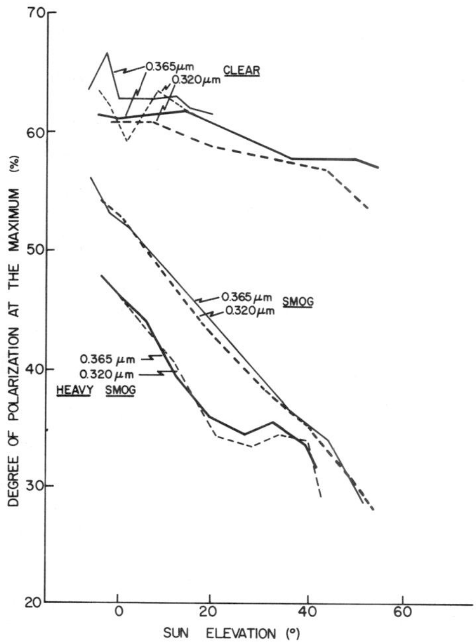

Figure 5.28 Magnitude of the polarization maximum as a function of sun elevation as measured at λ=0.320 and 0.365 μm for cases of moderate and heavy smog in Los Angeles, California and, for cases of clear conditions in Los Angeles and Davis, California. (Coulson et al., 1974.)

low sun elevations, but a different dependence on sun elevation resulted in large discrepancies in general. The Monte Carlo technique of Plass and Kattawar (1970) produced large statistical uncertainties in the results and, more importantly, their results were widely different from the measurements in polarization magnitude and in dependence on both sun elevation and wavelength. Hitzfelder et al. (1976) used a matrix operator method to compute the polarization and intensity of skylight for a turbid atmosphere with the normal and one-third the normal amount of aerosols. The results conform reasonably well with measurements in a relatively clear atmosphere, but they are not well adapted to comparisons for cases of smog. It is likely that the difficulty in modeling polarization effects in smog is in obtaining a realistic characterization of the particles (size distribution, index of refraction, concentration), rather than in inadequacies of the methods of computation.

An opportunity occurred during the Los Angeles observations to measure polarization and intensity of skylight under conditions of a heavy and persistent overcast of the typical stratus cloud which occurs along the California coast. The cloud is confined to the marine layer of air in the first 400 to 500 m of altitude, and is very uniform in thickness and droplet characteristics. By taking the data of Carrier et al. (1967) on the optical characteristics of California stratus and assuming a cloud thickness of 400 m, Coulson (1971) computed the normal optical thickness of the cloud to be about 28 in the early morning. The degree of polarization of the light from the cloud at $\lambda = 0.652$ μm in the sun's vertical is shown for five different sun elevations in Fig. 5.29. The lack of any appreciable dependence on scattering angle for most of the period shows multiple scattering to have been the dominant mode of radiative transfer, as would be anticipated for such a high optical thickness with little absorption. It was only during the late forenoon, at which time the stratus had decreased in thickness and started to break up, that the degree of polarization exceeded 1% to 2%. It is interesting also that only for the thin cloud did skylight exhibit any negative polarization.

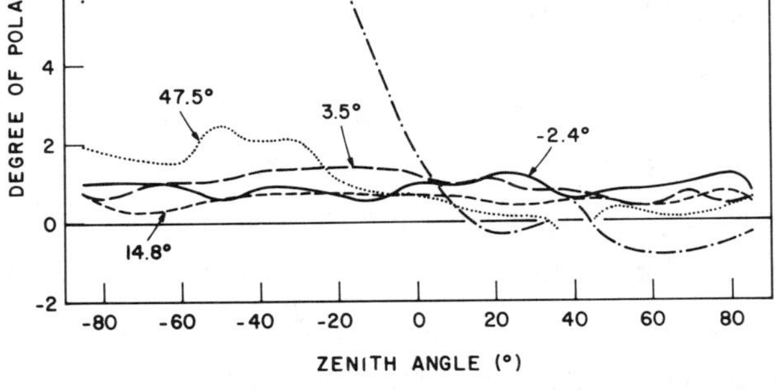

Figure 5.29 Degree of polarization as a function of zenith angle in the sun's vertical as measured at $\lambda > 0.652$ μm at five different sun elevations for the case of a stratus overcast in Los Angeles, California. (After Coulson, 1971.)

Intensity measurements made simultaneously with those of polarization showed the intensity over the sky to be essentially independent of azimuth, with a gradual decrease from zenith to horizon and the zenith being two to three times as bright as the horizon. Only in the later stages of cloud breakup did the intensity assume its more normal distribution over the dome of the sky.

5.5.6 Polarization during a partial solar eclipse

An opportunity to make skylight measurements during a partial (approximately 80%) eclipse of the sun occurred at Davis on February 26, 1979. Two polarimeters were operated for some time before, during, and after the eclipse, the eight-wavelength polarimeter being oriented in the direction of the zenith and that with six wavelengths being oriented at 90° from the sun in the sun's vertical. Both relative intensity and degree of polarization were measured in all of the available wavelengths, so a relatively complete set of data were obtained. The instrument oriented at the 90° point suffered from a noisy signal on that day, and the polarization values appeared to be approximately 2% too low in comparison with those of the more accurate eight wavelength instrument during the sunrise period. An arbitrary 2% was added to all of the polarization data taken with that instrument during the measurements.

Presumably this discrepancy persisted throughout the eclipse period, but the changes which occurred should be faithfully represented by both instruments. The atmosphere on the day of the eclipse was very clear after a cold front passage during the night, the sky was very blue, and the only clouds were wisps of stratocumulus along the western horizon. An additional indication of high atmospheric clarity were the high polarization values (approximately 82% at $\lambda=0.70$ μm) measured during the early morning.

The relative intensities (actually photon count rate) at $\lambda=0.40$ and 0.70 μm at 90° from the sun in the sun's vertical are shown as a function of sun elevation in Fig. 5.30. The eclipse began at a sun elevation of approximately 7°, was most complete (about 80% of disk obscured) at a sun elevation of 14.3° (the position of the arrows in the diagram), and ended at approximately 22° sun elevation. At the time of maximum coverage, the intensity was decreased by something less than an order of magnitude at both wavelengths, as would be expected for 80% totality.

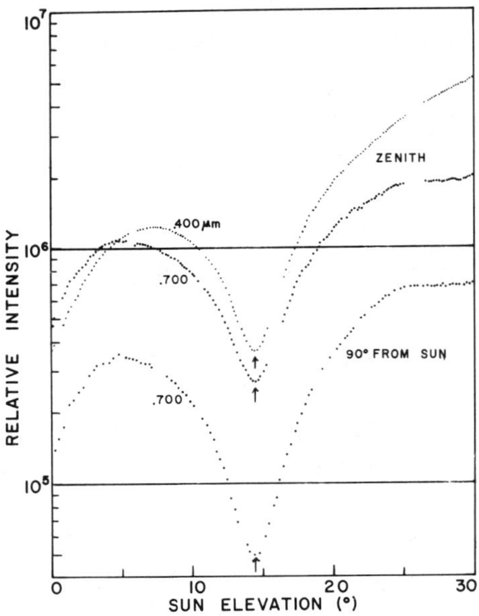

Figure 5.30 Relative intensities of skylight for λ=0.40 and 0.70 μm in the direction of the zenith, and for λ=0.70 μm at 90° from the sun in the sun's vertical, for the case of a partial eclipse of the sun at Davis, California, on February 26, 1979. The arrows mark the time of maximum obscuration. (Coulson et al., 1974.)

On the basis of observations at 90° from the sun by deBary et al. (1961) during the total eclipse of February 15, 1961, in Italy, it was not expected that the skylight polarization would decrease significantly for this case of partial eclipse. Because the direct sunlight is cut off during a total eclipse, the skylight is produced entirely by secondary and higher order scattering, in which case the degree of polarization should be very low. DeBary et al. observed it to reach zero at the time of totality. It remained low, however, only during totality. At 1 minute before and 1 minute after totality the polarization was essentially normal, even though the intensity reacted throughout the course of the eclipse. Shaw (1975), observing skylight at λ=0.40 μm in the sun's vertical 90° from the sun during the 1973 total eclipse in Kenya, found the degree of polarization to decrease from a pre-eclipse value of 45% to 4% during totality. The polarization decrease began some 6 to 7 minutes before totality and the depolarization effect disappeared in a like amount of time following totality. The intensity at 90° from the sun was decreased by a factor of 7.6×10^3 during the period of totality. Dandekar and Tuttle (1971) observed an intensity decrease of slightly more than four orders of magnitude at a point 90° from the sun during the total eclipse of March 1970. At the same time the polarization reached a minimum of 4.0% at $\lambda=0.475$ μm and less than 0.6% at $\lambda=0.60$ μm, and the sky color was shifted toward the blue during the period of totality. A summary of zenith skylight intensities observed in a number of different eclipses was given by Sharp et al. (1971). They found a variation in the ratio of zenith intensities during complete totality to those at 98.7% obscuration from 0.01 to 0.04, depending on sun elevation and other parameters.

A virtual lack of polarization response is shown by Fig. 5.31 for the partial eclipse at Davis. This is to be expected, as the 20% of the solar disk unobscured results in the direct flux being greatly dominant over the diffuse flux. During a total eclipse, of course, the only illumination would be diffuse. While the skylight intensity was strongly decreased in the partial eclipse, as shown in Fig. 5.31, the relative magnitudes of primary and secondary scattering were probably not much affected by a part of the disk being obscured.

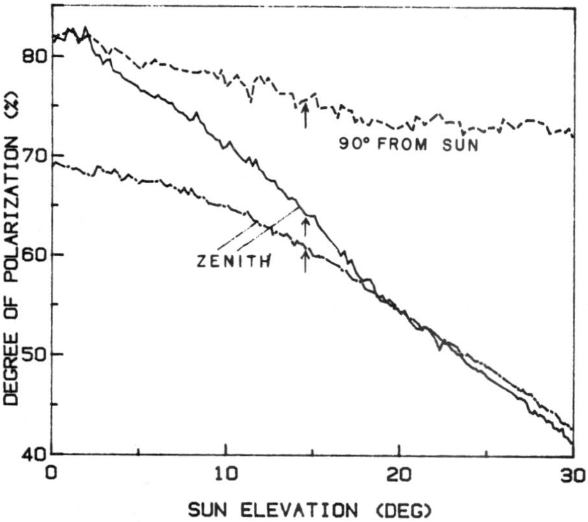

Figure 5.31 Degree of polarization of skylight for $\lambda=0.40$ and 0.70 μm in the direction of the zenith, and for $\lambda=0.70$ μm at 90° from the sun in the sun's vertical, for the case of a partial eclipse of the sun at Davis, California, on February 26, 1979. The arrows mark the time of maximum obscuration. (Coulson et al., 1974.)

5.6 Skylight polarization in the sun's almucantar

It has been standard practice by many authors, particularly in Europe and the Soviet Union, to concentrate on measurements of polarization in the almucantar of the sun (scans in azimuth at the elevation angle of the sun). The main advantage of choosing the sun's almucantar is that the optical pathlength seen by the instrument at a given sun elevation is a constant, so the only parameter which varies during one set of measurements is scattering angle Θ. An auxiliary advantage is that the relative magnitudes of primary and multiple scattering, while not constant, show less variation with Θ in the almucantar than in other directions in the sky. Since much of scattering theory uses Θ as a principal parameter, data for the sun's almucantar are amenable to relatively simple theoretical analysis. For the sun's vertical, on the other hand, the main advantage is that there is more basic information in the polarization distribution in that plane than any other. Not only is the polarization maximum P_{max} in or close to the sun's vertical, but so also are the neutral points and the negative branch of the polarization curves. Another consideration is that a polarization maximum exists in the sun's almucantar only at sun elevations less than 45°.

For purposes of comparison, a few measurements in the sun's almucantar were made in the clean atmosphere from an altitude of 3400 m at the Mauna Loa Observatory. The results of such measurements in low turbidity conditions on May 23, 1977, are shown for a wavelength $\lambda = 0.80$ μm and seven different sun elevations in Fig. 5.32. Not only does the azimuthal position of the maximum change with sun elevation (as it must to maintain a 90° scattering angle), but the curves become more asymmetric with increasing sun elevation. At a sun elevation of approximately 45°, the P_{max} is located at $\Theta = 180°$, which, of course, is the sun's vertical. At that time the maximum applies for both sun's almucantar and sun's vertical. At lower sun elevations the two are separate, and at higher sun elevations no maximum exists in the sun's almucantar. As for the case of the sun's vertical, Fig. 5.32 shows that the magnitude of P_{max} is greatest at the lowest sun elevation also for the sun's almucantar. Measurements at other wavelengths show the same general configuration as that of Fig. 5.32, and the change of magnitude with wavelength is about the same as that for the sun's vertical.

A comparison of the change of P_{max} with sun elevation as measured in low turbidity conditons on May 23, 1977, is shown for the two different wavelengths in Fig. 5.33. In order to decrease the time requirements for the measurements and achieve as much simultaneity as possible, the measurements in the sun's vertical were made at a point 90° from the sun. At $\lambda=0.80$ μm the measured values may be up to 1% too low. The data for the solar almucantar were taken by actual scans of the sky. At $\lambda=0.80$ μm the curves show generally similar behavior, a decrease of P_{max} with increasing sun elevation being evident in both cases. The magnitude of P_{max} is somewhat lower in the sun's almucantar than in the sun's vertical for the mid range of solar elevations. At the lower and higher sun elevations, however, the discrepancy is probably within measurement error. The behavior of P_{max} at $\lambda=0.365$ μm for the sun's almucantar has a quite different character from that of the other curves, particularly at low sun elevations. The long atmospheric pathlengths seen for short wavelengths when the sun, and therefore the point of observation, is low on the horizon results in an initially low value of P_{max}. As the pathlength rapidly decreases with increasing sun elevation, a commensurate rapid increase of P_{max} occurs. In the mid-range of solar

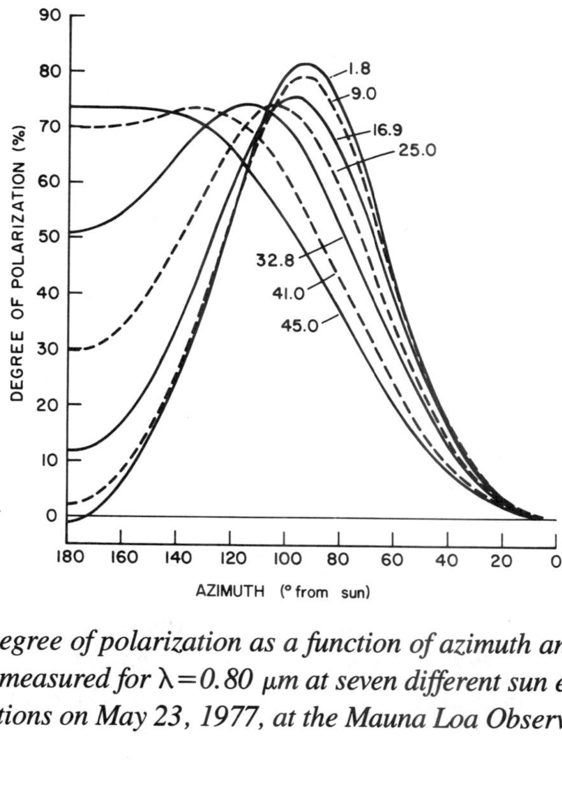

Figure 5.32 Degree of polarization as a function of azimuth angle in the sun's almucantar as measured for $\lambda=0.80$ μm at seven different sun elevations in low turbidity conditions on May 23, 1977, at the Mauna Loa Observatory. (Coulson et al., 1974.)

elevations the point of P_{max} has migrated away from the sun's vertical for the short wavelengths, resulting in P_{max} being a few percent higher in the sun's almucantar than in the sun's vertical. The two curves converge to within measurement error when both maxima appear in the sun's vertical at sun elevation 45°. Equivalent data at other wavelengths show similar features, although details vary with conditions.

The polarization in the sun's almucantar at lower altitudes and in more turbid conditions have been measured by several different authors. For instance, early measurements were made by Pyaskovskaya-Fesenkova (1958, 1960) at altitudes of 1450, 500, and 200 m, the first two locations being near Alma Ata in the Soviet Republic of Kazakhstan, and the third in the desert of Egypt. The instrument was a visual photometer equipped with a yellow Schott filter. The polarization was determined by a method devised by V.G. Fesenkov (the husband of the observer) in which the brightness at a point in the sky was measured

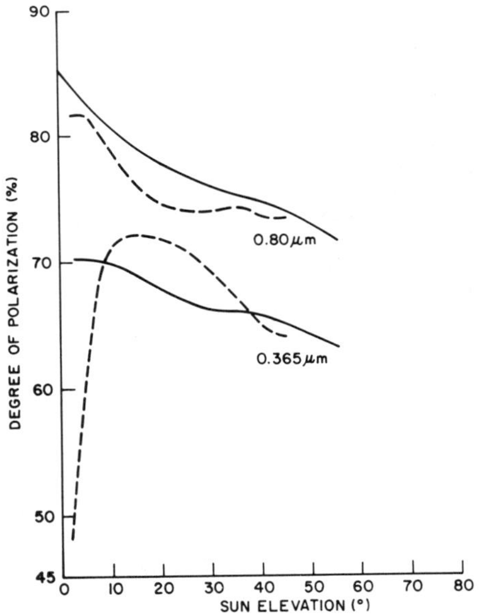

Figure 5.33 Comparison of the polarization maximum in the sun's almucantar (dashed curves) with that in the sun's vertical (solid curves) as a function of sun elevation from measurements at $\lambda=0.365$ and 0.80 μm on May 23, 1977, at the Mauna Loa Observatory. (Coulson et al., 1974.)

through a polarizer oriented at three different angles of 60° separation. The degree of polarization was calculated from the equation

$$P\Theta = \frac{2[B_1(B_1 = B_2) + B_2(B_2 - B_3) + B_3(B_3 - B_1)]^{1/2}}{B_1 + B_2 + B_3}$$

where $B_i(i=1,3)$ are the brightnesses obtained at the three positions of the polarizer, and Θ is the scattering angle. Data from the three sets of measurements are shown in Fig. 5.34, in which the solid curves are from an observatory at 1450 m altitude near Alma Ata and the dashed curve from an altitude of 200 m on the Egyptian desert. The very low polarization from the 1500-m altitude was observed in a dense dry haze, whereas the transparency was high for the clear case. It is somewhat surprising that the clear summer day at the mountain obser-

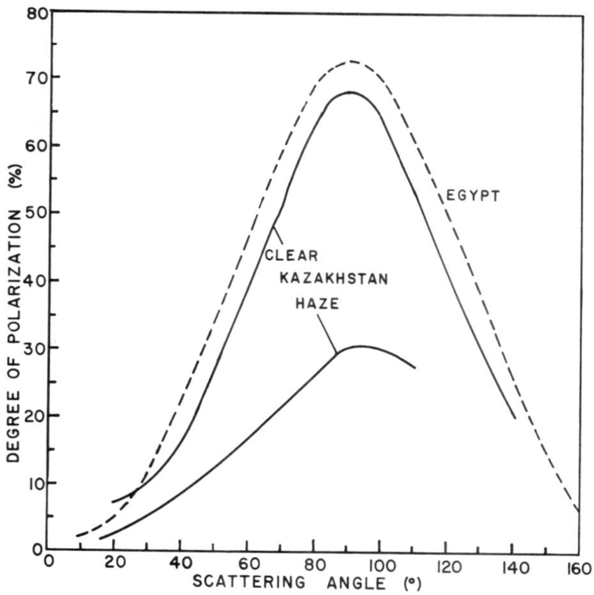

Figure 5.34 Degree of polarization of skylight as a function of scattering angle in the sun's almucantar as observed visually with a yellow filter by Pyaskovskaya-Fesenkova (1960) at two locations in Kazakhstan (solid curves) and at a site in the Egyptian desert (dashed curve).

vatory, where the surface albedo is presumably low, gave a polarization maximum of less than 70%, whereas equivalent measurements from 200-m altitude in a highly reflecting desert setting yielded a more reasonable value of 74%. The author did not explain this apparent anomaly.

Data for the sun's almucantar from the measurements of Bullrich (1964) at two different geographical locations are shown by the curves of Fig. 5.35. Haze in the atmosphere of S'Agata, a city near Naples in southern Italy, limited the polarization maximum to quite low values at both wavelengths, and the polarization maximum was observed to lie outside the sun's vertical. The Italy data are unusual in that the polarization was significantly higher at $\lambda = 0.709$ μm than at $\lambda = 0.405$ μm. This, of course, is the reverse of the situation under most hazy conditions. Even lower values of polarization are shown by the data of Mainz, West Germany in spite of the fact that the atmosphere appeared to be relatively clear at the time of the measurements with a visual range of 40 km. The aerosol burden must have been distributed through a thick layer of the atmosphere in order to produce such low polarization values with apparently clear conditions near the surface. The polarization maximum at Mainz was observed to be in a broad band which crossed the sun's vertical at a scattering angle of about 90°. Comparison of the data with theoretical computations showed the particle distribution to be rich in small sizes with relatively few large particles.

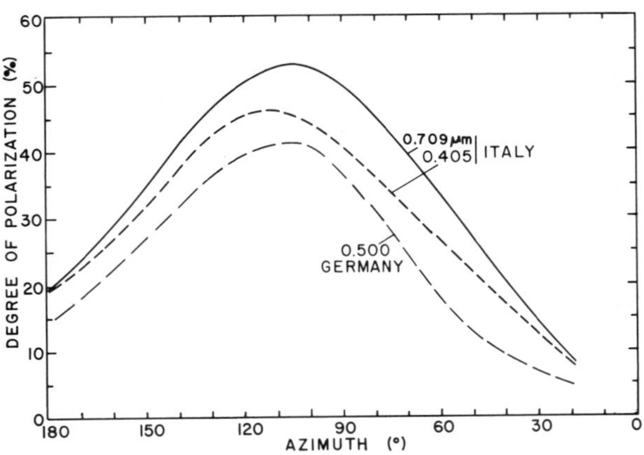

Figure 5.35 Degree of polarization of skylight as a function of azimuth angle in the sun's almucantar as measured at $\lambda = 0.405$ and 0.709 μm at S'Agata, Italy and at $\lambda = 0.500$ μm at Mainz, West Germany. (Data from Bullrich, 1964.)

The possibility of determining the average albedo of a large area of the earth's surface by measurements of the degree of polarization at a point on the sun's almucantar at 90° from the sun was investigated by Sivkov (1968), Ivanov et al. (1971), and others. The latter authors compared albedo determinations by the skylight polarization method with direct measurements of albedo in four wavelength ranges from 0.405 μm to 0.650 μm in a semi-desert region of Kazakhstan. The conclusion was that the method can be used to give albedo values to a precision of 10% to 15%, at least for summer observations. As might be expected, winter observations gave mean albedo values for a large area significantly less than those of a simple snow surface, the implication being that rocks and vegetation extending above the snow surface decreased the average albedo from that of the snow itself. The method is based on sound physical principles, but it has not been exploited to its fullest potential for practical application in other situations.

References

Bullrich, K. (1964): Scattered radiation in the atmosphere and the natural aerosol, *Adv. in Geophys., Vol. 10,* Academic Press, pp. 99–242.

Bullrich, K., R. Eiden, and W. Nowak (1966): Sky radiation, polarization, and twilight radiation in Greenland, *PAGEOPH, 64,* 220–242.

Carrier, L.W., G.A. Cato, and K.J. von Essen (1967): The backscattering and extinction of visible and infrared radiation by selected major cloud models, *Appl. Opt., 6,* 1209–1216.

Cornu, M.A. (1884): Observations relatives a la couronne visible actuellement autour du Soleil, *Comp. Rend., 99,* 488–493.

Cornu, A. (1890): Sur l'application du photopolarimetre a la meteorologie, C.R. Ass. franc. *Av. Sci., Session a Limoges,* pp. 267–270.

Coulson, K.L. (1971): On the solar radiation field in a polluted atmosphere. *J. Quant. Spectrosc. Radiat. Transfer, II,* 739–755.

Coulson, K.L., J.J. Carroll, and B.W. Fitch (1974): Unpublished data.

Coulson, K.L., J.V. Dave, and Z. Sekera (1960): *Tables Related to Radiation Emerging from a Planetary Atmosphere with Rayleigh Scattering,* University of California Press, Berkeley and Los Angeles.

Coulson, K.L., and D.W. Reynolds (1971): The spectral reflection of natural surfaces, *J. Appl. Met., 10,* 1285–1295.

Coulson, K.L., and R.L. Walraven (1972): A high-precision computer-controlled dual-channel polarizing radiometer, Preprints, Conf. on Atm. Rad., Fort Collins, Colorado, pp. 161–162.

Coulson, K.L., R.L. Walraven, and B.W. Fitch (1975): Atmospheric turbidity in the Antarctic as indicated by the intensity and polarization of skylight, I.U.G.G. General Assembly, Grenoble, France, August 28, 1975.

Dandekar, B.S., and J.P. Tuttle (1971): Day sky brightness and polarization during the total eclipse of 7 March 1970, *Appl. Opt., 10,* 1220–1224.

Dave, J.V. (1956): On the intensity and polarisation of the light from the sky, *Part II, Proc. Indian Acad. Sci., XLIII, No. 6, Sect. A,* 336–358.

Dave, J.V., and K.R. Ramanathan (1956): On the intensity and polarisation of the light from the sky during twilight, *Proc. Indian Acad. Sci., XLIII, No. 2, Sect. A.,* 67–78.

deBary, E. (1964): Influence of multiple scattering of the intensity and polarization of diffuse sky radiation, *Appl. Opt., 3,* 1293–1303.

deBary, E., K. Bullrich, and D. Lorenz (1961): Messungen der himelsstrahlung und deren Polarisationsgrad wahrend der Sonnenfinsternis AM 15.2. 1961 in *Viareggio (Italian), Geofis. Pura e Applicata, 48,* 193–198.

Dorno, C. (1919): Himmelshelligkeit, Himmelspolarisation und Sonnenintensitat in Davos (1911 bis 1918), *Meteor. Zeit., 36,* 109–124, 181–192.

Ellis, H.T., and R.F. Pueschel (1971): Solar radiation: Absence of air pollution trends at Mauna Loa, *Science, 172,* 845–846.

Elterman, L. (1968): UV, visible and IR attenuation for altitudes to 50 km, 1968, AFCRL–68–0153, Env. Res. Pap. No. 285, U.S. Air Force.

Fitch, B.W. (1976): Albedo of the snow-covered Antarctic Plateau, M.S. Thesis, Dept. of Land, Air, and Water Resources, University of California, Davis.

Fitch, B.W., and K.L. Coulson (1983): Effects on skylight at South Pole Station, Antarctica by ice crystal precipitation in the atmosphere, *Appl. Opt., 22,* 71–74.

Fraser, R.S. (1968): Atmospheric neutral points over water, *J. Opt. Soc. Am., 58,* 1029–1031.

Gehrels, T. (1962): Wavelength dependence of the polarization of the sunlit sky, *J. Opt. Soc. Am., 52,* 1164–1173.

Gockel, A. (1920): Beitrage zur Kenntnis von Farbe und Polarisation des Himmelslichtes, *Ann. Phys., Leipzig (4), 62,* 283–292.

Hitzfelder, S.J., G.N. Plass, and G.W. Kattawar (1976): Radiation in the earth's atmosphere: its radiance, polarization, and ellipticity, *Appl. Opt., 10,* 2489–2500.

Holzworth, G.C., and C.R.N. Rao (1965): Studies of skylight polarization, *J. Opt. Soc. Am., 55,* 403–408.

Ivanov, A.I., G.Sh. Livšhitz, and E.L. Tem (1971): Seasonal variation of albedo of a region by skylight polarization measurements, *Scattering and Absorption of Light in the Atmosphere,* Nauka Pub. House, Kazakh SSR, Alma Ata, 56–58 (in Russian).

Jensen, C. (1899): Beitrage zur photometrie des Himmels, *Meteor. Zeit. 14,* 488–499.

Jensen, C. (1928): Die Himmelsstrahlung, *Handbuch der Physik, Bd. 19, Springer, Berlin,* 70–152.

Jensen, C. (1932): Normale, gestorte, und pseudonormale polarisationserscheinungen der Atmosphare, *Meteor. Zeit., 49,* 419–430.

Kano, M. (1968): Effect of a concentrated turbid layer on the polarization of skylight, *J. Opt. Soc. Am., 58,* 789–797.

Kimball, H.H. (1903): Observations of solar radiation with Angstrom pyrheliometer at Asheville and Black Mountain, N.C., *Mo. Wea. Rev., 31,* 320–334.

Kimball, H.H. (1913a): The effect upon atmospheric transparency of the eruption of Katmai volcano, *Mo. Wea. Rev., 41,* 153–159.

Kimball, H.H. (1913b): The effect of the atmospheric turbidity of 1912 on solar radiation intensities and skylight polarization, *Bul. Mt. Weather Observatory, V(5),* 295–312.

Pickering, E.C. (1884): A new form of polarimeter, *Proc. Am. Acad. Arts and Sciences, XIII, New Series,* 294–302.

Plass, G.N., and G.W. Kattawar (1970): Polarization of the radiation reflected and transmitted by the earth's atmosphere, *Appl. Opt., 9,* 1122–1130.

Pyaskovskaya-Fesenkova, E.V. (1958): On scattering and polarization of light in desert conditions, *Dokl. Akad. Nauk, USSR, 123(6),* 1006–1009.

Pyaskovskaya-Fesenkova, E.V. (1960): Some data on the polarization of atmospheric light, *Doklady Akad. Nauk, USSR, 131(2),* 297–299.

Sekera, Z. (1956): Recent developments in the study of the polarization of skylight, *Adv. in Geophys., Vol. III,* Academic Press, New York, 43–104

Sekera, Z., and Collaborators (1955): Investigation of the Polarization of Skylight, Final Report, Contr. AF19(122)-239, Dept. of Meteorology, Univ. of California, Los Angeles, 68 pp.

Shah, G.M. (1969): Enhanced twilight glow caused by the volcanic eruption on Bali Island in March and September 1963, *Tellus, XXI,* 636–640.

Sharp, W.E., S.M. Silverman, and J.W.F. Lloyd (1971): Summary of sky brightness measurements during eclipses of the sun, *Appl. Opt., 10,* 1207–1210.

Shaw, G.E. (1975): Sky brightness and polarization during the 1973 African eclipse, *Appl. Opt., 14,* 388–394.

Shaw, G.E. (1982): Solar spectral irradiance and atmospheric transmission at Mauna Loa Observatory, *Appl. Opt., 21,* 2006–2011.

Sivkov, S.I. (1968): Determinations of the albedo of large areas of the earth's surface from measurements of polarization of the sky, *Actinometry, Atmospheric Optics, and Ozonometry, Hydromet. Pub. House,* Leningrad, 17–20 (in Russian).

Stamov, D.G. (1961): Determination of atmospheric turbidity in various directions by polarimetry, actinometry and atmospheric optics, *Gidromet. Publishing House,* Leningrad, 115–124 (in Russian).

Tichanowski, J.J. (1926): Resultate der Messungen der Himmel polarisation in verschiedenen Spektrumabschnitten, *Meteor. Zeit., 46,* 288–292.

Tichanowski, J.J. (1927): Die Bestimmung des optischen Anisotropiekoeffizienten der Luftmolekulen durch Messungen der Himmelspolarisation, *Phys. Zeit., 28,* 252–260.

CHAPTER 6

INTENSITY OF THE SUNLIT SKY

The intensity of skylight is of interest in the natural illumination of buildings, the response of plants to diffuse radiation, and other practical problems, as well as in studies of atmospheric properties. The theory of scattering is well developed, and a large body of literature on the application of the theory to various models of the clear and turbid atmosphere has been accumulated. Typical results of such studies will be outlined below. Preliminary to that discussion, however, is a brief survey of daylight illumination and sky luminance.

6.1 Illumination and luminance

Because of their direct applicability to engineering and architectural design, the photometric quantities, luminance of, and illumination from the sunlit sky have received more attention than have the equivalent radiometric quantities radiance (intensity) of, and irradiance (flux) from the sky. Efforts to find a method of conversion from the photometric system of units to the radiometric system have been unsuccessful because the photometric units are referenced to the response of the human eye, whereas the radiometric units are in terms of absolute energy. The conversion factor between the illumination and solar flux depends on the spectral distribution of the light, which, of course, changes with sun elevation, atmospheric conditions, and other factors. Furthermore, the conversion for direct sunlight is different from that for diffuse skylight. According to Walsh (1961), the factor by which the flux of diffuse skylight on a horizontal surface expressed in cal cm^{-2} min^{-1} has to be multiplied to give the illumination in lm ft^{-2} ranges from about 6300 for sun elevation at 20° elevation to about 7000 for the sun at 60° elevation. For cases of direct sunlight without any contribution from skylight, the factor is about 6600, and somewhat higher values apply for overcast skies.

Because a minimum of illumination occurs under conditions of overcast skies, building designers consider this as a worst case, and design accordingly. In addition, the illumination is relatively steady under a stratus or thick altostratus overcast, and many locations of the world are subject to extensive periods of full cloudiness. For all of these reasons, a representation of the luminance distribution for an average or typical overcast sky is of practical interest. The simple empirical formula proposed by Moon and Spencer (1942), originally developed from observations in Washington, D.C., has been found to yield reasonable approximations for locations in England and Sweden (Walsh, 1961). The Moon and Spencer formula is written

$$L_\theta = \frac{1}{3} L_z (1 + 2 \sin \theta) \tag{6.1}$$

where L_θ is luminance of the sky at angle θ above the horizon and L_z is luminance at the zenith. The lack of azimuthal dependence corresponds to the commonly observed uniformity in azimuth under a solid overcast. The formula gives the luminance at the horizon as one-third that at the zenith, whereas Walsh (1961) states that 0.4 is a better approximation to reality. In this case the formula becomes

$$L_\theta = \frac{1}{5} L_z (2 + 3 \sin \theta) \tag{6.2}$$

At any rate, the original version was adopted officially by the Commission Internationale de l'Eclairage at the Paris meeting of 1955, and has enjoyed widespread use in building design. The ratio of luminance at elevation angle θ to that at the zenith, as computed from Eq. (6.1), is shown at 5° intervals of θ in Table 6.1.

TABLE 6.1 Luminance distribution of an overcast sky (Walsh, 1961).

θ	L_θ	θ	L_θ	θ	L_θ
0	0.333	35	0.716	70	0.960
5	0.391	40	0.762	75	0.977
10	0.449	45	0.805	80	0.990
15	0.506	50	0.844	85	0.997
20	0.561	55	0.879	90	1.000
25	0.615	60	0.911		
30	0.667	65	0.938		

Another type of sky which has a more or less stable luminance distribution is the clear blue sky. Pokrowski (1929) developed from a combination of theory and measurements the formula for luminance from a clear sky written as

$$L_{\theta\alpha} = L_{90}\left(\frac{1 + \cos^2\alpha}{1 - \cos^2\alpha}\right)\left[1 - \exp\left(\frac{-P}{\sin\theta}\right)\right] \tag{6.3}$$

where θ and α are elevation and azimuth angles, respectively, L_{90} is the luminance on the horizon at an angular distance 90° from the sun, and P is an empirical constant. The change of the magnitude of luminance with sun elevation is accounted for by L_{90}. From measurements of sky illumination in Sweden, Hopkinson (1954) found the best values of P to be 0.32. The luminance distribution for a clear sky for a sun elevation of 40°, as given by Walsh (1961), is shown by Fig. 6.1. The pattern is essentially equivalent to that for skylight intensity given, for example, in Figs. 6.4 and 6.5.

For cases in which the sky is neither clear nor overcast, the distribution of illumination is complicated and depends on the type and extent of cloudiness. No simple relation is available for the distribution of luminance over the sky, but the total illumination on a horizontal surface should vary in a fashion similar to that of total diffuse radiation. Many recent studies of this quantity are available (e.g., Carroll, 1985; Pinker and Ewing, 1985).

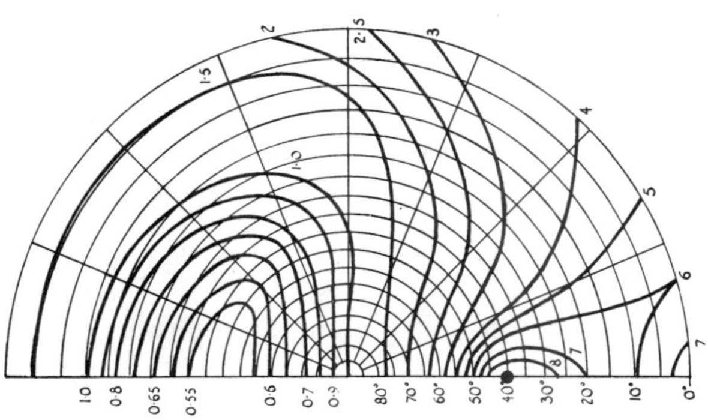

Figure 6.1 Distribution of luminance over the sky for a sun elevation of 40°. (After Walsh, 1961.)

6.2 Intensity of skylight

Although illumination and skylight intensity have qualitatively similar distributions, quantitatively they may be quite different. Steven (1977) found values may differ by as much as a factor of 3 at some points of the sky when both quantities are referenced to zenith values. This is an extreme, however, and generally the photometric and radiometric fields of skylight exhibit similar variations with sun elevation, atmospheric turbidity and other parameters.

6.2.1 Standard distributions of clear sky intensity

It is useful in solar energy computations to define some standard intensity distributions for representative values of solar zenith angles. A set of four such distributions has been developed by Steven (1977) from 67 sets of skylight intensity measurements in England, each set consisting of individual measurements at a grid of 34 points over the dome of the sky. The measurements, made with a Linke-Feussner actinometer, were for wavelength-integrated white light. The results of Steven's measurements are summarized in Table 6.2, in which the entries, termed normalized intensity, are the ratio of intensity measured at the grid points to the total diffuse flux on a horizontal surface. Thus the units of the data are $(\pi st)^{-1}$. The data are segregated into four ranges of solar zenith angle θ_o centered at 35°, 45°, 55°, and 65°, and tabulated according to zenith and azimuth angles with respect to those of the sun. The number of sets of measurements in each range of θ_o varied from 11 to 13, and the mean Linke turbidity factors varied from 0.22 to 0.38. According to the author, the distribution of the intensity normalized in this way is remarkably constant from day to day if the sky is clear. However, Steven's attempts to develop an empirical relation by which the observed distributions might be represented were unsuccessful.

6.2.2 Skylight intensity in a Rayleigh atmosphere

While distributions of luminance or of total intensities are useful tools for practical applications, they are not well adapted to providing a clarification of the physical processes of scattering in the atmosphere. More quantitative models are required for this basic understanding and the simplest model for the purpose

TABLE 6.2 Mean values of normalized intensity of skylight according to the measurements of Steven (1977).

Zenith angle (°)	Azimuth (°)	Mean solar zenith angle (°)			
		35	45	55	65
−10	0	3.12	3.54	4.04	4.93
+10	0	3.25	3.93	4.72	6.44
0	–	3.14	3.04	3.25	3.65
−10	±20	2.51	2.55	2.58	2.87
+10	±20	2.53	2.84	3.17	3.87
+20	0	2.49	2.98	3.83	–
+30	0	1.98	2.52	3.83	–
+40	0	1.57	–	–	–
−20	0	1.86	1.92	2.05	2.46
−30	0	–	1.21	1.26	1.36
−40	0	–	–	0.82	0.88
30	180	0.41	0.35	0.36	0.31
60	180	0.40	0.44	0.48	0.49
75	180	0.54	0.58	0.66	0.71
30	±90	0.69	0.66	0.54	0.43
60	±90	0.60	0.67	0.63	0.58
75	±90	0.68	0.76	0.76	0.76
30	±45	1.61	1.30	0.93	0.71
60	±45	1.20	1.37	1.44	1.33
75	±45	1.15	1.41	1.60	1.74
30	±135	0.45	0.43	0.38	0.32
60	±135	0.44	0.46	0.49	0.47
75	±135	0.54	0.58	0.64	0.64
0	–	0.84	0.69	0.55	0.41
Mean coefficient of variation		0.105	0.12	0.13	0.14

is that of the clear Rayleigh atmosphere. Generally, scattering by aerosols may be considered as a perturbation superimposed on the Rayleigh distribution, but, of course, under certain conditions aerosol effects dominate the intensity fields. This aspect will be discussed further below.

Examples of the intensity distribution of skylight in a plane-parallel model of the Rayleigh atmosphere, as computed by Coulson et al. (1960), is shown for six different wavelengths in the sun's vertical for a sea-level location with no surface reflection by Fig. 6.2. The position of the sun at a zenith angle of $\theta_o = 53.1°$ is indicated by the arrow. Although the ordinate is in relative units,

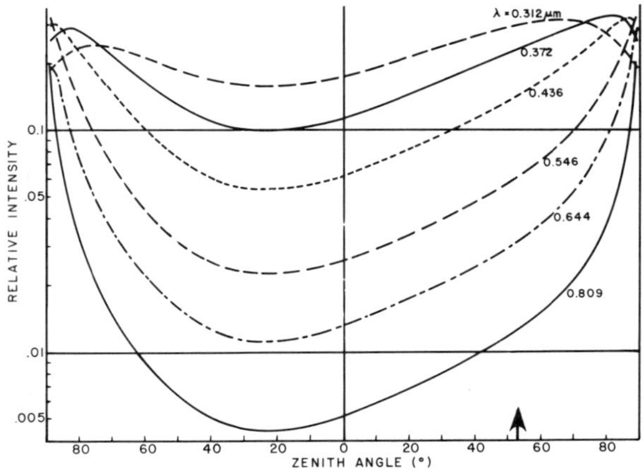

Figure 6.2 Relative intensity of skylight in the sun's vertical at six different wavelengths for a Rayleigh atmosphere with no surface reflection. The position of the sun at a zenith angle of $\theta_o = 53.1°$ is indicated by the arrow. (Data from Coulson et al., 1960.)

it can readily be changed to absolute energy units by multiplying by F_o/π, where F_o is the flux of solar radiation, expressed in the desired units, which is incident at the top of the atmosphere. Strong horizon brightening is a characteristic feature of skylight at the longer wavelengths, whereas attenuation of the scattered radiation through the long optical pathlengths at large zenith angles causes horizon darkening at the shorter wavelengths. The increase of scattering efficiency with decreasing wavelength according to the λ^{-4} relationship is responsible for the general increase of intensity with decreasing wavelength shown by the curves. The intensity minimum occurs at an angle of approximately 90° from the sun, the minimum being particularly pronounced and the intensity gradients particularly strong for the longer wavelengths. For the shorter wavelengths, however, multiple scattering creates a more nearly isotropic intensity distribution across the sky. For instance, for $\lambda = 0.312$ μm the difference between maximum and minimum intensities is only about a factor of 2, whereas for $\lambda = 0.809$ μm, the horizon intensity is over 45 times as great as that in the region of the minimum.

The low intensity at long wavelengths is a characteristic that has practical applications in remote sensing. Since Rayleigh scattering is very weak, it is other parameters that mainly control the radiation field. For skylight, the controlling parameter is scattering by aerosols, so for aerosol studies by means of skylight, the longer wavelengths should be emphasized. For remote sensing of the surface from space, the background intensity is low and atmospheric transmission is good at the long wavelengths, so here also, other things being equal, the long wavelengths are most advantageous for the purpose.

The distributions for other wavelengths and other solar zenith angles are similar to those shown, except that in general the pattern shifts across the sky with changing solar zenith angles in such a manner as to keep the minimum intensity at approximately 90° from the sun and maintain intensity gradients near the horizon similar to those shown. This characteristic is well illustrated by the solid curves of Fig. 6.3, for which the solar zenith angle is $\theta_o = 84.3°$. Multiple scattering at both wavelengths is insufficient to prevent a change of intensity of more than an order of magnitude from horizon to zenith. The slight horizon darkening shown for $\lambda = 0.495$ μm may be an artifact due to the assumption of a plane-parallel atmosphere. The dashed curves of Fig. 6.3 are for the sun at the zenith, in which case the strong tendency for horizon brightening overrides that for the minimum at 90° from the sun. In fact, the minimum intensity

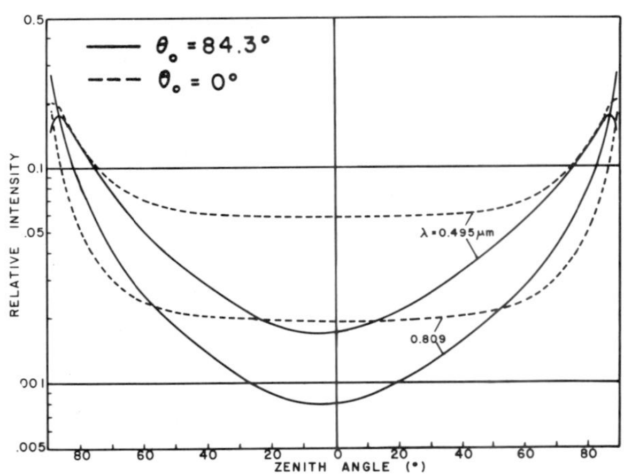

Figure 6.3 Relative intensity of skylight in the sun's vertical at two different wavelengths ($\lambda = 0.495$ and 0.809 μm) and two different solar zenith angles ($\theta_o = 0.0°$ and $84.3°$) for a Rayleigh atmosphere with no surface reflection. (Data from Coulson et al., 1960.)

of scattered light for that case is precisely at the zenith, and not at the horizon. The two opposing effects combine to give a remarkably constant intensity over much of the sky for the sun at the zenith. Realizing that for a zenith sun the intensity field is independent of azimuth, it is evident from the dashed curves that for a Rayleigh atmosphere the intensity would be essentially constant across the sky out to a zenith angle of nearly 60° in all directions. In the real atmosphere, of course, aerosol scattering would destroy this uniform distribution. However, it is an interesting effect that shows up in the Rayleigh computations and there are many cases in nature in which the Rayleigh model is not a bad approximation. Furthermore, the sun reaches the zenith at all tropical locations. Thus the uniform intensity distribution that occurs at high sun elevations should be a useful feature for studies of aerosol effects.

The intensity distribution over the entire dome of the sky at a wavelength of $\lambda = 0.546\ \mu m$ is shown by Fig. 6.4. The position of the sun at $\theta_o = 66.4°$ is shown by the circle. Although the minimum intensity is in the region at approximately 90° from the sun, there is not a strong tendency for this characteristic to persist over the whole hemisphere, as there is for the polarization (see Fig. 4.11). Here the dominant effect is that of horizon brightening at all azimuths. A different distribution is seen for ultraviolet wavelengths ($\lambda = 0.3112\ \mu m$) in Fig. 6.5. Although a minimum intensity occurs in a general region about 90° from the sun, the tendency for horizon darkening at this short wavelength causes an equally low minimum at the horizon in a wide angle centered at an azimuth

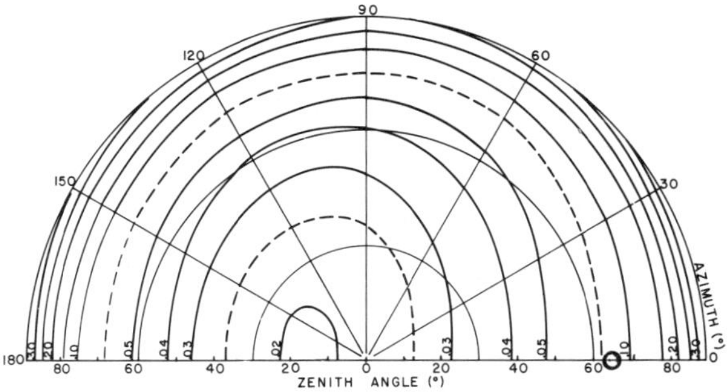

Figure 6.4 The distribution of skylight intensity over the dome of the sky at $\lambda = 0.312\ \mu m$ and $\theta_o = 66.4°$ for a Rayleigh atmosphere with no surface reflection. (Data from Coulson et al., 1960.)

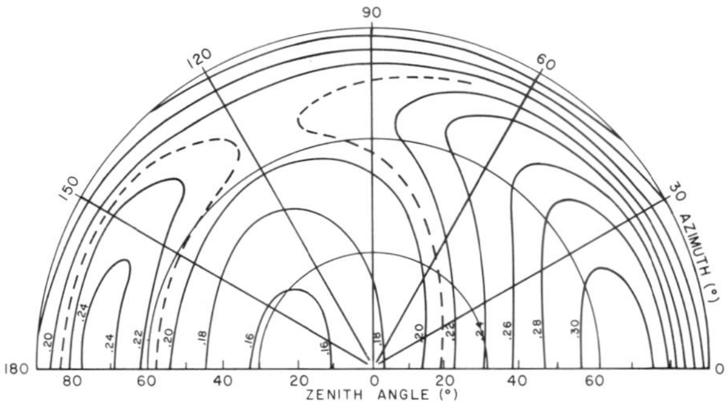

Figure 6.5 The distribution of skylight intensity over the dome of the sky at $\lambda = 0.312$ μm and $\theta_o = 53.1°$ for a Rayleigh atmosphere with no surface reflection. (Data from Coulson et al., 1960.)

of 90°. At the same time a pronounced intensity maximum occurs at 20° to 30° above the horizon at all azimuths.

6.3 Flux of skylight in a Rayleigh atmosphere

As shown by Deirmendjian and Sekera (1954), the total flux of diffuse skylight on a horizontal surface at the ground is given by the relation

$$F_s = \frac{\pi \mu_o F_o A \bar{s}}{2(1 - A\bar{s})} \left[\gamma_e(\mu_o) + \gamma_r(\mu_o)\right] \tag{6.4}$$

where A is Lambert surface albedo and $\bar{s}$, γ_e, and γ_r are functions for which tabulations have been made [see Eq. (4.82) of Chapter 4]. Coulson (1968) has computed F_s for eight values of optical thickness from $\tau = 0.01$ to 1.0 for seven positions of the sun from $\theta = 0°$ to 84.3°. The resulting flux of diffuse skylight for the Rayleigh atmosphere is shown for seven solar zenith angles and no surface reflection ($A = 0$) by Fig. 6.6. For these computations F_o has been set to unity, thereby making the flux incident on a horizontal surface at the top of the atmosphere $\pi\mu_o$. The numerical values of Fig. 6.6 are taken relative to the flux assumed at the top of the atmosphere. For instance, if F_o is in units of W m^{-2},

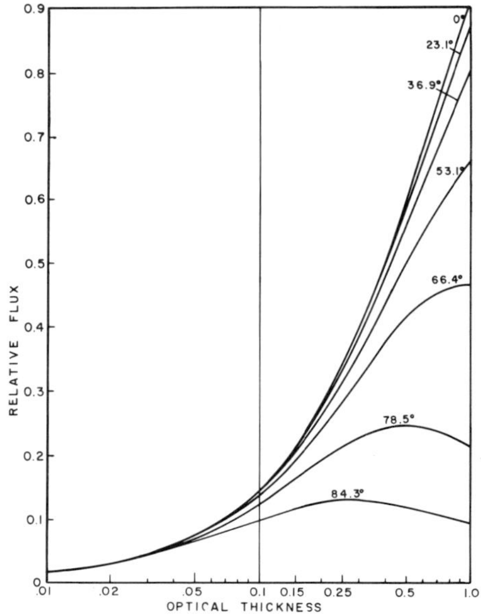

Figure 6.6 Relative flux of skylight on a unit horizontal surface at the ground due to Rayleigh scattering for seven different zenith angles of the sun and no ground reflection. The flux incident on a unit horizontal surface at the top of the atmosphere is assumed to be $\pi\mu_o$.

the values shown can be changed to W m^{-2} by multiplying by F_o/π. It should be noted that Deirmendjian and Sekera (1954) set $F = 1/\pi$, which makes their values less than these by a factor of π.

The curves of Fig. 6.6 show the dramatic increase of scattering efficiency with increasing optical thickness (decreasing wavelength) typical of Rayleigh scattering. It is of interest, however, that for large solar zenith angles the strong attenuation through the long pathlength through the atmosphere, with the consequent strong backscatter of the radiation to space, results in the flux for the atmosphere of large optical thickness being reduced from that for a somewhat thinner atmosphere. Generally, however, the diffuse flux increases with increasing optical thickness.

By solving Eq. (6.4) for various values of A, it is a simple matter to compute the surface flux due only to the downward scatter of the radiation reflected from the surface. Coulson (1968) has tabulated this quantity for the same optical thicknesses and solar zenith angles as in Fig. 6.6. The resulting data for the sun at a zenith angle of 53.1 ° are shown for ten different values of surface albedo in Fig. 6.7. The curves show that the surface albedo would have to be over 0.90 for the flux due to surface reflection to contribute one-half of the total flux of skylight in a Rayleigh atmosphere. This is much higher than the albedo of most natural surfaces, although the albedo of a thick cover of new snow may approach or exceed this value. For instance, Carroll and Fitch (1981) observed the albedo of the snow surface at the geographical South Pole to be higher than 0.90 for large zenith angles of the sun, but, of course, the solar zenith angle never gets as low as 53.1 ° at the South Pole. Computations for other solar zenith angles show that a highly reflective surface contributes more than one-half of the total surface flux due to Rayleigh scattering for solar zenith angles less than 53.1 °,

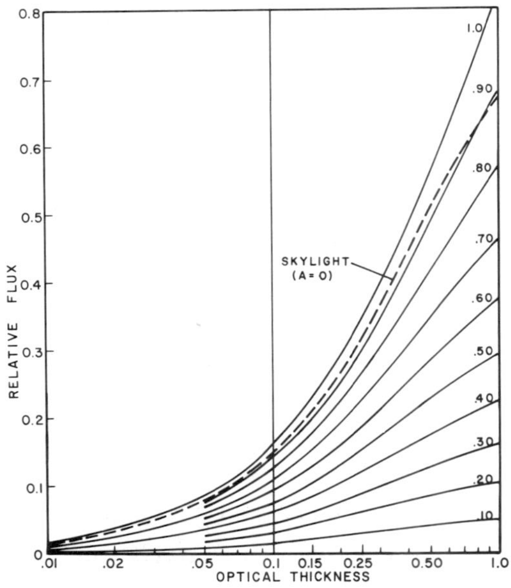

Figure 6.7 Relative flux of skylight on a unit horizontal surface at the ground due to backscatter by the overlying Rayleigh atmosphere of radiation reflected from Lambert surfaces with ten different values of surface albedo and a solar zenith angle of 53.1 °. The dashed curve, which is transferred from the previous figure, represents atmospheric scattering of the incident light with no reflection from the ground.

but for still larger solar zenith angles surface reflection cannot contribute one-half of the diffuse flux in a Rayleigh atmosphere.

6.4 Skylight intensity computed for turbid atmospheres

A great number of theoretical studies have been carried out by various authors for computations of the skylight intensity distribution for more or less realistic models of the aerosol content of a cloudless atmosphere. The main parameters necessary for models of the turbid atmosphere are size-frequency distributions and concentration of particles, their index of refraction, and their variation with altitude. Various models of polydisperse aerosols were outlined above in section 3.2.7. Almost universally the particles have been assumed to be spheres, thereby permitting use of the Mie theory for computations, and most authors, though not all, have assumed the plane-parallel model of the atmosphere. Some examples of the many theoretical studies of skylight intensity in the turbid atmosphere are Feigelson et al. (1960), Bullrich (1964), de Bary and Bullrich (1964), Hansen (1969), Dave (1970), Tanaka (1971), Blattner and Wells (1973), Kattawar et al. (1976), Hitzfelder et al. (1976), and Nazaraliyev and Pavlov (1980). In addition, Elterman (1968) has developed a model of the turbid atmosphere which has found wide application in radiative transfer calculations.

Results of computations on the intensity of skylight in the sun's vertical for five different solar zenith angles and a wavelength of $\lambda = 0.55$ μm by Hitzfelder et al. (1976) are shown in Fig. 6.8. Rayleigh scattering, scattering and absorption by aerosols, and absorption by ozone were taken into account in the computations, the total optical thickness being 0.378. The altitude distribution of these components used was that of Elterman (1968). The aerosol size distribution was given by the Haze L model, and the index of refraction of aerosols was assumed to be $m = 1.55\text{–}0.05i$. Surface reflectance was assumed to be Lambertian with a value of $A = 0.15$. The extraterrestrial solar flux across a unit area normal to the solar beam was taken to be unity for the computations.

The curves of Fig. 6.8 show a strong intensity maximum in the region of sky surrounding the sun. This brightening is due entirely to scattering by aerosols, as no such feature is produced by Rayleigh scattering (see Figs. 6.2 and 6.3). Although the incident flux is proportional to $\cos\theta_o$, the intensity maximum does

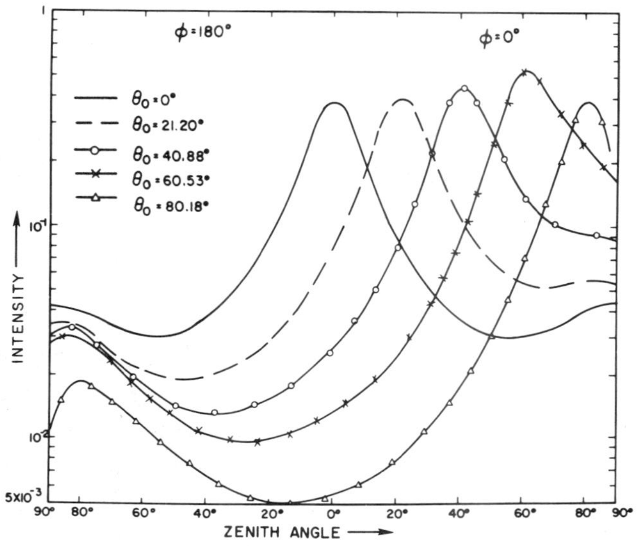

Figure 6.8 Intensity of skylight in the sun's vertical in a turbid atmosphere at a wavelength of $\lambda=0.55$ μm and five different solar zenith angles according to the computations of Hitzfelder et al. (1976). The ordinate label "Radiance" is equivalent to the term "Intensity" used here. (After Hitzfelder et al., 1976.)

not respond to the decreases of incident solar flux with increasing θ_o, but in fact the maximum increases with increasing solar zenith angle out to $\theta_o=60.53°$. The number of scatterers encountered by the direct beam in its traverse of the atmosphere increases with increasing θ_o, thereby leading to the intensity increase as shown. The maximum for $\theta_o=80.18°$ is diminished somewhat by attenuation of the scattered light through the long pathlength near the horizon. The minima are located at approximately 90° from the sun, and they do respond to the combination of decreased incident flux and decreased numbers of scattering particles in the direction of the minimum as the sun approaches the horizon. Horizon darkening near the antisolar horizon is not necessarily an aerosol effect, as it also occurs for a Rayleigh atmosphere of large optical thickness.

Tanaka (1971) has used a version of the adding method to compute the intensity and degree of polarization of light at two wavelengths emerging from a plane-parallel atmosphere overlying a Lambert surface. In addition to the

Rayleigh case, two models T_1 and T_2 of the turbid atmosphere were considered. The aerosol and Rayleigh optical thicknesses, τ^A and τ^R, at the two wavelengths for the two atmospheric models are listed in Table 6.3. The size frequency distribution of particles was similar to that of Haze *C* proposed by Deirmendjian (1964), the vertical distribution was according to the Elterman (1968) profile, and the index of refraction was taken as 1.33.

The computations yielded the intensity distributions in the sun's vertical shown in Fig. 6.9 for $\lambda = 0.45$ μm and two values of surface albedo. Tanaka assumed the solar flux on a horizontal surface at the top of the atmosphere to be $\pi F_o \mu_o$ and set $F_o = 1.0$. Thus by multiplying the relative intensities given by the actual F_o, one obtains the intensity in absolute units (e.g., W m^{-2} sr^{-1}). The data show that skylight intensity increases with increasing turbidity, except in the region of horizon darkening at the antisolar azimuth. In that direction the mean optical pathlength of scattered photons is so large that attenuation overcompensates the additional scattering by the more numerous particles, which is the basic reason for horizon darkening, and the effect is amplified by increasing turbidity. The strong intensity gradients extending right up to the position of the sun are a more realistic distribution than that of the previous figure, due no doubt to the smaller angular resolution of Tanaka's computations.

The intensity distribution over the dome of the sky for $\lambda = 0.45$ μm and $\theta_o = 55°$ for turbidity model T_1 and no ground reflection is shown in Fig. 6.10. The solar aureole with strong intensity gradients is pronounced throughout the region surrounding the sun, but appreciable horizon darkening exists only within about 30° of the solar azimuth. The antisolar half hemisphere has a very flat intensity distribution, with a change of only slightly more than a factor of 2 over the entire region.

TABLE 6.3 Atmospheric models used for the computations of Tanaka (1971).

Model	$\lambda(\mu)$	τ^R	τ^R	Total τ
T_1	0.45	0.221	0.219	0.440
	0.70	0.037	0.141	0.178
T_2	0.45	0.221	0.437	0.658
	0.70	0.037	0.281	0.318

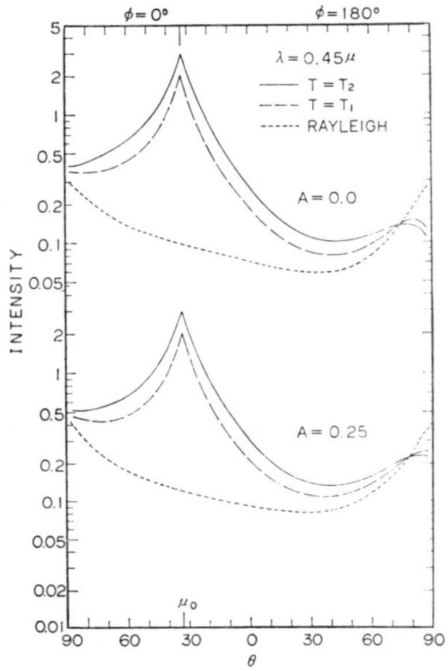

Figure 6.9 Intensity of skylight in the sun's vertical for a Rayleigh atmosphere and for two models of a turbid atmosphere at a wavelength of $\lambda = 0.45$ μm solar zenith angle of 53°, and surface albedo values of $A = 0$ and 0.25. (After Tanaka, 1971.)

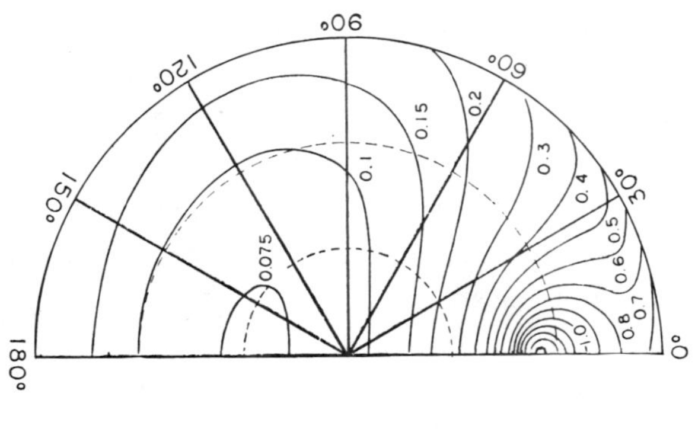

Figure 6.10 Distribution of skylight intensity over the hemisphere of the sky for a moderately turbid atmosphere, wavelength of $\lambda = 0.45$ μm, and no surface reflection. (After Tanaka, 1971.)

In addition to the provision of illumination by skylight mentioned in section 6.1, one practical aspect of skylight intensity is in the use of solar energy as an alternative energy source. Concentrating solar collectors by necessity have an aperture somewhat larger than that subtended by the disk of the sun, and thus include a certain amount of diffuse light from the solar aureole which adds to the total energy received. The same situation applies to pyrheliometers, but in this case the diffuse light received represents a source of error in the pyrheliometric measurement. In both of these cases, interest is in the total wavelength-integrated energy flux represented by the diffuse light, as opposed to that in narrow spectral bands required for some applications.

In a study oriented toward concentrating solar collectors, Thomalla et al. (1983) have computed the distribution of total wavelength-integrated intensity in the solar aureole for four different aerosol models and two solar zenith angles. They took into account the finite angular diameter of the sun, limb darkening of the sun, and absorption of radiation by ozone, carbon dioxide, and water vapor. Aerosol models were chosen as being typical of continental background conditions, urban areas, maritime surface conditions, and aerosols over the desert, all with a refractive index of 1.50–0.02i. Aerosol optical thicknesses were taken as $\tau^A = 0.05$, 0.20, and 0.40 for clear, average, and turbid atmospheres, respectively.

The intensities in the solar aureole, as computed by Thomalla et al. (1983), are given for the continental background model and the average urban model for all three aerosol optical thicknesses of the computations and two sun elevations in Table 6.4. The data are of reasonably low precision, having been obtained by graphical interpolation from diagrams in the original publication, but they should be adequate for practical application. The angles listed from 0.3° to 10° are with respect to the center of the solar disk. The data are strictly applicable only in the solar almucantar, as differences of intensity of the sky above and below the sun were not considered.

The data of Table 6.4 were used to construct the intensity distribution curves of Fig. 6.11. Continental background and average urban aerosol models for low ($\tau^A = 0.05$) and high ($\tau^A = 0.40$) turbidity conditions are represented for two different solar zenith angles. The curves show intensity gradients to be particularly strong near the disk of the sun, with decreases of factors between about 5 and

TABLE 6.4 Intensities of skylight in the region of the solar aureole as a function of angular distance from the center of the sun for two models of atmospheric aerosols, two positions of the sun, and three different aerosol optical thicknesses, as determined by graphical interpolation from the diagrams of Thomalla et al. (1983).

Angle from sun (°)	$\theta_o = 70°$			$\theta_o = 30°$		
$\tau^A =$	0.05 ($\times 10^3$)	0.20 ($\times 10^3$)	0.40 ($\times 10^3$)	0.05 ($\times 10^2$)	0.20 ($\times 10^3$)	0.40 ($\times 10^3$)
	Continental background aerosol model					
0.3	1.2	3.2	3.95	7.3	2.6	4.1
0.4	1.25	3.1	3.80	7.1	2.5	4.0
0.5	1.17	3.02	3.65	6.8	2.4	3.9
1.0	0.90	2.30	2.80	5.3	1.85	2.9
2	0.56	1.40	1.80	3.3	1.05	1.8
3	0.38	0.95	1.15	2.1	0.70	1.15
4	0.265	0.65	0.80	1.5	0.50	0.79
5	0.190	0.49	0.59	1.05	0.36	0.57
6	0.142	0.36	0.44	0.80	0.27	0.41
7	0.114	0.28	0.34	0.65	0.20	0.32
8	0.095	0.23	0.28	0.55	0.17	0.26
9	0.082	0.19	0.23	0.47	0.15	0.22
10	0.075	0.17	0.20	0.42	0.13	0.19
	Average urban aerosol model					
	($\times 10^2$)	($\times 10^2$)	($\times 10^3$)	($\times 10^2$)	($\times 10^2$)	($\times 10^2$)
0.3	3.5	8.8	1.3	1.9	6.1	9.9
0.4	3.4	8.6	1.21	1.85	6.0	9.8
0.5	3.2	8.4	1.1	1.8	5.9	9.6
1.0	2.6	7.1	0.91	1.5	4.9	7.9
2	1.8	5.0	0.60	0.05	3.3	5.0
3	1.4	3.8	0.47	0.78	2.5	4.0
4	1.1	2.9	0.36	0.61	1.9	3.0
5	0.90	2.4	0.30	0.50	1.6	2.6
6	0.77	1.9	0.26	0.45	1.4	2.1
7	0.70	1.7	0.22	0.41	1.3	1.9
8	0.65	1.6	0.20	0.39	1.2	1.7
9	0.61	1.5	0.18	0.38	1.1	1.6
10	0.60	1.4	0.17	0.37	1.0	1.5

20 within the first 10° from the sun's center. Gradients are stronger for the continental background aerosol than for the average urban model because of the higher number of submicron-sized particles assumed for the latter. Intensities are high and show only a modest dependence on sun position for the highly turbid atmosphere, whereas the clear atmosphere yields lower intensities and a stronger dependence on solar zenith angle. The equivalent curves for a moderately turbid atmosphere fall between the two families shown here.

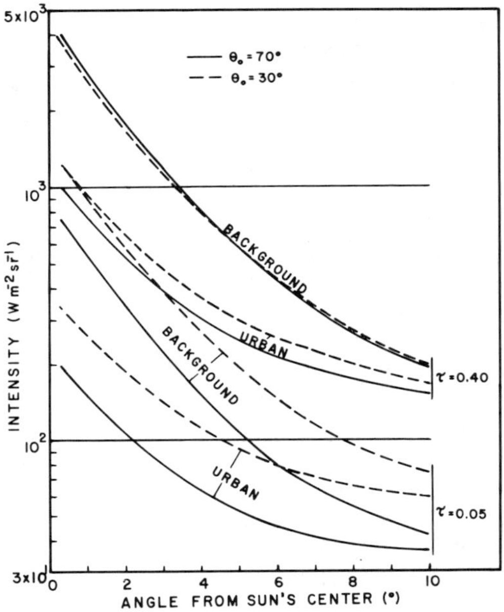

Figure 6.11 Distribution of skylight intensity in the region of the solar aureole for two models of the atmospheric aerosol and two solar zenith angles, according to the computations of Thomalla et al. (1983).

6.5 Skylight intensity measured at various locations

Since skylight intensity for the clear atmosphere is dependent on both atmospheric turbidity and surface albedo, it is useful to look at the skylight measurements that have been made at various geographic locations. We have chosen cases in which cloud interference was absent or minimal and atmospheric conditions were relatively stable over the period of measurements. In most cases, the data are given in relative units only, as radiation sources and other calibration facilities are generally inadequate to yield accurate and reliable calibrations of narrow-band radiometers, particularly those based on photomultiplier tubes, over the extremely wide dynamic range required for skylight measurements. Consequently, the data, with few exceptions, have generally been expressed on a relative or arbitrary scale of intensity. This is normally the case in the diagrams that follow.

6.5.1 Skylight intensity at the Mauna Loa Observatory, Hawaii

As discussed in Chapter 5, atmospheric conditons at the Mauna Loa Observatory, located at an altitude of 3400 m on the island of Hawaii, are among the cleanest to be found anywhere on the planet. Not only is the island at a mid-Pacific location thousands of miles from continents and thereby largely exempt from continental pollution, but also the temperature inversion at an altitude of 2 to 2.5 km which is associated with the trade wind regime keeps most of the locally produced particulates at levels below the Observatory. Thus the atmosphere at and above the Observatory usually has very low turbidity, making skylight intensity there approximate that of a Rayleigh atmosphere as closely as possible under real conditions.

The distribution of skylight intensity through the plane of the sun's vertical, as measured by the present author but previously unpublished, is shown for two wavelengths and a number of sun elevations in Fig. 6.12. The data were

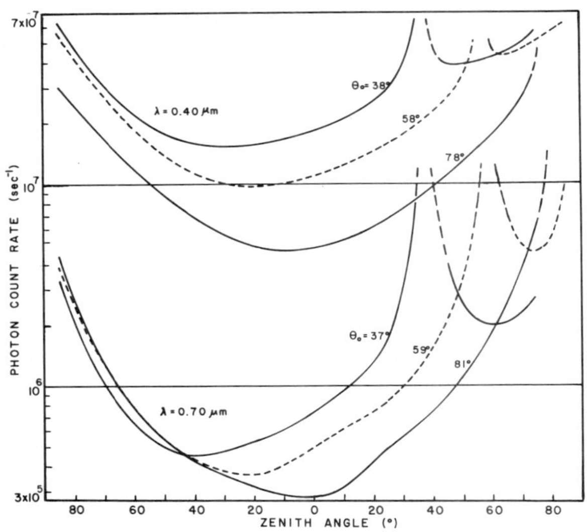

Figure 6.12 Photon count rate as a function of zenith angle in the sun's vertical for different zenith angles of the sun at the two wavelengths as measured on a very clear day (February 10, 1977) at the Mauna Loa Observatory (altitude 3400 m). The long dashes in the region near the sun indicate extrapolations. (Coulson, unpublished data.)

taken under very clear atmospheric conditions on February 10, 1977, with the dual-channel polarizing radiometer briefly described by Coulson et al. (1974). The ordinate scale of the diagram is the actual photon count rate of the instrument. Since measurements at the two different wavelengths were taken on different channels, with different photomultiplier tubes, different neutral density filters, and different optics in general, the relative magnitudes of the count rates at the two wavelengths have no significance. The families of curves for each wavelength, however, are internally consistent, and the shapes of the curves accurately depict the intensity distribution in the sky.

Several features of the curves deserve comment. First, the change of intensity across the sky is much greater at $\lambda=0.70$ μm than at 0.40 μm. For instance, the intensity difference between the antisolar horizon and the minimum at $\theta_o=57.9$ is a factor of 9.6 for $\lambda=0.40$ μm, whereas for a comparable solar position ($\theta_o=59.4$) the difference at $\lambda=0.70$ μm is a factor of 12.5. The smaller amplitude of the curves for $\lambda=0.40$ μm than for $\lambda=0.70$ μm is brought about by the greater effect of multiple scattering of the shorter wavelength radiation. A second feature of note is the much sharper and more pronounced solar aureole for the longer wavelength. Unfortunately, in order to protect the photomultiplier tubes from damaging radiation, measurements could not be made much closer than 2.5° to 3° from the disk of the sun, so details of the aureole region were not well determined. The main curves in the diagram were constructed from actual data points and the portions with long dashes are extrapolations. It is well established from computations that the solar aureole is due mainly to primary scattering by the larger particles, and the fact that primary scattering is more pronounced at $\lambda=0.70$ μm than at $\lambda=0.40$ μm is responsible for the aureole being sharper and more pronounced at the longer wavelength.

One feature of the curves at $\lambda=0.70$ μm that appears anomalous is an apparent intensity bulge just sunward of the minimum, which makes the minimum sharper than it would be otherwise. The bulge decreases in importance with decreasing solar zenith angle, and it does not appear at $\lambda=0.40$ μm. The cause of the feature is not known. Computations show that intensity maxima occur at various scattering angles for monodispersions of small absorbing spheres (Plass, 1966), but there is no reason to believe the aerosol component of the atmosphere above Mauna Loa was monodisperse in nature. Polydispersions of particles normally exhibit a smooth phase function through scattering angles of 90° and more. (See for example Fig. 3.12). There may be some particle distribution, such as a bimodal distribution or a distribution of irregularly shaped particles, which could explain the observed feature, but as yet it remains unexplained.

Horizon brightening occurs at both wavelengths, but is more pronounced at $\lambda = 0.70$ μm than at 0.40 μm. The long pathlength through the atmosphere near the horizon causes significant amounts of multiple scattering at both wavelengths, but at $\lambda = 0.40$ μm it is relatively strong all across the sky. At $\lambda = 0.70$ μm, however, the contribution by multiple scattering is minor across most of the sky, so the increase near the horizons is relatively more important for this wavelength than at $\lambda = 0.40$ μm.

Skylight intensity data as measured in the sun's vertical on the very clear day of February 10, 1977, at the Mauna Loa Observatory for a solar zenith angle of approximately 66° and three different wavelengths are shown by the solid curves of Fig. 6.13. Similar data as computed for the Rayleigh atmosphere with

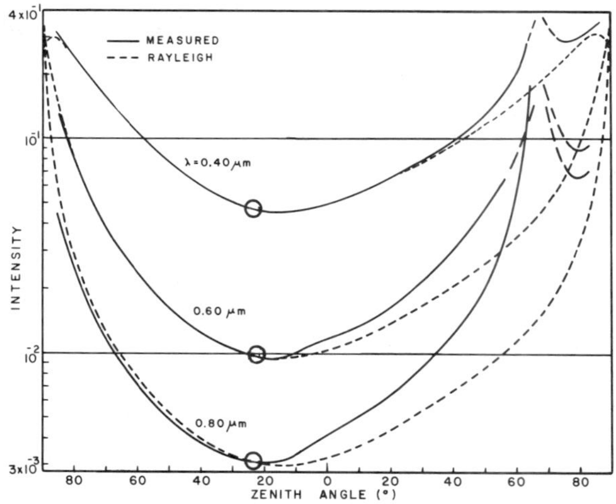

Figure 6.13 Distribution of skylight intensity in the sun's vertical as measured at the Mauna Loa Observatory (solid curves) and as computed for the Rayleigh atmosphere (dashed curves) at three different wavelengths for a solar zenith angle of approximately 67°. The measurements have been normalized to give the same intensity values as those of the Rayleigh atmosphere at a point 90° from the sun (circles). (Coulson, unpublished data.)

no surface reflection by Dr. Bruce Fitch for equivalent optical thicknesses at the altitude of Mauna Loa, are shown by the dashed curves of the diagram. Data from the measurements have been arbitrarily normalized to correspond to the Rayleigh intensities at a point 90° from the sun. This normalization point is indicated by circles located approximately 23° from the zenith at the antisolar azimuth. To the extent that this normalization is valid, the measured data may be interpreted as absolute intensities, in spite of a lack of absolute calibration of the measuring instrument. As mentioned previously, the Rayleigh computations were based on the assumption of the solar energy incident, per unit time and unit wavelength interval, on a unit surface normal to the solar beam at the top of the atmosphere having the value of πF_o, and F_o was set to unity. Thus absolute intensity values are obtained, in units of W cm^{-2} $(\mu m)^{-1}$ sr^{-1}, by multiplying values on the diagram by the quantity F_o/π, where F_o is the actual energy flux at the top of the atmosphere expressed similarly in units of W cm^{-2} $(\mu m)^{-1}$. Data on the extraterrestrial flux throughout the solar spectrum are available in various publications (e.g., Thekaekara and Drummond, 1971).

Returning to Fig. 6.13, we see that even in the clear atmosphere at a high altitude there is sufficient aerosol scattering to produce a significant solar aureole, the aureole being particularly pronounced at the longer wavelengths. For instance, the measured increase of intensity from 20° above the sun to the vicinity of the sun is more than an order of magnitude greater than the increase over the same angle in the Rayleigh atmosphere. Unfortunately, the instrument used was not well adapted to measurements very near the disk of the sun. The data do show, however, that the aureole is relatively minor at $\lambda = 0.40$ μm due to the strong scattering by the Rayleigh atmosphere in comparison to that produced by the minor component of turbidity in such clear conditions.

The data of Fig. 6.13 show a very close congruence between the measured and computed intensities in the region of the sky opposite the sun. In fact, for $\lambda = 0.40$ μm and 0.60 μm, there is no detectable difference between the curves throughout the range of angles of greater than 90° from the sun, and for $\lambda = 0.40$ μm the two are equivalent to within 35° of the sun. Even the discrepancies at $\lambda = 0.80$ μm may be within measurement error. This situation is gratifying in several respects. First, it attests to the clarity of the atmosphere over Mauna Loa, a result which was already deduced from visual observation and polarization measurements and led to the choice of this date as having a very clear atmosphere. Second, the agreement tends to bolster confidence in the measurement technique and the computational program. Finally, it illustrates that neglecting the reflection from the dark lava surface of the mountain is a valid procedure for at least this one case at Mauna Loa.

To show the effect of dust in the atmosphere, we choose the date of May 2, 1977, on which an influx of dust carried by westerly winds from the Asian continent appeared over Hawaii. (Polarization data taken on this date were discussed in Chapter 5, sections 5.8 and 5.9.) Skylight intensities measured in four different wavelengths on May 2, 1977, are shown by the solid curves of Fig. 6.14 for solar zenith angles of 15° to 20°. Comparison data taken in clear atmospheric conditions on May 16, 1977, are shown for the same wavelengths by the dashed curves. The photon count rate shown for the same wavelengths are directly comparable, as the instrument settings remained unchanged over the 2-week period

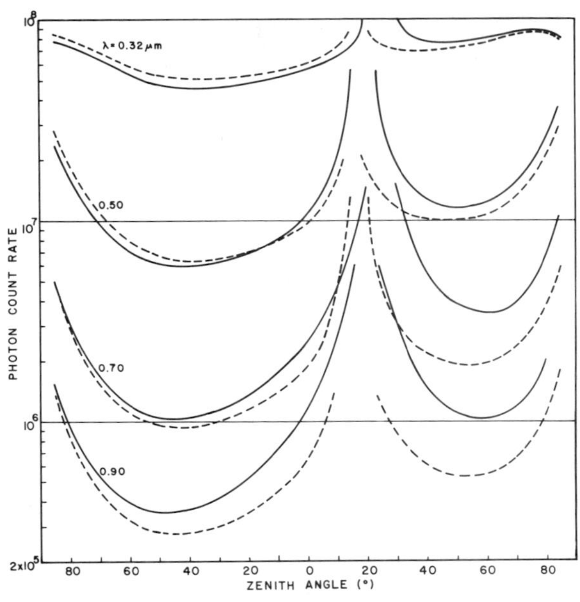

Figure 6.14 Distribution of skylight intensity in the sun's vertical as measured in four different wavelengths at the Mauna Loa Observatory during an episode of Asian dust in the atmosphere over Hawaii (May 2, 1977–solid curves) and in a very clear atmosphere (May 16, 1977–dashed curves).

separating the measurements. As mentioned previously, however, the data for different wavelengths bear no direct relation to each other.

The data show that the dust produced a significant increase of skylight intensity, particularly in the region of the sky below the sun, at the longer wavelengths. For instance, in the region 30° to 35° above the solar horizon, the intensity at $\lambda = 0.90$ μm was about 94% higher in the turbid than in the clear atmosphere. At $\lambda = 0.70$ μm the analogous increase was about 89%. The change of intensity above the sun was less dramatic but still significant at the longer wavelengths. At $\lambda = 0.50$ μm, however, the change of intensity due to dust was relatively minor everywhere, as it was at $\lambda = 0.32$ μm. In fact, at this latter wavelength the intensity in the sky above the sun and throughout the antisolar half of the hemisphere was slightly decreased by the dust. The decrease was no doubt due to slight absorption by the dust particles, the absorption being more significant at the shorter than at the longer wavelengths because many more scattering events occurred as the short wavelengths were subjected to multiple scattering. Single scattering was predominant at the long wavelengths, so a slight absorption at each scattering event was not effective in changing the total intensity. Another effect of multiple scattering at $\lambda = 0.32$ μm was to make the intensity almost isotropic across the sky in both clear and turbid atmospheres, and to produce a small amount of darkening at the horizon below the sun. The presence of the dust did not change these features significantly.

6.5.2 Skylight intensity in industrial areas

6.5.2.1 Mainz, West Germany

Bullrich et al. (1965, 1968) measured the intensity of skylight in the industrial city of Mainz, West Germany, under various turbidity conditons during the summer and fall months of 1964. The Linke turbidity factor during the measurements varied between 2.8 and 7.2, and the horizontal visual range from 5 to 15 km. A typical example of the intensity distribution in the sun's vertical for four different wavelengths as measured in Mainz on September 9, 1964, is shown in Fig. 6.15. The ordinate of the diagram is in relative units in spite of the fact that Bullrich and his colleagues indicated that their intensity (radiance in their terminology) was in absolute energy units. The unit of measure they use

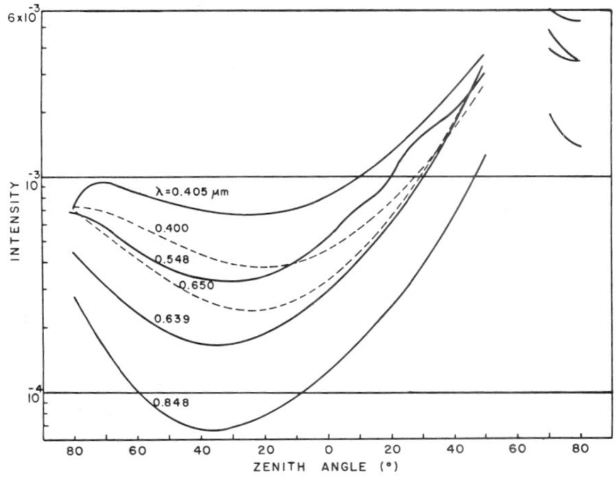

Figure 6.15 Distribution of skylight intensity in the sun's vertical as measured at four different wavelengths (solid curves) by Bullrich et al. (1965) in the industrial city of Mainz, West Germany, on September 9, 1964. The dashed curves represent computations for a model of atmospheric aerosols thought to be similar to the aerosol prevailing at the time of the measurement.

(cal cm^{-2} min^{-1}) is a simple flux for some unknown spectral interval and solid angle. Without those factors specified, the data cannot be considered spectral intensities, as they indicate. The horizontal visual range at the time of the measurements of Fig. 6.15 was estimated as 6 km, and the Linke turbidity factor was a moderate 3.8. The general configuration of the curves is typical of intensity distributions observed in industrial areas elsewhere, but, of course, the details vary. For instance, the region of the solar aureole is broad and intensity gradients are not so pronounced in the turbid atmosphere as in clearer atmospheric conditions, and some optical instability, produced no doubt by varying particle concentrations, is evident. The spectral distribution shows normal variations, however, with multiple scattering causing high intensity of short wavelengths and less multiple scattering of long wavelengths yielding low intensities. The position of the minimum intensity, while difficult to define precisely, appears to be at its normal position of approximately 90° from the sun. The horizon darkening in the region below the sun indicated in the diagram could be expected to give way to horizon brightening as the solar elevation increases, and indeed measurements at higher solar elevations in the same series show horizon brightening at the longer wavelengths. At $\lambda = 0.405$ μm, however, horizon darkening occurs at all sun elevations at which measurements were made, and it shows up at the antisolar horizon as well.

The dashed curves in Fig. 6.15 represent attempts by Bullrich et al. (1968) to interpret the data by computational models of aerosols. The aerosol model was based on the Junge size-frequency distribution with $\gamma^*=3.0$, the Linke turbidity factor was taken as 4.0, and the surface was assumed to be nonreflecting. While the main features were approximated by the computations, it is obvious that the scattering pattern was more complex than could be represented by the simple model.

6.5.2.2 Los Angeles, California

The well known problem of air pollution in Los Angeles has generated considerable activity in measurements of skylight under different atmospheric conditions in the area. A number of meteorological factors, in addition to the large number of automobiles, power plants, and oil refineries, combine to produce this serious problem. The two most important of the meteorological factors are the low-level temperature inversion that limits vertical motions and dispersion of pollutants, particularly in summer and fall, and the intense solar radiation that promotes chemical reactions and gas-to-particle conversions creating numerous aerosols which limit horizontal visual range and cause changes of polarization and intensity of skylight. Contributing factors are the light winds typical of high pressure systems and a diurnal sea breeze cycle that transports polluted air parcels out to sea at night and back to land after sunrise.

A realization of the diurnal cycle of smog is important for interpreting the skylight measurements from the Los Angeles area. In the early morning the main smog layer is confined to the very low levels of the atmosphere, as evidenced by low values of horizontal visual range with blue skies above. As automobile emissions pick up during the morning commute and solar radiation intensity rapidly increases as the sun rises above the horizon, the smog thickens and increases in height, reaching its maximum density and vertical extent at about noon. During the afternoon the aerosol component normally diminishes, although the oxidant level, a major index of smog, may continue to increase until midafternoon or later. As will be seen, the diurnal cycle is very evident in skylight data.

At least four series of skylight measurements have been made in Los Angeles during the last three decades (Sekera and Collaborators, 1955; Holzworth and Rao, 1965; deBary, 1964; and Coulson, 1971). The first two of these did not include skylight intensity; selected data from the last two are discussed next.

The measurements by deBary (1964) were made in May and June 1963 with the polarizing radiometer developed in Mainz and described by Bullrich (1964). Data were taken in three wavelengths at angular intervals of 10° in the sun's vertical for various zenith angles of the sun. Data from June 19, 1963, are shown for a solar zenith angle of 66° in Fig. 6.16. The ordinate here also must be considered to be in relative units of intensity, as the units given by deBary are for flux instead of intensity. The visual range at the time of the measurements was estimated at only 1.6 km, but the observed blueness of the sky showed the pollution to be of restricted vertical extent. The horizon darkening at $\lambda = 0.405$ μm, with a hint of the same feature at the longer wavelengths, is evidence that the aerosol layer was shallow, and the pronounced minima and

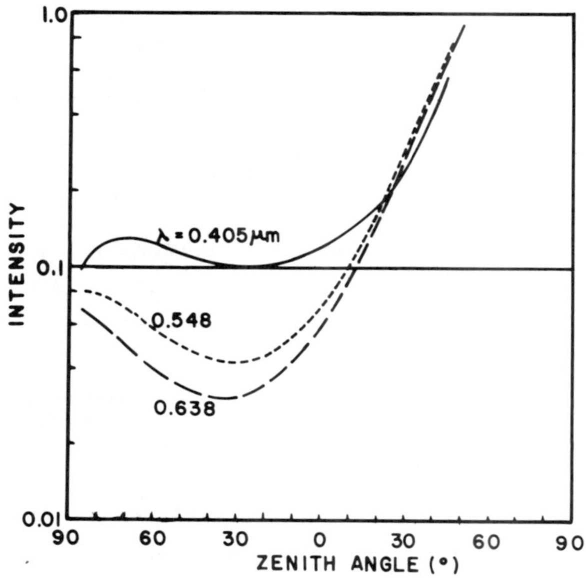

Figure 6.16 Distribution of skylight intensity in the sun's vertical as measured by deBary (1964) in three different wavelengths in the polluted atmosphere of Los Angeles on June 19, 1963.

sharp solar aureole at all three wavelengths indicate relatively little multiple scattering throughout the region of the sky above the darkened horizon. DeBary estimated, on the basis of visual observation and Mie theory computations, that the Linke turbidity factor was only about 2.0 at the time, in spite of the restricted visual range. The efforts of deBary to fit the measurements by computed intensity for the Rayleigh atmosphere in combination with a model of atmospheric aerosols following the Junge size-frequency distribution with exponential factors of 3 and 4 and various particle concentrations were only moderately successful.

Since the solar ultraviolet plays a crucial role in the formation of the photochemical smog of Los Angeles, Coulson (1971) emphasized wavelengths of $\lambda = 0.320$ μm and 0.365 μm in the skylight measurements made there. It was fortunate for the study that a smog-free day (September 23, 1968) was followed 3 days later by an episode of heavy smog. Both intensity and polarization of skylight throughout the sun's vertical and in vertical planes at azimuths 45°, 90°, 135° were measured with the polarizing radiometer briefly described by Coulson (1971), on these and several other days of the period. The polarization data were shown in Chapter 5; here we give intensity data for both clear and smoggy conditions.

The intensity distribution in the sun's vertical as measured on the day of heavy smog (September 26, 1968) at $\lambda = 0.365$ μm during six scans of the sky at various sun elevations between sunrise and noon are shown in Fig. 6.17. During the early morning the pollution appeared, from a typical brown-colored sky near the horizons with a blue sky overhead, to be confined to the low levels of the atmosphere. The lower curves of the diagram show a definite minimum of intensity near 90° scattering angles, as occurs in clearer atmospheres. Intensity gradients in the solar aureole, however, are not strong at any sun elevation. Visual observations showed a progressive buildup of both concentration and height of the smog layer as traffic increased and solar heating became a factor. The brown color near the horizons intensified and the blue color of the sky changed to a gray-white color by midforenoon. By that time the solar aureole had spread out to cover the whole sky, thereby erasing the intensity minimum, causing a monotonic increase of intensity from horizon to sun, and creating very gentle intensity gradients everywhere. The sky was bright enough to cause discomfort to the eye, which combined with irritation due to oxidants in the smog to yield the unpleasant sensation normally experienced in smog.

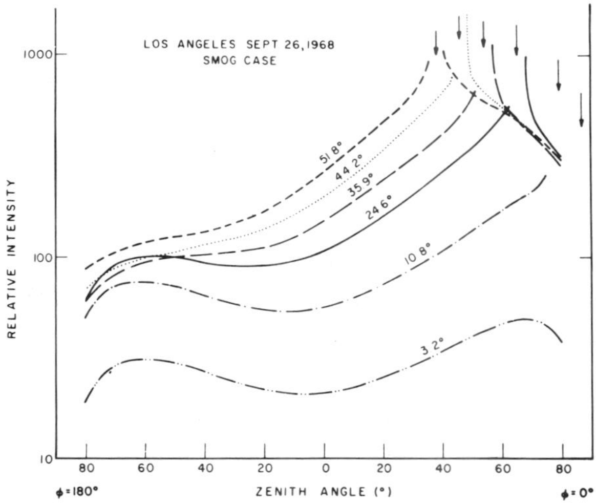

Figure 6.17 Distribution of skylight intensity in the sun's vertical as measured in the polluted atmosphere of Los Angeles at six different sun elevations and a wavelength of $\lambda=0.365$ μm. (After Coulson, 1971.)

The intensities measured at two different sun elevations on a smoggy day are compared with similar measurements on a clear day, and with theoretical computations for a Rayleigh atmosphere and the atmospheric model of Kano (1964), in Fig. 6.18. The lower curves are for a time in the early forenoon, in which case the smog was confined to the low levels and manifested itself mainly in horizon darkening and decreasing gradients in the solar aureole. The overall intensity was not greatly different from that in a clear atmosphere. The upper family of curves, however, show a strongly modified intensity distribution. Spreading of the solar aureole erased the intensity minimum, weakened intensity gradients near the sun, and caused at least a doubling of skylight intensity over much of the sky. Some of the observed features, including a broad solar aureole and distinct horizon darkening are reproduced in the computations of Kano (1964) for a concentrated turbid layer underlying a Rayleigh atmosphere, but their relative magnitudes are greatly different from those observed. As would be expected, there is little relation between intensities in the smog-laden atmosphere and those of a Rayleigh atmosphere.

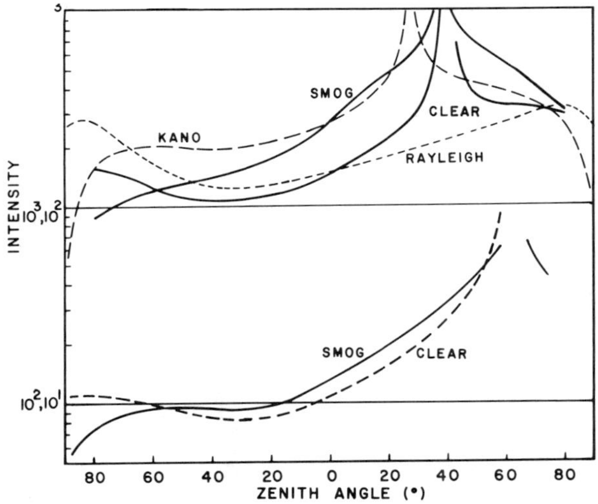

Figure 6.18 Distribution of skylight intensity in the sun's vertical as measured in clear and polluted atmospheric conditions in Los Angeles at two different sun elevations and a wavelength of $\lambda = 0.365$ *μm. Comparison curves are given for a Rayleigh atmosphere and for the case of a concentrated turbid layer underlying the Rayleigh atmosphere according to the computations of Kano (1964). (Adapted from Coulson, 1971.)*

The distribution of intensity over the entire hemisphere of the sky is shown for clear and polluted conditions at a sun elevation of approximately 57° in Fig. 6.19. For the polluted atmosphere (solid curves) the monotonic increase of intensity from horizon to sun occurs at all azimuths, and the configuration of the field shows the controlling parameter to be scattering angle. This is not surprising in view of the strong angular dependence of the Mie scattering pattern. For the clear case (dashed curves), however, the Rayleigh pattern is evident in the broad minimum about 90° from the sun, and the small amount of horizon darkening. Only in the bright solar aureole are aerosol effects the dominant feature in the clear atmosphere.

6.5.3 Skylight intensity over snow surfaces

Three series of skylight measurements over extensive areas of snow are those of Bullrich et al. (1966) in Greenland, a short series at McMurdo, Antarctica

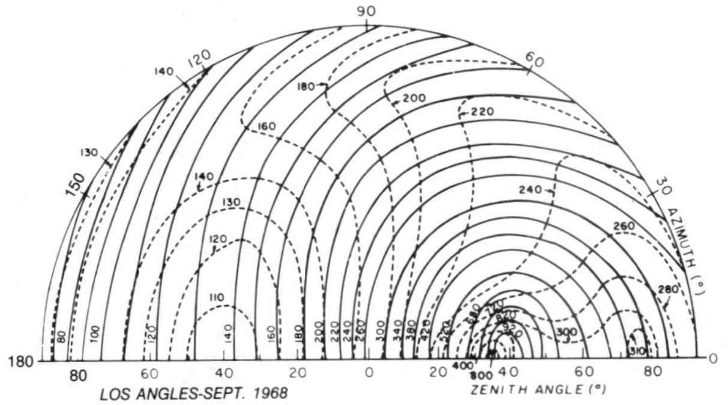

Figure 6.19 Distribution of skylight intensity over the dome of the sky as measured at a wavelength of $\lambda=0.365$ *μm in clear (dashed isopleths; sun elevation 55°) and polluted (solid isopleths; sun elevation 52°) conditions in Los Angeles. (After Coulson, 1971.)*

by Coulson, Carroll, and Fitch in 1974 (unpublished), and a long series at the geographic South Pole during the Austral summer periods of 1974 and 1975. The measurements of this last series were made by B.W. Fitch during an over-winter year at the South Pole; the data have been tabulated by Fitch et al. (1978), and reported in more detail by Fitch and Coulson (1977, 1983).

The measurements at both McMurdo and the South Pole were made at six different wavelengths ($\lambda=0.321$, 0.398, 0.550, 0.700, 0.805, and 0.872 μm) by a high-precision polarizing radiometer similar to that described by Walraven (1977). Operation of the device was controlled by computer, and the data were reduced to the physical quantities of interest on a real time basis. The usual mode of operation was to make measurements at 5° intervals in vertical planes across the dome of the sky. Emphasis was on measurements in the sun's vertical, but since the scan pattern was arbitrary, scans in the sun's almucantar or scans at various azimuths sufficient for making complete maps of the sky were sometimes included in the observational routine. Measurements at McMurdo were confined to the sun's vertical.

The measurement site at McMurdo is at a sea level location on the coast of Antarctica. At the time of the skylight measurements (February 6, 1974) the ice pack extended several miles out to sea and the surrounding land surface was completely snow covered, the sky was very clear, and the visual range was at least 130 km. Data taken at sun elevations of 23° to 28° in five wavelengths are shown by the solid curves of Fig. 6.20. The dashed curves are the intensities in a Rayleigh atmosphere at roughly equivalent wavelengths and sun elevation of 23.6°, as given by the tabulations of Coulson et al. (1960) for a surface albedo of $A = 0.80$. Since the measurements were recorded simply as photon count rates of the instrument, they can be expressed only in relative magnitudes. For a comparison, the measurements have been normalized to the Rayleigh values at the zenith in Fig. 6.20. An element of most interest is the astonishingly similar distribution of the two sets of data in the antisolar half of the plane, particularly at the longer wavelengths. The slight indication of horizon darkening is probably within measurement error, although it does appear at all three of the longer wavelengths. In contrast, at the shorter wavelengths, the measurements indicate less horizon darkening than that shown in the Rayleigh model. This is probably the result of scattering by very small particles in the long pathlengths near the horizon, the particles being too small to contribute appreciably to intensity at the longer wavelengths.

A pronounced feature of the McMurdo data is the appearance of a distinct solar aureole at all wavelengths. The weak intensity gradients in the aureole

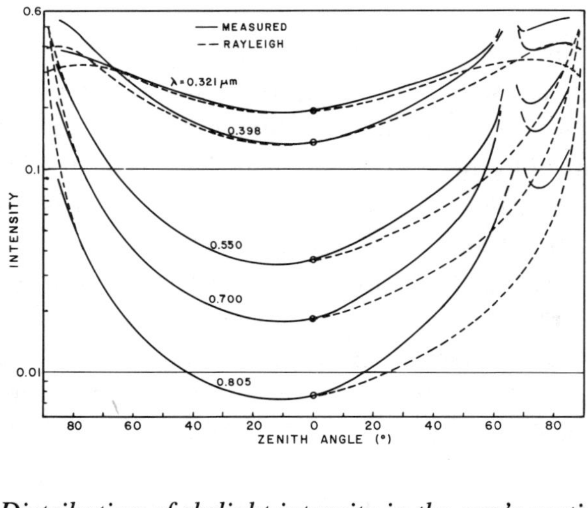

Figure 6.20 Distribution of skylight intensity in the sun's vertical as measured in five different wavelengths over a snow surface at McMurdo, Antarctica (solid curves) and as computed at equivalent wavelengths for a Rayleigh atmosphere overlying a Lambert surface with albedo A=0.80 (dashed curves).

indicate that the scatterers are of predominantly small sizes, as would be expected in this remote location. It is not unreasonable to expect the particles responsible for the aureole at McMurdo to be mostly in the low levels of the atmosphere, and perhaps of marine origin.

As mentioned, the long series of skylight measurements at the South Pole was made with the same polarizing radiometer as used at McMurdo. The South Pole station is on an extensive snow field at an altitude of approximately 2800 m on the Antarctic plateau. Being over 1000 km from the nearest coastline and far removed from major sources of turbidity, it is one of the cleanest locations on the planet. Skylight measurements repeated at several year intervals at this unique site will be useful for indicating any secular trends in turbidity of the planetary atmosphere.

Measurements of the intensity distribution in the sun's vertical at the South Pole are shown for four different wavelengths by the solid curves of Fig. 6.21. The dashed curves give the best approximation available for the Rayleigh atmosphere. Two problems with the Rayleigh approximation are that the corresponding wavelengths are somewhat different due to the high altitude of the

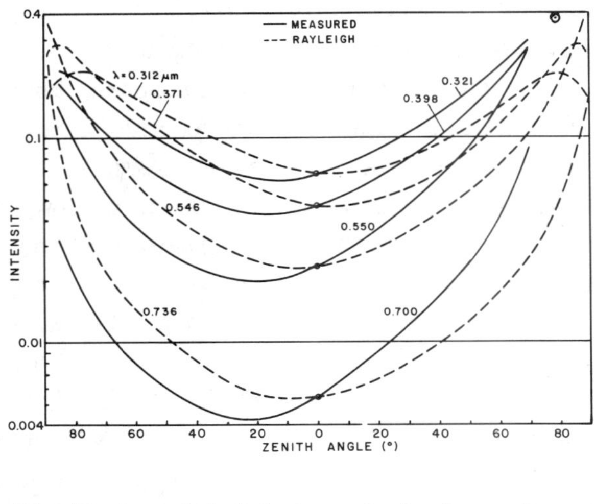

Figure 6.21 Distribution of skylight intensity in the sun's vertical as measured in four different wavelengths over a snow surface at the South Pole (solid curves) and as computed at similar wavelengths for a Rayleigh atmosphere overlying a Lambert surface with albedo A=0.80 (dashed curves).

South Pole station, and that the actual albedo of the snow surface was greater than the $A = 0.80$ assumed in the computations. Neither of these should produce major changes of the angular distribution, however, and discrepancies in magnitude are obviated by the measurements being normalized on the Rayleigh values at the zenith. An additional problem is that the actual reflection from the snow is not necessarily Lambertian in character, as assumed in the Rayleigh computations.

One feature of the intensity distributions at all four wavelengths, which appears to be unique to the South Pole data, is the altered character and shift of position of the intensity minimum to angles more than 100° from the sun. Data from other locations, including McMurdo, show the intensity minimum to be roughly 90° from the sun for clear atmospheric conditions and farther than that for turbid conditions in which the solar aureole is the dominant feature. The sun elevations in the diagram are closely analogous for the two sets of curves (10.7° versus 11.50°), and turbidity was minimal. One feature of the surface that is not accounted for in the computations are the sastrugi (wind-blown ridges of snow), but it is not clear how reflection from sastrugi could cause a shift of the pattern as indicated.

A second feature of the South Pole data that, as far as is known, has not been observed elsewhere is a pronounced shift and distortion of the intensity pattern by ice crystals in the lower levels of the atmosphere. Fitch and Coulson (1977, 1983) have discussed this feature which occurred on several occasions. Under conditions of a relatively warm airmass rising over the Antarctic plateau from the Weddell Sea, ice crystals sometimes form in the lowest 1 km of the atmosphere and slowly sift down to the surface. They do not constitute a cirrus cloud in the usual sense, as the sky often appears to be completely clear during the ice crystal episodes. From collections of these crystals, Ohtaki (1978) observed the crystals to be thin hexagonal plates and prisms of dimensions less than 0.2 mm with concentrations of up to 100 crystals per liter of air. Skylight intensities measured during such an ice crystal episode (January 6, 1974) at the South Pole are shown for six different wavelengths in Fig. 6.22. The absolute magnitudes of the curves, being in terms of photon count rates, have no significance, but their relative distributions are valid. The sun elevation for this case was 22.5°, as indicated by the circle. As shown by the diagram, the ice crystals produced a shift of the intensity minimum for $\lambda = 0.70\ \mu m$ from its normal 90° position

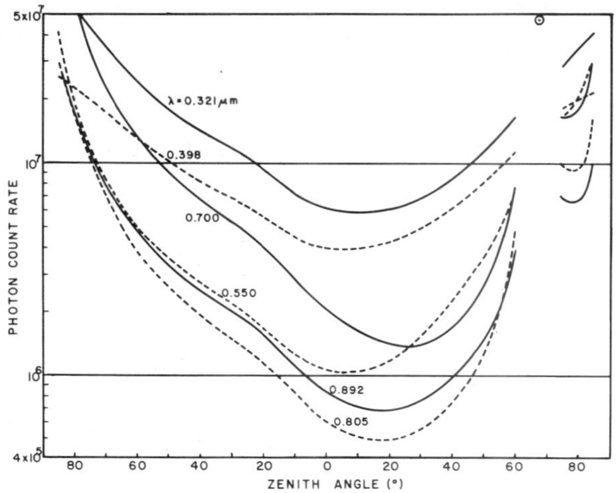

Figure 6.22 Distribution of skylight intensity in the sun's vertical as measured at six different wavelengths during a period of ice crystal precipitation at the South Pole on January 6, 1974. ⊙ represents position of the sun.

to as close as 32° from the sun. The shift is somewhat less at other wavelengths, but it is evident at all wavelengths. At the same time, a pronounced solar aureole developed. The intensity gradients were particularly strong at the longer wavelengths due to forward scattering by the large ice crystals. A second distortion of the pattern was due to an intensity increase in the region of the sky centered about 20° from the zenith at the antisolar azimuth, a scattering angle of approximately 88° to 90°. The intensity increase is evident at all wavelengths, but is most pronounced in the near infrared.

The effects of ice crystals on the intensity distribution are seen most clearly by the plot of Fig. 6.23, in which the measurements at $\lambda = 0.70$ μm for clear and ice crystal conditions are both compared with the distribution for a Rayleigh atmosphere. The net result of the ice crystals was to distort the intensity pattern by increasing the intensity in the antisolar half of the sky by a factor of approximately 3, decreasing intensity between sun and zenith by a similar amount, and sharpening the intensity gradients in the solar aureole from their clear-sky values. As far as is known, no complete theoretical explanation of these effects has been developed. The calculations of ice crystal scattering by Jacobowitz (1971), Liou (1972), Coleman and Liou (1981), and others all indicate strong and peaked

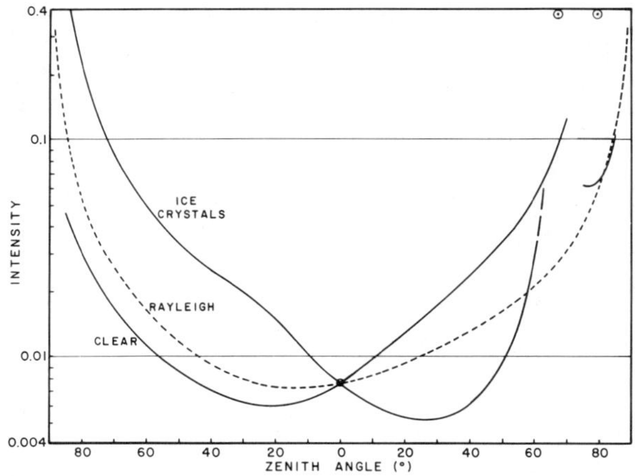

Figure 6.23 Distribution of skylight intensity in the sun's vertical as measured at λ=0.70 μm during clear sky conditions and in an ice crystal episode at the South Pole. The dashed curve is for computations for a similar wavelength for a Rayleigh atmosphere overlying a Lambert surface of albedo A=0.80. Circles (⊙) represent two different sun elevations.

forward scattering with reasonably strong backscattering. The first of these explains well the short intensity gradients in the solar aureole, and the second may explain the intensity increase in the region opposite the sun. However, neither can account for the pronounced intensity minimum between sun and zenith as observed at the South Pole.

The intensity distribution in the sun's almucantar at the South Pole as measured at four wavelengths on March 11, 1974, is shown by the solid curves in Fig. 6.24. The sun elevation at the time had decreased to 3.7° as the polar night approached. A few cirrus clouds were observed to extend an estimated 2° above the horizon, a fact which was no doubt responsible for the apparent optical instability in the data. The general behavior is evident, however, with relatively sharp intensity gradients in the region near the sun, a broad and ill-defined minimum intensity at azimuths 90° to 130°, and a flat intensity field near the antisolar azimuth. At this low sun elevation, the scattering angle is almost the same as the azimuth. As in the case of the sun's vertical, the intensity gradients near the sun are sharpest at the longer wavelengths, as is the overall intensity difference between maximum and minimum. This behavior is in accord with the pattern already seen for the intensity in the sun's vertical.

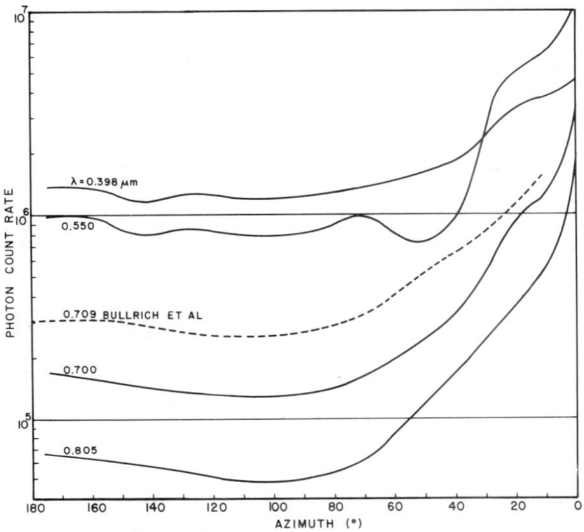

Figure 6.24 Distribution of skylight intensity in the sun's almucantar as measured at four different wavelengths over a snow surface at the South Pole on March 11, 1974 (solid curve) and as measured at λ=0.709 μm by Bullrich et al. (1966) over a snow surface in Greenland.

The dashed curve in Fig. 6.24 is from measurements at $\lambda=0.709$ μm taken by Bullrich et al. (1966) over the snow surface in Greenland. Although the relative magnitudes of the curves have no significance, the intensity distribution in the Greenland data corroborates that for the South Pole. The gradient near the sun in Greenland was somewhat less short than at the South Pole, but the higher sun elevation (6.5° versus 3.7°) might account for the difference. The optical pathlength through the atmosphere is, of course, very sensitive to sun elevation when the sun is low on the horizon. The broad intensity minimum has the same character in Greenland as at the South Pole, as does the flat intensity field in the region of the antisolar azimuth.

References

Blattner, W.G., and M.B. Wells (1973): Monte Carlo studies of sky radiation, Report AFCRL-TR-73-0613, Air Force Cambridge Research Laboratories, Bedford, MA.

Bullrich, K. (1964): Scattered radiation in the atmosphere and the natural aerosol, *Adv. in Geophys., Academic Press, 10,* 99–260.

Bullrich, K., E. deBary, R. Eiden, and K. Heger (1965): Research on atmospheric optical radiation transmission, Sci. Rept. No. 3, Johannes Gutenberg—Universitat, Mainz, Ger. AFCRL–65–109.

Bullrich, K., R. Eiden, and W. Nowak (1966): Sky radiation, polarization, and twilight radiation in Greenland, *PAGEOPH, 64,* 220–242.

Bullrich, K., R. Eiden, R. Jaenicke, and W. Nowak (1968): Solar radiation extinction, sky radiation, sky light polarization, and aerosol particle total number and size distribution on the Island of Maui (Hawaii), *Pure Appl. Geophys., 69,* 280–319.

Carroll, J.J. (1985): Global transmissivity and diffuse fraction of solar radiation for clear and cloudy skies as measured and predicted by bulk transmissivity models, *Solar Energy, 35,* 105–118.

Carroll, J.J., and B.W. Fitch (1981): Effects of solar elevation and cloudiness on snow albedo at the South Pole, *J. Geophys. Res., 86,* 5271–5276.

Coleman, R.F., and K.N. Liou (1981): Light scattering by hexagonal ice crystals, *J. Atmos. Sci., 38,* 1260–1271.

Coulson, K.L. (1968): Effect of surface reflection on the angular and spectral distribution of skylight, *J. Atm. Sci., 25,* 759–770.

Coulson, K.L. (1971): On the solar radiation field in a polluted atmosphere, *J. Quant. Spectrosc. Radiat. Transfer, 11,* 759–770.

Coulson, K.L., J.V. Dave, and Z. Sekera (1960): *Tables Related to Radiation Emerging From a Planetary Atmosphere With Rayleigh Scattering,* University of California Press, Berkeley and Los Angeles.

Coulson, K.L., R.L. Walraven, G.I. Weight, and L.B Soohoo (1974): Photon-counting polarizing radiometer, *Appl. Opt., 13,* 497–498.

Dave, J.V. (1970): Intensity and polarization of radiation emerging from a plane parallel atmosphere containing monodispersed aerosols, *Appl. Opt., 9,* 2673–2684.

Dave, J.V., and P.M. Furukawa (1966): Intensity and polarization of the radiation emerging from an optically thick Rayleigh atmosphere, *J. Opt. Soc. Am., 56,* 394–400.

deBary, E. (1964): Influence of multiple scattering of the intensity and polarization of diffuse sky radiation, *Appl. Opt.*, *3*, 1293–1303.

deBary, E., and K. Bullrich (1964): Effects of higher-order scattering in a molecular atmosphere, *J. Opt. Soc. Am.*, *54*, 1413–1416.

Deirmendjian D. (1964): Scattering and polarization properties of water clouds and hazes in the visible and infrared, *Appl. Opt.*, *3*, 187–196.

Deirmendjian, D., and Z. Sekera (1954): Global radiation resulting from multiple scattering in a Rayleigh atmosphere, *Tellus*, *4*, 382–398.

Elterman, L. (1968): UV, visible and IR attenuation for altitudes to 50 km, 1968, AFCRL-68-0153, Env. Res. Pap. No. 285. U.S. Air Force.

Feigelson, E.M., M.S. Malkevich, S.Ya Kogan, T.D. Koronatova, K.S. Glazova, and M.A. Kuznetsova (1960): Calculation of the Brightness of Light in the Case of Anisotropic Scattering, Consultants Bureau, Inc., New York (Trans. from Russian)

Fitch, B.W., and K.L. Coulson (1977): Polarizing radiometer measurements of skylight at South Pole Station, Antarctica, *Proc. Soc. Photo-Optical Inst. Eng.*, *112*, 184–190.

Fitch, B.W., and K.L. Coulson (1983): Effects on skylight at South Pole Station, Antarctica by ice crystal precipitation in the atmosphere, *Appl. Opt.*, *22*, 71–74.

Fitch, B.W., K.L. Coulson, and R.H. Hamilton (1978): Skylight measurements at the South Pole, Contrib. in Atm. Sci., No. 15, University of California, Davis.

Hansen, J.E. (1969): Exact and approximate solutions for multiple scattering by cloudy and hazy planetary atmospheres, *J. Atm. Sci.*, *26*. 478–487.

Hitzfelder, S.J., G.N. Plass, and G.W. Kattawar (1976): Radiation in the earth's atmosphere: Its radiance, polarization, and ellipticity, *Appl. Opt.*, *15*, 2489–2500.

Holzworth, G.C., and C.R.N. Rao (1965): Studies of skylight polarization, *J. Opt. Soc. Am.*, *55*, 403–408.

Hopkinson, R.G. (1954): Measurements of sky luminance distribution at Stockholm, *J. Opt. Soc. Am.*, *44*, 455–459.

Jacobowitz, H. (1971): A method for computing the transfer of solar radiation through clouds of hexagonal ice crystals, *J. Quant. Spectros. Rad. Trans.*, *11*, 691–695.

Kano, M. (1964): Effect of a turbid layer on radiation emerging from a planetary atmosphere, Doctoral Dissertation, University of California, Los Angeles.

Kattawar, G.W., G.N. Plass, and S.J. Hitzfelder (1976): Multiple scattered radiation emerging from Rayleigh and continental haze layers. I. Radiance, polarization and neutral points, *Appl. Opt.*, *15*, 632–647.

Liou, K.N. (1972): Light scattering by ice clouds in the visible and infrared: A theoretical study, *J. Atmos. Sci.*, *29*, 524–536.

Moon, P., and D.E. Spencer (1942): Illumination from a non-uniform sky, *Trans. Illum. Eng. Soc.*, *37*, 707–726.

Nazaraliyev, M.A., and V.Ye. Pavlov (1980): On a determination of the absorptivity of the atmospheric aerosol from the sky brightness in the ultraviolet spectral region, *Atmospheric and Oceanic Physics, Izvestia, Akad. Nauk*, *15*, 422–427.

Ohtake, T. (1978): Atmospheric ice crystals at the South Pole in summer, *Antarctic J. of U.S.*, *13*, 174–175.

Pinker, R.T., and J.A. Ewing (1985): Modeling surface solar radiation: Model formulation and validation, *J. Clim. Appl. Met.*, *24*, 389–401.

Plass, G.N. (1966): Mie scattering and absorption cross sections for absorbing particles, *Appl. Opt.*, *5*, 279–285.

Pokrowski, G.I. (1929): *Phys. Zeit*, *30*, 697.

Sekera, Z., and Collaborators (1955): Investigation of the Polarization of Skylight, Final Report, Contr. AF19(122)-239, Dept. of Meteorology, Univ. of California, Los Angeles, 68 pp.

Steven, M.D. (1977): Standard distributions of clear sky radiance, *Q.J. Roy. Met. Soc.*, *103*, 457–465.

Tanaka, M. (1971) Radiative transfer in turbid atmospheres. II. Angular distribution of intensity of the solar radiation diffusely reflected and transmitted by turbid atmospheres, *J. Met. Soc., Japan*, *49*, 321–332.

Thekaekara, M., and A.J. Drummond (1971): Standard values of the solar constant and its spectral components, *Nature Physical Sci.*, *229*, 6–9.

Thomalla, E., P. Kopke, H. Muller, and H. Quenzel (1983): Circumsolar radiation calculated for various atmospheric conditions, *Solar Energy, 30,* 575–587.

Walraven, R.L. (1977): Reduction of polarizing radiometer data, *Rev. Sci. Inst.*, *48,* 613–616.

Walsh, J.W.T. (1961): *The Science of Daylight,* Macdonald, London.

CHAPTER 7

THE NEUTRAL POINTS OF SKYLIGHT POLARIZATION

7.1 Introduction

The general behavior of points in the sky at which the degree of polarization is zero, the neutral points, was discussed in Chapter 4. Concentration there was on neutral point location as computed for the Rayleigh atmosphere, with a wide range of optical thicknesses and for different laws of reflectance for the underlying surface. In this chapter we emphasize observations of the neutral points and their deviations from their theoretical positions under various types of atmospheric conditions and surface reflection.

The first of the three normally occurring neutral points, the Arago point, was discovered by the French astronomer, D.F.J. Arago, a few years after his discovery of skylight polarization in 1809. Arago developed a new type of polarimeter in 1814 and 1815, and it is logical to assume that the neutral point was discovered with his new instrument. However, no published account of either of these discoveries has been found by the present author, despite a concentrated search of Arago's publications and of the general literature of the period. The fact that he was the discoverer of the phenomena appears to be well established, however, as several writers in the field (e.g., Brewster, 1861) have stated as much. Perhaps the information was disseminated by private correspondence, as Arago is reported to have been a prolific letter writer. In any case, the Arago point, normally located 20° to 30° above the antisolar point, has been observed more frequently than either of the other neutral points. Its popularity has been bolstered, no doubt, because it is easy to see, and also because its position has been found to be more sensitive to the effects of atmospheric turbidity than the position of either the Babinet or Brewster point. The variations of its position with changes of turbidity will be discussed more fully below.

It is of interest to note that Arago was one of the several French scientists who was responsible for the tremendous impetus given science in France following the French Revolution. The National Institute was established in Paris during this period, and the names of Biot, Cauchy, Malus, Fresnel, and Laplace joined that of Arago as scientists making great strides in discovering the nature and properties of light in the atmosphere. This is, of course, in addition to advances in fields such as chemistry, biology, and other sciences made by members of the National Institute in France. Across the English Channel were such stars as Brewster, Herschel, Wollaston, Young, and Airy in the field of optics, but on balance the French were more successful in advancing the knowledge of optics during the first quarter of the 19th century than were other European scientists. It is interesting to speculate on whether this drive for excellence among the French was a result of the terrible days of the French Revolution.

The Babinet point, the second neutral point to be discovered, was first observed by the French meteorologist J. Babinet in the year 1840. (Babinet came after the primary impulse of French science mentioned above.) The discovery came during a trip by Babinet to the seacoast for making observations of the Arago point in the clear atmosphere far from industrial pollution. However, on discovering this second point, Babinet abandoned the Arago point observations to concentrate on his discovery (Babinet, 1840). He found it at about the same angular distance above the sun as the Arago point is above the antisolar point. Its position above the sun makes the Babinet point observable any time the sun is above the horizon and for considerable periods of time both before sunrise and after sunset. Being closer to the sun, and hence in a brighter part of the sky than the Arago point, the Babinet point is somewhat less readily seen than the Arago point. Even so, it can be observed without difficulty using the normal Savart polariscope or other types of polarization devices. Numerous observations, with both visual and electronic instruments, have been accumulated over the decades, although the detailed observations of Babinet himself do not appear to have been published.

After making enough observations to convince himself that his discovery was not an artifact of the instrument, Babinet submitted a note on the discovery to the Academy of Sciences, Paris (Babinet, 1840). According to Babinet's report, the new neutral point was observed at about the same time by one M. Guerard in the Pyrenees. Even though this fact was not communicated until after Babinet's note was delivered to the Academy, Babinet did not consider this discrepancy in time sufficient to establish any priority in his own favor. Nevertheless, the name of Babinet has always been associated with the discovery of this second neutral point.

The most difficult of the three neutral points to observe visually is the one below the sun—the Brewster point. Since he already knew of the Arago point and on hearing about Babinet's discovery, the Scottish physicist Sir David Brewster thought that whatever the cause of the neutral points, "it was manifest that the same cause ought to produce a neutral point below the sun" (Brewster, 1847). However, the region between the sun and horizon is the brightest area of the sky, and haze near the horizon yields generally low polarization values. Thus the polarization bands of the Savart polariscope are swamped by the excessive glare in the instrument, making the Brewster point difficult to see. In searching for the point at the low sun elevations prevailing at the high latitude location of St. Andrews (56.3° N), Brewster was "perplexed beyond measure with the feeble and uncertain indications of the polariscope." Finally, by positioning himself inside a long, dark passage from the end of which a small area below the sun was visible, he got a distinct view of the new neutral point located 15° to 16° below the sun. The date was February 28, 1842. This discovery was communicated to Babinet, who also began searching for the point. After several unsuccessful attempts over a 4-year period, Babinet confirmed the existence of this third neutral point on July 23, 1846. The same type of difficulty in seeing the Brewster point was experienced by Rubenson (1864) observing in the clear atmosphere of Italy in 1861 and 1862. At no time in that 2-year period was he able to definitely verify the existence of a neutral point below the sun.

Before discussing observations of the neutral points, it is worth describing briefly the appearance of a neutral point in the Savart polariscope. This simple device was widely used for well over a century for observing neutral points, and still has not been entirely supplanted by electronic polarimeters. Briefly, in a polariscope, two plates of rock crystal are mounted inside a metal tube 15- to 20-cm long and 3 cm in diameter. These plates are backed up by an analyzer, such as a tourmaline or Nicol prism, or, more recently, a disk of sheet polarizer. Polarized light passing through this arrangement yields a system of parallel black and white fringes in the field of view. The fringes created by positively polarized light are shifted laterally by one-half fringe spacing from those of negatively polarized light. Thus a neutral point is observed as an abrupt displacement of one set of bands with respect to the other set in the field of view. By mounting the device on a clinometer of some type, the elevation angle at which the band shift occurs, the elevation angle of the neutral point, can be determined. Being a visual instrument, most observations with the Savart polariscope have been in white light, although some observers have also employed wide-band color filters to determine low-resolution spectral variations of neutral point positions. Modern electronic polarimeters normally utilize narrow band interference filters for measurements of neutral point positions at different wavelengths, yet more neutral point observations have been made with the Savart polariscope than with any of the more sophisticated instruments. Unfortunately, the Savart polariscope is not well adapted to determining degree of polarization, making other types of instruments necessary.

For a neutral point to occur, the normal positive polarization must be exactly matched by an equal quantity of negative polarization. Considerable discussion over several decades was devoted to explaining the source of negative polarization. One obvious source is light reflected from the earth's surface, but relatively little was known about reflection from natural surfaces except for reflection by still water bodies, to which Fresnel's laws apply. However, it was not clear how scattering by the atmosphere could introduce negative polarization. Brewster, having investigated polarization of light reflected and refracted by transparent materials, considered the negative polarization to be introduced by reflection by water droplets suspended in the atmosphere, but the details could not be worked out. Lord Rayleigh (Strutt, 1871) showed how scattering of light by molecular size particles may introduce polarization, but the polarization due to single or primary scattering is only positive in nature. It was thought at the time that the negative polarization must result from scattering by atmospheric aerosols. However, from computations of multiple scattering in the atmosphere initiated by Chandrasekhar's treatise (1950), it became obvious that multiple scattering of light can cause negative polarization. The stronger the multiple scattering, the more negative polarization is introduced, and the more the neutral point is displaced from its exact forward or backward scattering position applicable to primary scattering. Since the amount of multiple scattering is affected by atmospheric aerosols, the positions of the neutral points are also affected by, and may be used as an index of, atmospheric aerosols. Herein lies their utility for practical applications.

7.2 Neutral point observations

7.2.1 The Arago point

Numerous observations of the position of the Arago (AR) point have been acquired during the 170 years since Arago's discovery. They have been principally in the form of concentrated groups of observations, as opposed to extensive series covering long periods of time at any one location. Usually, though not universally, studies have been initiated by the eruption of a major volcano injecting copious quantities of debris into the upper atmosphere and causing major changes of skylight polarization. (The effects of volcanic eruptions on the intensity and polarization fields of skylight are the subject of Chapter 9.) Pickering (1884) developed a new type of polarimeter to observe the effects of Krakatoa in 1883, and used the new instrument for skylight measurements in 1884 and 1885. It does not appear that any special observations followed the gigantic eruption of Mount Tamboro in 1815, and of course, polarization observations appear to have been confined to those of Arago. Kimball's (1903, 1913a, 1913b) observations followed the eruptions of Pelee in 1903 and Mount Katmai in 1912, while those of Dorno (1919) were triggered by the Mount Katmai eruption, as were some of the many observations made by Jensen (e.g., 1915). The observations of Volz (1970, 1975) have tended to be directly or indirectly motivated by the existence of high-level volcanic clouds, and Coulson (1983) made a concentrated series of skylight measurements after the 1982 eruption of El Chichon. Though many sets of observations were made under these unique atmospheric conditions, there were others as well. Somewhat extended series of polarization observations, including neutral point positions, under more normal atmospheric conditions include those of Brewster (1861, 1864) from 1841 to 1845, Rubenson (1864) in 1861 and 1862, several series by Jensen (1899, 1928, 1942), and an extensive study devoted specifically to the behavior of the Arago point by Neuberger (1940, 1950). Other important series of measurements have already been listed in Chapter 5.

Returning to observations of the Arago point specifically, we note the 5-year series by Brewster (1861, 1864) taken at St. Andrews, Scotland, from 1841 to 1845. In view of what is known now about the Arago point, there are some puzzling aspects to Brewster's data, which are tabulated in detail in his publications. The angular distance between the Arago point and the antisun, the AR distance according to what Brewster considered the most reliable of his observations, is plotted as a function of sun elevation in Fig. 7.1. For this plot, it was necessary to compute the sun elevation for each observation, as this information was not included in the original tabulations. I am indebted to Dr. R.S. Fraser of Goddard Space Flight Center for these computations. Brewster's observations were made with the Savart polariscope in white light. Mean values of observations made with an equivalent instrument in white light in Pennsylvania by Neuberger (1940) are shown by the comparison curve of Fig. 7.1. Shown also is a curve of the Arago point distances computed for a wavelength of 0.546 μm in a pure Rayleigh atmosphere by Coulson (1954).

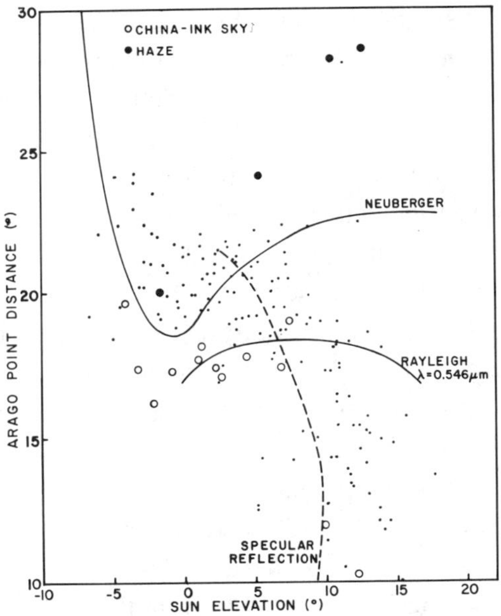

Figure 7.1 Angle between the Arago point and the antisolar point (the Arago point distance) as a function of sun elevation according to the observations of Brewster (1861) at St. Andrews, Scotland (points and circles), with curves derived from observations by Neuberger (1940) and computations for the Rayleigh atmosphere by Coulson (1954). The dashed curve is for computations at λ=0.495 μm by Fraser (1968) for the case of a Rayleigh atmosphere overlying a specularly reflecting surface.

Although Brewster's data show a general decrease of AR distance with increasing sun elevation in the twilight period just as later measurements indicate they should, the dependence is not pronounced and it is embedded within a large scatter of observations. At positive sun elevations, however, Brewster's data are not only highly variable but the dependence with sun elevation seems to be opposite that observed by Neuberger (1940). The reasons for these anomalies in Brewster's observations are not known, but they may be due to widely varying atmospheric conditions at the high latitude location of St. Andrews. The observations in which atmospheric haze was specifically mentioned, indicated by circles in the diagram, do not show any coherent dependence on atmospheric haze. However, haze effects are responsible for the observations of both Brewster and Neuberger showing considerably larger AR distances than those computed by Coulson (1954) at a wavelength in the middle of the visible spectrum ($\lambda=0.546$ μm) for the Rayleigh atmosphere. Interpretation is made even more difficult by descriptive terms Brewster used that have no modern counterpart. For example, it is not clear what the term "China-ink sky" means, as apparently the term is no longer in use. Presumably, it refers to a sky made uniformly dark by atmospheric aerosols. In one case Brewster mentioned a "muddy China-ink sky," and in another a "China-ink sky without any blue." In any case, the data of Fig. 7.1 indicate the aerosols of the China-ink sky to introduce positive polarization sufficient to cause a significant shift of the Arago point in the direction of the antisolar point.

A second puzzling feature of Brewster's observations is his statement that in the normal state of the atmosphere the AR distance is 18.5° when the sun is on the horizon. As seen in Fig. 7.1, this corresponds almost precisely with the observations of Neuberger (1940) taken in central Pennsylvania, but it does not correspond well with Brewster's own observations. Brewster's data show 20° to be a more representative value. Furthermore, he concluded that the Arago point was on the horizon at a sun elevation of 11° or 12°, which means that the AR distance was likewise only 11° or 12°. The data do not show this clearly, and it is certainly at variance with what would be expected from either Neuberger's observations or the Rayleigh atmosphere. There is no obvious explanation of these discrepancies, but these same statements were repeated by Rubenson (1864).

One aspect of Brewster's observations that should not be overlooked is that they were made at a location near the sea. St. Andrews is near the Firth of Tay on the northeast coast of Scotland, and the effects of sea-surface reflection may be an important aspect of Brewster's data. The strong decrease of AR distance at sun elevations of 5° to 10° may be associated with the introduction of positive polarization into the skylight by the reflection of positively polarized light from the water surface. This effect was observed on the Island of Sylt in the North Sea by Neuberger (1950). In this case, decreases of AR distances of 1° to 6° occurred at sun elevations of 4.5° to 12.5°, with little or no changes at the lower sun elevations with less sea surface reflection. A much larger effect of sea-surface reflection is shown by the computations of Sekera (1961) and Fraser (1968) for the case of specular (mirror) reflection from the underlying surface. While sea-surface reflection is not necessarily specular, there is certainly a specular component. This component has the capability of introducing positively polarized light into the atmosphere, which causes very significant changes in neutral point behavior. The position of the Arago point at $\lambda=0.495\ \mu m$ for a Rayleigh atmosphere overlying a specularly reflecting surface, according to the computations of Fraser, is shown by the dashed curve of Fig. 7.1. It is obvious that such reflection can explain the small Arago point distances in the region of sun elevations 8° to 14°, a feature which is difficult to explain otherwise.

A number of parameters associated with the position of the Arago point have been proposed as indices of atmospheric turbidity. Neuberger (1940, 1950) summarized Arago point observations taken in Hamburg before and after the eruption of Mount Katmai and showed that not only the absolute magnitude of the AR distance was strongly affected by volcanic debris, but so too was the so-called AR amplitude. This latter quantity is the difference in angle between the maximum AR distance occurring shortly after sunrise (or before sunset) and the minimum distance occurring shortly before sunrise (or after sunset). The AR amplitude increases with increasing atmospheric turbidity. One problem in using the AR amplitude as a turbidity index is that the observations necessary to determine the entire AR curve, and thus its amplitude, cover a period of several hours and become too burdensome for frequent use. In an effort to circumvent this problem, Neuberger (1950) proposed the use of a single measurement of AR distance taken at a sun elevation of 10.5°. His analysis indicates that the major variations of AR amplitude are due primarily to variations of the AR distance at positive sun elevations, so the single observation at a sun elevation such as 10.5° should be about as sensitive to turbidity changes as the AR amplitude. The economy of observer time is a strong point in favor of the single measurement.

The measurements of Brewster, Neuberger, and others before about 1950 were made visually, mainly in white light. However, with the advent of electronic polarimeters, narrow-band measurements became possible, and considerable numbers of such data are available. For instance, Sekera and Collaborators (1954) observed Arago point behavior at sun elevations from 0° to 20° in four different wavelengths (λ=0.365, 0.460, 0.515, and 0.625 μm) and at four different observing sites in southern California. The data were plotted as deviations of observed neutral point distances from those computed for a Rayleigh atmosphere. In general, the observed AR distances were greater than the theoretical by an amount which varied from 0° to 2° in the ultraviolet to 4° to 8° at the longest wavelength, with data at the intermediate wavelengths lying between these extremes. The relative insensitivity of the AR distance to turbidity at ultraviolet wavelengths was further confirmed by the measurements of Coulson (1971). According to the data of Sekera and Collaborators, the more turbid atmospheric conditions prevailing in Los Angeles and Pasadena generally produced larger AR distances than those in the clearer conditions of Table Mountain and Cactus Peak, but this was not universally the case. Furthermore, the neutral point positions were not well correlated with the commonly observed meteorological parameters such as horizontal visibility or wind direction. The positions were found to be quite different on different days with apparently similar atmospheric conditions. The explanation for this anomalous result is not known, but it may be a result of aerosols being dispersed more in the vertical direction on some days than on others. Visual manifestations observed at the surface would be different for the two types of distributions, but the neutral point positions, being affected by the entire atmospheric path of the incident sunlight, would not respond only to conditions in the surface layers. Thus the lack of correlation between neutral point distances and horizontal visibility should not be taken to indicate a lack of reliability of neutral point distances as an index of atmospheric turbidity. In reality, such a behavior would more logically indicate the neutral point distances to be an index of overall atmospheric turbidity, whereas visual effects at the surface would be primarily indicators of conditions in the lower atmosphere.

In summarizing information on the position of the Arago point as an index of atmospheric turbidity, the evidence available supports the following conclusions:

1. The position of the Arago point is very sensitive to the existence of volcanic debris in the upper atmosphere. (This aspect will be discussed more fully in Chapter 9.)

2. Under more normal atmospheric conditions, the changes in Arago point position with atmospheric turbidity are much more subtle. The changes are greater in the longer visible wavelengths than in blue or ultraviolet wavelengths.

3. A single measurement of AR distance taken at a sun elevation of about 10° is about as effective for indicating atmospheric turbidity as are more numerous observations taken over several hours at both positive and negative sun elevations. The economy of observer time tips the scales in favor of the single observation for routine use.

4. It is known that reflection of light from snow or water surfaces may affect the Arago point position, a fact which must be considered in using the Arago point as an index of atmospheric turbidity.

5. In spite of its sensitivity to atmospheric turbidity, observations of the Arago point, or for that matter any of the neutral points, have never been pursued to the stage of routine use as a quantitative index of turbidity.

7.2.2 The Babinet point

Normally located in the sun's vertical at some angular distance above the sun, the Babinet point is visible throughout the day from before sunrise until after sunset. Its angular distance from the sun, termed here the Babinet point distance, varies from 25° to 30° at low sun elevations to a coincidence with the sun when the sun is at the zenith. The Babinet point distances, as computed at various wavelengths for a pure Rayleigh atmosphere, were shown above in Chapter 4.

As with the Arago point, the position of the Babinet point is strongly affected by the existence of volcanic debris in the upper atmosphere (see Chapter 9), and its behavior in such cases has been studied by many authors. For more normal cases of atmospheric turbidity, the Babinet point has received much less attention than has the Arago point. Brewster (1861, 1864) included Babinet point distances with his Arago point observations in the 1840s. Brewster's Babinet point observations are shown, along with a curve computed by Coulson (1954) for a Rayleigh atmosphere at visible wavelengths, in Fig. 7.2. There is a large amount of scatter with little coherence in this data. The variability of atmospheric conditions at the high latitude of Scotland which is responsible for the Arago point scatter is likewise effective in varying the position of the Babinet point. Jensen (1932) found the behavior of the Babinet point to be more complicated than that of the Arago point, and concluded that the Babinet point was more sensitive to conditions higher in the atmosphere.

Be that as it may, the measurements of Sekera and Collaborators (1954) do indicate a sensitivity of Babinet point distances to low level turbidity. Fig. 7.3 is a plot of the deviation of Babinet point distances from their theoretical positions for a Rayleigh atmosphere (see Chapter 4) as a function of sun elevation for four different wavelengths according to the measurements of Sekera and Collaborators at various locations in southern California. The atmosphere was least turbid at the higher altitude sites on Table Mountain and Cactus Peak, most turbid at the California Institute of Technology in Pasadena, with turbidity between the two extremes on the campus of the University of California, Los Angeles. The curves do not extend to the same sun elevations at all wavelengths, as the Babinet point distance decreases with increasing wavelength and was located

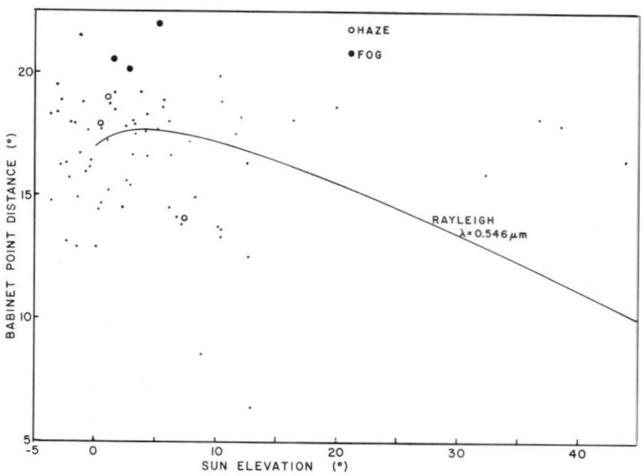

Figure 7.2 Angle between the Babinet point and the sun as a function of sun elevation according to the observations of Brewster (1861) (points and circles), with a curve computed for a Rayleigh atmosphere by Coulson (1954).

too close to the sun to observe at the higher sun elevations in the longer wavelengths. The most important points of Fig. 7.3 are the following:

1. The Babinet point in the less turbid atmosphere (Cactus Peak and Table Mountain) is generally closer to its theoretical position than in the turbid atmospheres of Pasadena and Los Angeles.

2. At low sun elevations, the Babinet point distance is relatively insensitive to atmospheric turbidity, the curves being closely grouped and not systematically arranged.

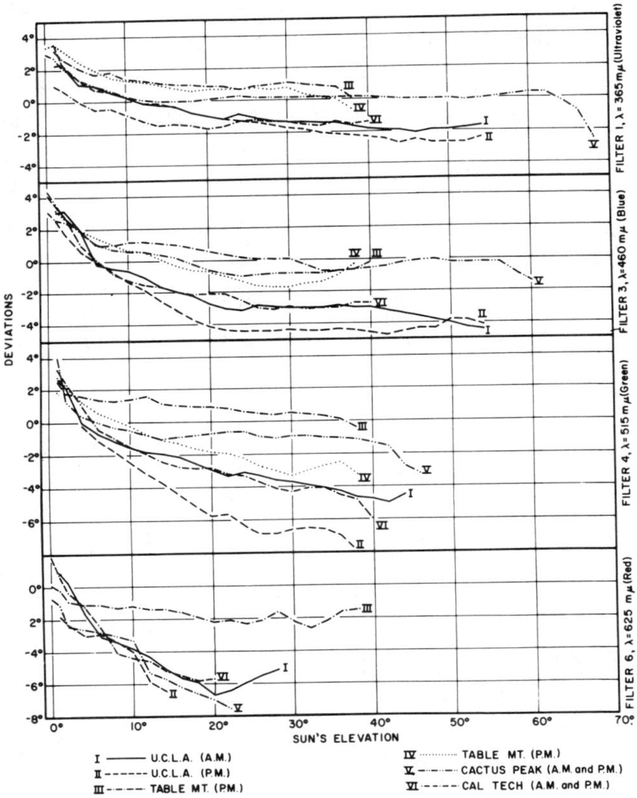

Figure 7.3 Deviations of the observed position of the Babinet point from its theoretical position in a Rayleigh atmosphere as a function of sun elevation according to the observations at four different wavelengths taken by Sekera and Collaborators (1954) at four different locations in southern California (after Sekera and Collaborators, 1954).

3. At the higher sun elevations the effect of turbidity is to shift the Babinet point deviations in the negative direction, which is the same as shifting the point closer to the sun. The division is particularly clear at blue and green wavelengths. The data for Cactus Peak in the red wavelengths appear anomalous, the reasons of which are unknown but may be associated with reflection of light from the red sands and soils surrounding the site.

Babinet point observations made by Coulson et al. (1974) in four different wavelengths at the Mauna Loa Observatory are shown for the clear but optically unstable day of March 15, 1973, in Fig. 7.4. Although measurements were also taken at longer wavelengths, the point was too close to the sun for observations at $\lambda=0.60$ μm. This was the usual situation observed on most days at this high altitude (3400 m) site. At near-infrared wavelengths both Babinet and Brewster points disappeared, leaving only positive polarization in the vicinity of the sun. The data of Fig. 7.4 show a great variability of Babinet point position at all four wavelengths, the behavior at $\lambda=0.50$ μm being particularly erratic. The at-

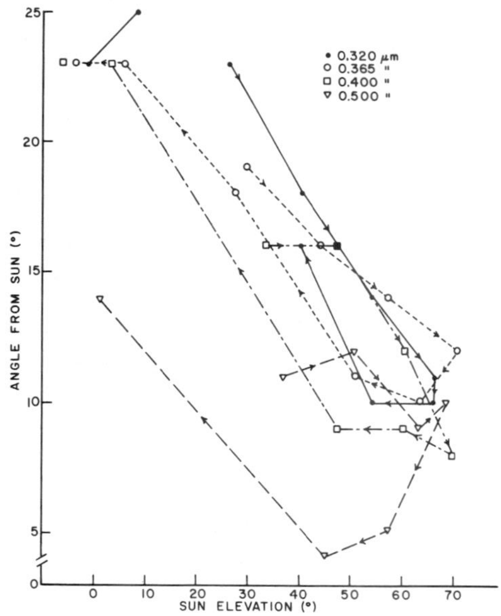

Figure 7.4 Angle between the Babinet point and the sun as a function of sun elevation according to the observations of Coulson et al. (1974) at four different wavelengths in optically unstable atmospheric conditions at the Mauna Loa Observatory. Arrows show the effect of an increase in turbidity throughout the day.

mosphere appeared hazy on this day, the haze possibly consisting of an influx of dust from the deserts of Northern China (Braaten and Cahill, 1986). By following the arrows superimposed on the curves, the effect of an increase of turbidity over the course of the day can be seen.

Although observations of the Babinet point are less extensive than those of the Arago point, the data available indicate that the position of the Babinet point responds to changes of atmospheric turbidity, though its usefulness as a turbidity index is limited under normal conditions. Only in cases when volcanic debris is in the upper atmosphere, the subject of Chapter 9, does the Babinet point position provide a sensitive and reliable indicator of the atmospheric situation.

7.2.3 The Brewster point

After his discovery of a neutral point below the sun on February 28, 1842, Sir David Brewster (1861) began efforts to characterize its behavior during the same time he was measuring the Babinet and Arago points. However, the brightness of the sky and low polarization values in the region of the sky below the sun make the Brewster point very difficult to detect visually, and during the period 1842 to 1845 Brewster was able to see the neutral point on a total of only 23 days. One factor limiting the observations of the Brewster point at St. Andrews, Scotland, is that throughout the winter months the sun is relatively low on the horizon. A second factor is the variable nature of atmospheric turbidity at the site mentioned above in connection with Brewster's Arago point observations. However, these factors had no influence on the observations Rubenson (1864) made in Italy in 1861 and 1862, in which he tried many times to see the Brewster point, but always in vain.

The anomalous nature of either the atmospheric conditions or observing site at St. Andrews is further indicated by the limited angle between the Brewster point and sun according to Brewster's observations. Theoretical computations of neutral point positions (see Chapter 4) for the Rayleigh atmosphere show the Brewster point to be located several degrees farther from the sun than the Babinet point at visible wavelengths. However, 26 observations by Brewster give an average Brewster point distance of only 12.5 °, whereas analogous observations of the

Babinet point yielded a mean distance of 18.5°. One possible explanation of such a discrepancy may be the positive polarization introduced by reflection from the sea surface in the vicinity of St. Andrews. The polarization field near the horizon would be more sensitive to such reflection than that at higher elevation angles in the sky.

Later authors such as Kimball (1903, 1913b) Dorno (1919), Jensen (1928, 1942), Neuberger (1940), Dahlkamp and Kantus (1946–1947), and others, observing through the Savart polariscope, concentrated on the Babinet and Arago points to the virtual exclusion of the Brewster point. As a result, relatively little is known about the response of the Brewster point position to atmospheric turbidity.

The advent of electronic polarimeters obviated the difficulty of detecting the Brewster point, however, and a number of Brewster point observations have been made since about 1950. For instance, Sekera and Collaborators (1954) combined Brewster point observations with those of the Babinet point, shown in Fig. 7.3, and obtained the deviations of the observed neutral point positions from their theoretical counterparts of Fig. 7.5. The curves do not extend to equally high sun elevations at all wavelengths because the neutral point was closer to the sun in longer wavelengths than in shorter wavelengths and interference from stray sunlight in the instrument terminated the observations at sun elevations that were lower for the longer wavelengths than for shorter wavelengths. The curves show that the Brewster points are closer to their theoretical positions in the less turbid atmosphere (Cactus Peak and Table Mountain) than in the more turbid conditions at Pasadena and Los Angeles, and that the effect of turbidity is to shift the positions in the direction toward the sun. A considerable asymmetry between forenoon and afternoon occurred at UCLA, probably because of that site's location near the ocean. The sunlight reflected from the sea surface acts as a source of positively polarized light which serves to shift the Babinet and Brewster points toward the sun. This effect is even more pronounced at UCLA during the afternoon than earlier in the day because the ocean lies in a westerly direction from the site, and the atmosphere in the vicinity receives most light from the sea surface when the sun is in that direction. The effect of sea surface reflection was mentioned previously in connection with Brewster's observations on the coast of Scotland.

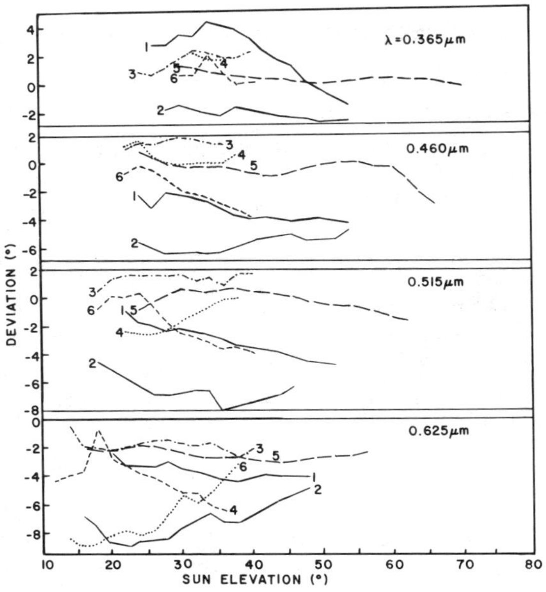

Figure 7.5 Deviations of the observed position of the Brewster point from its theoretical position in a Rayleigh atmosphere as a function of sun elevation according to the observations at four different wavelengths taken by Sekera and Collaborators (1954) at four different locations in southern California. Curves 1: UCLA (a.m.); Curves 2: UCLA (p.m.); Curves 3: Table Mountain (a.m.); Curves 4: Table Mountain (p.m.); Curves 5: Cactus Peak (a.m. and p.m.); Curves 6: Pasadena (a.m. and p.m.).

Data typical of the observations of the Brewster point made by Coulson (1974) at the Mauna Loa Observatory are shown in Fig. 7.6. These data were taken on the same day (March 25, 1973) as the Babinet point observations of Fig. 7.4, and indicate generally the same effects. In interpreting the curves it should be realized that the Brewster point, being located below the sun, is not observable until the sun elevation is 20° to 25°, so that fewer measurements are available for the Brewster point than for the Babinet point. Furthermore, the gradients in the polarization field near the horizon are weaker than those higher in the sky, so it is to be expected that observational errors are greater for the Brewster point than for the Babinet point. Only the shorter wavelengths are represented in Fig. 7.7 because at $\lambda=0.60$ and 0.70 μm the Brewster point, when it exists in the clear atmosphere of Hawaii, is too close to the sun to observe, and it is

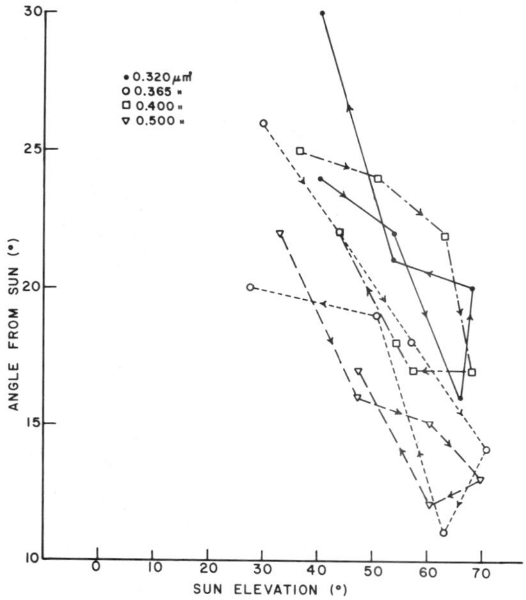

Figure 7.6 Angle between the Brewster point and the sun as a function of sun elevation according to the observations of Coulson et al. (1974) at four different wavelengths in optically unstable atmospheric conditions at the Mauna Loa Observatory.

generally nonexistent at still longer wavelengths. For the wavelengths shown, however, the pattern illustrates the expected movement of the Brewster point toward the sun with increasing sun elevation and a scatter of the data which is consistent with the errors of observation. The absolute magnitude of the angle between neutral point and sun is generally much greater than the 12.5° average found by Brewster (1861), but it is consistent, considering the decreased optical thickness at the 3400-m altitude of Mauna Loa, with the observations of Sekera and Collaborators (1954).

7.3 Secondary and anomalous neutral points

Under normal clear atmospheric conditions, the only neutral points of skylight polarization are the Arago, Babinet, and Brewster points, all of which are single points located in the plane of the sun's vertical. Under unusual conditions, however, multiple neutral points have been observed, and in some cases these

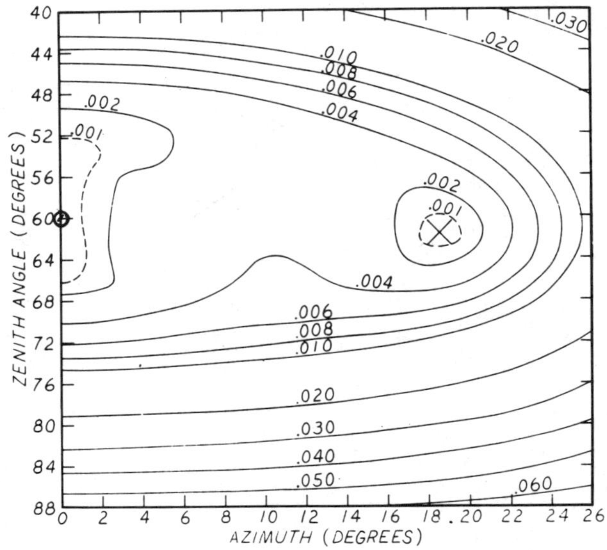

Figure 7.7 Contour lines of degree of polarization over a region of the sky surrounding the sun as computed by Fraser (1981) for a model of the turbid atmosphere overlying a still water surface at a sun elevation of 60° and wavelength 0.55 μm (after Fraser, 1981).

neutral points are located outside of the sun's vertical. On 22 different occasions on which the sea horizon was terminated by a dark purple belt extending to about 1.5° above the horizon at the azimuth of the antisolar point, Brewster (1864) observed the belt to have strong positive polarization, instead of the usual negative polarization existing between the Arago point and horizon. The positive polarization of the purple band then gave way to the normal negative polarization in the region, creating a secondary Arago point at the edge of the belt. In one case when the primary Arago point coincided with the edge of the band, a strange situation occurred in which a neutral point had positive polarization on both sides of it. More frequently the normal Arago point and the band of positive polarization near the horizon were separated by a band of negative polarization, with the primary and secondary Arago points well defined. While the phenomenon was most often seen above a sea horizon, it occasionally occurred above a land horizon as well.

A different type of neutral point anomaly was observed by Soret (1888) above the surfaces of lakes in Switzerland. At times when the sun was seen low over the lake, resulting in the reflected image of the sun forming a second source of positively polarized illumination of the atmosphere in the vicinity, the normally negative polarization in the region of the sun was replaced by positive polarization. At the same time two new neutral points appeared, one on each side of the sun at an angle of 15° to 20° horizontally from the sun. From many observations of the phenomenon, Soret found the effect to depend on the areal extent and surface roughness of the lake, the altitude of the sun, and the existence of fog or haze over the lake. From 13 observations, the angle between the sun and its lateral neighbor was 18.3°. From these observations, Soret concluded that the appearance of neutral points located laterally from the sun is not a rare phenomenon. Perhaps not, but in the century since Soret's observations no other reports of visual sightings of these anomalous neutral points have appeared. As will be seen below, however, numerical modeling of a turbid atmosphere overlying a still-water surface shows very small polarization values outside the sun's vertical, and their positions correspond reasonably well to those observed by Soret.

In theoretical investigations of the effects of surface reflection on the polarization of skylight, Sekera (1961) and Fraser (1981) computed the polarization field of skylight for an atmosphere overlying a Fresnel reflector such as a still-water surface. Of particular interest are Fraser's results for an atmospheric model containing aerosols such as were likely to exist at the time Soret (1888) observed the anomalous neutral points in the sky over the Swiss lakes. Fig. 7.7 shows the contours of the degree of polarization in the region of the sky near the sun's vertical for the case of a Fresnel surface and sun elevation 60°. The polarization is very small throughout a large area surrounding the sun with an absolute minimum at some 18° from the sun near the solar almucantar. At the minimum point, the light is partially circularly polarized, the value of parameter V being $V=1.2\times10^{-4}$. The circular polarization is due to scattering by aerosols, and would occur with a value of $V=0.4\times10^{-4}$ for no surface reflection. Thus reflection by the Fresnel surface increases slightly the circular polarization, but even so, the amount of circular polarization remains very small. It is unlikely that Soret, observing with the Savart polariscope, would have been able to detect the small amounts of polarization shown by the computations out to an angle of at least 25° from the sun's vertical. The fact that he was able to observe the polarization field sufficiently well to detect the location of the anomalous neutral points shows that the model calculations of Fraser did not reproduce in all details the skylight characteristics observed by Soret, though they certainly point to Fresnel reflection as having the capability of causing such anomalous polarization features.

References

Babinet, J. (1840): Sur un nouveau point neutre dans l'atmosphere, *Comptes Rendus, 11,* 618–620.

Babinet, J. (1845): Note de M. Babinet sur la polarisation de la lumiere atmospherique, *Comptes Rendus, 20,* 801–804.

Braaten, D.A., and T.A. Cahill (1986): Size and composition of Asian dust transported to Hawaii, *Atmos. Environ., 20,* 1–5.

Brewster, Sir David (1847): On the polarisation of the atmosphere, *Phil. Mag. and J. Sci., XXXI,* 444–454.

Brewster, Sir David (1861): Observations of the polarisation of the atmosphere, made at St. Andrews in 1841, 1842, 1843, 1844, and 1845, *Trans. Roy. Soc. Edinburgh, XXIII,* 211–240.

Brewster, Sir David (1864): Additional observations on the polarisation of the atmosphere, made at St. Andrews in 1841, 1842, 1843, 1844, and 1845, *Trans. Roy. Soc. Edinburgh, XXIV,* 247–286.

Chandrasekhar, S. (1950): *Radiative Transfer,* Clarendon Press, Oxford.

Cornu, M.A. (1884): Observations relatives a la couronne visible actuellement autour du Soleil, *Comptes Rendus, 99,* 488–493.

Coulson, K.L. (1954): Neutral points of skylight polarization in a Rayleigh Atmosphere, Sci. Rept. No. 7, Contr. AF 19(122)-239, University of California, Los Angeles.

Coulson, K.L. (1971): On the solar radiation field in a polluted atmosphere, *J. Quant. Spectrosc. Radiat. Transfer, 11,* 739–755.

Coulson, K.L. (1977): Atmospheric turbidity determinations by skylight measurements at the Mauna Loa Observatory, *Contrib. in Atm. Sci. No. 13,* University of California, Davis.

Coulson, K.L. (1983): Effects of the El Chichon volcanic cloud in the stratosphere on the polarization of light from the sky, *Appl. Opt., 22,* 1036–1050.

Coulson, K.L., R.L. Walraven, and L.B. Soohoo (1974): Polarization of skylight at an altitude of 3461 meters on Mauna Loa, Hawaii, *Contrib. in Atm. Sci. No. 9,* University of California, Davis.

Dahlkamp, V., and H. Kantus (1946–1947): Untersuchengen uber den reinheitsgrad der atmosphare mit hilfe der neutralen punkte der atmospharischen polarisation, *Zeit. fur Meteor.*, *1,* 253–263.

Dorno, C. (1919): Himmelselligkeit, Heinmelspolarisation, und Sonnenintensitat in Davos (1911 bis 1918), *Meteor. Zeit.*, *36,* 109–124, 181–192.

Fraser, R.S. (1968): Atmospheric neutral points over water, *J. Opt. Soc. Am.*, *58,* 1029–1031.

Fraser, R.S. (1981): Atmospheric neutral points outside of the principal plane, *Contrib. Atm. Phys.*, *54,* 286–297.

Jensen, C. (1899): Beitrage zur photometrie des Himmels, *Meteor. Zeit.*, *14,* 488–499.

Jensen, C. (1928): Die Himmelsstrahlung, *Hanbuch der Physik, Bd. 19, Springer, Berlin,* 70–152.

Jensen, C. (1932): Normale, gestorte und pseudonormale polarisationserscheinungen der atmosphare, *Meteor. Zeit.*, *49,* 419–430.

Jensen, C. (1942): Die polarisation des himmelslichts, *Handbuck der Geophysik, Bd. 8, Berlin,* 527–620.

Kimball, H.H. (1903): Observations of solar radiation with the Angstrom pyrheliometer at Asheville and Black Mountain, N.C., *Mo. Wea. Rev.*, *31,* 320–334.

Kimball, H.H. (1913a): The effect upon atmospheric transparency of the eruption of Katmai Volcano, *Mo. Wea. Rev.*, *41,* 153–159.

Kimball, H.H. (1913b): The effect of the atmospheric turbidity of 1912 on solar radiation intensities and skylight polarization, *Bul. Mt. Weather Observatory, V* (5), 295–312.

Neuberger, H. (1940): Studies in atmospheric turbidity in Central Pennsylvania, Penn. State Coll. Studies No. 9, 36 pages.

Neuberger, H. (1950): Arago's neutral point: A neglected tool in meteorological research, *Bul. Am. Meteor. Soc.*, *31,* 119–125.

Pickering, E.C. (1884): A new form of polarimeter, *Proc. Am. Acad. Arts and Sciences, XIII, New Series,* 294–302.

Rubenson, R. (1864): Memoire sur la polarisation de la lumeire atmospherique, Nova Acta Regiae Societatis Scientiarum, *Seriei Tertiae Vol. V, Upsal,* 1–145.

Sekera, Z. (1961): The effect of sea surface reflection on the sky radiation, in Jerlov (Ed.) *Monograph No. 10, Paris l'Institut Geophysique National,* pp. 66–72.

Sekera, Z., and Collaborators (1954): Investigation of polarization of skylight, Final Rept., Contr. AF 19(122)-239, University of California, Los Angeles.

Soret, M.J.L. (1888): Influence des surfaces d'eau sur la polarisation atmospherique et observation de deux points neutres a droite et a gauche de Soleil, *Comptes Rendus, 107,* 867–870.

Strutt, J.W. (Lord Rayleigh) (1871): On the light from the sky, its polarisation and coulour, *Phil. Mag.*, 107–120, 274–279.

Volz, F. (1970): Atmospheric turbidity after the Agung eruption of 1963 and size distribution of the volcanic aerosol, *J. Geophys. Res., 75,* 5185–5193.

Volz, F., (1975): Distribution of turbidity after the 1912 Katmai eruption in Alaska, *J. Geophys. Res., 80,* 2643–2648.

CHAPTER 8

SKYLIGHT INTENSITY AND POLARIZATION DURING TWILIGHT

8.1 Intensity of the twilight sky

Observations of light from the sky during periods of twilight have long been used to deduce the turbidity profile of the upper atmosphere. The method is very convenient from an observational standpoint, as no artificial light sources are required, solar energy is intense enough that instrument sensitivity problems are minimized, the source is essentially constant from day to day, and the position of the sun at any given instant is readily calculated. The main problems in interpreting the observational data from a physical standpoint arise from the spherical geometry of the system, the long atmospheric paths through which the solar beam passes during twilight, and the strong asymmetric diffuse radiation field resulting from primary and higher order scattering of the incident energy. Most authors have emphasized only the radiant intensity in twilight, even though the polarization field is often a more sensitive indicator of aerosols in the atmosphere than is the intensity. Because of the large number of measurements by various authors of twilight intensity, we begin this section on twilight by reviewing studies of twilight intensity, and postpone for the time being the discussion of polarization during twilight.

Thus far, the most extensive work on twilight is that of the late Professor G.V. Rozenberg (1963) and his colleagues and students in the Soviet Union (Rozenberg and Mikirtumova, 1976; Nazaraliyev and Rozenberg, 1977). Systematic observations of twilight were begun in 1941 at the Abastumani Observatory, located at an altitude of 1700 m in the clear atmosphere of the Soviet Republic of Georgia. An analysis of data covering the period from 1942 to 1952, with a total of 373 twilight series of intensity measurements at $\lambda = 0.527$ μm and 0.397 μm and 185 series of polarization observations at $\lambda = 0.527$ μm, was reported by Megrelishvili et al. (1978), and in more detail in the monograph of Megrelishvili (1981). Values of zenith intensity at the two wavelengths obtained from this extensive data set are shown as a function of solar depression angle in Fig. 8.1. Unfortunately, their measurements were terminated at a solar depression of 4.5°, which corresponds to a tangent ray at about 20-km altitude, so no information on scattering in the lower stratosphere or troposphere was obtained. Their data extended, however, to the phenomenal altitude of 160 km, which corresponds to a solar depression of 12.7°, and thus were appropriate to studies of atmospheric turbidity in the region of the mesosphere. According to previous measurements of Rozenberg and Mikirtumova (1976) and theoretical work by Nazaraliyev and Rozenberg (1977), the effects of secondary scattering are so dominant at and above 80 to 100 km that twilight data gives little information on turbidity above the mesosphere. At 100 km the intensity due to secondary scattering is about equal to that from primary scattering, and the errors of turbidity determinations from measurements of total intensity are about equivalent to those of the natural variability of the atmosphere.

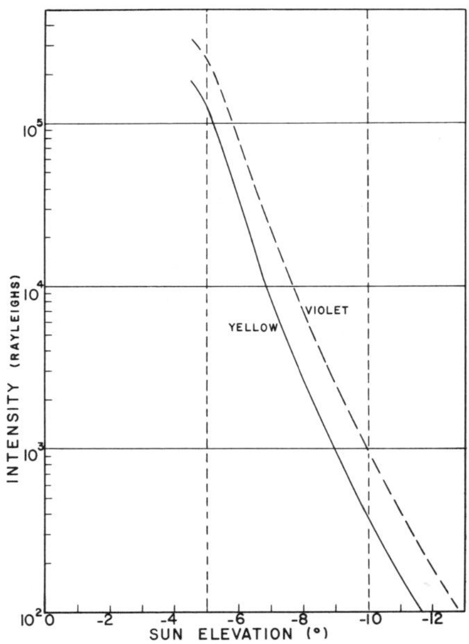

Figure 8.1 Average values of twilight brightness at the zenith as a function of sun elevation for yellow and violet portions of the spectrum as measured by Megrelishvili et al. (1978) at the Abastumani Observatory.

From analyses of the color index and autocorrelation matrix of the data at the two wavelengths of Fig. 8.1, Megrelishvili et al. (1978) conclude that (1) a mesospheric layer of high turbidity exists between 40 and 60 km, (2) the mesospheric scattering coefficients for the violet and yellow spectral regions are practically the same as for Rayleigh scattering, indicating the particles were small compared to these wavelengths, (3) the turbidity has a distinct increase with height above 70 km, and (4) the increasing blueness of the sky as the solar depression increases from 4° to 9° is due only to vertical variations of atmospheric turbidity.

By a correlational analysis of the same set of data as reported by Megrelishvili et al. (1978), Mel'nikova et al. (1980) computed autocovariance matrices from measured intensities at the two wavelengths for the turbidity coefficient vector $\sigma(h)$ at 17 altitudes in the 40- to 120-km range. The profile of the second eigenvector of the matrix confirmed the existence of a turbid layer between 40 and 70 km, and the third eigenvector indicated a layer in the 55- to 85-km range due to the breakup of meteoric material. Twilight methods are ineffective for characterizing turbidity above 85 km due to the effects of secondary scattering, according to these authors. They concluded, in contrast to other authors, that the polarization of twilight contains practically no significant information on the scattering substance at different altitudes in the mesosphere, and that indications of correlations between polarization anomalies and state of the ionosphere are due to secondary scattering.

Although the results reported by Megrelishvili et al. and by Mel'nikova et al. represent important and pioneering efforts, it is in some respects unfortunate that the measurements did not extend to longer wavelengths. As seen by Fig. 5.6 of Chapter 5, measurements at eight different wavelengths at the Mauna Loa Observatory showed that wavelengths $\lambda \geq 0.70$ μm were subject to fewer multiple scattering effects than those of $\lambda \leq 0.60$ μm. In fact, the polarization of twilight at Mauna Loa under ordinary atmospheric conditions is essentially wavelength independent for the near-infrared spectral region, whereas a very significant spectral dependence up to and including $\lambda = 0.60$ μm appeared in the data. This can be taken to mean that multiple scattering played a minor role in the near-infrared region, but not in the visible and ultraviolet. This result is not surprising in view of the critical dependence of Rayleigh scattering on wavelength. Megrelishvili et al. concluded that mesospheric aerosols have scattering characteristics similar to those of Rayleigh particles, in which case extension of the measurements to longer wavelengths would have greatly reduced contamination of the data by secondary and higher order scattering. A step toward the use of longer wavelengths was made by Rozenberg and Mikirtumova (1976), in which data in the region $\lambda=0.59$ to $0.66\ \mu m$ were analyzed, but further extension in wavelength would have been advantageous.

Coulson (1981) reported on zenith intensity measurements taken during morning twilight on 43 days during the spring of 1977 at the Mauna Loa Observatory. The 1977 period was characterized by the absence of any appreciable contamination of the stratosphere by volcanic eruptions. The measurements were made in eight different wavelength intervals from $\lambda=0.32\ \mu m$ to $0.90\ \mu m$. In analysis of the data, 9 days of extremely clear atmospheric conditions were designated as low turbidity days, and measurements for those days were averaged to obtain data representing low turbidity conditions over Hawaii. Average zenith intensities at the various wavelengths are plotted as a function of sun elevation in Fig. 8.2 and tabulated at one-half degree intervals in Table 8.1. The 9 days of 1977 on which the measurements were made were February 10, 15, 19, 21, 23, and March 4, 9, 10, 11. The data have been normalized to a count rate of 1.0×10^6/sec at 0° sun elevation, and a certain amount of graphical interpolation was necessary. The sun elevation scale on the abscissa of the figure applies to the curve for $\lambda=0.90\ \mu m$ only, and the other curves have successive displacements by spaces of 1° to the right for curve separation. In addition, curves of sun elevation are superimposed and labeled on the right. The dotted curve in the diagram is the relative air mass (as given by the U.S. Standard Atmosphere, 1976) above the solar ray which is tangent to the earth's surface at the various solar elevation angles. It is shifted one division to the left for clarity.

An obvious feature in Fig. 8.2 is the effect of ozone absorption. The normalization process compensated for any ozone absorption which took place at 0° sun elevation, but the curves show greater absorption occurred in the longer atmospheric paths for negative sun elevations and lesser absorption after sunrise. The most pronounced absorption was for $\lambda=0.60\ \mu m$, but the wings of the Chappuis band were responsible for slight absorption at $\lambda=50\ \mu m$ and $0.70\ \mu m$. A second feature of Fig. 8.2 is the similarity of the curves for $\lambda=0.80$ and $0.90\ \mu m$. This similarity is an indication that for these longer wavelengths primary scattering is the dominant mechanism by which the incident sunlight is directed

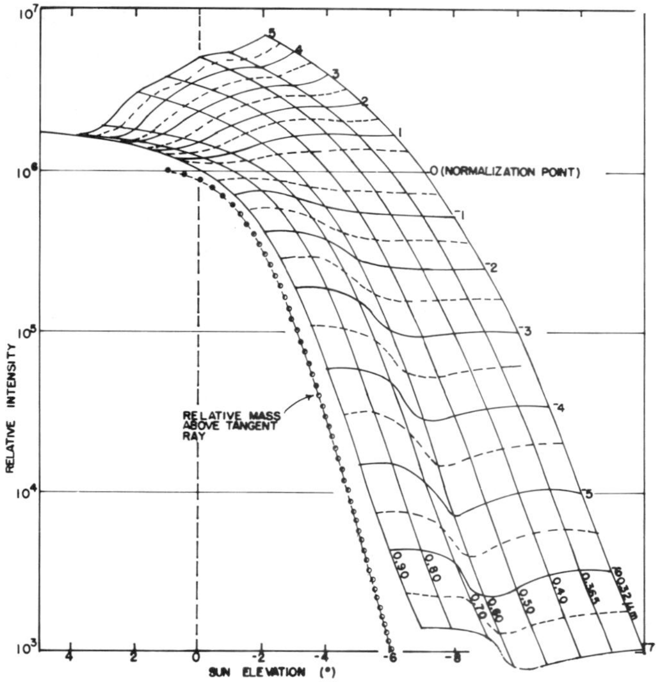

Figure 8.2 Average zenith intensities as a function of sun elevation for 9 low turbidity days at the Mauna Loa Observatory. Data have been normalized to a photon count rate of 10^6/sec at a 0° sun elevation. The sun elevation scale on the abscissa applies only for $\lambda = 0.90$ μm (the other curves have successive displacements of 1° to the right), but sun elevation curves have also been superimposed and labeled on the right. Included is a curve of the mass, on a relative scale, of the amount of air above the solar ray, which is tangent to the earth's surface at the various solar depressions. (After Coulson, 1981.)

into the nadir direction by the atmosphere above Mauna Loa. If this is a valid conclusion, then the zenith intensity at any solar depression angle should be approximately proportional to the number of scattering centers illuminated by the direct solar beam. For a pure gaseous and cloudless atmosphere, the intensity should thus be approximately proportional to the pressure at the level of the tangent ray. This quantity, as given by the U.S. Standard Atmosphere 1976, is indicated by the dotted curve of Fig. 8.2. Throughout the range of solar depressions of 2° to 7°, the measured intensity does not decrease as rapidly with

TABLE 8.1 Average photon count rate at eight different wavelengths in the direction of the zenith as a function of sun elevation from observations on 9 very clear days at the Mauna Loa Observatory. The data have been normalized to a count rate of 1.0×10^6/sec at 0° sun elevation.

Sun elevation (°)	Wavelength (μm)							
	0.320	0.365	0.400	0.500	0.600	0.700	0.800	0.900
5.0	6.88(6)	5.27(6)	5.06(6)	3.80(6)	3.05(6)	1.87(6)	1.62(6)	1.72(6)
4.5	6.17 ″	4.73 ″	4.64 ″	3.56 ″	2.85 ″	1.81 ″	1.58 ″	1.68 ″
4.0	5.36 ″	4.22 ″	4.22 ″	3.32 ″	2.68 ″	1.76 ″	1.54 ″	1.64 ″
3.5	4.58 ″	3.74 ″	3.77 ″	3.08 ″	2.49 ″	1.70 ″	1.49 ″	1.58 ″
3.0	3.89 ″	3.26 ″	3.35 ″	2.80 ″	2.31 ″	1.63 ″	1.45 ″	1.54 ″
2.5	3.26 ″	2.82 ″	2.94 ″	2.52 ″	2.11 ″	1.55 ″	1.40 ″	1.47 ″
2.0	2.67 ″	2.40 ″	2.48 ″	2.22 ″	1.90 ″	1.46 ″	1.33 ″	1.40 ″
1.5	2.14 ″	2.01 ″	2.05 ″	1.92 ″	1.71 ″	1.36 ″	1.27 ″	1.32 ″
1.0	1.66 ″	1.65 ″	1.66 ″	1.60 ″	1.46 ″	1.26 ″	1.19 ″	1.22 ″
0.5	1.31 ″	1.31 ″	1.32 ″	1.30 ″	1.24 ″	1.14 ″	1.10 ″	1.12 ″
0.0	1.00 ″	1.00 ″	1.00 ″	1.00 ″	1.00 ″	1.00 ″	1.00 ″	1.00 ″
−0.5	7.19(5)	7.40(5)	7.34(5)	7.48(5)	7.75(5)	8.47(5)	8.68(5)	8.62(5)
−1.0	5.12 ″	5.36 ″	5.33 ″	5.42 ″	5.84 ″	6.83 ″	7.43 ″	7.19 ″
−1.5	3.59 ″	3.71 ″	3.71 ″	3.71 ″	4.13 ″	5.21 ″	5.75 ″	5.63 ″
−2.0	2.46 ″	2.47 ″	2.48 ″	2.42 ″	2.74 ″	3.77 ″	4.31 ″	4.16 ″
−2.5	1.62 ″	1.60 ″	1.56 ″	1.52 ″	1.69 ″	2.50 ″	2.96 ″	2.56 ″
−3.0	1.20 ″	1.01 ″	9.76(4)	9.13(4)	9.97(4)	1.57 ″	1.92 ″	1.84 ″
−3.5	6.38(4)	6.11(4)	5.78 ″	5.21 ″	5.42 ″	9.10(4)	1.16 ″	1.13 ″
−4.0	3.71 ″	3.47 ″	3.32 ″	2.88 ″	2.75 ″	4.79 ″	6.29(4)	6.05(4)
−4.5	2.04 ″	1.95 ″	1.86 ″	1.57 ″	1.40 ″	2.41 ″	3.14 ″	2.96 ″
−5.0	1.30 ″	1.07 ″	1.03 ″	8.65(3)	6.98(3)	1.68 ″	1.48 ″	1.45 ″
−5.5	5.69(3)	5.81(3)	5.57(3)	4.67 ″	3.83 ″	6.53(3)	7.37(3)	7.51(3)
−6.0	3.14 ″	3.17 ″	3.34 ″	2.58 ″	2.17 ″	3.65 ″	3.92 ″	4.22 ″
−6.5	1.90 ″	1.68 ″	3.07 ″	1.44 ″	1.24 ″	2.15 ″	2.19 ″	2.34 ″
−7.0	1.04 ″	9.37(2)	9.40(2)	8.11(2)	7.72(2)	1.30 ″	1.33 ″	1.39 ″

increasing solar depression as does the pressure at the tangent ray. There may be two reasons for this. First, there is always a certain amount of multiple scattering, even though primary scattering is dominant, and thus multiple scattering would tend to increase the intensity. In addition, the model of the standard atmosphere is only for gases, whereas the actual atmosphere includes some aerosols. Aerosol scattering would contribute to the divergence of the curves at the larger solar depressions since the effective number and efficiency of scatterers is actually higher than for atmospheric gases alone. Finally, astronomical refraction, not taken into account in the diagram, lowers the geometric shadow height by a few kilometers.

Comparisons of the low turbidity twilight intensities at Mauna Loa with those determined by other authors are shown by Figs. 8.3 and 8.4. Mauna Loa data at $\lambda = 0.70$ and $0.50\ \mu m$ were chosen for the diagram of Fig. 8.3 so as to be comparable to the computations at the same wavelengths by Blattner et al. (1974) and as close as possible to the measurements of Koomen et al. (1952) for the light-adapted human eye. The measurements of Koomen et al. were made at a low-altitude site in Maryland and at an altitude of 2800 m on Sacramento Peak in New Mexico. The computations of Blattner et al. were for a spherical shell model of the Rayleigh atmosphere. All of the data have been normalized to 10^6/sec at 0° sun elevation. Except in the region of positive and small negative sun elevations, at which time aerosols in the lower troposphere are illuminated, the slopes of the curves are all very similar. This similarity is evidence that the amount of aerosol scattering in the higher levels of the atmosphere must have been of small magnitude, or the turbidity consisted of very small particles as

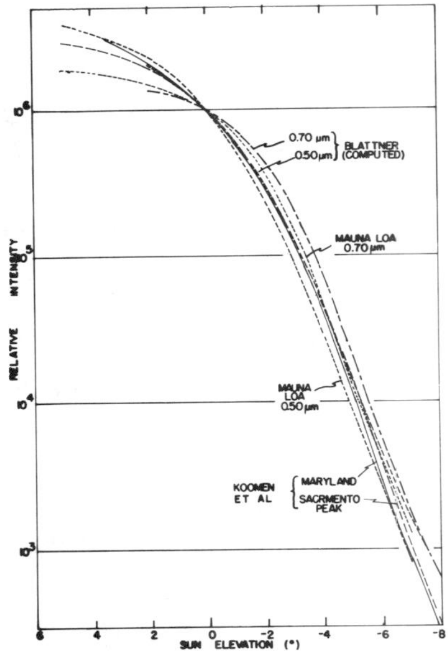

Figure 8.3 Zenith intensity as a function of sun elevation as measured by Koomen et al. (1952) at two different locations, as computed by Blattner et al. (1974) for a pure Rayleigh atmosphere, and as measured at the Mauna Loa Observatory. Data have been normalized to the same value at a 0° sun elevation. (After Coulson, 1981.)

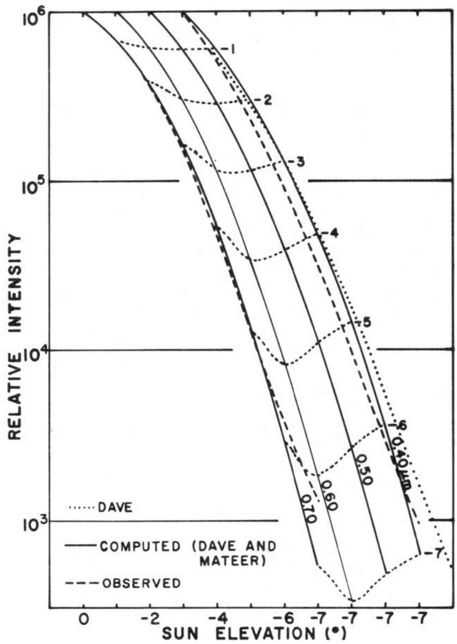

Figure 8.4 Normalized zenith intensity as a function of sun elevation as computed for a normal clear atmosphere by Dave and Mateer (private communication) at four different wavelengths, as observed on the clearest days at two wavelengths at the Mauna Loa Observatory, and as computed by Dave (1956) for primary and secondary scattering. Data have been normalized to the same value at a 0° sun elevation.

found by Megrelishvili et al. (1978), since the computations of Blattner et al. were for a pure Rayleigh atmosphere. The small differences in the curves are probably within measurement or computational error.

In a study of twilight colors, Dave and Mateer (1980) calculated the intensities of twilight at several different wavelengths in many directions, including the zenith, for a number of atmospheric models. The sphericity of the atmosphere was taken into account, as were ozone absorption and the inhomogeneous nature of aerosols in the various models. Atmospheric refraction was not considered, and the computations neglected multiple scattering and aerosol absorption. The normalized intensities computed by Dave and Mateer for the clear atmosphere containing 0.242 atm cm of ozone and the normal aerosol distribution (rapid decrease of concentration with height in the lower levels, with secondary maxima at 20 and 50 km). Details of the atmospheric model are given by Dave and Mateer (1968). The normalized intensities for this model as computed by Dave and Mateer are shown for four different wavelengths by the solid curves of Fig. 8.4. The dashed curves show the average low turbidity intensities at $\lambda=0.40$ and 0.70 μm made at Mauna Loa. The similarity of the curves of $\lambda=0.70$ μm is striking, particularly for solar depression angles less than about 6°. For $\lambda=0.40$ μm, however, the neglect of multiple scattering makes the slope of the computed curve too great at small solar depressions, resulting in artificially high normalized intensities at solar depressions of 2° to 5°. Increased multiple scattering as the solar depression angle passes 5° results in a larger slope for the observed than for the computed curves. The dotted curve, taken from an earlier study by Dave (1956) in which secondary scattering was taken into account, has essentially the same slope as the observed curves at the larger solar depressions. This indicates that although multiple scattering is important at $\lambda=0.40$ μm, a good approximation is achieved by accounting for only primary and secondary scattering.

In studies of twilight intensity at a point in the sky 20° above the horizon at the azimuth of the sun, several authors (Bigg, 1956; Volz and Goody, 1962; Volz, 1969, 1970) have deduced atmospheric turbidity profiles to altitudes of 50 to 60 km under undisturbed and volcanic conditions. The rationale for choosing a 20° elevation angle for the intensity measurements was to take advantage of the well-known forward scattering maximum for aerosol particles. This angle is not appropriate for polarization measurements during twilight, as it is in the general region of the Babinet point and the polarization there is weak and variable. For twilight polarization measurements the zenith direction is preferred. The 20° measurements, however, have been largely confined to intensities, in which case the 20° elevation angle is a logical choice.

From measurements at the 20° elevation angle, Bigg (1956) deduced the existence of a small turbidity maximum near the top of the mesosphere at 80 km, with a large scattering maximum near the tropopause at 15 km. The mesospheric maximum corresponds reasonably well with the conclusions of Megrelishvili et al. (1978), but the data of these latter authors do not extend down to 15 km. Bigg's data are unusual, however, because of his observation of sharp peaks in the intensity gradient (dI/Idt) as a function of time, which he found associated with the temperature inversions, with the presumed dust concentrations, detected by radiosonde. Other authors (Megrelishvili, 1958; Volz, 1969; Coulson, 1981) failed to find sharp changes of gradient at small height differences, Megrelishvili (1958) stating that in her observations of 1500 twilight curves sharp breaks such as those found by Bigg were never observed. Volz and Goody (1962) also pointed out that Bigg had, in effect, assumed the median scattering height to be the same as the height of the tangent ray, and thus his correlations of twilight intensity with atmospheric parameters such as temperature and humidity profiles had no physical basis.

8.2 Color ratios during twilight

It is generally recognized that the changes of intensity during twilight are so rapid, the range being over four orders of magnitude during a given twilight period, that intensity profiles themselves are difficult to interpret. More convenient parameters derived from the intensity profiles are intensity gradients or the ratios of intensities at two different wavelengths. This latter parameter is termed the color index or, preferably, the color ratio (CR). Volz (1970) found the red/green CR (wavelengths 0.715 and 0.477 μm) to be the most significant parameter for interpretation of his measurements. Coulson (1981), utilized several wavelength combinations for the CR $\lambda = 0.80/\lambda_i$, where λ_i assumed the other seven wavelength values of his measurements. He found the CR 0.80/0.90 and 0.80/0.70 to be very small and uninformative because of the similarity of scattering at these longer wavelengths in the clear atmosphere at Mauna Loa. Such might not be the case, however, in more turbid conditions. The CR 0.80/0.60 was contaminated by ozone absorption at $\lambda = 0.60$ μm. Perhaps the most useful CR in Coulson's measurements were 0.80/0.50 and 0.70/0.40, as all of these wavelengths are outside the main absorption band of ozone and there is enough of a wavelength spread to yield different scattering characteristics for the two. The computations of Blattner et al. (1974) showed that the CR 0.70/0.50 during twilight verified the feasibility of determining the characteristics of high altitude dust layers.

The CR $0.80/\lambda_i$ for the average of 9 low-turbidity days at Mauna Loa are shown in Fig. 8.5 (Coulson, 1981). The measurements have been normalized to unity at a sun elevation of 5°. For wavelengths less than 0.80 μm, positive values of the CR represent a shift of the color toward $\lambda = 0.80$ μm from its value at the normalization point. This would correspond to a reddening of the sky

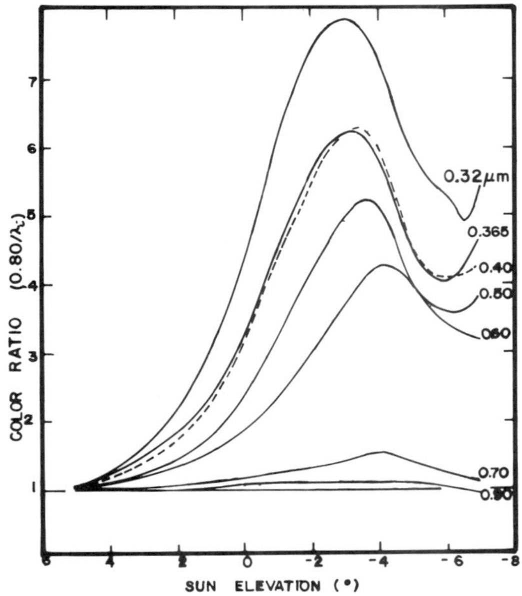

Figure 8.5 Color ratio 0.80 μm/λ_i of low turbidity average zenith intensity at 0.80 μm in conjunction with that at other wavelengths λ_i as a function of sun elevation at the Mauna Loa Observatory. Data were normalized to unity at 5° sun elevation. (After Coulson, 1981.)

for positive values in the diagram and explain the well-known feature of increased warm colors in landscape photographs taken at a time of low sun elevations. The CR 0.80/λ_i reaches a definite maximum during the period of sun elevation −2° to −5°, the precise time being somewhat dependent on wavelength. The maxima of CR 0.80/0.40, being at sun elevation −3° to −4° corresponds to the time of the "purple light" commonly observed as a phase of twilight and described in detail by Gruner (1942). The magnitude of the CR has a strong spectral dependence, the enhancement of radiation at 0.32 μm being a factor of nearly 8 at the maximum. Comparable, but progressively lower factors, occur at the other wavelengths. The minor influence of multiple scattering of near-infrared wavelengths in the clear atmosphere at Mauna Loa is responsible for the low values of CR 0.80/0.70 and 0.80/0.90 shown in the diagram.

The decrease of CR 0.80/λ_i, for solar elevations of −4° to −6° is probably due to greater relative attenuation at 0.80 μm by the normal aerosol layer in the 18- to 24-km altitude range than to lesser attenuation at shorter wavelengths. The pronounced minimum at sun elevation about $-6°$ is similarly a manifestation of aerosol effects. The prevailing increase of CR at still larger solar depressions, at least for $\lambda = 0.50$ μm, is probably due to multiple scattering, as it was predicted for a pure Rayleigh atmosphere by the computations of Blattner et al. (1974).

Appreciable amounts of atmospheric turbidity over Mauna Loa cause very significant changes of the behavior of the CR during twilight. As will be seen in Chapter 9, the volcanic cloud from the eruption of El Chichon in 1982 caused profound changes in the CR, but here we show the more modest perturbations which occurred in the spring of 1977. As judged by skylight polarization measurements and visual evidence, the atmosphere was unusually turbid on 7 days out of the 63 days of skylight measurements. The CR 0.80/0.50 μm during twilight was given for all 7 days by Coulson (1980), three of which are shown here, along with the low turbidity average, in Fig. 8.6. Generally, an increase of turbidity causes a decrease of CR 0.80/0.50 μm, but this is not universally true. The intensity deviations occurring under turbid conditions (see Coulson, 1980) indicate that the decrease of CR may be due to a decrease of intensity at 0.80 μm, an

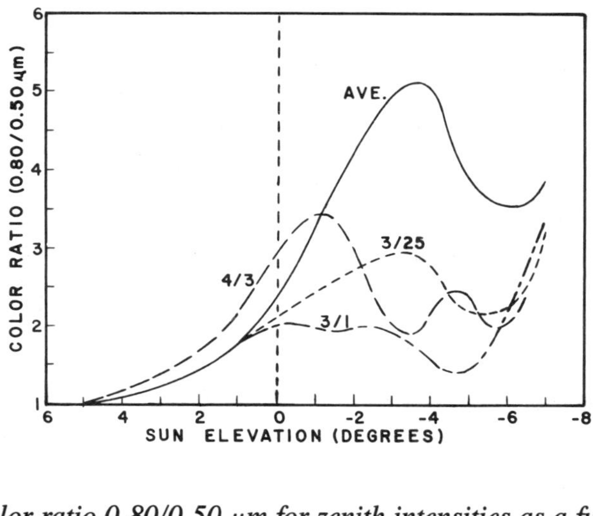

Figure 8.6 Color ratio 0.80/0.50 μm for zenith intensities as a function of sun elevation for various moderate to high turbidity conditions at the Mauna Loa Observatory. The average low turbidity curve is given for comparison.

increase at 0.50 μm, or both. Thus the CR alone is difficult to interpret physically, but still is a useful indicator of atmospheric conditions.

A period of disturbed weather occurred in Hawaii during the first days of April 1977, with a pronounced modification of the twilight CR resulting from this weather system. The most anomalous situation of the period occurred on April 3. In this case, intensities in the post-sunrise period were increased at 0.80 μm and decreased at 0.50 μm, leading to an anomalously high CR at that time. At solar depressions beyond about 1°, however, very low intensities at 0.80 μm resulted in low values of the CR, with a minimum occurring at the same solar depression as the usual maximum. The original data show that all of the near-infrared intensities were unusually low at this time, whereas the shorter wavelengths were affected to a much lesser extent. This feature was probably the result of attenuation by relatively large aerosol particles being produced by the condensation of water vapor on the usual aerosols in the marine air which was then carried to high levels by strong atmospheric turbulence.

The CR for two other anomalous twilight periods is represented in Fig. 8.6. On March 25, the atmosphere had a hazy appearance with dust layers extending to several degrees above the horizon. The sunrise sky was unusually yellow on that day and a definite solar aureole persisted throughout the forenoon. An even more anomalous condition occurred on March 1. Although no clouds appeared near the zenith, wisps of cirrus were observed elsewhere in the sky, and scattering by particles which had not yet formed visible clouds cannot be ruled out. There was considerable upslope motion carrying air from below the trade wind inversion to the mountain top. Both the CR and zenith polarization during twilight showed optical instability which may have been caused by either actual or nascent clouds at the higher levels. Whether or not such particles constitute turbidity depends on the definition of the term, but their existence is evident from light in the twilight sky.

In addition to the CR, a second useful index for interpreting twilight intensities is the deviation of the intensity from normal. As discussed above, a case of anomalous upper level turbidity occurred at the Mauna Loa Observatory on April 3, 1977. The deviations of the intensities for that day from their low turbidity averages, expressed as a fraction of these averages, are shown for six wavelengths in Fig. 8.7. The 20% to 60% reduction of intensity of the longer

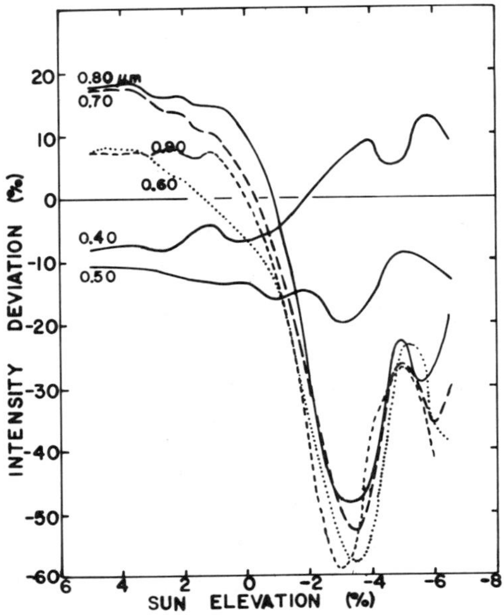

Figure 8.7 Deviation of the zenith intensity from the low turbidity average as a function of sun elevation for six different wavelengths as measured at the Mauna Loa Observatory on the high turbidity day of April 3, 1977. (After Coulson, 1981.)

wavelength radiation at sun elevations of $-3°$ to $-6°$ is evidence of a strongly absorbing aerosol layer in the upper troposphere or lower stratosphere, and some aerosol layering is evident in the data. The rapid return of the intensity to normal immediately before sunrise and the small overshoot after sunrise indicate more aerosol scattering than usual in the lower atmosphere. Similar, although less pronounced, twilight intensity deviations occurred on the other days of high turbidity during the measurement series. Volz (1981) has suggested that the anomalous twilight intensities at Mauna Loa may have been due to distant clouds or haze which caused either an anomalous attenuation of the direct solar beam or an increase of elevation of the effective horizon. While this type of explanation cannot be ruled out by the information available, the tentative explanations advanced here are to be preferred.

From 93 cases of twilight intensity measurements at an angle 20° above the horizon in the azimuth of the sun taken at the Blue Hill Observatory, Milton, Massachusetts (altitude 205 m), Volz and Goody (1962) endeavored to distinguish nine different twilight types to characterize the state of turbidity in the atmosphere. The criteria for classifying the types were position and strength of various features of color ratios and intensity gradients in the five different wavelengths of the measurements, type of weather situation in which the measurements were made, and some subjective judgment. Their general procedure for interpreting the measurements was to assume a turbidity profile for the low levels, use measured intensities to compute the turbidity of the upper atmosphere, and treat variations of low level turbidity as a source of error. Because of evidence that red light ($\lambda = 0.659$ μm) is not seriously disturbed by multiple scattering, much of their analysis utilized the red wavelengths only. (A similar conclusion, although for wavelengths of 0.70 μm, was reached by Coulson (1981) from the Mauna Loa measurements.) Volz and Goody (1962) concluded that the atmosphere over Massachusetts normally has a turbidity maximum in the 15- to 30-km altitude range, with a certain amount of dust at all levels up to 65 km. From 30 to 65 km the extinction coefficient for dust was almost constant, with a value of about one-fifth that for the Rayleigh atmosphere at the same level. Their data indicate a small seasonal variation of turbidity between 10 and 55 km with a maximum in winter at all levels. There was no correlation between upper atmospheric turbidity and meteor showers, contrary to what had been postulated by Giovanelli (1954), Grandmontagne (1941), and others.

In a series of twilight measurements at a 20° elevation angle over the period 1962–1968, Volz (1969) utilized the same grouping of twilight types as defined by Volz and Goody (1962) whereby the types were separated on the basis of intensity gradients, color ratios, date of measurement, and other parameters. Most of the 1962–1968 series was affected by volcanic debris from the March 1963 eruption of the volcano Agung in Indonesia, but a few measurements of undisturbed twilights were made before the eruption. An example of the red/green color ratio (CR 0.659/0.477 μm) for undisturbed twilights at the Blue Hill Observatory is shown in Fig. 8.8. For a comparison of different profiles, Volz normalized the CR to 1.0 at sun elevations of −1° and −4.5°, whereas Coulson's data were normalized at sun elevation 5.0°. Thus the different normalization procedure, with the different angles of observation and different wavelengths of the measurements, eliminates the feasibility of a direct comparison of the two sets of data.

The CR 0.659/0.477 μm as observed by Volz for undisturbed conditions is shown as a function of sun elevation by the solid curve in Fig. 5.18. The

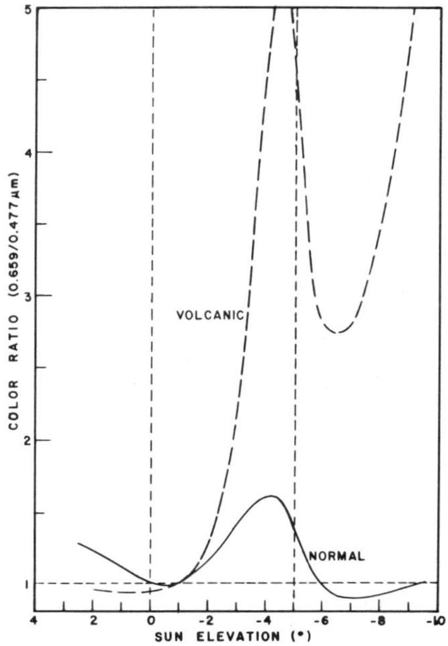

Figure 8.8 Color ratio of wavelengths 0.659/0.477 μm as a function of sun elevation as observed by Volz (1969) in normal and volcanically disturbed conditions at Blue Hill Observatory in a direction 20° above the solar horizon. The data were normalized to unity at a solar depression of 1°.

relatively pronounced maximum at solar depressions of 4.0° to 4.5° corresponds to approximately the time of the maximum reddening of the main purple twilight glow. The most intense or main purple light occurs at solar depression of 3.5° to 4°, and the faint secondary purple light occurs in the deeper twilight at solar depressions of 8° to 11°. The crimson glow, an intense red glow near the horizon, occurs simultaneously with the secondary purple light. The slight increase of the CR beyond −7° in Fig. 5.18 is an indication of redness at 20° elevation, but the main red glow is confined to the region nearer the horizon. The gradients of the ultraviolet light ($\lambda = 0.377$ μm), given by Volz but not reproduced here, show well the periods of main and secondary purple light. Measurements in the direction of the zenith by Coulson (1980) and others do not depict these color changes as well as those at 20° elevation, as the colors normally do not reach the zenith.

The strongest effects of the stratospheric dust cloud from the eruption of Agung in March 1963 caused the intense reddening of the sky indicated by the dashed curve of Fig. 8.8. This effect will be mentioned in some detail in Chapter 9 which is devoted to volcanic debris in the atmosphere, but the curve here shows the two periods of intense red light in deep and moderate twilight, along with the virtual disappearance of red color in the sky just at sunrise or sunset. In fact the data here show the sky in the immediate post-sunrise period to have less of a red color under volcanic conditions than in more normal circumstances.

8.3 Polarization at the zenith during twilight

As mentioned above, the zenith is the most logical direction in the sky for measuring polarization during twilight. Aside from the convenience of orienting the polarimeter in a fixed direction, the zenith is close to the polarization maximum when the sun is near the horizon. This minimizes measurement errors, and the region of weak and variable polarization which occurs at, for instance, 20° above the solar horizon is avoided.

Many authors have shown zenith polarization during twilight to be a sensitive indicator of aerosols in the upper atmosphere. For instance, Dave and Ramanathan (1956) reported a series of measurements showing changes of polarization for solar depression angles from 0° to nearly 20°, and Dave (1956) interpreted the polarization effects on the basis of primary and secondary scattering of the incident light by gases at various heights in the atmosphere. Shah (1969) found a strong effect on twilight characteristics produced by the volcanic debris from Agung in 1964, and Coulson (1983) observed an even stronger and persistent effect of the stratospheric cloud from the eruption of El Chichon in 1982. The monograph of Rozenberg (1963) and the computations of Blattner et al. (1974) have shown that some, but by no means all, twilight variations can be explained on the basis of multiple scattering and absorption in a molecular atmosphere. The complete polarization problem certainly involves aerosol effects. This conclusion is bolstered by the calculations of Divari (1967) and Blattner (1976), and by the observations of Ashburn (1952), Robley (1950), Dietz (1969), and others. Dietz and Kohl (1968) endeavored, with only moderate success, to correlate twilight polarization with the turbidity characteristics of various types of air masses. In a more extensive analysis, Steinhorst (1974, 1977) showed not only that lower stratospheric aerosols produce significant effects in twilight polarization, but he also developed an iterative method for deducing the vertical profile of aerosol concentration from twilight observations.

8.3.1 Polarization measurements at the Mauna Loa Observatory

Emphasis in the present discussion will be a series of twilight polarization measurements by the present author from an altitude of 3400 m at the Mauna Loa Observatory. The intensity measurements from the series were presented in the previous section. Part of the polarization data was discussed by Coulson (1980), but some of the results are presented here for the first time. The twilight polarization at the zenith was measured during the early morning hours of 43 days during the spring of 1977, at which time the stratosphere was clear of any volcanic debris. As will be seen, however, atmospheric turbidity was not constant throughout the period, and the polarization field responded to changes of turbidity.

The polarimeter used for the twilight measurements is the photon counting polarimeter described in Chapter 11. Briefly, it is a rotating analyzer type polarizing radiometer consisting of two channels with four interference filters each. The field of view is a 1° circular cone. In the twilight measurement mode, the instrument is oriented in the direction of the zenith, and the four filters of each channel are caused to cycle at a rate such that each measurement is repeated once every 48 seconds. The uncertainty associated with each data point was computed by the well-known chi-squared method and printed out immediately, as were the intensity and degree of linear polarization. The uncertainties varied somewhat between the two channels, among the four filters of each channel, and with intensity of the light. Generally, the uncertainties in degree of polarization were less than 0.5%, but during the early morning when the light was very dim the uncertainties were a factor of 2 or 3 greater than this. The scatter of data points typical of the measurements can be seen in Figs. 8.9 and 8.10.

Data from the shortwave channel of the polarimeter for a case of very low atmospheric turbidity (February 19, 1977) are shown as a function of sun elevation by the plotted points in Fig. 8.9. The curves are fitted to the points by eye. In the range of wavelengths 0.32 to 0.50 μm the polarization P increases monotonically with increasing wavelength, as the effects of multiple scattering

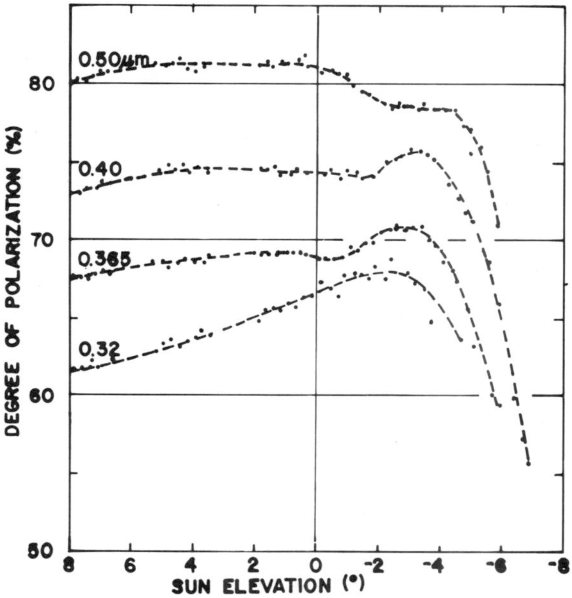

Figure 8.9 Degree of polarization at the zenith as a function of sun elevation as measured in four different wavelengths of the shortwave channel of the polarimeter during the sunrise period at the Mauna Loa Observatory in very clear atmospheric conditions on February 19, 1971. (After Coulson, 1980.)

diminish with increasing wavelength. There is evidence that P for the ultraviolet and short visible regions is controlled largely by scattering in the troposphere for solar depression angles represented here. Bullrich et al. (1969) interpreted a minimum of the polarization curves at solar depressions of 7 ° to 11°, as observed for $\lambda = 0.397$ μm at Davos, Switzerland, to indicate a dust layer in the 80- to 120-km altitude range. This seems questionable, however, in view of a great amount of multiple scattering in the long atmospheric pathlengths for the larger solar depressions. The polarization at $\lambda = 0.636$ μm in their measurements appeared to be unaffected by the high altitude dust.

The curves for $\lambda \leq 0.40$ μm in Fig. 8.9 show polarization maxima to occur at solar depressions of 2° to 4°, in contrast to the situation for the longer wavelengths and to the expectations from simple Rayleigh scattering. Dave (1956) obtained a somewhat similar effect from computations of primary and secondary scattering in a Rayleigh atmosphere, and Steinhorst (1974) obtained a polariza-

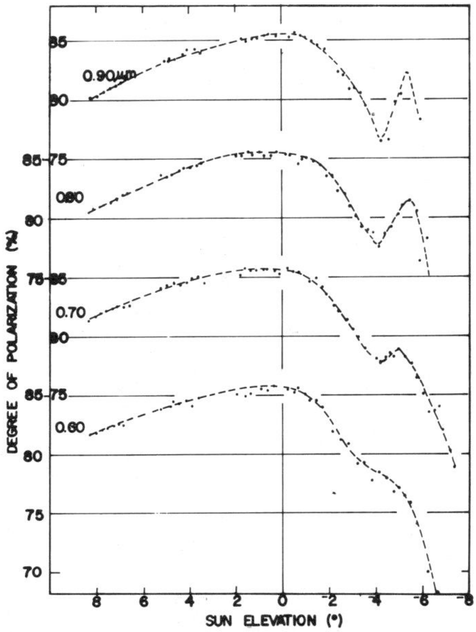

Figure 8.10 Degree of polarization at the zenith as a function of sun elevation as measured in four different wavelengths of the longwave channel of the polarimeter during the sunrise period at the Mauna Loa Observatory in very clear atmospheric conditions on February 19, 1977. (After Coulson, 1980.)

tion maximum at about the right solar angles for atmospheric models with aerosol layers in the 16- to 22-km altitude range. In both of these cases, however, the polarization magnitudes do not correspond well with the measurements.

Data for the longer wavelengths from measurements in the clear atmosphere of February 19, 1977, at Mauna Loa are shown by the points of Fig. 8.10. The curves have been drawn by eye and are separated by staggering of the ordinate scale. Three features of the curves indicate that primary scattering dominates the radiation field at these wavelengths. First, the polarization maximum is located approximately 90° from the sun, in contrast to its position at the shorter wavelengths which show the effects of significant multiple scattering. Second, the magnitude of the maximum is essentially independent of wavelength, a feature which was destroyed by multiple scattering effects at the shorter wavelengths. Finally, the high value of the maximum ($\sim 85.5\%$) shows that Rayleigh-type scattering was predominant in this case of clear atmosphere. The pronounced minimum of the curves at solar depression angles of 3° to 6° was almost certainly produced by aerosols in the so-called Junge layer in the 18- to 22-km altitude range. The magnitude of the minimum varied somewhat from day to day, but its position and general configuration was very constant in clear atmospheric conditions.

For purposes of comparison, Coulson (1980) selected the clearest days which occurred at Mauna Loa during the measurement period and used the twilight data for those days to compute the low turbidity average twilight curves for all eight wavelengths of the measurements. The principal criterion on which these days were selected for low turbidity was that the degree of polarization at $\lambda = 0.80$ μm in the direction of the zenith at sunrise had to be at least 85%. (It may be of interest to note that the highest polarization observed at Mauna Loa was 85.5%, which occurred at sunrise on February 19, 1977.) As mentioned in connection with the twilight intensities, 9 days out of the measurement series met the low turbidity criterion, and the data from those days were used to compute the low turbidity averages listed in Table 8.2. The standard deviations of the data for $\lambda = 0.40$ and 0.70 μm are entered in parentheses in the table. The presunrise minimum and secondary maximum which occur at the longer wavelengths are delineated well, although slight variations of the positions from day to day tend to broaden their outlines somewhat. Both the average data and the data for a single day shown in Figs. 8.9 and 8.10 indicate a zone of transition in the scattering regime between wavelengths of about 0.40 and 0.60 μm. For example, the polarization minimum at solar depression angles of about 4°, with a secondary maximum at a depression of about 5°, is very distinct at $\lambda = 0.60$ μm and all longer wavelengths at which measurements were made. In fact, all of the data for $0.60 \leq \lambda 0.90$ μm are closely congruent, indicating that the scattering process is virtually independent of wavelength in this range. As mentioned previously, the high value of the maximum, the position of the maximum corresponding to a scattering angle of 90°, and the wavelength independence all indicate the strong dominance of primary scattering at these wavelengths. At wavelengths of $\lambda \leq 0.40$ μm, on the other hand, the effects of multiple scattering are evident in the relatively low polarization magnitudes, the strong wavelength dependence, and the migration of the polarization maximum toward smaller scattering angles. The data show that the wavelength of $\lambda = 0.50$ μm, corresponding to a Rayleigh optical thickness of 0.098 at an altitude of 3400 m, is in the middle of the transition zone. The maximum polarization

TABLE 8.2 Average degree of polarization as measured in eight different wavelengths at the Mauna Loa Observatory on 9 very clear days in February and March 1977. The standard deviations of the data for $\lambda=0.40\ \mu m$ and $\lambda=0.071\ \mu m$ are given in parentheses.

Sun elevation (°)	Wavelength (μm)							
	0.320	0.365	0.400	0.500	0.600	0.700	0.800	0.900
−7.0	—	0.513	0.558(—)	0.625	0.676	0.689(—)	0.676	—
−6.5	0.507	0.527	0.584(0.016)	0.664	0.696	0.715(—)	0.720	0.700
−6.0	0.547	0.579	0.631(0.021)	0.691	0.724	0.745(0.006)	0.761	0.719
−5.5	0.577	0.613	0.669(0.021)	0.725	0.749	0.767(0.008)	0.793	0.772
−5.0	0.615	0.645	0.696(0.022)	0.749	0.763	0.778(0.007)	0.798	0.796
−4.5	0.627	0.667	0.715(0.022)	0.761	0.766	0.766(0.008)	0.775	0.774
−4.0	0.637	0.681	0.728(0.021)	0.765	0.769	0.763(0.007)	0.765	0.766
−3.5	0.654	0.688	0.733(0.020)	0.766	0.776	0.778(0.007)	0.780	0.779
−3.0	0.662	0.691	0.734(0.020)	0.765	0.790	0.796(0.005)	0.800	0.800
−2.5	0.666	0.691	0.731(0.20)	0.765	0.814	0.814(0.005)	0.818	0.817
−2.0	0.666	0.688	0.726(0.019)	0.768	0.824	0.828(0.003)	0.831	0.830
−1.5	0.664	0.682	0.722(0.024)	0.775	0.836	0.840(0.002)	0.841	0.840
−1.0	0.660	0.677	0.721(0.020)	0.782	0.843	0.848(0.002)	0.847	0.847
−0.5	0.655	0.673	0.722(0.020)	0.788	0.843	0.852(0.002)	0.851	0.849
0.0	0.649	0.672	0.723(0.020)	0.793	0.850	0.854(0.002)	0.852	0.850
0.5	0.645	0.672	0.725(0.020)	0.796	0.850	0.854(0.003)	0.851	0.850
1.0	0.641	0.673	0.726(0.019)	0.798	0.849	0.853(0.003)	0.849	0.848
1.5	0.637	0.674	0.727(0.019)	0.798	0.847	0.851(0.005)	0.846	0.844
2.0	0.633	0.674	0.728(0.019)	0.798	0.844	0.848(0.006)	0.843	0.841
2.5	0.630	0.674	0.728(0.018)	0.797	0.841	0.846(0.007)	0.840	0.837
3.0	0.626	0.674	0.728(0.018)	0.795	0.838	0.841(0.008)	0.835	0.833
3.5	0.623	0.673	0.727(0.018)	0.793	0.834	0.837(0.009)	0.831	0.828
4.0	0.620	0.672	0.726(0.017)	0.791	0.830	0.833(0.010)	0.826	0.823
4.5	0.617	0.671	0.724(0.017)	0.788	0.827	0.829(0.011)	0.821	0.818
5.0	0.615	0.669	0.723(0.017)	0.784	0.822	0.824(0.012)	0.816	0.813

is relatively high, indicating only a modest amount of multiple scattering, but the character of the curves during the twilight period is more representative of those for short wavelengths than for $\lambda \geq 0.60\ \mu m$. Thus it appears that measurements at $\lambda=0.50\ \mu m$ would not be well adapted to detecting aerosol layers in the upper troposphere and lower stratosphere, at least at the Mauna Loa location. At low-altitude sites, the increased multiple scattering would seem to make 0.50-μm radiation even less sensitive to high-altitude turbidity than it is at Mauna Loa.

The standard deviations of the data, listed for $\lambda=0.40$ and 0.70 μm in Table 8.2, show a significant variability of zenith polarization, even in low turbidity conditions on the 9 selected days. The high consistency of the data at $\lambda=0.70$ μm, in which the standard deviation is generally 1%, is probably an artifact arising from the method of selecting clear days. At $\lambda=0.40$ μm and for shorter wavelengths, the standard deviation of about 2% is an indication of variable conditions in either aerosols in the lower troposphere or reflection from the tops of clouds surrounding the mountain. It gives little information about turbidity in higher atmospheric layers.

As mentioned above, the atmosphere at Mauna Loa may be quite turbid at times. Neglecting for the moment the case of volcanic debris in the stratosphere, water vapor and particulates carried up from the surface layer as weather systems pass over the islands may result in considerable high tropospheric turbidity, and episodes of the transport of dust from the deserts of eastern Asia to Hawaii are not infrequent in spring. Coulson (1980) has discussed twilight measurements taken on 6 days with atmospheric turbidity, 2 of which are represented in Fig. 8.11. During the first week of April 1977, strong vertical motions in the atmosphere were associated with a low-pressure system over Hawaii. Large cumulus clouds penetrated the usual trade wind temperature inversion, scattered thunderstorms occurred, and clouds formed much earlier than usual over the observatory. Twilight measurements at $\lambda=0.70$ μm taken in the middle of the disturbed period (April 3) are shown by the solid curve in Fig. 8.11. The low-turbidity average is included for reference. The behavior of the presunrise polarization was clearly anomalous on April 3, the scatter of data points indicating optical instability, the presunrise minima and maxima being shifted a full 2° toward smaller depression angles, and the sunrise maximum being decreased by 3% from the low turbidity average. The data at the larger solar depression angles indicate turbid layers in the upper troposphere or stratosphere.

The data of May 2, 1977, taken during an episode of Asian dust over Hawaii show conditions in the higher levels of the atmosphere to have been near normal, with the principal anomaly being due to lower atmospheric turbidity. Although the twilight minimum and secondary maximum were displaced somewhat, it was only in the near-sunrise period that the major effect of the dust occurred. As the direct solar beam entered the turbid layers, the degree of polarization plummeted, reaching nearly 17% below the clear day average at a sun elevation

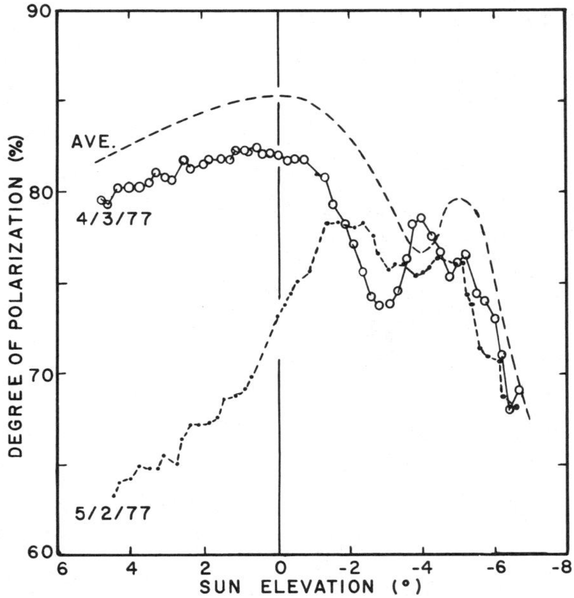

Figure 8.11 Degree of polarization at the zenith as a function of sun elevation as measured at 0.80 μm during the sunrise period on a day (April 3, 1977) of strong vertical motions in the atmosphere and a day (May 2, 1977) with moderate to heavy tropospheric turbidity at the Mauna Loa Observatory. The low turbidity average curve for λ=0.80 μm is given for comparison.

of 5°. The dust was evidently at much lower altitude and had different scattering characteristics than the moist aerosols injected into the upper atmosphere on April 3.

8.3.2 Spectral distribution of twilight polarization

The measurements made at eight different wavelengths from λ=0.32 to 0.90 μm at the Mauna Loa Observatory provide an opportunity to define the spectral distribution of polarization at the zenith for different sun elevations during the sunrise period. The average low turbidity distribution, from data for the 9 clearest days of the 1977 series of observations, is shown for eight different sun elevations in Fig. 8.12. The curves of the diagram are divided into two groups, the upper group, for which the ordinate scale on the right applies, is for sun

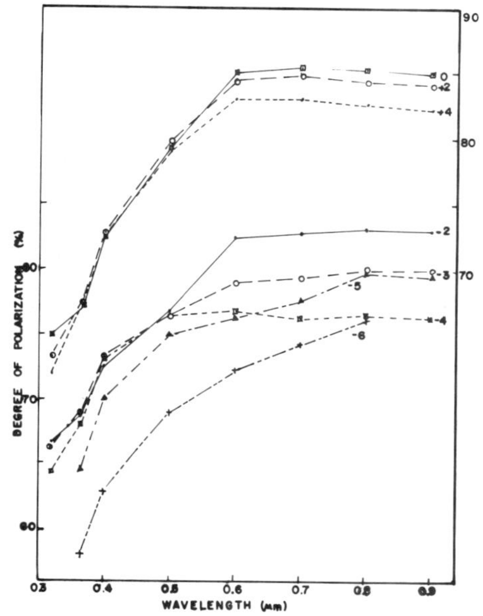

Figure 8.12 Spectral distribution of the degree of polarization at the zenith for eight different sun elevations in clear atmospheric conditions in the sunrise period at the Mauna Loa Observatory. The ordinate scale on the right applies to the upper three curves and that on the left to the lower family of curves. (After Coulson, 1980.)

elevations 0° to 4°, whereas the lower group, with the scale on the left, is for the presunrise period. The strong variation of polarization with wavelength in the range $0.32 \leq \lambda \leq 0.60$ μm is due mainly to secondary and higher order scattering by atmospheric gases. It will be remembered that the degree of polarization resulting from a single scattering event is essentially independent of wavelength, but that the efficiency of Rayleigh scattering is proportional to λ^{-4}. Thus radiation at the shorter wavelengths is depolarized more than that at longer wavelengths in the range in which multiple scattering is significant. The very small amount of dispersion at $\lambda > 0.60$ μm is evidence of mainly primary scattering. Only at solar depressions of 5° and 6° does multiple scattering, perhaps combined with scattering by the small number of aerosol particles in the long atmospheric pathlengths, operate to extend the polarization dispersion to 0.80 μm. For instance, the computations of Dave (1956) show that primary scattering from the 40-km altitude region is dominant over lower level secondary scattering at a 4° depression angle, but the reverse is true when the sun is at 8° below the horizon. It is likely that the spectral dependence of the primary/secondary scattering ratio is responsible for the dispersion shown for solar depressions of 5° and 6°, but further quantitative work is needed to clarify the point.

The change of dispersion from this low turbidity average by the increased upper tropospheric turbidity brought about by atmospheric mixing in the April 3, 1977, weather system over Hawaii is shown by Fig. 8.13. The ordinate scales are the same as in the previous diagram, and the low turbidity curve for solar elevation 0° is shown for comparison. The polarization at the longer wavelengths is much more sensitive to aerosol scattering than that at $\lambda \geq 0.50$ μm. Multiple scattering still dominated the shorter wavelength regime, the aerosols having little added effect. At wavelengths of $\lambda \geq 0.60$ μm, however, illumination of the

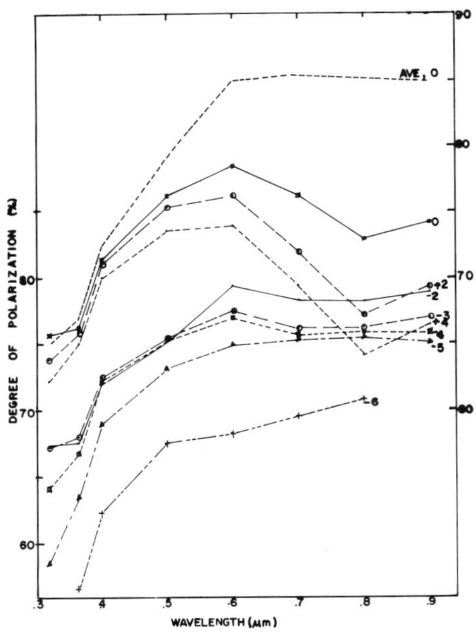

Figure 8.13 Spectral distribution of the degree of polarization at the zenith for eight different sun elevations in the sunrise period in the turbid atmospheric conditions of April 3, 1977, at the Mauna Loa Observatory. The ordinate scale on the right applies to the upper family of curves and that on the left to the lower family. The low turbidity average for 0° sun elevation is shown for comparison. (After Coulson, 1980.)

troposphere by the direct solar beam when the sun was near or above the horizon resulted in considerable depolarization by the tropospheric aerosols, with little change from the previous case at the larger solar depressions.

In an effort to better understand the causes of the observed variations of zenith polarization during twilight, Coulson (1980) computed the correlation coefficients between the maximum and minimum polarization magnitudes with other meteorological parameters such as precipitable water, amount of clouds below the Observatory, surface aerosol scattering coefficient, etc., but the results indicated no significant relationships among the data. This is not surprising in view of the decoupling between the lower and upper troposphere brought about by the trade wind inversion existing at the latitude of Hawaii. This decoupling would destroy any correlation between skylight polarization and air mass, such as that found by Dietz and Kohl (1968) at a higher latitude location.

8.3.3 Characterization of stratospheric aerosols from twilight polarization

The most important aspect of twilight measurements is their use in characterizing aerosols in the stratosphere. Theoretically it should be possible to determine atmospheric properties completely by skylight measurements as a function of time or solar depression during twilight. The problem is complicated, however, by effects of multiple scattering, illumination of the aerosol by light scattered upward from the underlying atmosphere, atmospheric inhomogeneities in the long path of the incident radiation traversing the atmosphere, effects of clouds, and other aspects. Because of these difficulties, the twilight method has not yet been developed to its fullest potential. However, a step in this direction was made by Steinhorst (1977), who interpreted zenith polarization measurements at several wavelengths made during twilight to deduce aerosol concentrations as a function of height for volcanic and nonvolcanic conditions. The main steps in Steinhorst's method are the following:

1. Obtain polarization and intensity data in the direction of the zenith as a function of time during twilight. Emphasis in the measurements is on red and near-infrared wavelengths, as shorter wavelengths are subject to more multiple scattering and more attenuation by the molecular atmosphere.

2. Use model calculations to estimate and subtract out the intensity components due to multiple scattering. If P is the measured degree of polarization and subscripts p and m indicate primary and multiple scattering, respectively, then we have the relation

$$P = \frac{P_p I_p + P_m I_m}{I_p + I_m} \tag{8.1}$$

which reduces to

$$P_p = P\left(1 + \frac{I_m}{I_p}\right) - P_m\left(\frac{I_m}{I_p}\right) \tag{8.2}$$

The quantities P_m and I_m/I_p are obtained from model calculations, and the resulting P_p is due only to primary scattering by the atmosphere. It is used in later steps to select the most appropriate model for the stratospheric aerosol.

3. Assume, on the basis of previous work, a size-frequency distribution and limiting sizes of the aerosol particles. Steinhorst assumed a lognormal distribution with limits of r_o=0.17 μm and r_o=0.5 μm.

4. From calculations based on several such distributions compute the polarization P_m of multiply scattered light and relative contribution of intensity due to multiple scattering I_m/I_p. On introduction of these quantities into Eq. (8.2), together with the measured polarization P as a function of solar depression angle, the polarization of the primary scattered light P_p can be determined for the different atmospheric models.

5. Determine the most probable median radius r_o for the lognormal distribution existing at the time of the measurement by selecting models with several values of r_o and increasing the stratospheric aerosol concentration for each until agreement between calculated and measured degrees of polarization is obtained for one wavelength.

6. For this wavelength subtract out, by means of Eq. (8.2), the effects of secondary scattering and obtain the polarization P_p due only to primary scattering for the several models and concentrations.

7. Compare calculations with measurements for one or more additional wavelengths to finally obtain the most probable value of r_o.

While this procedure defines the most probable aerosol concentration and size distribution at a given level in the atmosphere corresponding to a given solar depression angle, further efforts are necessary to develop the vertical profile of aerosol concentration $n(h)$. Steinhorst (1977) utilized an iterative method for approximating this profile. Modifying his notation somewhat, we have the intensity at the zenith during twilight for solar depression D due to primary scattering through the layer from h_1 to h_2 as

$$I_D = \int_{h_1}^{h_2} (f_R + f_A)\, e^{-\tau}\, dh \tag{8.3}$$

Here τ is total optical thickness between aerosol layer and measuring instrument. It is the sum of Rayleigh optical thickness (with ozone) τ^R and aerosol optical thickness τ^A. We now define $f_R(h,D)$ as the normalized Rayleigh scattering function for a unit volume of air at height h, and $f_A(h,D)$ as the analagous quantity for aerosols. If $f_A(0,D)$ is the scattering function of a typical particle, then in general $f_A(h,D) = f_A(0,D)n(h)$, where $n(h)$ is the number concentration of particles. For simplicity we further define $F_R(h,D) = f_R(h,D)\ e^{-\tau^R}$ and $F_A(h,D) = f_A(0,D)\ e^{-\tau^R}$. Then Eq. (8.3) becomes

$$I_D = \int_{h_1}^{h_2} F_R(h,D)\ e^{-\tau^A}\, dh + \int_{h_1}^{h_2} F_A(h,D)\ e^{-\tau^A} n(h)\, dh \tag{8.4}$$

If, as usual, we let $I_\perp$ and $I_\parallel$ represent the orthogonal components of intensity normal and parallel, respectively, to the scattering plane, and introduce them into Eq. (8.4), we have the polarization in the form

$$P_D = \frac{\int_{h_1}^{h_2} [F_{R\perp}(h,D) - F_{R\parallel}(h,D)]e^{-\tau^A} dh + \int_{h_1}^{h_2} [F_{A\perp}(h,D) - F_{A\parallel}(h,D)]e^{-\tau^A} n(h) dh}{\int_{h_1}^{h_2} F_R(h,D) e^{-\tau^A} dh + \int_{h_1}^{h_2} F_A(h,D) e^{-\tau^A} n(h) dh} \tag{8.5}$$

Since the degree of polarization for primary scattering in the Rayleigh atmosphere is a function of scattering angle only and for observations at the zenith during twilight the scattering angle varies little from 90°, the main changes in observed polarization are due to the effects of aerorols. Aside from simple attenuation, aerosol effects are represented by the second integrals of numerator and denominator of Eq. (8.5).

The aerosol profile, represented by $n(h)$, cannot be determined directly from Eq. (8.5), but as shown by Steinhorst (1977), an iterative method based on comparisons of the measured and calculated degrees of polarization at the various solar depression angles can be applied to yield a reasonably good approximation to the actual profile. The method requires that the height range for the solar beam for each solar depression angle be defined as an initial input. The height ranges given by Steinhorst for a wavelength of $\lambda = 0.85$ μm, as determined by the method of Grassl (1970), are given in Table 8.3. The aerosols were taken to extend to 35 km, with only molecular scattering above that level. The iterative procedure by which the final aerosol profile $n(h)$ is obtained is based on the relation

$$n_i(h) = \frac{P_D}{P_D'} n_{i-1}(h) \tag{8.6}$$

where P_D and P_D' are the measured and computed degrees of polarization, respectively, and i is the order of iteration. To start the process, an initial profile is assumed and used for the first computation of P_D'. On the basis of this computation a corrected profile is obtained for the second computation of P_D'. This

TABLE 8.3 Height ranges at the various solar depression angles applicable to scattering of radiation at a wavelength of 0.85 μm. Data from Steinhorst (1977).

Solar depression (°)	Height range (km)	Solar depression (°)	Height range (km)
1.0	5–8	4.5	21–23
2.0	8–11	5.0	23–28
2.5	11–13	6.0	28–35
3.0	13–15	6.5	35–40
3.5	15–17	7.0	40–45
4.0	17–21		

procedure is continued until the value of the ratio $|P'-P|/P'$ falls below a specified limit, at which time the best approximation to the actual profile has been obtained. Steinhorst tested the method on a given aerosol profile with the results shown in Fig. 8.14. The profile computed by the iteration method is an adequate approximation to that given, regardless of the type of initial profile assumed. The successive corrections eventually yield the main features of the given profile from 10-km to 30-km altitudes. Outside these limits the results obtained were not found to be useful.

The primary sources of error in the method are the uncertainties in eliminating multiple scattering in the computations, the poor height resolution of the twilight intensity profiles, unknown effects of clouds, and errors in the iteration method itself. Additional errors arise from unknown and variable sizes

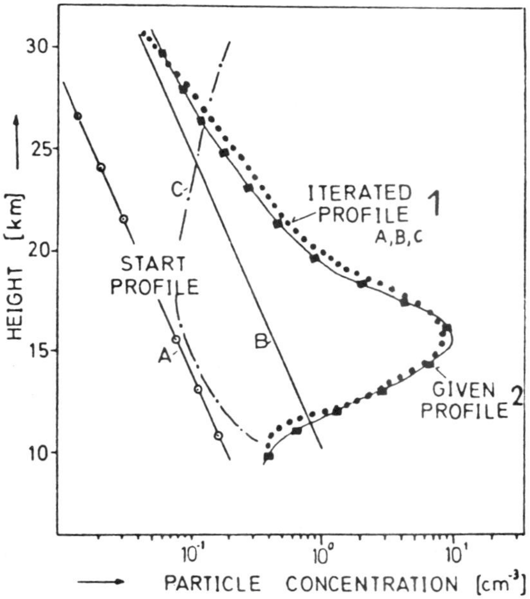

Figure 8.14 The height distribution of aerosol concentrations used by Steinhorst (1977) for testing the iteration method for determining aerosol profiles from twilight polarization measurement. Curve 2 is the assumed profile, curves A, B, and C are assumed initial profiles, and curve 1 is the profile determined by iterative calculations. (After Steinhorst, 1977.)

and indices of refraction of the aerosols. In spite of these difficulties, however, Steinhorst (1977) obtained at least reasonable profiles from twilight measurements made during three different periods at two different mountain sites and an urban location between 1968 and 1974. For instance, twilight polarization measurements made at Davos in 1968 and processed by the iteration method resulted in the aerosol profile shown in Fig. 8.15. The thin error bars show the estimated errors arising from inaccuracies in eliminating the effects of multiple scattering from the data, and the thick error bars are the estimated errors from all causes. The errors increase rapidly for altitudes above 25 km. Only Rayleigh scattering was assumed above 35 km, but this assumption had little effect on concentrations below 30 km. A greater number of height intervals than those of Fig. 8.15 would not improve the results because of the poor resolution of the twilight intensity profiles.

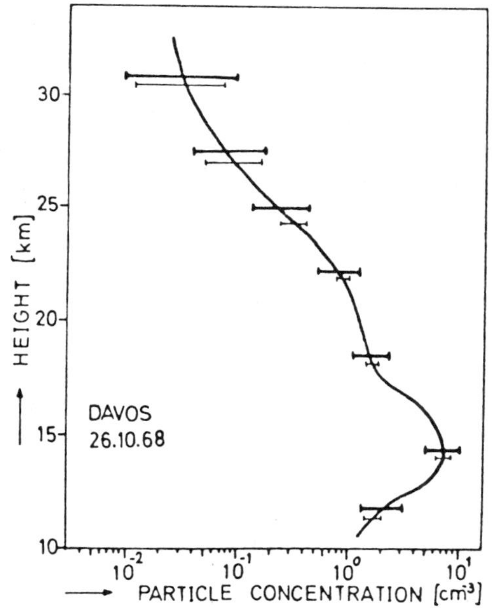

Figure 8.15 The height distributions of aerosol concentrations computed by the iteration method from zenith polarization measurements in Davos, Switzerland, in 1968. The thin error bars indicate the uncertainties arising from multiple scattering and the thick bars are for uncertainties from all causes. (After Steinhorst, 1977.)

The profile of Fig. 8.15, as well as others given by Steinhorst (1977), indicate changes of aerosol distributions with time. But, of course, there were no real measurements of the aerosols existing at the time. We can conclude that the method developed by Steinhorst has interesting possibilities but it has significant limitations, mainly in the assumed size-frequency distributions of the aerosol particles and in the neglect of scattering orders higher than the second.

References

Ashburn, E.V. (1952): The density of the upper atmosphere and the brightness of the twilight sky, *J. Geophys. Res.*, *57*, 85.

Bigg, E.K. (1956): Detection of atmospheric dust and temperature inversions by twilight scattering, *Nature, 177 (4498)*, 77–79.

Blattner, W.G.M. (1976): Light scattering in spherical-shell atmospheres, in Bolle (ed.) *Radiation in the Atmosphere,* Science Press, Princeton, 268–270.

Blattner, W.G., H.G. Horak, D.G. Collins, and M.B. Wells (1974): Monte Carlo studies of sky radiation at twilight, *Appl. Opt.*, *13,* 534–547.

Bullrich, K., et al. (1969): Research on atmospheric optical radiation transmission, Final Rept., Contr. F–61052–67–C–0046, Gutenberg University, Mainz, West Germany.

Coulson, K.L. (1980): Characteristics of skylight at the zenith during twilight as indicators of atmospheric turbidity. 1: Degree of polarization, *Appl. Opt.*, *19,* 3469–3480.

Coulson, K.L. (1981): Characteristics of skylight at the zenith during twilight as indicators of atmospheric turbidity. 2: Intensity and color ratio, *Appl. Opt.*, *20,* 1516–1524.

Coulson, K.L. (1983): Effects of the El Chichon volcanic cloud in the stratosphere on the polarization of light from the sky, *Appl. Opt.*, *22,* 1036–1050.

Dave, J.V. (1956): On the intensity and polarization of the light from the sky during twilight, *Proc. Indian Acad. Sci.*, *XLIII, No. 6, Sec. A,* 336–358.

Dave, J.V., and C.L. Mateer (1968): The effect of stratospheric dust on the color of the twilight sky, *J. Geophys. Res.*, *73,* 6897–6913.

Dave, J.V., and C.L. Mateer (1980): Private communication.

Dave, J.V., and K.R. Ramanathan (1956): On the intensity and polarization of the light from the sky during twilight, *Proc. Indian Acad. Sci., XLIII, No. 2, Sec. A,* 67–78.

Dietze, G. (1969): On the aerosol factor in the twilight method, *PAGEOP, 77,* 159–167.

Dietze, G., and G. Kohl (1968): Einfluss des Luftmassentyps auf die polarisation des Dammerungslichts, *Zeit. fur Meteor., 20,* 209–213.

Divari, N.B. (1967): Calculations of the polarization of the light from the twilight sky, *Atm. Oceanic Phys., 3,* 507–515.

Giovanelli, R.G. (1954): The attenuation of light by meteoric dust in the upper atmosphere, *J. Austral. Phys., 7,* 641–648.

Grandmontagne, R. (1941): Etudes photoelectriques sur la lumiere du ceil, *Ann. de Phys., Ser. 11, 16,* 253–305.

Grassl, H. (1970): Bestimmung der Grossenverteilung aus spektralen Transmissionsmessungen, *Beitr. Phys. Atmos., 43,* 255–284.

Gruner, P. (1942): Dammerungscheinungen, *Hdb. d. Geophys., VIII, Physik der Atmosphare I, Berlin, Borntraeger,* 432–526.

Koomen, M.J., C. Lock, D.M. Packer, R. Scolnik, R. Tousey, and E.O. Hulbert (1952): Measurements of the brightness of the twilight sky, *J. Opt. Soc. Am., 42,* 353–356.

Megrelishvili, T.G. (1958): On the possibility of investigating aerosol layers by the crepuscular method, *Izv. Akad. Nauk, USSR, Series Geofiz., (4),* 560–563.

Megrelishvili, T.G. (1981): *Regularities of Variations of Scattered Light and Emission of the Earth's Twilight Atmosphere,* Tbilisi, Georgian SSR, Metsniyereba Pub. House, 273 pp.

Megrelishvili, T.G., I.G. Mel'nikova, G.V. Rozenberg, and A.V. Khovanskiy (1978): Mean twilight at the zenith and vertical profile of scattering coefficient according to observations at the Abastumani Astrophysical Observatory (1942–1952), *Izv. Akad. Nauk, USSR, Atmospheric and Oceanic Physics, 14,* 805–812.

Mel'nikova, I.G., T.G. Megrelishvili, and G.V. Rozenberg (1980): Correlational analysis of light scattering coefficient stratification in the mesosphere, *Izv. Akad. Nauk, USSR, Atmospheric and Oceanic Physics, 16,* 344–348.

Nazaraliyev, M.A., and G.V. Rozenberg (1977): Monte Carlo simulation of twilight observations, *Izv. Akad. Nauk, USSR, Atmospheric and Oceanic Physics, 13,* 94–100.

Rozenberg, G.V. (1963): *Twilight,* Fizmatgiz, Moscow.

Rozenberg, G.V., and G.G. Mikirtumova (1976): Mesospheric turbidity from twilight sounding data, *Izv. Akad. Nauk, USSR, Atmospheric and Oceanic Physics, 12,* 631–636.

Robley, R. (1950): La diffusion multiple dans l'atmosphere deduite des observations crepusculaires, *Annales de Geophysique, 6,* 191–211.

Shah, G.M. (1969): Enhanced twilight glow caused by the volcanic eruption on Bali Island in March and September, 1963, *Tellus, XXI,* 636–640.

Steinhorst, G. (1971): Einfluss atmospharischer aerolschichten auf den Himmelsstrahlungspolarisationsgrad wahrend der dammerung, *Beitr. Phys. Atmos., 44,* 279–292.

Steinhorst, G. (1974): Recent measurements of twilight polarization, *Appl. Opt., 13,* 219–220.

Steinhorst, G. (1977): Stratospheric aerosol concentrations determined by an iterative method from twilight polarization measurements, *Contrib. to Atm. Phys., 50:4,* 508–523.

Volz, F.E. (1969): Twilights and stratospheric dust before and after the Agung eruption, *Appl. Opt., 8,* 2505–2517.

Volz, F.E. (1970): On dust in the tropical midlatitude stratosphere from recent twilight measurements, *J. Geophys. Res., 75,* 1641–1646.

Volz, F.E. (1981): Zenith polarization and intensity during twilight, *Appl. Opt., 24,* 4172–4174.

Volz, F., and R.M. Goody (1962): The intensity of twilight and upper atmospheric dust, *J. Atmos. Sci., 19,* 385–406.

CHAPTER 9

RADIATIVE EFFECTS OF VOLCANIC ERUPTIONS

9.1 Volcanoes in history

It is well known that the material injected into the upper atmosphere by large volcanic eruptions causes many optical effects, some of which are extremely beautiful and awe-inspiring, and volcanic eruptions are thought to be a principal cause of changes of climate. Unfortunately, this latter subject is beyond the scope of the present discussion, but reviews of volcanoes and climate are given by Mitchell (1973), Pollack et al. (1976), Toon and Pollack (1982), Budyko (1984), and Parker (1985), among many others. Optical effects usually take the form of intense colors, usually red and yellow, in the sky during twilight; the change of the normal deep blue of the daytime sky to a washed out whitish blue or even gray; and prolongation of sunrise or sunset colors of up to an hour beyond their normal period. Occasionally, the sky or the disk of the sun or moon is observed to have a green, blue, or violet cast, and, in some cases, a dry haze apparently of volcanic origin, has covered wide areas of Europe for weeks or months as a result of a major eruption.

The most completely documented eruption in history is the famous eruption of the Indonesian volcano Krakatoa in 1883. Because of the widespread destruction, worldwide optical effects, and generally intense interest generated by this gigantic cataclysm of nature, the Royal Society of London appointed the Krakatoa Committee to collect and synthesize information on the eruption and its aftereffects. The report of the Committee (Symons, 1888) is a unique document of 494 pages that summarizes not only the information on Krakatoa but also includes a listing of some of the effects of 247 other eruptions over the world dating back to the year 1510. Of these, 39 were recorded as being violent in character.

Cataclysmic as it was, Krakatoa was not the most violent eruption on record. At least three eruptions of more violent character preceded it, two of the three occurring in 1783—just 100 years before Krakatoa. The third was the gigantic eruption of Tambora in 1815, which has been credited with causing such cold conditions in the northeastern United States and Europe that the year 1816 has been termed "the year without a summer." Just whether the observed climatic anomaly was volcanically initiated is still a matter of controversy, however. "The most frightful eruption on record," according to the Krakatoa report (p. 388), was that of Asama in Japan in 1783. "Rocks from 40 to 80 feet in some of their dimensions were hurled through the air in all directions. Towns and villages were buried. One stone is said to have measured 264 feet by 120 feet; it fell into a river and looked like an island." A second major eruption of 1783 was that of Skaptar Jokull, Iceland. The main meteorological phenomenon resulting from these 1783 eruptions, presumably mainly from the latter, was a thick dry mist or fog which spread over Europe during the summer of 1783 and persisted for 1 to 4 months in different locations. The sun was often obscured by the haze at the time of sunrise and sunset, and even during the day it was either pale like the moon or red-colored and dim enough to be viewed with the naked eye. There appear to be no reports of twilight colors, such as were observed after Krakatoa, resulting from these gigantic 1783 eruptions.

Indices of the strength or severity of volcanoes are useful for geologic and climatic studies, as a gauge of real destruction or loss of life and other purposes. Two such indices are the dust veil index (DVI) and the volcanic explosivity index (VEI) developed by Lamb (1970, 1977) and Newhall and Self (1982), respectively. A scale of Tsuya (1955) is based on the total volume of ejecta resulting from the eruptions, but his approach is less useful for meteorological or climatic analyses than those mentioned above. For studies of optical effects, volcanic effects on solar radiation and climate, and similar meteorological or climatological investigations, the dust veil index of Lamb (1970) is most useful. Its shortcomings arise more from lack of data than from the basic concepts. The DVI has three simple forms, the choice among the three for any particular event being determined entirely on the basis of the available data. They are written in Lamb's nomenclature as:

$$\mathrm{DVI} = 0.97\, R_{D\max}\, E_{\max}\, t_{ms} \tag{9.1}$$

$$\mathrm{DVI} = 52.5\ T_{D\max}\ E_{\max}\ t_{ms} \tag{9.2}$$

$$\mathrm{DVI} = 4.4\ q\ E_{\max}\ t_{ms} \tag{9.3}$$

Here $E_{\max}$ is the maximum fraction of the globe covered by the dust veil and t_{ms} is the number of months between the time of the eruption and the last observation of the dust veil. $R_{D\max}$ is the greatest percent depletion, expressed as a monthly average, of the direct solar radiation, $T_{D\max}$ is the lowering of the average surface temperature in °C over the midlatitude zone, and q is the cubic kilometers of solid matter dispersed into the atmosphere as dust. The constants in the expressions are set to yield a DVI of 1000 for the 1883 eruption of Krakatoa.

Lamb (1970) considers the first of the relationships to be most reliable for those cases with the required input data, then the second, and finally the third. In all these cases the paucity of data forces much estimation of the parameters, and the final value is highly nonprecise and indicative only. However, even gross approximations, if developed in a consistent manner, are useful in ranking the many volcanic eruptions as to their probable atmospheric effects.

Lamb (1970) computed the DVI for almost 400 volcanic eruptions between 1500 and 1963, which gave DVI values of a minimum of 100 to a maximum of 4000. These are relative to the value of 1000 for Krakatoa. The eruptions with DVI of 800 or greater are listed, along with additional information on the eruption or its effects, in Table 9.1. Although Krakatoa is the most famous of the eruptions listed, 12 others have DVI the same as that of Krakatoa and five are given as having produced greater quantities of atmospheric dust than Krakatoa. It is notable that the "most frightful eruption on record" (Asama, Japan, 1783) yielded a DVI of only 600.

Because of the great scientific, sociological, and economic importance of volcanoes, as well as the spectacular optical phenomena and unleashed fury of the eruption that fascinates the populace, several sources of information are presently available (Simkin and Fiske, 1983; Rampino and Self, 1982; Wexler, 1951). However, none can compare in scope and synthesis with the report of the Krakatoa Committee (Symons, 1888). The most current information on volcanoes and other geophysical events is provided by the *Bulletin of the Scientific Event Alert Network (SEAN)* of the Smithsonian Institution, Washington, D.C.

TABLE 9.1 Volcanic eruptions and dust veil indices (DVI) as given by Lamb (1970) for all known eruptions since 1955 giving DVI of 800 or greater. The DVI for the 1883 eruption of Krakatoa was set at 1000. Numbers in parentheses are estimates.

Year	Volcano	Location	DVI	Remarks
1500	Java	6–8 °S, 106–114 °E	1000	
1586	Kelud, Java	8 °S, 112.5 °E	1000	
1593	Riuggit, Java	8 °S, 114 °E	1000	
1614	Little Sunda Islands	8 °S, 118 °E	1000	
1625	Katla, Iceland	63.5 °N, 19 °W	(800)	Quantities of ash fell up to 1000 km distance
1641	Awu, Celebes	2.5 °N, 125.5 °W	1000	
1660	Omate, Peru	16.5 °N, 71 °W	1000	
1660	Katla, Iceland	63.5 °N, 19 °W	800	Quantities of ash fell up to 1000 km; sky a ''bath of lurid fire'' in Denmark
1673	Gamma Kunorra, Moluccas	1.5 °N, 127.5 °E	1000	
1680	Tongkoko, Celebes	1.5 °N, 125 °E	1000	
1752	Little Sunda Islands	8 °S, 118 °E	1000	Possibly Tambora
1755	Katla, Iceland	63.5 °N, 19 °W	1200	Haloes around sun and moon; fiery red sunsets in Europe and Atlantic; red glow till 2½ hours after sunset in Ireland.
1766	Mayon, Philippines	13.5 °N, 123.5 °E	2300	Classed as violent
1768	Cotopaxi, Ecuador	1 °S, 78 °W	900	
1783	Eldeyjar, Skapar, Jokull, Iceland	63.5–64 °N, 18–23 °W	2300	Dust destroyed crops in Scotland; thick dry haze from North Africa to Scandinavia, sun obscured at elevations below 17° in France; sun dim and red or bluish-white; in England sun coppery colored until it was 20° high; colored sunsets and unusual glare of sky at night
1795	Pogrumni, Aleutian Is.	55 °N, 165 °W	1000	
1815	Tambora, Little Sunda Islands	8 °S, 118 °E	3000	Eruptions continued until 1819; remarkable sunsets in London; responsible for ''year without a summer'' (see text)
1835	Coseguina, Nicaragua	13 °N, 87.5 °W	(4000)	Great explosion blew away whole top of mountain
1846	Amargura Is., S. Pacific	18 °S, 174 °W	1000	It was suggested that dust reached N. Hemisphere after delay of more than a year
1861	Makjan, Moluccas	0.5 °N, 127.5 °E	(800)	Fine sunsets with unusual colors in Europe, 1862 and 1863.

TABLE 9.1 (Continued)

Year	Volcano	Location	DVI	Remarks
1875	Askja, Iceland	65°N, 17°W	1000	Much of the 3- to 4-km^3 ejecta in form of dust; dust deposited in Scandinavia; air over Britain hazy up to great heights
1883	Krakatoa	6°S, 105.5°E	1000	Much of island disappeared and new islands formed; dust fell up to 2000-km distance; dust column observed up to 27 km; direct solar beam diminished 20% to 30% in mean monthly values; up to 20% for 2–3 years; optical effects over much of globe
1886	Tarawer, New Zealand	38.5°S, 176.5°E	(800)	
1963	Agung, Bali	8.5°N, 115.5°E	800	Dust veil encircled earth and spread to all latitudes; remarkable sunsets and prolonged twilight glows over N. America and Europe; direct solar beam decreased 30% in some locations; possible 0.4°C decrease of global temperatures

A particularly useful feature of this Bulletin is a listing, in international scope, of individuals from whom additional information may be obtained. A second source of current information is the publication *EOS* of the American Geophysical Union. The more detailed survey and research reports are, of course, published in scientific journals, of which an unprecedented number are presently in existence. Unfortunately, a valuable source of data on volcanoes in the past, namely eyewitness accounts, has largely dried up. An observer, no matter how astute, is not likely to write a letter about a spectacular sunset in Massachusetts if he has already seen on the evening news a week earlier a comparable scene in Arizona. This situation not only robs the record of information on the spread of the dust cloud, but it also tends to diminish the thrill and exultation the nature lover receives from discovering for himself the awe-inspiring sequence of reds, yellows, violets, and other colors of a volcanic sunset. We are all the poorer for it.

9.2 Volcanoes of the last century

The most recent period of just over 100 years has seen four eruptions that are considered major with respect to the dust and attendant gases they have injected into the atmosphere: Krakatoa, 1883; Katmai, 1912; Agung, 1963; and El Chichon, 1982. No fewer than 49 additional eruptions during the period have injected substantial volumes of material into the troposphere, with definite injections into the stratosphere, according to the listing of Newhall and Self (1982). Although the individual eruptions in this latter category probably are too small to be climatically significant, their cumulative effect cannot be so readily neglected, and the four major eruptions are each thought to have had climatic consequences.

The optical and other radiative aspects of the major eruptions of the last century are outlined in the following sections.

9.2.1 Krakatoa, 1883

This most famous eruption of the mountain of the same name in the Sunda Straits between Sumatra and Java was most violent on August 26 and 27, 1883. Copious amounts of information gathered from eyewitness accounts of the event and from observers of atmospheric effects over the world was synthesized and published by the Krakatoa Committee of the Royal Society of London (Symons, 1888) only 5 years after the event. Several radiative effects of the dust cloud thrown violently upward into the statosphere by this gigantic cataclysm deserve comment.

9.2.1.1 Sunrise and sunset glows

The spectacular colors that lit up the sky during twilight for several months after the August 26, 1883, eruptions of Krakatoa generated comments from essentially all parts of the civilized world well into the following year. During the first month to 6 weeks following the eruption, the dust cloud was confined mainly to the low-latitude regions of the earth. According to the report of the Krakatoa Committee (Symons, 1888), the dust was caught up in the high-altitude easterly winds and made two complete circuits of the globe in 26 days for an average speed of 115 to 133 km/hr (72 to 82 mph). The existence of such strong easterly circulation was unknown before the discovery of these "Krakatoa winds." With changing atmospheric circulation due to the onset of fall and winter of 1883, the dust cloud was fragmented and dispersed to higher latitudes. In the winter and spring of 1884, twilight glows and other optical effects produced by the stratospheric cloud were reported from most parts of the world, but an analysis of the observations shows that after a few months the fragmented nature of the cloud caused the phenomena to have only variable and sporadic effects at the higher latitudes.

The specific colors constituting the twilight glow varied widely with location and time. Not only did the cloud vary in density and height, but, in addition, colors are difficult to describe verbally and their perception is highly subjective. Thus there is wide diversity in the descriptions collected by the Krakatoa Committee. Most observers report a predominance of red colors, the tints including flame red, crimson, brick red, etc., the last of these being the last glow on the horizon as twilight fades into darkness. Others colors frequently seen are yellow, silvery white, copper color, and, more rarely, various shades of blue and green. Violet or delicate pink colors are common, and in fact a large portion of the sky is suffused with a beautiful pink veil during the latter stages of many twilight spectaculars. Such was the case in the sketch of the colored sky by F.A.R. Russell in Surrey, England, in November 1883, which is reproduced here as Fig. 9.1. The colors at that time were somewhat atypical because of the

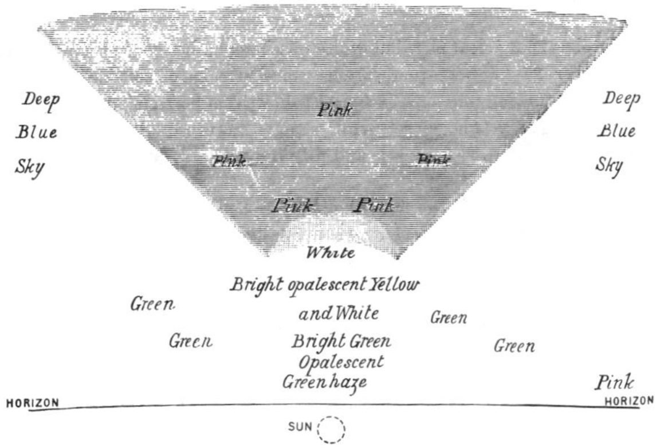

Figure 9.1 Sketch of the twilight colors seen by F.A.R. Russell in Surrey, England, after the eruption of Krakatoa (Russell and Archibald, 1888).

absence of various shades of red in the scene. Some of the spectaular skies seen in London from late 1883 to 1886 were captured in over 530 paintings of the evening twilight by William Ashcroft (South Kensington Museum). The finest colors were in midwinter of 1883–1884 in the "Blood Afterglows." At later times "Amber Afterglows" were more common and their appearance was "fitful."

9.2.1.2 Blue and green sun and moon

The blue and green colors of sun and moon, being rare, as indicated by the expression "once in a blue moon," generated much comment. Actually, these effects may be observed in nonvolcanic clouds as well (e.g., the blue moon seen in Europe as smoke from Canadian forest fires was driven across the Atlantic). After Krakatoa these exotic colors of sun and moon were most common in the tropics before dispersal of the cloud to higher latitudes. In the first few days, the luminaries were reported as being blue more often than green, whereas the reverse prevailed in observations of later dates and higher latitudes. Green colors were seen most often when the sun was near the horizon. One observation of the green sun viewed through a spectroscope showed strong general attenuation in the red region beyond $\lambda=0.621$ μm, with lesser attenuation at $\lambda=0.428$ μm. If the attenuation were due mainly to absorption, solid particles would be indicated, but the ubiquitous red colors of the twilight glows would show a large amount of scattering of these wavelengths. Thus it is not clear whether the scattering particles were solid or liquid, but we can say in the light of later knowledge that they were probably a mixture, with liquid predominant in the later stages.

The normal optical thickness of the stratospheric cloud had been deduced on the basis of largely qualitative evidence by Deirmendjian (1973). By subtracting out the optical thickness of the Rayleigh atmosphere and that of a normal background aerosol, Deirmendjian computed an optical thickness of the cloud to be $\tau^C=0.63$ at $\lambda=0.5$ μm. While this seems to be a reasonable value for the newly generated cloud, the assumptions on which it is based do not correspond well with the situation existing at the time. For instance, the optical thickness which is independent of wavelength in the visual range (Deirmendjian, p.272) does not correspond with the red absorption mentioned above, and it is difficult to see how the cloud acting as a neutral filter could produce a blue-green color of the sun. A neutral filter would make the sun appear either in its normal yellow color, or white if the extra-atmosphere spectrum were applicable. In neither case would it be green. In reality it is likely that the optical thickness was similar to that of the El Chichon cloud of 1982 (see section 9.2.4), in which case τ^C at $\lambda = 0.490$ μm reached 0.49 to 0.60 in the young cloud. This is not too different from $\tau^C = 0.63$ of Deirmendjian for Krakatoa. A very significant decrease at longer and shorter wavelengths was observed in the case of El Chichon, but, of course, the El Chichon data were not available at the time of Deirmendjian's analysis.

9.2.1.3 Altitude of the Krakatoa cloud

One feature that is universal in volcanic sunsets is a prolongation of the twilight glow much beyond that of a normal sunset. For instance, several twilight glows persisted to solar depressions of more than 20° during the period August to November 1883. It will be remembered that astronomical twilight ends at solar depression 18°; at this time the sky is supposed to be devoid of sunlight. This criterion is not met in many cases of volcanic twilights. From 29 observations considered reliable by the Krakatoa Committee, the average solar depression for the end of twilight glow during August 27 to November 30, 1883, was 17.8°, with a range of angles 13.4° to 21.8°. During the following 2 months a decrease of height and spreading of the cloud caused an earlier termination of the glow, the last remnants disappearing at an average solar depression of 14.8° for 25 observations. A positive correlation between the apparent intensity of the colors and the duration of the glow was noted in the summary.

The altitude of the strata causing twilight glows was the subject of much analysis and comment by the Krakatoa Committee. The concept adopted was that the entire glow phenomenon consisted of a primary glow produced by primary scattering of the direct solar beam and a secondary glow resulting from secondary and higher order scattering of light from the primary glow when the solar depression angle was too great for illumination of the cloud by direct sunlight. The secondary glow was considered responsible for the prolongation of twilight by periods of up to an hour beyond the time of cessation of normal twilights. According to this analysis, the height of the upper part of the stratum was initially at 121,000 ft (36.9 km), or some 20 km above the tropical tropopause, and the lower portion was at about 104,000 ft (31.7 km). These upper and lower limits decreased over the succeeding 5 months to about 64,000 ft (19.5 km) and 56,000 ft (17.1 km), respectively. By that time the mean duration of twilight yielded a height of 70,000 ft (21.3 km), with the longer durations indicating 87,500 ft (26.7 km). The decrease in height was attributed to gravitational fall of the larger particles, but it was thought that from the development of Bishop's ring (see next section) that a larger portion of the finer particles remained at about their original altitude for some time. Fragmentation of the cloud by varying winds in the upper atmosphere caused sporadic optical effects in different locations over the world.

9.2.1.4 Bishop's ring

The first mention of a "faint halo" around the sun as seen through the Krakatoa dust cloud was in the Japanese *Gazette,* but the first detailed observation was that of Sereno Bishop in Honolulu just 10 days after the August 26 eruption. Bishop observed the ring to exist more or less continuously in Honolulu over periods into 1886, and it was observed, although less regularly, in other locations. Since then the Bishop's ring is recognized as a fairly common feature associated with volcanic clouds.

Although descriptions of the phenomenon vary widely, the ring is a diffuse white or faintly colored band, circular except when near the horizon, extending from an inner radius of 10° to 12° to an outer radius of 22° to 23°. Bishop himself gave limits of 20° to 30°. The inner border is white and the outer edge is faint reddish or brownish in color. This order of colors is that of a corona, and opposite that of the ordinary halo arising by refraction in ice crystals. Being a diffraction phenomenon, the corona is sharpest and the colors brightest if made by particles of uniform size; the smaller the drops, the larger the corona. If different sizes are present, the corona of one sized drop overlaps that of a different sized drop, thereby making the corona diffuse and ill-defined. The diffuse nature of Bishop's ring indicates a dispersion of sizes for the particles of the dust, but the dispersion cannot be too broad or the corona would not appear.

The Krakatoa Committee considered Bishop's ring to be due to first-order diffraction by the particles. In this case the modal radius r_m of particles responsible for the corona can be computed from the relation (Johnson, 1954)

$$r_m = 1.22 \frac{\lambda}{\sin\Delta} \tag{9.4}$$

where Δ is the angle from the sun to the first red border, and λ and r_m are in the same units. If the red border is taken as 23° in Bishop's ring and $\lambda = 0.69$ μm, then $r_m = 0.94$ μm. This is based on the assumption that the particles are water droplets, which is probably unrealistic for the volcanic cloud, but it should give the correct order of magnitude. Particles of this size fall slowly, thereby accounting for the long duration of the observed phenomenon. In addition, there may have been a renewal of the particles with time as some of the gaseous material such as sulfur dioxide or hydrogen sulfide reacted with water vapor to form sulfuric acid droplets, as is thought to have occurred in the case of El Chichon.

9.2.1.5 Polarization of light from the Krakatoa cloud

Polarimetric observations in the volcanic cloud appear to have been meager in spite of the interest the cloud generated otherwise. Pickering (1886) had a polarimeter of new design completed about a year after the eruption and used it for skylight observations in white light in August and September of 1884, and in the following summer. Although the polarization measured by the new instrument was very low, with a maximum of 48.2% in August 1884, Pickering did not comment on the cause of such low values, though they were almost certainly connected with the Krakatoa cloud. The polarization remained low at least through September 1885, but the maximum had risen to 60.5% by that time. Cornu (1884) reported on polarization observations taken in France in which the maximum polarization at 90° from the sun in the sun's vertical was reduced from the normal 75% on clear days to only 4.8% at the time the Krakatoa cloud was over the observation site. Observations with red and cobalt blue filters showed that the degree of polarization was lower in red than in blue light. The normal situation in the clear atmosphere is the reverse of this.

In connection with his skylight measurements, Cornu (1884) described a corona, which is undoubtedly the same phenomenon as Bishop's ring, that had been seen previously by two French observers between March and September 1884. According to Cornu's observations, the Arago, Babinet, and Brewster points were significantly displaced at the time of the corona, but, more interestingly, he found four other neutral points lying outside the sun's vertical. A pair of neutral points was found in the region of the corona, one on either side of the sun at about the same elevation as the sun. They were situated outside the red corona, which means at an angle of something more than 20° from the sun if the corona was really a Bishop's ring. Observations through a green filter rendered the neutral points indistinct, and with a blue filter they were moved close to the sun. Cornu considered the perturbation to be equivalent to the superposition of a beam of negatively polarized light on the normal distribution of skylight. The other two neutral points were situated symmetrically on either side of the antisolar point and at about the same elevation. These two were much more difficult to see than those near the sun, and they were visible for only a few minutes before sunset. The instrument used for neutral point observations was the normal Savart polariscope with and without colored filters.

As mentioned in Chapter 7, Soret (1888) observed pairs of neutral points somewhat similarly located to those of Cornu, but they appeared only in observations through haze overlying a water surface. There do not appear, however, to have been water surfaces involved with Cornu's neutral point observations in the volcanic cloud.

9.2.2 Mount Katmai, 1912

The Alaskan volcano Mount Katmai (58°16'N, 155°W) suddenly exploded in a major eruption on the afternoon of June 6, 1912, continued in a violent state for about 3 days, and in a reduced level of activity for 5 more months. A descriptive narrative of the eruption was issued by the National Geographic Society (Anonymous, 1912), and the event was witnessed from several ships that happened to be in the vicinity. Copious ashfalls covered a wide area, and the visibility in the Shelikof Strait dropped to zero for some 40 hours. Detonations, while heard at distances up to 600 miles, apparently did not compare in violence with those of Krakatoa, although the different atmospheric conditions would affect the transmission of sound waves. The National Geographic Society report indicated that the total amount of material ejected by Katmai may have surpassed that of Krakatoa, but the amount that ended up as dust in the atmosphere is not known. Likewise there were no reliable estimates of the altitude reached by the dust, although Gruening (1963) thought it may have reached 40 km. The dust veil index of Lamb (1970) was set at 600, compared to 1000 for Krakatoa.

Information on the Katmai eruption is neither as extensive nor detailed as that of the Royal Society report on Krakatoa, but two sources constitute a summary of the main atmospheric effects of Katmai. Kimball (1913a) circulated a letter to weather bureau stations, observatories, and other scientific establishments in the United States requesting eyewitness accounts of atmospheric phenomena seen during the summer and fall of 1912, and summarized the results within a few months after the event. A similar summary of about 40 reports from Europe was developed by Maurer and Dorno (1914).

9.2.2.1 Twilight glows and colored sun and moon

Relatively colorful twilight glows were seen during the summer and fall of 1912 in various locations of North America and Europe, but they were sporadic and generally did not persist beyond solar depressions of 10° to 12°. It is apparent that the twilight phenomena after Katmai were but poor cousins to the spectacular displays of twilight colors after Krakatoa. There were several reports of dusty daytime skies as the Katmai cloud overspread British Columbia and the northwestern part of the United States within the first week after the eruption. Similar reports from Europe and the Middle East into the fall, and even into December of 1912, indicate the widespread nature of the cloud. There seem to be no observations to reliably indicate Katmai dust in tropical latitudes, and the cloud distribution at higher latitudes shows no such concentrations and consistency in time as those of the Krakatoa cloud in the tropics during 1882 and 1884. The more irregular circulation of the high-latitude atmosphere would tend to fragment and disperse the Katmai cloud more quickly than the steady tropical wind system did the cloud of Krakatoa.

No observations of blue or green sun or moon were reported after Katmai, but there were several reports of the sun becoming dim and even disapearing in haze just before sunset. Greenish or bluish colors in the sky near sunset were confined to a few days after the eruption. Also at this time, Bishop's rings were observed around the sun and moon in some locations.

9.2.2.2 Atmospheric transparency

Several stations were measuring incident solar radiation in 1912, and the resulting data may be interpreted in terms of atmospheric transparency or solar radiation attenuation. Among such observing stations were the Mount Weather Observatory, Virginia; Madison, Wisconsin; Lincoln, Nebraska; Santa Fe, New

Mexico; Washington, D.C.; and several stations in Europe (Potsdam, Warsaw, Paris, and St. Petersburg). In addition, the solar constant program of the Smithsonian Institution was well under way in 1912, the two stations in operation at the time being Mount Wilson, California, and Bassour, Algeria (Abbot et al., 1913, 1922). Data from the Smithsonian stations are particularly reliable, and show well the decrease of atmospheric transparency, for both total solar radiation and for different wavelengths, by the Katmai cloud.

Data from observations at the Mount Wilson Observatory (34.9°N, 118.10°W) as tabulated by Abbot et al. (1913) for the summer months of 1912 are shown in Fig. 9.2a. The dust apparently reached Mount Wilson only in July, in which case the data for June should represent normally clear atmospheric conditions at that location. The transmission for wavelength-integrated radiation

(a)

(b)

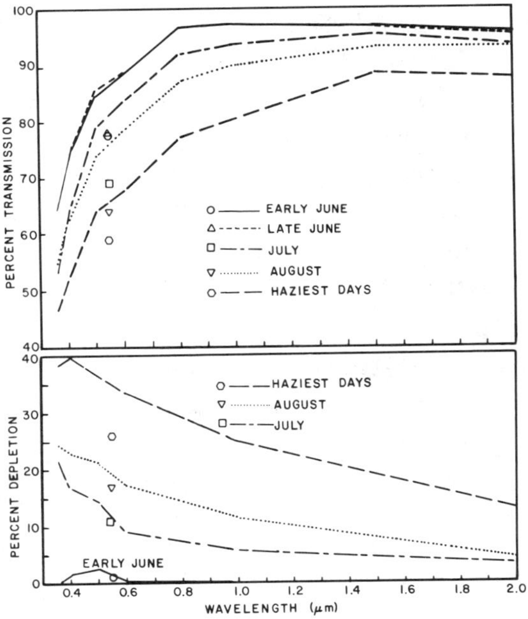

Figure 9.2 (a) Transmission of solar radiation as a function of wavelength by the volcanic cloud from the Katmai eruption as observed at the Mount Wilson Observatory by Abbot et al. (1913); (b) Spectral depletion of solar radiation by the Katmai volcanic cloud as observed at the Mount Wilson Observatory by Abbot et al. (1913).

in the direct beam corresponding to the sun's zenith distance of 48° (airmass 1.5) is shown by the single points of the diagram, and its spectral distribution corresponding to the sun at the zenith is given by the curves. The dust was over the station through July and August, being thickest in August. The average total transmission for the entire month of August was only 64%, for a decrease of 13% from the June average. On the haziest days the decrease was over 18%. The curves show much better transmission of near-infrared wavelengths than those of the visible, indicating the small sizes of the dust particles. The wavelength dependence is more readily seen on the plot of percent depletion versus wavelength shown in Fig. 9.2b. In order to exhibit the effects of only the volcanic cloud, the data are taken as deviations from the early June averages, which are set for zero. The difference between late June and early June is probably an indication of natural variability, and is unrelated to volcanic dust. The lesser effect of dust at the longer wavelengths is shown well by the curves, the fractional depletion at $\lambda = 2.0$ μm being a third less of that in the short visible and ultraviolet. For total direct beam radiation, depletion by dust on the haziest days was fully 26%, and the average for the whole month of August was 17%.

A second Smithsonian station was operating in Bassour, Algeria (36°N, 3°E) in 1912, and data similar to those of Mount Wilson were taken there (Abbot et al., 1913). The same general characteristics are shown by the Bassour data, except that the dust cloud had about the same opacity in July as in August, and the overall opacity was about the same as that at Mount Wilson. For the haziest days, however, the Algerian cloud was slightly more opaque than that over California. Abbot et al. (1913) concluded that the maximum decrease of solar radiation attributable to the Katmai dust reached about 20% at both stations. This, of course, is for the direct solar beam, and part of the decrease was counteracted by the flux due to scattered light. Abbot et al. estimated that the net decrease of solar energy due to dust was about 10% of the solar constant.

Measurements of wavelength integrated solar energy at Mount Weather, Virginia, by Kimball (1913a) showed the Katmai dust cloud to have been thickest in July and August 1912. The average depletion of the direct beam at airmass 1.5 was approximately 23% for each of these 2 months. The cloud diminished with time so that the analogous depletion in October was less than 5% in Virginia.

A survey of the distribution of the Katmai dust by Volz (1975) indicates a highly variable cloud over Europe in the summer of 1912, and it had decreased to low-turbidity values by early 1913. In the United States the cloud continued to be relatively dense through the fall of 1912 and into the following winter. Colorful twilights were seen in Europe as late as November and December 1913, but they may have been from other causes. As seen in the next section, however, skylight polarization did not return to pre-Katmai conditions until at least 1914.

9.2.2.3 Skylight polarization

Skylight measurements by Kimball (1913a, 1913b) in the United States and by Jensen (1928, 1942) and Dorno (1919) in Europe show that the volcanic cloud from the Katmai eruption had a pronounced effect on the polarization of skylight, both in decreasing the polarization maximum and in shifting the positions of the neutral points.

Observations of the degree of polarization P_{max} at a point 90° from the sun in the sun's vertical made by Kimball (1913b) at Mount Weather, Virginia, during the last 8 months of 1911 and 1912 are shown in Fig. 9.3. The measurements were in white light at a sun elevation of 30°. The dashed curves of the diagram show the maximum and minimum values of P_{max} in the various months of 1911, and the solid curves are for equivalent data taken in 1912. From these and other data it is evident that the Katmai cloud reached Virginia about June 19, 1912, and continued to thicken during July and August. While the cloud diminished in thickness during the fall of 1912, it did not disappear until well into 1913, and fragments were observed in some locations until 1914.

In order to show more explicitly the relation of P_{max} in 1912 with that in the normal case, the difference between P_{max} for 1912 and the average for 1908, 1909, and 1911 (i.e., $[(1908+1909+1911)/3]-1912$) is shown by the dotted curve of Fig. 9.3. In the densest part of the cloud P_{max} was decreased by 27% as the average for the whole month of August, and by 23% throughout the following 2 months. Even the December average showed a decrease of 16%. No data are available for P_{max} after the end of 1912, but Kimball's neutral point observations were continued into 1913.

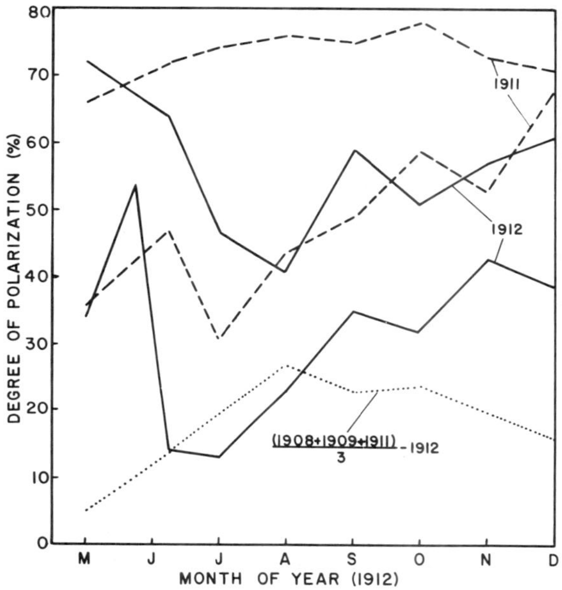

Figure 9.3 Degree of polarization of skylight at a point 90° from the sun in the sun's vertical as a function of time according to the observations of Kimball (1913b) at Mount Weather, Virginia. The dashed curves show the maximum and minimum values of the polarization maximum P_{max} for 1911, and the solid curves are the same quantities for 1912. The dotted curve is the difference between the average P_{max} for 1908, 1909, and 1911 and that of 1912.

Supplementary to the measurement of $P_{\max}$, a large number of observations of neutral point positions were taken in the United States and Europe both before and after the Katmai eruption. Although the Arago point is somewhat sensitive to high-level turbidity, its response is greatest to turbidity in the lower layers of the atmosphere. In fact, as shown in Chapter 7 for this latter case, the Arago point is more sensitive to low-level turbidity than is the Babinet point. For volcanic dust in the stratosphere, however, the Babinet point is much the more sensitive of the two. This is shown well by the data for 1911 to 1914 given by Jensen (1928) and plotted in Fig. 9.4. As before, the Babinet point is referenced to the sun and the Arago point to the antisun. The 1911 data show the normal behavior of both Babinet and Arago points. Under normal conditions, the angle of the Babinet point with respect to the sun changes only about 2° through the whole range of sun elevations from +6° to −6°, and the Arago point distance changes

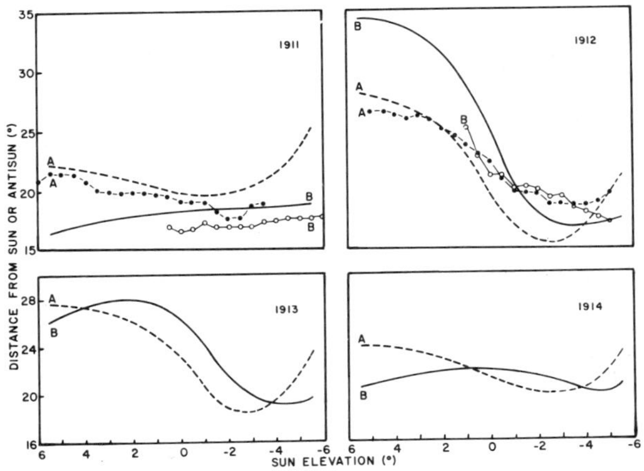

Figure 9.4 Distance of the Babinet point (Curves B) and Arago point from the antisun (Curves A) for various periods before and after the 1912 eruption of Mount Katmai as observed by Jensen (1928)—smooth curves—and Kimball (1913b)—plotted points.

slightly more than 5° over the same range. Furthermore, the Arago point distance is always greater than the Babinet point distance.

The introduction of volcanic dust in 1912 caused profound changes of neutral point behavior, as shown by the 1912 data of Fig. 9.4. The amplitude of the Arago point changed from only 5° in 1911 to over 12.5° in 1912, and the position of the minimum was shifted to larger negative sun elevations. The most pronounced change, however, was in the behavior of the Babinet point. In fact, the whole character of the Babinet point curve was transformed by the Katmai dust. Negatively polarized light was added to the region of the sky near the sun sufficient to shift the Babinet point by up to 17° farther from the sun at positive sun elevations. The source of such negative polarization is not clear as multiple scattering, often a source of negative polarization, should be stronger for the long atmospheric pathlengths corresponding to the sun at the horizon than in positive sun elevations. In addition, the data of 1912 show a slight shift of the Babinet point toward the sun at negative sun elevations, which is a time of similarly long atmospheric pathlengths and much multiple scattering of the incident sunlight. Negative polarization may be introduced by single scattering of absorbing submicron size particles (Dave, 1970), but normally the polarization is of positive sign throughout the first few tens of degreees of scattering angle for a polydispersion such as would be expected in the volcanic cloud. The neutral point positions migrated nearer to their pre-Katmai positions over the subsequent 2 years, but even in 1914 there was still some evidence of lingering volcanic dust over Europe.

Neutral point behavior as measured by Kimball (1913b) in white light at Mount Weather Observatory is shown by the plotted points in Fig. 9.4. Data in the 1911 section are for the fall months of 1911 and spring of 1912 before the appearance of the Katmai cloud over the site, and the 1912 data were taken in the fall of 1912 and extended into the following winter. It is clear from the plot that the response of both neutral points to the volcanic cloud was qualitatively the same in Virginia as in Europe, although quantitatively the modification was more extreme in Europe. The extreme sensitivity of the Babinet point should make it particularly useful for detecting stratospheric dust for whatever cause.

9.2.3 Mount Agung, 1963

On March 17, 1963, the most sacred volcano on the island of Bali, Indonesia (8°25′S, 115°30′E), broke into a violent eruption after being dormant for 120 years. The entire eastern half of the island was devastated by fallout from the ejection, and 1500 people were killed (Booth et al., 1963). Twilight glows were observed in many locations of both the northern and southern hemispheres, and solar radiation fluxes were diminished at many stations over the world. The dust veil index assigned for the Agung cloud by Lamb (1970) was 800, compared to 1000 for Krakatoa, but this estimate was made with a relative paucity of data.

It is surprising, in this day of worldwide communication and scientific interest, that so few observations of the Agung cloud have been reported. The answer to this puzzle must lie in the fact that most of the cloud went to the southern hemisphere, with its small land area, thin population, and restricted scientific acumen. To counter the restricted observational material, two new measurement systems (lidar and high-altitude aircraft with particle collectors) never before available for observations of volcanic dust were utilized for the Agung cloud. The principal types of data available, and the deductions therefrom, are summarized in the following sections.

9.2.3.1 Twilight glows

Widely scattered observations of twilight glows in both hemispheres show that while the Agung cloud was generally not as dense as that from Krakatoa, it was widespread and persistent. Volz (1969) observed volcanic twilights in Germany in November and December 1963, and lingering effects were still noticed into 1967. Steinhorst (1977), also in Germany, ascribed anomalous twilight polarization measurements taken in 1968 to the Agung cloud. In this case, however, it is likely that some less violent eruptions between 1963 and 1968 (Villarca, 1964; Taal, 1965; Redoubt, Soputan, Kelund, and Awa, 1966; and others) contributed more to Steinhorst's results than did the 5-year-old Agung cloud. Twilight observations by Meinel and Meinel (1963, 1964) show the cloud reached Arizona (33.3 °N) within 6 months after eruption. Agung dust was observed over Minneapolis (45 °N) in January 1964 and at College, Alaska (65 °N), in the summer of 1964. The arrival date of the cloud over India is unknown because of monsoon cloudiness, but twilight glows were observed in many locations in late September (Shah, 1969). It is likely, however, that the cloud had arrived earlier than this. Solar radiation records show that Agung dust arrived over Hawaii (19.5 °) in the first few weeks after the eruption. Observations in the southern hemisphere (Weinert, 1967) show the cloud over Australia as early as May 1965 (Dyer and Hicks, 1968; Mossop, 1964; Hogg, 1963). It started affecting atmospheric transparency at the South Pole in November 1963 (Flowers and Viebrock, 1965).

Red/green color ratio measurements during twilight at a point in the sky 20 ° above the solar horizon, a ratio thought by Volz (1969) to be a sensitive indicator of high-level turbidity, indicated the arrival of the Agung cloud over Weissenau in southern Germany as early as July 1963. The ratio showed a sharp peak in the winter of 1963–64 and a lesser peak a year later, with decreases during the intervening summer months. This observation would bolster the concept of Dyer and Hicks (1968) that the breakdown of predominantly zonal circulation of summer in the tropics into increased meridional flow with the approach of winter tapped the low-latitude turbidity reservoir for aerosol injections into the higher latitude stratosphere in the winters of both hemispheres.

The duration of twilight glows from Agung dust were generally similar to those from other volcanoes. For instance, the glow in Arizona (Meinel and Meinel,

1963) persisted until about 1 hr 30 min after sunset in October 1963, and Shah (1969) measured twilight enhancements up to solar depressions of 11° to 13° in India in early 1964. The observations of Volz (1964) in Germany, however, indicated the glows terminated at solar depression angles of 8° to 9°. It will be remembered that twilight glows after Krakatoa could still be seen at solar depression angles of 17° to 19° in some cases. The lesser persistence of the Agung glows may be ascribed to the lesser density of the Agung cloud, although a difference in altitude cannot be ruled out from the available data.

9.2.3.2 Altitude of the cloud layer

While the worldwide dispersion of the Agung dust is well established, the altitude of the stratum is less well known. Mossop (1964) collected particles from an altitude of 20 km over Australia. Lidar observations by Grams and Fiocco (1967) in Massachusetts in the summer of 1964 showed dust layers of variable density between 16.5 and 18 km, and similar measurements in Alaska gave 14 km. From an analysis of sunlight reflected from a large balloon floating at 30 km during twilight, Pittock (1966) indicated the Agung cloud over Colorado to be at 20- to 32-km altitude. In Massachusetts the maximum density was at about 19 km (Grams and Fiocco, 1967). Meinel and Meinel (1964) estimated from visual observations during twilight that the mean altitude over Arizona was 17.8 km, with the top at 20 km. This estimate was about 2-km lower then their original estimate of a year earlier. Dyer and Hicks (1968) summarized data from 10 different locations in the northern hemisphere and showed a latitudinal dependence of altitude of the dust layer. For instance, the altitude at 80°N was less than 10 km, rising to 15 to 20 km in midlatitude and to over 20 km in the tropics.

9.2.3.3 Optical thickness of the Agung dust

The normal optical thickness of the atmosphere $\tau(\lambda)$ is composed of that due to Rayleigh scattering $\tau^R(\lambda)$, aerosols $\tau^A(\lambda)$, ozone $\tau^O(\lambda)$, water vapor $\tau^W(\lambda)$, and other trace gases $\tau^T(\lambda)$. Thus we have

$$\tau(\lambda) = \tau^R(\lambda) + \tau^A(\lambda) + \tau^O(\lambda) + \tau^W(\lambda) + \tau^T(\lambda) \qquad (9.5)$$

The most reliable data on $\tau(\lambda)$ for the Agung dust cloud appears to be the stellar extinction values reported from the Mount Stromolo Observatory in Australia (Hogg, 1963) and from the Boyden Observatory in South Africa (Irvine and Peterson, 1970). Unfortunately, measurements on solar spectral extinction, were terminated by closure of the Smithsonian solar constant stations in the 1950s. As mentioned below, broadband solar radiation data from different sources are available, but this type of data is not adaptable to determinations of $\tau(\lambda)$.

The stellar extinction data from the observatories in Australia and South Africa were summarized by Deirmendjian (1973) and compared with similar data from model calculations. At the wavelengths of the stellar data, $\tau^W(\lambda)$, and $\tau^T(\lambda)$ are negligibly small and $\tau^R(\lambda)$ and $\tau^O(\lambda)$ may be evaluated and subtracted out from $\tau(\lambda)$. This leaves only the aerosol component $\tau^A(\lambda)$, which itself is composed of that due to a normal background aerosol plus that of the Agung cloud. The change of $\tau^A(\lambda)$ produced by the arrival of the Agung cloud over the Observatory should, other components being unchanged, yield a measure of cloud optical thickness. The results of Deirmendjian's analysis of the Observatory data are summarized in Fig. 9.5. Here $\tau^A(\lambda)$ for the Agung cloud

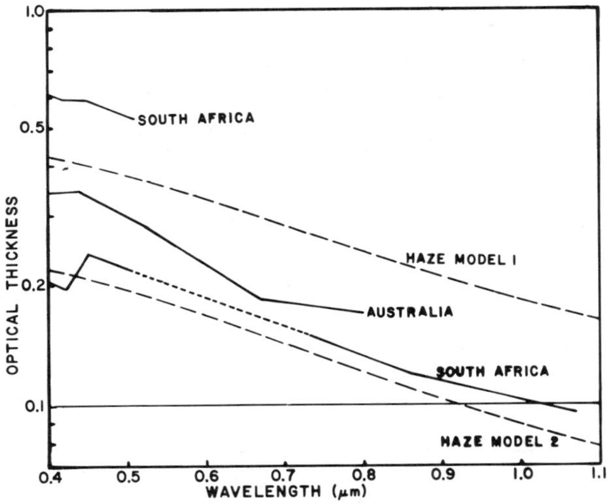

Figure 9.5 Optical thickness of Agung dust as a function of wavelength as determined from stellar extinction data at observatories in South Africa and Australia and according to computations based on two models of atmospheric haze. (Data from Deirmendjian, 1973.)

is shown as a function of wavelength for the average of 5 nights with moderate dust and one with high extinction in blue and green wavelengths, both from South Africa, and a curve for data taken at Mount Stromolo in southern Australia. It should be realized, of course, that data taken far from the site of the eruption and at widely separated locations may not be representative of the cloud at other places and times. However, the measurements appear to be consistent in spectral dependence at least, the larger $\tau^A(\lambda)$ at short wavelengths indicating the generally small sizes of cloud particles in accord with the particles collected by Mossop (1964) at 20-km altitude in the southern hemisphere. Included in the diagram are two curves of optical thickness computed for haze model *H* with refractive index equivalent to that of silicates (1.55), curves 1 and 2 being for horizontal ranges of 14 and 7 km, respectively (Deirmendjian, 1969).

Broadband measurements of direct solar radiation were under way at the environmental monitoring stations at Mauna Loa in Hawaii and at the South Pole at the time of the eruption. Flowers and Viebrock (1965) reported that the Agung cloud reached the South Pole in several bursts in November and December 1963 as the atmospheric circulation changed from winter to summer. At the time of these incursions the normal incidence solar radiation was reduced by values ranging from 23% to 78% for sunlit periods at the Pole at least through November 1964. Solar noon values of direct solar radiation at the Mauna Loa Observatory were reduced by about 5% to 6% in April and May 1963 (Mendonca et al., 1978). They gradually increased back to normal over a period of 8 years, but it is not definite that the reduction over such a long period was all due to the Agung cloud.

It appears, surprisingly enough, that the only polarization measurements made from the Agung cloud were those of Shah (1969) from November 1963 to March 1964 at Ahmedabad, India (23 °N). The measurements at λ=0.66 μm were taken from the direction of the zenith for solar depression angles of 0° to 12°–14°. The principal change of polarization was the appearance of a pronounced polarization minimum occurring at a solar depression angle of approximately 4°, the magnitude being depressed from the normal 60% to 65% in the pre-Agung twilight to as low as 44% in the Agung cloud. (One of Shah's polarization curves is shown below in Fig. 9.8 for comparison with the El Chichon measurements.) Shah interpreted this polarization minimum as indicating a stratospheric dust layer at 18-km altitude over Ahmedabad in the winter of

1963–64. Concurrent measurements of twilight intensity at 20° above the solar horizon tend to confirm the existence of such a layer.

9.2.4 El Chichon, 1982

Ninety-nine years after the eruption of Krakatoa, an apparently insignificant mountain in southern Mexico broke out in a convulsion which rivaled, and perhaps even exceeded, that of its famous predecessor. El Chichon (17.33 °N, 93.20 °W, 1260 m high) underwent a series of eruptions from March 28 to April 4, 1982 (Weintraub, 1982), at least two of which injected material into the stratosphere, according to satellite images of the area (Matson and Robock, 1984) and as indicated by the production of strong atmospheric gravity waves received as far away as Dallas, Texas, and McMurdo, Antarctica. The smaller of the two occurred in the late evening of March 28. The satellite-derived temperature −75.2 °C at the top of the cloud indicated a height of either 16.3 km or 18.5 km, the uncertainty arising by −75.2 °C corresponding to two different heights on the radiosonde plot. A more forceful explosion occurred in the early morning of April 4. The temperature of the top of this largest cloud was measured by satellite at −83.0 °C, corresponding to some unknown height in the stratosphere (Matson and Robock, 1984).

The April 4 cloud was caught up in the easterly wind of the stratosphere, and daily satellite images showed that it had circled the globe in 21 days. The first land area it encountered after leaving Mexico was the island of Hawaii—the site of the Mauna Loa Observatory (MLO). The Observatory, the primary purpose of which is monitoring atmospheric parameters, was in full operation at the time, and a complete record of the first appearance and subsequent changes of the El Chichon cloud over Hawaii has been compiled. These Mauna Loa observations will be emphasized in this discussion, not necessarily because they are the best (although they have some unique features), but because they are very complete, they have been widely quoted, and the present author was involved with their acquisition and analysis. A short description of our initial detection of the cloud at MLO and our subsequent monitoring of its properties may be of interest for the record.

Our regular measurement program at MLO included weekly lidar observations for monitoring the turbidity of the atmosphere, and April 9, 1982, happened to be on the lidar schedule, so Tom DeFoor (an engineer at MLO) and I drove up the mountain in the evening, arriving at the site after dark. Before starting the lidar we noticed that the moon looked fuzzy, and assumed there was a layer of cirrus through which we would have to shoot the lidar. This was not particularly unusual, but we tried to avoid cirrus as it degrades the lidar signal. On the first lidar shot, however, we saw that there was no cirrus, but instead a monstrous signal from the level of the stratosphere. Having recently seen signals from the so-called mystery cloud (a volcanic cloud of unknown origin which had appeared in January 1982), we knew this was a volcanic cloud, but we had never seen anything even approaching this magnitude of lidar return. We spent the next several hours decreasing power and gain settings to get the maximum signal without saturation of the system, and in estimating the height and thickness of the cloud. On terminating the observations and driving back down the mountain we knew we had seen a whale of a cloud but we did not know where it had been spawned. It was only the next morning when we reported the discovery of the cloud to the headquarters in Boulder that we learned that a volcano in Mexico had erupted a few days previously, and that El Chichon was the probable source of the cloud. As a result of this major atmospheric phenomenon, we doubled the frequency of lidar observations, initiated a concentrated project of measuring polarization and intensity of skylight, made more frequent sunphotometer measurements, and generally stepped up the observational program at MLO to get as much information as possible on the properties and effects of the El Chichon cloud over Hawaii. Some of the results are shown in following sections.

Although eyewitness accounts of twilight colors and associated phenomena, such as were most complete for the Krakatoa eruption, are extremely scanty from El Chichon, they are not completely lacking. For instance, Peterson (1983) described a sequence of color changes during sunrise in Texas, as did Verschure (1984) in the Netherlands. The vivid twilight colors photographed by the present author and reproduced as the Frontspiece of this book is another of the few eyewitness accounts of El Chichon twilights. But other types of data on the El Chichon cloud are sufficient to show the spread of the cloud, so the absence of widespread sightings is not as serious now as it once would have been (McCormick and Swissler, 1983), though it must be admitted that eyewitness descriptions of twilight colors and Bishop rings have a fascination that can never be quite compensated for by satellite or lidar measurements. Be that as it may, systems of satellites, surface and airborne lidar, particle counters and collectors carried aloft by balloons, solar radiation instruments, and other devices make up a very powerful observation network, and the combination has certainly yielded more information on the El Chichon cloud than has been compiled for any such phenomenon in the past. In addition, several major effects directed toward modeling the event have added greatly to the understanding of the physics and chemistry involved. Unfortunately, the scope of the present discussion is too restricted to even mention much of this material. Two collections of reports, however, provide valuable summaries of the various aspects, and are readily available. First Dr. Ignacio Galindo with guest editors Dr. D.J. Hofmann and Dr. M.P. McCormick, initiated a special issue of *Geofisica Internacional* (Vol. 23, Nos. 2 and 3) in which 11 major articles are included, and Dr. J.B. Pollack and Dr. O.B. Toon were guest editors of a special issue of *Geophysical Research Letters* (Vol. 10, No. 11, 1983) containing 18 reports on the El Chichon cloud. These collections are in addition to numerous publications in the regular scientific journals.

El Chichon has been compared in size and amount of ejected material with Krakatoa, although there is little firm basis for such a comparison. Early estimates of the mass of the cloud generated such terms as "weather-maker of the century" (Mitchell, 1982) and "dust cloud of the century" (Robock, 1983). McCormick and Swissler (1983) estimated that the total stratospheric dust cloud contained 12 megatonnes of solid and gaseous material, most being at 21- to 27-km altitude in the early stages. The total amount of sulfur dioxide injected into the stratosphere was put at 13.4 megatonnes by Evans and Kerr (1983). Within the first month or two the top had risen in height to at least 30 km. Particulate material in the young cloud is thought to have consisted mainly of volcanic ash, but a later conversion of the gases, principally sulfur dioxide and hydrogen sulfide, by reaction with water vapor is thought to have produced submicron-sized sulfuric acid droplets. Such droplets were the main component in later periods. Lidar observations in Hawaii showed the cloud had gone through three distinct phases of development. Profiles of the backscatter ratio for the cloud in the three phases, as observed by ruby lidar at MLO, are shown in Fig. 9.6. In the initial or transition phase, the young cloud was characterized by strong variability in space and time, with pronounced layering in its vertical profile. This first phase lasted 1 to 2 months, and gradually gave way to the second or stabilized phase. The profile in the stabilized phase, while containing a definite peak of concentration, was much more homogeneous and varied only slowly with time. In this phase the top of the cloud had risen and filled through the region 27 to 32 km, presumably

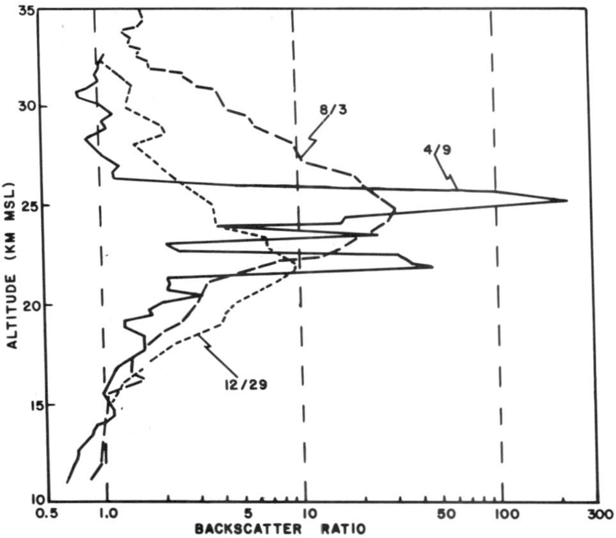

Figure 9.6 Altitude profiles of the lidar backscatter ratio during three different stages of El Chichon volcanic cloud at the Mauna Loa Observatory: April—transition phase; August 3—stabilized phase; December 29—dissipation phase.

by sulfuric acid droplets formed by gas-to-particle conversion. The stabilized cloud lasted with almost constant height and concentration until October, at which time the dissipation phase set in. To initiate the third phase, the summer type of wind regime, with principally zonal flow, gave way to the stronger meredional circulation of fall and winter, causing movement of cloud particles to higher latitudes. The first evidence of the cloud at Fairbanks, Alaska, was on November 15, 1982 (Wendler, 1984). Once the third phase had been initiated, the cloud, at least over Hawaii, continued to decrease in concentration and decrease in altitude. After a period of 2 years most of the cloud had dissipated. During the 2 years, however, it had spread to all parts of the globe and made a worldwide dust veil. The beautiful blue skies of Hawaii had been turned to a washed-out milky appearance, and only since 1984 have they reacquired their normal blue color.

Because the primary interest here is on radiation in the atmosphere, the remainder of the El Chichon discussion is on optical thickness of the cloud, its modification of solar flux reaching the surface, and the changes of polarization and intensity of skylight which occurred at the Mauna Loa Observatory.

9.2.4.1 Optical thickness of the El Chichon cloud

As indicated previously in Eq. (9.5), the total optical thickness $\tau\ (\lambda)$ at any wavelength is made of components due to the Rayleigh atmosphere, aerosols, ozone, water vapor, and trace gases. By measuring the energy $I(\lambda,\theta_o)$ of the direct solar beam at some solar zenith angle, introducing known corrections for the Rayleigh atmosphere and ozone, and avoiding the major absorption bands of water vapor and trace gases, it is a simple matter to determine the optical thickness due to aerosols. If $I(\lambda,\theta_o)$ is the measured energy in the direct solar beam and $I_o(\lambda)$ that at the top of the atmosphere, the optical thickness is given by the relation.

$$I(\lambda,\theta_o) = I_o(\lambda) \exp[-\tau(\lambda) \sec\theta_o] \tag{9.6}$$

By combining this with Eq. (9.5), we can obtain $\tau^A(\lambda)$ as

$$\tau^A(\lambda) = -\cos\theta_o ln\left(\frac{I}{I_o}\right) - \tau^R(\lambda) - \tau^O(\lambda) \tag{9.7}$$

The usual instrument for measuring $I(\lambda,\theta)$ is the well-known sunphotometer. Sunphotometer observations are taken on a routine basis at several environmental monitoring stations, including the Mauna Loa Observatory. Being at an altitude of about 3400 m and protected from low-level turbidity by the trade wind temperature inversion, this Observatory was in an advantageous location for monitoring the El Chichon cloud, and a large number of optical thickness measurements were obtained.

The optical thickness of the cloud during its first week or so of residence over Hawaii was generally above 0.60 at $\lambda=0.425\ \mu m$, and on one occasion it reached a phenomenal value of 0.77 (DeLuisi et al., 1983). The normal background optical thickness of aerosols at MLO is only about 0.02, and just prior to the appearance of the El Chicon dust the optical thickness had been increased to the 0.03 to 0.05 range by the mystery cloud over Hawaii. The El Chichon cloud was extremely variable through the month of May, generally showing values of τ^A in the 0.30 to 0.40 range, but with one case of $\tau^A = 0.05$. By June the cloud had become stabilized with $\tau^A = 0.20$ to 0.30. The spectral distribution τ^A for the stablized cloud of June and July at MLO is shown by dotted curves of Fig. 9.7. The curves were fitted to sunphotometer data at five different wavelengths. The maximum value of τ^A of about 0.25 was in the $\lambda = 0.6$- to 0.7-μm range in June and at somewhat shorter wavelengths in July, the wavelength shift probably resulting from the formation of smaller mean particle sizes as the original dust component gave way to smaller size droplets of the sulfuric acid aerosol. Inversion of these data for July by King et al. (1984) to

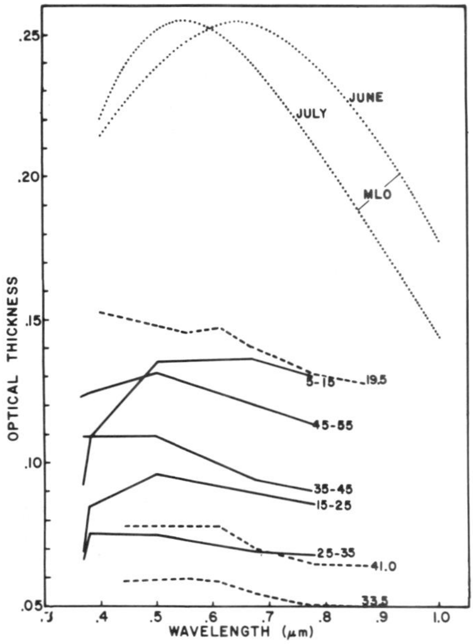

Figure 9.7 Optical thickness as a function of wavelength of the El Chichon volcanic cloud as observed at the Mauna Loa Observatory in June and July 1982 (dotted curves), as measured at three different latitudes by Spinhirne (1983) from aboard an aircraft with flight path over the central area of North America and the west coast of South America (dashed curves) in October 1982, and mean values for various latitude bands according to measurements by Dutton and DeLuisi (1983) on an aircraft flight along meridian 120°W in December 1982 (solid curves).

yield a size-frequency distribution showed particle radii in the 0.1- to 1.0-μm range, with a pronounced maximum at 0.3 to 0.4 μm. This is in general agreement with the balloon-borne measurements of Hofmann and Rosen (1984) in Wyoming and Texas in October 1982, although they tended to get a bimodal distribution with main mode radii of 0.3 and 0.15 μm.

In addition to its much-reduced magnitude, the latitudinal distribution of the cloud in October 1982 over central North America and west coast of South America (Swissler, 1983) was different from that along longitude 120°W in December 1982 (Dutton and DeLuisi, 1983), as shown by the solid and dashed curve of Fig. 9.7. Such variability is to be expected, of course, as the cloud is moved from its previous low-latitude reservoir to higher latitude regions with the onset of meridional-type circulations. Very near the equator the cloud remained relatively dense, but midlatitude values were low for one flight and high for the other. Generally, however, optical thicknesses remained in the 0.05 to 0.15 range, which is several times background levels, at least in the northern hemisphere. Although less in the southern hemisphere, an optical thickness of about 0.03 prevailed to at least 37.5°S, the southern terminus of the measurements (Swissler, 1983).

9.2.4.2 Solar energy attenuation by the El Chichon cloud

From a climatic standpoint, the most important aspect of the El Chichon cloud is its attenuation of solar energy reaching the earth's surface. Major environmental monitoring stations and some ordinary weather stations measure on a routine basis the energy of the direct solar beam, energy of diffuse skylight, and the sum of these—the global radiation. Representative values of the change of these three quantities by the El Chichon cloud at various stages are given in Table 9.2. The Mauna Loa data were taken on June 6, 1982, and compared with a reference day with insignificant stratospheric dust, June 5, 1981. Depletion of the global flux is of most interest, being 5.6% and 7.7% at solar noon and for the whole day, respectively. If such large depletion values were to continue over a sufficiently long time there would certainly be climatic consequences. The attenuation of the direct beam by 21.3% more than normal would be a great detriment in the use of focusing solar energy collectors and similar applications, but its climatic significance is less than that of the decrease of global radiation.

TABLE 9.2 Percentage changes of direct, diffuse, and global solar radiation reaching the earth's surface brought about by the El Chichon volcanic cloud at various locations.

Station	Lat. (°N)	Direct	Diffuse	Global	References
Mauna Loa	19.5				DeLuisi et al. (1983)
All λ		−21.3	—	−5.6±1.0	
λ>0.695 μm	—	—		−7.3±1.0	
Whole day total	—	—		−7.7±1.0	
Ann Arbor, Michigan	42.2				Baker et al. (1984)
All λ		−30	+88	−7	
Davis, California	38.5				Carroll (1984)
All λ, θ_0=60°		−17.7	+59.7	−3.6	
Tucson, Arizona	32.1	−5.2	—	—	Szymber and Sellers (1985)
Corvallis, Oregon	44.3	−27	+213	−13	Rao and Bradley (1983)
Fairbanks, Alaska	64.4	−29.4	—	−5.6	Wendler (1984)

The data for Ann Arbor, Michigan, were taken at the time of first arrival of the cloud, October 26, 1982. Equivalent measurements taken over the following 3 months show roughly equivalent radiative changes, but a large variability occurs normally in this industrial location and separation of the different effects is difficult. A similar difficulty exists at Davis, California, except in that case the variability arises from agricultural operations. The values shown in Table 9.2 for Davis at a sun elevation of 30° represent the average change for the entire first 6 months of 1983 from the previous 3-year mean, and is thought to be due principally to the volcanic cloud. The global flux in the shorter wavelengths (0.295–0.630 μm) was particularly sensitive to stratospheric dust, whereas that from 0.630 to 3.00 μm was affected more by atmospheric water vapor.

Values of the Linke turbidity factor T measured at Tucson, Arizona, by Szymber and Sellers (1984) were used to compute the change of atmospheric transmission for direct solar radiation given in Table 9.2. According to these authors, the all-wave transmission is closely approximated by the exponential $\exp(-0.08T)$. The average value of T for August and September 1982 was 3.37, and that for the same months over the period June 1956 to March 1982 was 2.48. From these, the transmission was decreased from 82.0% for the long-term average to 76.4% in 1982, a decrease of 5.6%. This seems like a small change in comparison to the analogous changes at other stations, but the authors point out that Tucson normally has both high turbidity and high values of precipitable water in summer, so the added attenuation by the El Chichon cloud is proportionally not as great as it otherwise would be. The Corvallis, Oregon, data were taken about solar noon during November and December 1982. They are representative of observations over a wide geographical area of southern Oregon after the arrival of the cloud in late 1982. The cloud first arrived over Fairbanks, Alaska, on November 15, 1982, and it caused a decrease of the direct solar beam of 19.4% from normal during the next 5 months (Wendler, 1984). A decrease in cloud density in April and May 1983 resulted in a decrease of the direct beam of only 15.4%. The average decrease of global flux amounted to 4.9% over the period November 1982 to May 1983. The low sun elevations prevailing at this high-altitude location result in unusually large attenuation of the direct beam, and give a somewhat distorted picture of the actual thickness of the cloud.

9.2.4.3 Polarization and intensity of skylight in El Chichon cloud

A concentrated program of skylight observations was carried out by the present author at the Mauna Loa Observatory over a period of about 18 months beginning at the time of the first appearance of the El Chichon cloud over Hawaii (April 9, 1982). Polarization and intensity, mainly in the sun's vertical, were measured in eight narrow spectral intervals from $\lambda=0.32$ μm to $\lambda=0.90$ μm. The distribution of these parameters across the sky at many positions of the sun, during both twilight and the sunlit part of the day, the magnitude of the polarization maximum, and the positions of the neutral points were all included in the measurements. Fortunately, a previous observational program (Coulson, 1980, 1981) conducted with the same instrument at the same observatory during the clean-stratosphere period of 1977 yielded baseline data with which El Chichon data may be compared.

Before discussing the Mauna Loa measurements, it is worth mentioning some measurements of skylight intensity made with a hand-held photometer, and of neutral point positions observed with a Savart polariscope, by Volz (1984) near Boston, Massachusetts, during 1983. The optical thickness was determined to be about 0.17 at that location during the winter 1982-83. Volz interpreted the $\lambda\lambda$ 0.850/0.500 color ratio in the solar almucantar to indicate a predominance of particles of 0.80-μm radius in the volcanic cloud. This is a variance with the study of King et al. (1984), which gave a pronounced maximum in the $r = 0.3$- to 0.4-μm range, and with direct measurements of Hofmann and Rosen (1984) showing maxima at 0.3 μm and 0.15 μm. Volz's measurements of neutral point positions show the very sensitive behavior of the Babinet point to volcanic aerosols. An equivalent sensitivity was observed by Kimball (1912b) in the cloud from Mount Katmai and measured by Coulson (1983a) for the El Chichon cloud over Hawaii.

The data of skylight polarization and intensity for the El Chichon cloud over Hawaii as measured at the Mauna Loa Observatory are too extensive for a complete discussion here. They have, however, been documented (Coulson, 1983a, 1983b), and a brief summary is given below. The data show that while the intensity and color ratio of skylight were both sensitive to the volcanic effect, the most profound modification was reserved for the degree of polarization and the behavior of the neutral points. In fact, during the densest period of the cloud, a new neutral point, which had not been observed previously, appeared at the zenith at the time the sun was only 3° above the horizon. This extreme value of Babinet point distance of 87° was observed only on one occasion (May 14, 1982), but the behavior of the plane of polarization in conjunction with the observation leaves no doubt as to the existence of the anomalous neutral point on that one occasion.

The degree of polarization at the zenith for the presunrise and immediate post-sunrise period at MLO at the time the cloud had become stabilized (July 13, 1982) is compared with analogous measurements by various authors in Fig. 9.8. The measurements of Jensen (1942) and Tichanowski (1926) were in the industrial area of Frankfurt and in the clear atmosphere of Kazakhstan, respectively, both being in nonvolcanic conditions. The dashed curve of Coulson (1980) is likewise for a clean-stratosphere case, as is the upper curve of Steinhorst (1977). Steinhorst considered the depression of his second curve to be due to lingering remnants of the Agung cloud, but this seems questionable in view of the fact that they followed the Agung eruption by some 5 years. A stratospheric contribution from some minor eruption in the interim seems more likely. The Agung cloud was definitely over the station at the time of Shah's (1969) twilight observations in India on February 1, 1964, however, which yielded the polarization curve shown in the diagram. Shah's data show essentially the same configuration as those at Mauna Loa taken on July 13, 1982, especially the primary minimum occurring at about 4° solar depression, but the Mauna Loa cloud was

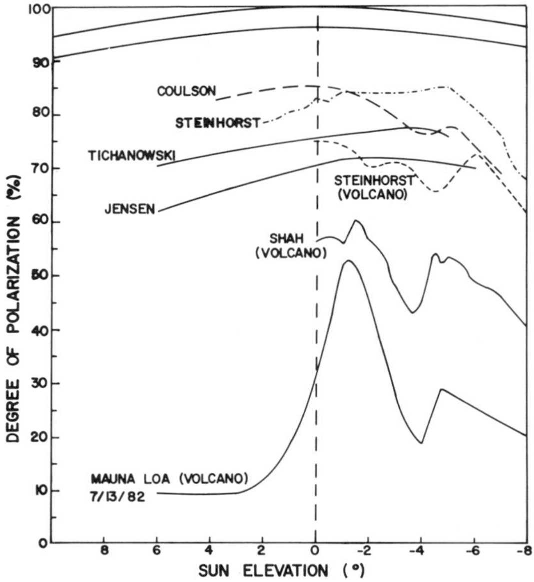

Figure 9.8 Degree of polarization of zenith skylight as a function of sun elevations at the Mauna Loa Observatory on July 13, 1982, compared with that observed by various authors in the past. The unlabeled curves at the top are for the Rayleigh atmosphere with primary and multiple scattering.

more extreme. Unfortunately Shah's measurements were terminated at sunrise, and the rapid decrease of polarization centered on the time of sunrise is the most dramatic feature of the zenith polarization at MLO.

Curves typical of the course of zenith polarization throughout the sunrise period at MLO are shown in Fig. 9.9. The principal features are the pronounced polarization maximum just prior to sunrise, the minimum and secondary maximum at solar depressions of 4° and 6°, respectively, and the plummeting of the polarization just at sunrise. After this dramatic decrease, the polarization over the entire sky remained low throughout the day. For instance, the polarization distribution in the sun's vertical given by Coulson (1983a), but not reproduced here, shows that in the stabilized phase of the cloud the polarization maximum at $\lambda=0.70$ μm in the sunlit period never reached 20% whereas in normal periods it was 80% or more. The data show that the polarization field over the whole sky was affected in a typical pattern, a fact that led Coulson (1983a) to suggest

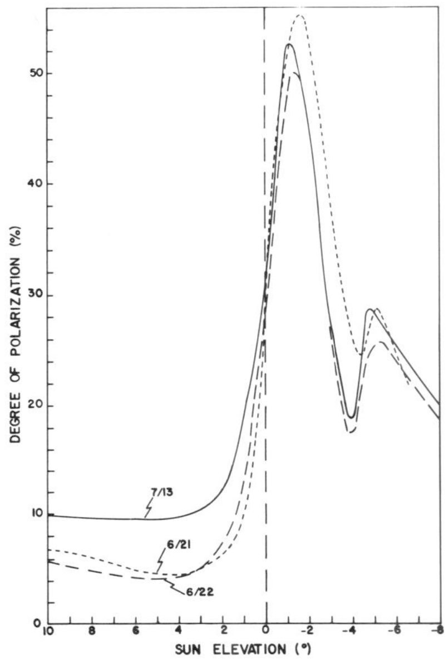

Figure 9.9 Degree of polarization of zenith skylight as a function of sun elevation as observed at λ=0.70 μm at the Mauna Loa Observatory on three different days during the stabilized phase of the El Chichon stratospheric cloud.

that measurements in one direction, such as the zenith, can be a good index of the behavior of the whole field. This would alleviate the need for many measurements for characterizing the polarizing effect of the cloud.

It must be realized for an inversion of the polarization field to give the total optical thickness and vertical distribution of the cloud, much less the particle size distribution, is not an easy task. However, some success in this regard was achieved by King and Fraser (1983) in reproducing the principal features of polarization and intensity of the daytime sky at Mauna Loa by a radiative transfer model of a plane-parallel atmosphere containing a volcanic cloud between 22- and 28-km altitude, a total aerosol optical depth of 0.233, and a strongly peaked aerosol size distribution with a mode radius of 0.4 μm. Not only was the measurement distribution of polarization and intensity in the sun's vertical fairly faithfully reproduced, but they also obtained multiple neutral points in positions roughly corresponding to those observed. Based on the plane-parallel atmospheric model, the method was not amenable to studies of twilight phenemona.

The displacement of the neutral points at Mauna Loa by El Chichon, particularly the Arago and Babinet points, was much more extreme than that produced by the Mount Katmai cloud of 1912. The behavior of the Arago point is shown in Fig. 9.10. The normal curve and the 1912 curve were obtained by Neuberger (1950) from the observations of Jensen (1928) in Hamburg before and after all the Katmai eruption. Similar results were given by Kimball (1913b) from observations in Virginia. In normal conditions the Arago distance is 17° to 18° and undergoes little change with sun elevation. The behavior changed markedly in 1912, with a pronounced minimum developing in the presunrise period, followed by a rapid increase of Arago distance by over 11° by 4° sun elevation. A comparison of these results with the Mauna Loa observations taken about 4 months after the El Chichon eruption, given by the other three curves in Fig. 9.10, shows the variations as much more extreme in 1982 than in 1912. Possible reasons for the differences are (1) the El Chichon cloud may have been denser than the

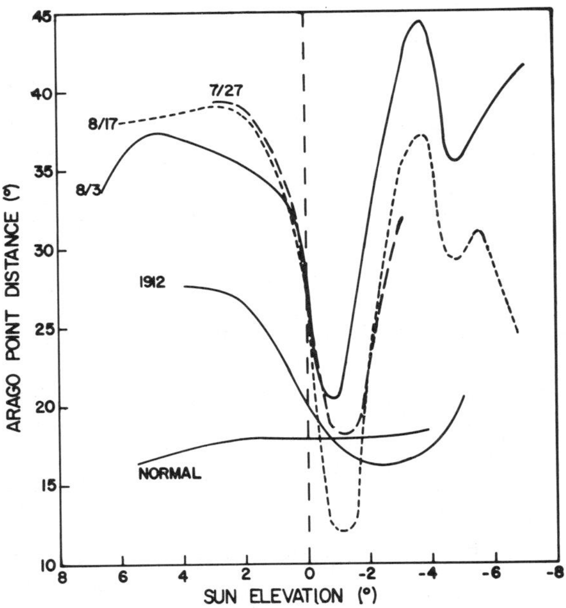

Figure 9.10 Angular distance of the Arago point from the antisun as a function of sun elevation as observed at λ=0.70 μm in the stabilized cloud at the Mauna Loa Observatory. The annual mean values from Jensen's observations in normal atmospheric conditions and after the eruption of Mount Katmai in 1912 are given for comparison.

Katmai cloud, (2) the altitude distribution may have been different for the two, and (3) the Mauna Loa observations were at $\lambda=0.70$ μm, whereas those of 1912 were in white light. The first of these possibilities is the most logical choice, as lidar observations show the El Chichon cloud to have been dense and deep, and other evidence indicates that El Chichon was a very large eruption. The wavelength difference between the two sets of measurements is probably of minor significance, as a few observations by Jensen indicate the Arago distance in red to have been less than 2° different from that in blue.

The Babinet point position is known, from Jensen's observations of the Katmai cloud, to be even more sensitive to volcanic debris than is the Arago point. Jensen's observations, as summarized by Sekera (1950), are shown for normal and 1912 conditions in Fig. 9.11. These data are qualitatively similar to analogous observations of Kimball (1913b). There is, however, little correspondence between the 1912 and Mauna Loa data, whereas the Mauna Loa data for 3 different days shows a common and characteristic pattern. The presunrise maximum of Babinet distance of over 50° is associated with the zenith polarization minimim at about

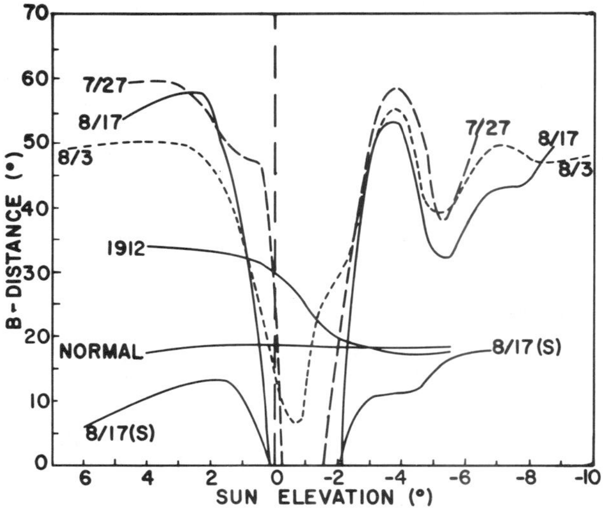

Figure 9.11 Angular distance of the Babinet point from the sun as a function of sun elevation as observed at $\lambda=0.70$ μm in the stabilized cloud at the Mauna Loa Observatory. The annual mean values from Jensen's observations in normal atmospheric conditions and after the eruption of Mount Katmai in 1912 are given for comparison.

the same solar depression angles, and the major minimum of the curves in Fig. 9.11 occurred at the time of the zenith polarization maximum just before sunrise. Finally, the dramatic rise in Babinet distance just after sunrise corresponds closely in time with the equally dramatic fall of zenith polarization through the sunrise period. This quasi-simultaneity of Babinet point shifts, Arago point shifts, and zenith polarization changes indicates that polarization features across the whole sky tend to occur in concert.

Two features of Fig. 9.11 are of particular interest. First, the Babinet point at $\lambda=0.70$ μm completely disappeared on 2 of the 3 days of observations. This is not a particularly unusual occurrence for dense turbidity in the lower atmosphere, but as far as is known it has not previously been observed in volcanic clouds. A more unique phenomenon is the appearance on August 17 of a second neutral point between the normal Babinet point and the sun. The existence of this anomalous distribution of polarization, with positive polarization near the sun, a region of negative polarization at some distance above the sun, and again positive polarization still farther above the sun, was confirmed by observations in four scans of the sky on August 17 and by several other scans 5 days later.

Skylight polarization measurements of the El Chichon cloud over Mauna Loa were continued from the time of its first appearance on April 9, 1982, through the remainder of 1982 and until late fall of 1983. Although both skylight and lidar measurements showed a lingering effect of the cloud, the situation had returned more or less to normal by the end of 1983.

The intensity of skylight at MLO was measured simultaneously with the polarization, and some significant changes in intensity and color of the sky were brought about by the El Chichon cloud (Coulson, 1983b). Instead of being a deep blue color as is normal for the sky at MLO, the volcanic debris produced a washed-out blue cast over the entire sky. The boundaries of the aureole were not sharply defined as they normally are in clear conditions, but instead there was a gradual diminution of intensity in all directions with increasing angle away from the solar disk. The Bishop ring was not observed at MLO at any time during the El Chichon episode. The intensity minimum, normally located about 90° from the sun, was shifted to 105° to 110° in the cloud, and the minimum was not as sharply defined as it is in normally clear conditions.

Perhaps the most striking intensity changes associated with the cloud occurred during twilight (Coulson, 1983b) in the longer wavelength range. The vivid color changes of twilight would indicate preferential scattering in the red, and data at $\lambda=0.70$ μm were emphasized in the Mauna Loa measurements. The change of zenith intensity at this wavelength as measured in the mature cloud is shown as a function of sun elevation in Fig. 9.12. Included for comparison are two curves representing clear conditions, one at MLO as an average of 9 clear days in 1977 and the other obtained by Ashburn (1952) at an altitude of 1653 m in the Mojave Desert of California. Ashburn's data has been normalized to the same intensity as that of MLO at sunrise. The similarity of the curves for clear conditions and those for El Chichon at solar depression angles greater than 6° indicates that the atmosphere in the 36- to 65-km range, which is above the top of the El Chichon cloud, was not greatly different in 1982 than it was in

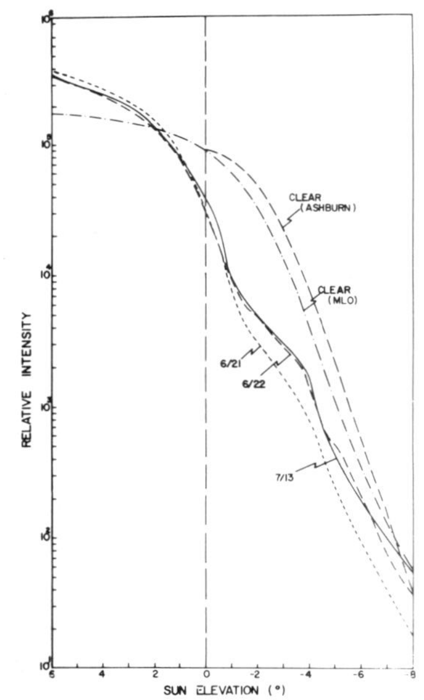

Figure 9.12 Relative intensity of zenith skylight as a function of sun elevation at $\lambda=0.70$ μm measured on three different days in the stabilized volcanic cloud at the Mauna Loa Observatory. Reference curves for the clear atmosphere, as observed at MLO in 1977 (Coulson, 1981) and as observed by Ashburn (1952) at an altitude of 1658 m in the Mojave Desert of California are included in the diagram.

in the clear periods of 1952 and 1977. As the solar beam descended through the cloud, however, the intensity in 1982 became anomalously low in the range of solar depression angles 4.5° to 6°, probably due to attenuation of the direct beam and light scattered downward from above. At the time the solar beam had descended to below the cloud ($D \approx 3.5°$ to $4.5°$), the similarity of the slopes of the curves indicates an addition of light by scattering in the clear tropospheric air, while their remaining low values are still the result of attenuation by the cloud of higher levels.

The change in intensity by nearly four orders of magnitude that occurs during twilight makes interpretation of the data more difficult than if less rapidly changing parameters were available. One such parameter, and the one advocated as a useful index by Volz (1969), is the ratio of intensities of two different wavelengths—the color ratio (CR)—during the twilight period. Whereas Volz concentrated on measurements at a point in the sky 20° above the solar horizon, Coulson (1981) determined color ratios in the direction of the zenith, and found them to be similarly responsive to atmospheric conditions. The CR (0.70/0.32 μm) in the zenith direction at MLO is shown as a function of sun elevation for 2 days of the El Chichon volcanic episode and for the average of 9 clear days of 1977 in Fig. 9.13. The large discrepancies correspond well to visual appearances of the sky in different conditions. The intense red color observed from early to mid-twilight is responsible for the large CR values in the D=5° to 8° range of solar depressions. The shortwave intensities during twilight are affected much less strongly than are the longer wavelengths. By the time D reaches 4°, the red color in the sky has faded and been replaced by yellow. In the period just before and just after sunrise, the color of the sky has changed to a strange bluish-white color, and, at slightly higher sun elevations, the sky color is almost entirely white. The strong diminution of red light during the immediate sunrise period results in the observed negative values of CR (0.70/0.32 μm) shown in the diagram, while the virtual lack of color in the sky after sunrise is reflected in the CR ≈ 1.0.

This color sequence is in strong contrast to that in clear conditions, in which case the red color is strongest at small solar depression angles. Since the bluish-white color regime does not occur in clear conditions, the CR remains positive through the sunrise period.

(a)

(b)

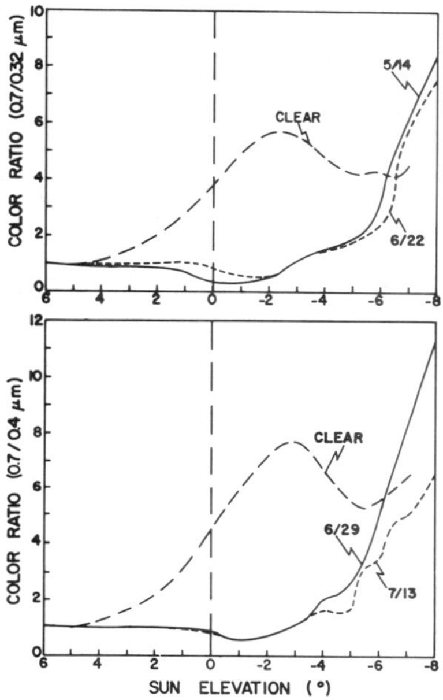

Figure 9.13 Color ratios at the zenith as a function of sun elevation as observed in the volcanic cloud compared with those for the average of 9 clear days in 1977 at the Mauna Loa Observatory for two wavelength pairs: (a) λ=0.70 *μm and* 0.32 *μm; and (b)* λ=0.70 *μm and* 0.40 *μm.*

The CR (0.70/0.40 μm) of Fig. 9.13b differs from that of 0.70/0.32 μm only in details. The cloud being somewhat thinner in late June and July may be mainly responsible for the small differences shown, but the intense reds and yellows typical of the large solar depressions could be expected to be dominant over both blue and ultraviolet light during the period.

It is clear that an adequate understanding of the measurements presented here will require the development and use of models of the volcanic cloud, both cloud particles and cloud geometry, in conjunction with a complete model of the unperturbed spherical and refracting atmosphere. As far as is known, no such comprehensive models have been utilized, although efforts in that direction are under way. Unfortunately, sampling and analysis of particles from the main body of the cloud are restricted in scope, and uncertainties in particle characteristics remain an impediment in modeling activities.

9.3 Discussion

The eruptions of volcanoes, and their atmospheric and climatic effects, constitute one of the most interesting and important subjects in geophysics. The effects of volcanoes in triggering the onset of ice ages has long been a favorite thesis of many scientists, and the possibility of worldwide temperature changes of at least several tenths of a degree in the lower atmosphere and several degrees in the stratosphere caused by volcanic debris from large eruptions is now beyond question. Such changes are difficult, however, to distinguish from the much greater natural variability of the atmosphere, and the extent to which the overall circulation of the atmosphere, and the resulting climate, is affected by volcanic clouds is still an active area of research. For instance, it has been suggested that the unprecedented strength of the 1982–83 El Nino, and perhaps even its initiation, was the result of the stratospheric cloud from the El Chichon eruption. If climatologists eventually find this to be the case, then the potential for major disruptions of climate by volcanoes has not been fully appreciated heretofore. The unusually cold summer experienced in New England and in Europe in 1816 has been ascribed to the gigantic eruption of Mount Tamboro in 1815, and there is evidence that the cloud from Krakatoa caused climatic anomalies in some regions.

Regardless of possible climatic effects, more immediate effects of volcanic eruptions are widespread. In the negative column must be listed the diminution of direct solar radiation that drives the solar power plants in California and elsewhere, the reduced response of photovoltaic and thermal solar collectors, diminished energy for photosynthesis, possible changes of atmospheric ozone, and a host of other aspects of response to solar radiation over the world. Offsetting, though only to a small degree, the general devastation and loss of life resulting from a major eruption are the positive effects of beautiful sunsets, Bishop's rings, and other optical phenomena which are viewed by millions of inhabitants of the planet.

Gauges of the size or severity of volcanic eruptions, such as the dust veil indices or severity index, remain crude because of lack of data. This situation is changing, however, as new observation tools—satellites, high-altitude aircraft, ground-based lidar—are brought into play. For instance, the El Chichon cloud has been observed, probed, collected, photographed, and generally examined as no previous cloud has been, and no doubt the trend will continue for future volcanoes. The one area which seems to have been unduly neglected is, in this writer's opinion, observations of skylight. The foregoing discussions in this chapter, as well as many studies elsewhere, have demonstrated that much information on volcanic clouds is simply, quickly, and inexpensively obtained by selected observations of the intensity and polarization of light from the sunlit sky.

References

Abbot, C.G., F.E. Fowle, and L.B. Aldrich (1913): *Annals of the Astrophysical Observatory,* Smithsonian Institution, Washington, D.C., Vol. III.

Abbot, C.G., F.E. Fowle, and L.B. Aldrich (1922): *Annals of the Astrophysical Observatory,* Smithsonian Institution, Washington, D.C. Vol. IV.

Anonymous (1912): Volcanoes of Alaska, *National Geographic Mag., 23(8).*

Ashburn, E.V. (1952): The density of the upper atmosphere and the brightness of the twilight sky, *J. Geophys. Res., 57,* 85–93.

Baker, C.B., W.R. Kush, and E. Rysnar (1984): Effects of the El Chichon volcanic cloud on direct and diffuse solar irradiances, *J. Clim. Appl. Met., 23,* 449–452.

Booth, P.W., S.W. Matthews, and R.E. Sisson (1963): Disaster in Paradise, *National Geographic Mag., 124(3),* 436–438.

Budyko, M.I. (1984): Effects of volcanic eruptions on climate, *Meteorologiya i Gidrologiya, No. 3,* 5–11.

Carroll, J.J. (1984): Effects of stratospheric aerosol on measured short-wave radiation incident at the ground, *J. Geophys. Res., 89(D),* 2553–2561.

Cornu, A. (1884): Observations relatives a la couronne visible actuellement autour du soleil, *Comptes Rend., Acad. Sci., Paris, 99,* 488–493.

Coulson, K.L. (1980): Characteristics of skylight at the zenith during twilight as indicators of atmospheric turbidity. 1: Degree of polarization, *Appl. Opt., 19,* 3469–3480.

Coulson, K.L. (1981): Characteristics of skylight at the zenith during twilight as indicators of atmospheric turbidity, 2. Intensity and color ratio, *Appl. Opt., 20,* 1516–1524.

Coulson, K.L. (1983a): Effects of the El Chichon volcanic cloud in the stratosphere on the polarization of light from the sky, *Appl. Opt.*, *22*, 2265–2271.

Coulson, K.L. (1983b): Effects of the El Chichon volcanic cloud in the stratosphere on the intensity of light from the sky, *Appl. Opt.*, *22*, 2265–2271.

Dave, J.V. (1970): Intensity and polarization of the radiation emerging from a plane-parallel atmosphere containing monodisperse aerosols, *Appl. Opt.*, *9*, 2673–2684.

Deirmendjian, D. (1969): *Electromagnetic Scattering on Spherical Polydispersions,* Elsevier, New York.

Deirmendjian, D. (1973): On volcanic and other turbidity anomalies, *Adv. in Geophys.*, *16*, 276–296.

DeLuisi, J.J., E.G. Dutton, K.L. Coulson, T.E. DeFoor, and B.G. Mendonca (1983): On some radiative features of the El Chichon volcanic stratospheric dust cloud and a cloud of unknown origin observed at Mauna Loa, *J. Geophys. Res.*, *88(C11)*, 6769–6772.

Dorno, C. (1919): Himmelshelligkeit, himmelspolarisation und sonnenintensitat in Davos (1911 bis 1918), *Meteorol. Zeit.*, *36*, 5/109–128, 181–192.

Dutton, E., and J. DeLuisi (1983): Spectral extinction of direct solar radiation by the El Chichon cloud during December 1982, *Geophys. Res. Lett.*, *10*, 1013–1016.

Dyer, A.S., and B.B. Hicks (1968): Global spread of dust from the Bali eruption of 1963, *Roy. Meteor. Soc. Quart. J.*, *94(402)*, 545–554.

Evans, W.F.J., and J.B. Kerr (1983): Estimates of the amount of sulphur dioxide injected into the stratosphere by the explosive volcanic eruptions: El Chichon, mystery volcano, Mount St. Helens, *Geophys. Res. Lett.*, *10(11)*, 1049–1051.

Flowers, E.C., and J.H. Viebrock (1965): Solar radiation: An anomalous decrease of direct solar radiation, *Science*, *148*, 493–494.

Grams, G., and G. Fiocco (1967): Stratospheric aerosol layer during 1964 and 1965, *J. Geophys. Res.*, *72*, 3523–3542.

Gruening, E. (1963): Lovely wonders of Katmai, *National Geographic Mag.*, *123(6)*, 800–831.

Hofmann, D.J., and J.M. Rosen (1984): Balloonborne particle counter observations of the El Chichon aerosol layers in the 0.01–1.8 μm radius range, *Geof. Int., 23(2)* 155–185.

Hogg, A.R. (1963): The Mount Agung eruption and atmospheric turbidity, *Austr. J. Sci., 26(4),* 119–120.

Irvine, W.M., and F.W. Peterson (1970): Observations of atmospheric extinction from 0.315 to 1.06 microns, *J. Atm. Sci., 27,* 62–69.

Jensen, C. (1928): Die Himmelsstrahlung, *Handbuch der Physik,* Bd. 19, Berlin, Springer, pp. 70–152.

Jensen, C. (1942): Die polarisation des himmelslichts, *Handbuch der Geophysik, 8,* 527–620.

Johnson, J.C. (1954): *Physical Meteorology,* M.I.T. Press and John Wiley, New York.

Kimball, H.H. (1913a): The effects upon atmospheric transparency of the eruption of Katmai volcano, *Mo. Wea. Rev., 41,* 153–159.

Kimball, H.H. (1913b): The effect of the atmospheric turbidity of 1912 on solar radiation intensities and skylight polarization, *Bul. Mt. Weather Observatory, V(5),* 295–312.

King, M.D., Harshvardhan, and A. Arking (1984): A model of the radiative properties of the El Chichon stratospheric aerosol layer, *J. Clim. Appl. Meteor., 23,* 1121–1137.

King, M.D., and R.S. Fraser (1983): The effect of El Chichon on the intensity and polarization of skylight at Mauna Loa, *Fifth Conf. on Atmospheric Radiation,* Baltimore, MD, Oct. 31–Nov. 4, 1983, Preprint Volume, pp. 370–373.

Lamb, H.H. (1970): Volcanic dust in the atmosphere; with a chronology and assessment of its meteorological significance, *Phil. Trans. R. Soc. London, 266,* 1121–1137.

Lamb, H.H. (1977): Supplementary volcanic dust veils assessments, *Clim. Monit., 6,* 56–67.

Matson, M., and A. Robock (1984): Satellite detection of the 1982 El Chichon eruptions and stratospheric dust cloud, *Geof. Int., 23(2),* 117–127.

Maurer, A., and C. Dorno (1914): Uber den verlauf und die geographische verbreitung der atmospharisch-optischen storung 1912–1913, *Meteor. Zeit., 31,* 49–62.

McCormick, M.P., and T.J.S. Swissler (1983): Stratospheric aerosol mass and latitudinal distribution of the El Chichon eruption cloud for October 1982, *Geophys. Res. Lett., 10,* 877–880.

Meinel, M.P., and A.B. Meinel (1963): Late twilight glow of the ash stratum from the eruption of Agung volcano, *Science, 142,* 582–583.

Meinel, M.P., and A.B. Meinel (1964): Height of the glow stratum from the eruption of Agung on Bali, *Nature, 201,* 657–658.

Mendonca, B.G., K.J. Hanson, and J.J. DeLuisi (1978): Volcanically related secular trends in atmospheric transmission at Mauna Loa Observatory, Hawaii, *Science, 202,* 513–515.

Mitchell, J.M. (1973): A preliminary evaluation of the atmospheric pollution as a cause of the global temperature fluctuation of the past century, in F. Singer (ed.) *Global Effects of Atmospheric Pollution,* Reidel Dordrecht, Netherlands, pp. 139–155.

Mitchell, J.M. (1982): El Chichon: Weather maker of the century, *Weatherwise, 35,* 252–259.

Mossop, S.C. (1964): Volcanic dust collected at an altitude of 20 km, *Nature, 203(4947),* 824–827.

Neuberger, H. (1950): Arago's neutral point: A neglected tool in meteorological research, *Bul. Am. Meteor. Soc., 31,* 119–125.

Newhall, C.B., and S. Self (1982): The volcanic explosivity index (VEI): An estimate of explosive magnitude for historical volcanism, *J. Geophys. Res., 87(C2),* 1231–1238.

Parker, D.E. (1985): Climatic impact of explosive volcanic eruptions. *The Meteorological Mag., 114(1355),* 1231–1238.

Peterson, E.C. (1983): Striking southern sunsets, *Weatherwise, 36,* 241–243.

Pickering, E.C. (1886): A new form of polarimeter, *Proc. Am. Acad. Arts and Sciences, XIII,* 294–302.

Pittock, A.B. (1966): A thin stable layer of anomalous ozone and dust content, *J. Atm. Sci., 23,* 538–542.

Pollock, J.B., O.B. Toon, C. Sagan, A. Summers, B. Baldwin, and W. Van Camp (1976): Volcanic explosion and climatic change: A research assessment, *J. Geophys. Res., 81,* 1071–1083.

Rampino, M.R., and S. Self (1982): Historic eruptions of Tambora (1915), Krakatau (1883), and Agung (1963): Their stratospheric aerosols and climate impact, *Quarternary Research, 18,* 127–143.

Rao, C.R.N., and W.A. Bradley (1983): Effects of the El Chichon volcanic dust cloud on insolation measurements at Corvallis, Oregon, *Geophys. Res. Lett., 10,* 389–391.

Robock, A. (1983): The dust cloud of the century, *Nature, 301,* 373–374.

Russell, F.A.R., and E.D. Archibald (1888): On the unusual optical phenomena of the atmosphere, 1883–1886, including twilight effects, coronal appearances, sky haze, coloured suns, moons, etc., Part IV of Symons, G.J. (Ed.) *The Eruption of Krakatoa and Subsequent Phenomena,* Krakatoa Committee, Royal Society, London, 494 pp.

Sekera, Z. (1950): Polarization of skylight, in *Compendium of Meteorology,* pp. 79–90, Am. Meteor. Soc., Boston, MA.

Shah G.M. (1969): Enhanced twilight glow caused by the volcanic eruption on Bali Island in March and September, 1963, *Tellus, XXI(5),* 636–640.

Simkin, T., and R.S. Fiske (1983): *Krakatau 1882: The Volcanic Eruption and Its Effects,* Smithsonian Institution Press, Washington, D.C.

Soret, J.L. (1888): Influences des surfaces d'eau sur la polarisation atmospherique et observation de deux points neutres a droite et a gauche de soleil, *Comptes Rend., Acad. Sci. Paris, 107,* 867–870.

South Kensington Museum, London (1888): Catalogue of Sky Sketches by William Ashcroft—Illustrating Optical Phenomena attributed to the Eruption of Krakatoa in the Java Straits August 27, 1883, 18 pages.

Spinhirne, J.D. (1977): El Chichon eruption cloud: Latitudinal variation of the spectral optical thickness for October 1982, *Geophys. Res. Lett., 10,* 881–884.

Steinhorst, G. (1977): Stratospheric aerosol concentration determined by an iterative method from twilight polarization measurements. *Beit. zur Phys. Atm., 50,* 508–523.

Swissler, T.J., M.P. McCormick, and J.D. Spinhirne (1983): El Chichon eruption cloud: Comparison of lidar and optical thickness measurements for October 1982, *Geophys. Res. Lett., 10,* 885–888.

Symons, G.J. (Ed.) (1888): *The Eruption of Krakatoa and Subsequent Phenomena,* Krakatoa Committee, Royal Society, London, 494 pp.

Szymber, R.J., and W.D. Sellers (1985): Atmospheric turbidity at Tucson, Arizona, 1956–83: Variations and their causes, *J. Appl. Met., 24,* 725–734.

Tichanowski, J.J. (1926): Resultate der messungen der himmelspolarisation in verschiedenen spektrumabschnitten, *Meteor. Zeit., 43,* 288–292.

Toon, O.B., and J.B. Pollack (1982): Stratospheric aerosols and climate, in R.C. Whitten (Ed.) *Topics in Current Physics,* Springer Verlag.

Tsuya, H. (1955): Geological and petrological studies of volcano Fuji, *Tokyo Daigaku Jishin Kenkyusho Iho, 33,* 341–382.

Verschure, P.P.H. (1984): Optical effects of El Chichon's volcanic aerosol, *Weather, 39,* 243–246.

Volz, F.E. (1964): Twilight phenomena caused by the eruption of the Agung volcano, *Science, 144,* 1121–1122.

Volz, F.E. (1969): Twilights and stratospheric dust before and after the Agung eruption, *Appl. Opt., 8,* 2505–2517.

Volz, F.E. (1975): Distribution of turbidity after the 1912 Katmai eruption in Alaska, *J. Geophys. Res., 80,* 2643–2648.

Volz, F.E. (1984): Volcanic turbidity, skylight scattering functions, sky polarization, and twilights in New England during 1983, *Appl. Opt., 23,* 2589–2593.

Weinert, R.A. (1967): The movement and dispersion of volcanic dust from the eruption of Mount Agung, Bali, 17 March 1963, *Aust. Meteor. Mag., 15,* 225–229.

Weintraub, B.N., and R.I. Tilling (1982): The disaster of El Chichon, *National Geographic Mag., 162(5),* 654–684.

Wendler, G. (1984): Effects of the El Chichon volcanic cloud on solar radiation received at Fairbanks, Alaska, *Bul. Am. Meteor. Soc.*, *65*, 216–218.

Wexler, H. (1951): Spread of the Krakatoa volcanic dust cloud as related to the high-level circulation, *Bul. Am. Meteor. Soc.*, *32*, 48–51.

CHAPTER 10

REFLECTING AND POLARIZING PROPERTIES OF NATURAL SURFACES

10.1 Background

Remote sensing of the earth's surface from aircraft or spacecraft is of scientific interest in many disciplines, including agriculture, hydrology, oceanography, geology, glaciology, and others. The economic value of remote sensing is now well recognized, as evidenced by the widespread use of observational data from the LANDSAT program of the United States, and the development of analogous satellite systems in Europe, Japan, and elsewhere. The data are of great value in mineral exploration, location of oil fields, areal extent and production potential of agricultural crops, forestry, grazing land inventory, land use planning, and similar uses, as well as in the many meteorological applications of satellite data.

The quantity that has received the lion's share of attention in remote sensing is the intensity of the reflected light, including its spectral distribution, angular dependence, and time variations, and a tremendous volume of data on light intensity at remote locations has been and continues to be accumulated. For instance, large banks of archival data tapes have resulted from earth observation programs of many types, but, thus far, the only observations of the earth from space that have included polarization measurements are those in an exploratory project aboard the Space Shuttle (Coulson et al., 1986). These data will be mentioned further below. By the nature of the problem, however, solar radiation directed to space at the level of the Shuttle and other spacecraft contains a significant component due to scattering by the atmosphere, not just that due to surface reflection. For surface characterization and discrimination, however, such atmospheric contamination of the radiation field should be minimized or corrected for by use of radiative transfer models applicable to the conditions of observation. For maximum information content, of course, both intensity and state of polarization of the reflected light should be included.

The traditional method of sensing the earth's surface from a remote platform has been by the use of photographic images. The first images obtained in aerial photography were with black and white film, but since about World War II color film with sensitivity extending to 0.90 μm has been available for this purpose. Photo interpretation of such images yields information on size, shape, texture, brightness, color, shadows, and other attributes of surface features. The space program has seen the development of new types of electronic sensors for the acquisition of images in several different spectral bands. An overview of the development and use of meteorological satellites over the last 25 years has recently been given by Smith (1986), as has a summary by Yates et al. (1986) of the use of satellites for observations of the earth's surface. Examples of the type of earth observations made from space are the measurements of solar insolation reaching the ground (Justus et al., 1986), and the monitoring of agricultural crops by the Advanced Very High Resolution Radiometer (AVHRR). Other instruments that have enjoyed widespread use for earth observations are the Multispectral Scanner (MSS) and Thematic Mapper (TM). A comprehensive discussion of sensing systems so far deployed in space is that of Chen (1985). The use of magnetic tape for recording signals from space systems permits computer processing of the data and the development of correlations and other statistical measures for maximum discrimination of the parameters describing the surface. In addition, repeated satellite passes over a given area give information on the time history of the developing crop, which can be interpreted in terms of growth stage, the onset of diseases or other problems in the crop, and even potential yield from the fields.

Basic to the whole problem of remote sensing of the earth's surface from space is the requirement for "ground-truth" data—i.e., reflection data taken at ground level for the various types of surfaces of interest. A very large body of literature has been built up on the reflection properties of various natural and artificial surfaces. Perhaps the most extensive observations of the spectral reflectance of natural surfaces is that of Krinov (1953), in which measurements made on natural surfaces show total reflectance values from as little as 0.02 for black sandy loam to as much as 0.35 to 0.75 for highly reflecting clay, limestone, and shale. A large number of reflection data for various surfaces have been compiled by Earing and Smith (1966); Coulson et al. (1965) and Coulson and Reynolds (1971) measured the reflection properties of various natural surfaces, as did Rao and Chen (1969), Egan (1968) and others.

Before discussing the large number of reflection measurements reported by various authors, we outline the properties of the general reflection matrix and show some unpublished data obtained by the present author for nine of the matrix elements characterizing the reflection from selected surfaces.

10.2 The reflection matrix

Emphasis in studies of remote sensing has generally been on the intensity of the reflected light as a function of wavelength and angles of incidence and reflectance, although some studies have included the polarizing properties as well. In principle, more information may be contained in the field of polarization than in the intensity field, and the polarization field is also dependent on wavelength and geometrical parameters. Thus by simultaneous measurements of both the intensity and state of polarization, the most powerful tool for extracting information from a field of radiation is obtained.

The relation between the light of vector intensity $\mathbf{I}'(\lambda;\theta',\phi')$ of wavelength λ incident on a surface from zenith angle θ' and azimuth ϕ' and intensity I $(\lambda;\theta,\phi)$ of the same wavelength reflected in direction (θ,ϕ) is given by the matrix equation

$$\mathbf{I}\ (\lambda;\theta,\phi) = \mathbf{M}\ (\lambda,\theta,\phi;\theta',\phi')\ \mathbf{I}'(\lambda;\theta',\phi') \tag{10.1}$$

where $\mathbf{M}\ (\lambda;\theta,\phi;\theta',\phi')$ is the reflection matrix. The elements of the reflection matrix are surface dependent, both individually and collectively, and data on those elements provide a complete representation of the reflecting properties of a surface. On expansion we have the relation

$$\begin{vmatrix} I \\ Q \\ U \\ V \end{vmatrix} = \begin{vmatrix} M_{II} & M_{IQ} & M_{IU} & M_{IV} \\ M_{QI} & M_{QQ} & M_{QU} & M_{QV} \\ M_{UI} & M_{UQ} & M_{UU} & M_{UV} \\ M_{VI} & M_{VQ} & M_{VU} & M_{VV} \end{vmatrix} \begin{vmatrix} I' \\ Q' \\ U' \\ V' \end{vmatrix} \tag{10.2}$$

It is of interest to note that the scalar intensity, the quantity normally used in remote sensing, is given in terms of the matrix elements by the equation

$$I = M_{II}I' + M_{IQ}Q' + M_{IU}U' + M_{IV}V' \tag{10.3}$$

It is generally assumed that the remotely sensed light is unpolarized, in which case $Q = U = V = 0$. If this is not the case, the polarization may be a source of some error in the intensity measurements.

For practical purposes, a significant simplification of Eq. (10.2) is possible. Observations show that for natural sunlight, the ellipticity V' of the incident light is vanishingly small. This means that the reflected light carries no information on the four matrix elements, M_{IV}, M_{QV}, M_{UV}, and M_{VV}. These elements may well be valuable for characterizing a reflecting surface in the laboratory, but they are not observable under natural sunlit conditions. Thus we will neglect them in the present discussion. This leaves 12 of the 16 elements still available for surface characterization in sunlit conditions. Laboratory measurements have shown that reflection by most natural materials, with the exception of birefringent crystals, introduces vanishingly small amounts of elliptic polarization, in which case the additional elements M_{VI}, M_{VQ}, and M_{VU} may be neglected. Thus for linear polarization, M reduces to a nine element matrix rewritten as

$$\begin{vmatrix} I \\ Q \\ U \end{vmatrix} = \begin{vmatrix} M_{II} & M_{IQ} & M_{IU} \\ M_{QI} & M_{QQ} & M_{QU} \\ M_{UI} & M_{UQ} & M_{UU} \end{vmatrix} \begin{vmatrix} I' \\ Q' \\ U' \end{vmatrix} \tag{10.4}$$

As seen below, many authors have made laboratory measurements of scalar intensity I, degree of plane polarization P, and angle χ of the plane of polarization of light reflected from natural materials. It is well to look at the method of extracting values of the individual matrix elements from the three quantities measured. If the incident light is unpolarized, $Q' = U' = 0$, and the Stokes parameters of the reflected light yield three of the matrix elements as

$$I = M_{II} I'$$

$$Q = M_{QI} I' \tag{10.5}$$

$$U = M_{UI} I'$$

If, further, the incident light is rendered plane polarized, such as by passing it through a polaroid sheet placed over the source, we have a number of simple combinations of the parameters of the incident light, depending on the angle of the plane of polarization.

$$\text{For } \eta = 0°,\ Q'_0 = I'_0,\ U'_0 = 0$$

$$\text{For } \eta = 45°,\ Q'_{45} = 0,\ U'_{45} = I'_{45}$$

$$\text{For } \eta = 90°,\ Q'_{90} = I'_{90},\ U'_{90} = 0 \tag{10.6}$$

$$\text{For } \eta = 135°,\ \mathrm{Q}'_{135} = 0,\ U'_{135} = -I'_{135}$$

By performing the algebra indicated in Eq. (10.4) we obtain the relations

$$I_\eta = M_{II} I'_\eta + M_{IQ} Q'_\eta + M_{IU} U'_\eta$$

$$Q_\eta = M_{QI} I'_\eta + M_{QQ} Q'_\eta + M_{QU} U'_\eta \tag{10.7}$$

$$U_\eta = M_{UI} I'_\eta = M_{UQ} Q'_\eta + M_{UU} U'_\eta$$

by which the individual matrix elements can be determined.

From the definition of the Stokes parameters we have for the reflected light

$$Q = IP \cos 2\chi$$

$$U = IP \sin 2\chi \tag{10.8}$$

where the degree of polarization P and angle of the plane of polarization χ are determined by the measurements.

By substituting appropriate values from Eq. (10.7), it is easy to obtain the following relations in which the measurements are used to compute the nine matrix elements. It is assumed that the incident intensity I' is independent of its plane of polarization; i.e.

$$I'_0 = I'_{90} = I'_{45} = I'_{135} \tag{10.9}$$

The individual elements are determined explicitly by the following relations:

$$\begin{aligned}
M_{II} &= \frac{I_0 + I_{90}}{2I'} = \frac{I_{45} + I_{135}}{2I'} \\
M_{IQ} &= \frac{I_0 - I_{90}}{2I'} \\
M_{IU} &= \frac{I_{45} - I_{135}}{2I'} \\
M_{QI} &= \frac{I_0\, P_0 \cos 2\chi_0 + I_{90}\, P_{90} \cos 2\chi_{90}}{2I'} \\
M_{QQ} &= \frac{I_0\, P_0 \cos 2\chi_0 - I_{90}\, P_{90} \cos 2\chi_{90}}{2I'} \\
M_{QU} &= \frac{I_{45}\, P_{45} \cos 2\chi_{45} - I_{135}\, P_{135} \cos 2\chi_{135}}{2I'} \\
M_{UI} &= \frac{I_{45}\, P_{45} \sin 2\chi_{45} + I_{135}\, P_{135} \sin 2\chi_{135}}{2I'} \\
M_{UQ} &= \frac{I_0\, P_0 \sin 2\chi_0 - I_{90}\, P_{90} \sin 2\chi_{90}}{2I'} \\
M_{UU} &= \frac{I_{45}\, P_{45} \sin 2\chi_{45} - I_{135}\, P_{135} \sin 2\chi_{135}}{2I'}
\end{aligned} \tag{10.10}$$

Thus all of the elements can be determined by measurements with the incident light plane polarized and the plane of polarization oriented at four different angles.

This method was used, in collaboration with Dr. R.L. Walraven, for laboratory measurements of reflection in the principal plane at $\lambda=0.70$ μm by desert sand from the Mojave Desert, California and by a green leaf of the philodendron plant (*Philodendron bipinnatifidum*). Values of the normalized matrix elements for desert sand at four different angles of incidence ($\theta_0=0°$, $20°$, $40°$, $60°$) are shown by the nine sections of Fig. 10.1. Element M_{II} is the same quantity as bidirectional reflectance discussed by many authors. It was convenient for plotting purposes to normalize the other elements to M_{II}. The ratio M_{QI}/M_{II} is the degree of plane polarization of the reflected light, which for this case of highly reflecting desert sand at $\lambda=0.70$ μm reaches only about 12%. The position of the neutral points on either side of the source position are evident, and the general configuration of the curves is almost independent of illumination angle.

The ratio M_{UI}/M_{II} is the Stokes parameter U, and for this case of the principal plane it appears, within the error of measurement, to be identically zero, as would be expected. The physical significance of the other elements is not so simple as for M_{II}, M_{QI}, and M_{UI}. Elements M_{UQ} and M_{IU} are zero within measurement error. Element M_{IQ} is very similar to element M_{QI} for this case of the principal plane, but this may not be true at other azimuths. M_{UU} is large and mostly negative, whereas M_{QQ} is large and always positive.

Of perhaps more interest in the context of remote sensing are the analogous curves for a green philodendron leaf shown in Fig. 10.2. Errors of measurement are larger in this case because of the low intensity of the reflected light, but the curves show some of the same general features as those of the previous diagram. There is obviously more variation among the elements for the green leaf than for desert sand, a feature of considerable importance for potential application in remote sensing.

10.3 Reflection by various surfaces

10.3.1 Reflection by green vegetation

A very complete and detailed work on the radiation regime of plants has been issued by Ross (1981), with a list of over 450 references on the subject.

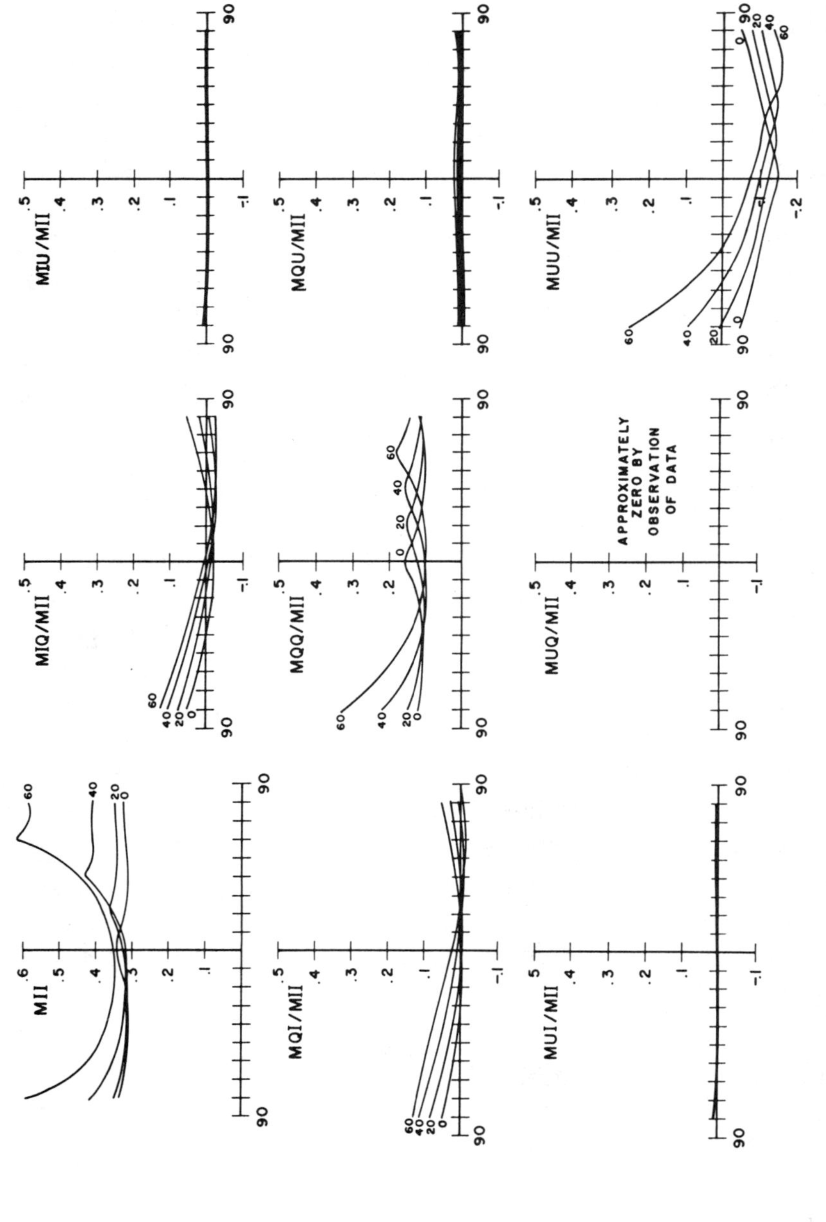

Figure 10.1 Measured values of the normalized elements of the reflection matrix for desert sand, as a function of angle in the principal plane, for four different angles of incidence. ($\theta_o = 0°, 20°, 40°, 60°$; $\lambda = 0.70\ \mu m$.)

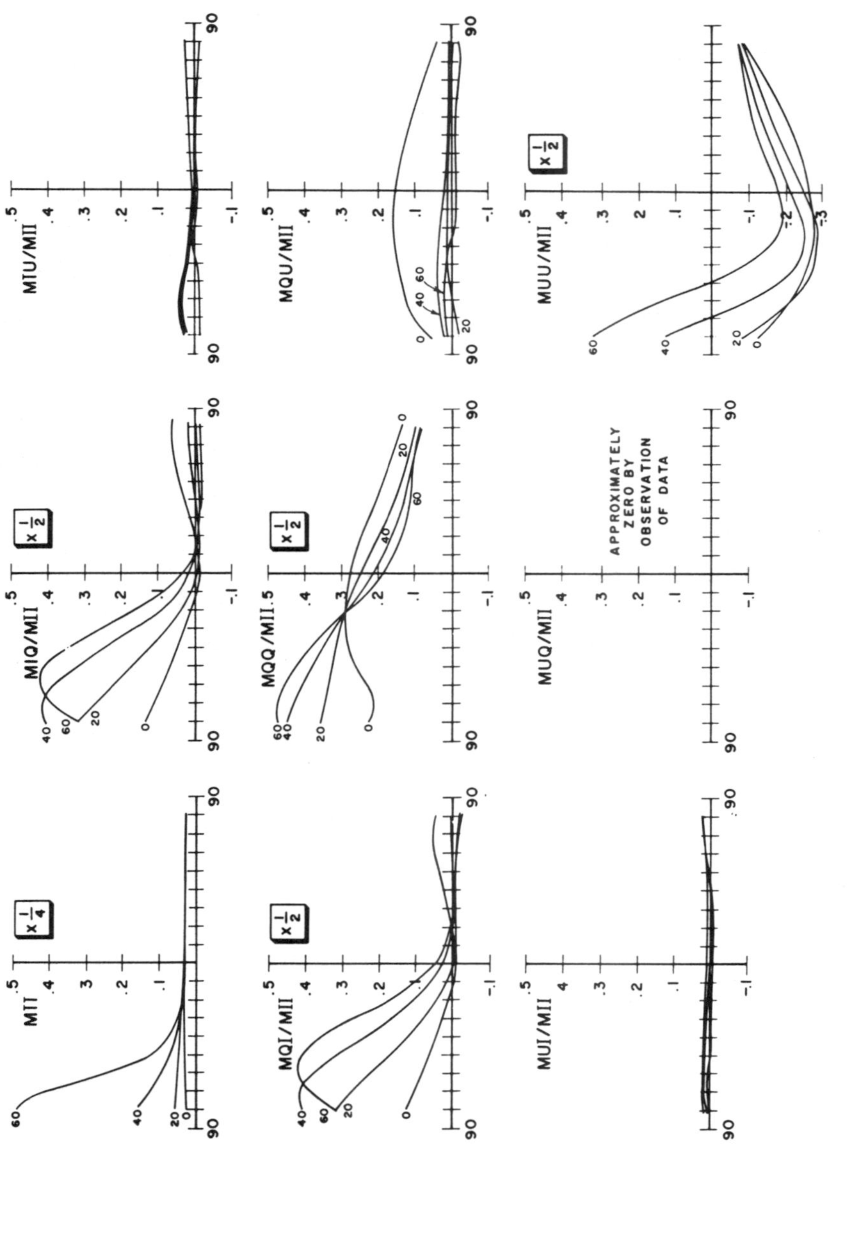

Figure 10.2 Measured values of the normalized elements of the reflection matrix of a green leaf of the philodendron plant, as a function of angle in the principal plane, for four different angles of incidence. ($\theta_o=0°, 20°, 40°, 60°$; $\lambda=0.70\ \mu m$.)

Ross used the simplified Schwarzschild equation to solve the radiative transfer problem in the plant canopy, including reflection, absorption, and scattering by the phytoelements. A similarly complete survey of the reflectance of shortwave radiation by agricultural crops was developed in a doctoral dissertation by Bunnick (1978). In addition to giving observational data on the reflection, absorption, and scattering of radiation by plants, Bunnick summarizes the modeling of the radiative regime of plant stands by seven different methods (Willstatter and Stoll, 1918; Kubelka and Munk, 1931; Allen and Richardson, 1968; Ross and Nilson, 1974; Suits, 1972; Smith and Oliver, 1974; Goudriaan, 1977). According to Bunnick, the last three of these models yield results which agree well with field measurements. Unfortunately, none of these models includes polarization effects. However, a newer model by Vanderbilt (1980) does include polarization of the reflected light.

Light reflected from normal green vegetation has a well defined spectral signature, as shown in Fig. 10.3. The wavelength interval λ=0.40 to 2.1 μm contains more than 85% of energy in the solar spectrum. While reflection spectra vary somewhat among plant species, this curve, as measured by Gates et al. (1965) by the intergrating sphere method for leaves of the oleander plant (*Nerium oleander*), is typical of the reflectance of green plants in general. The strong absorption bands of chlorophyll and other plant pigments are responsible for

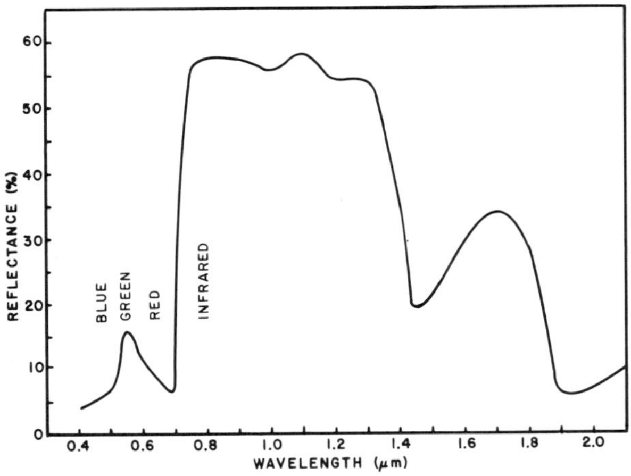

Figure 10.3 Spectral reflectance of green leaves of the oleander plant, according to measurements of Gates et al. (1955).

the low reflectance values in the $\lambda=0.4$ to $0.5\ \mu m$ and $\lambda=0.65$ to $0.69\ \mu m$ regions. As will be seen below, these are also the regions of polarization maxima. Lower absorption values at about $\lambda=0.55\ \mu m$ are responsible for higher reflectances for green light and predominance of green colors in growing plants. The dramatic increase of reflectance at about $\lambda=0.70\ \mu m$ is evidence of less absorption by plant pigments in the near infrared, with much of the radiation entering the leaf and being backscattered by the cellular structure, principally the spongy mesophyl, of the interior of the leaf. Finally, absorption by liquid water in the leaf is responsible for the low reflectance at wavelengths beyond $\lambda=1.3\ \mu m$.

The distribution of the reflected radiation in the outward hemisphere above plant canopies has been measured less frequently than has total reflectance, but the general outlines have been established. The distribution is a strong function of leaf type and orientation. For instance, vertically oriented leaves, such as in a growing stand of wheat or barley, present a different average aspect to the incident radiation and more mutual shadowing than do more horizontally oriented leaves of sugar beets or soybeans. An additional difference in the two types of orientations is the relative specularity of the reflected radiation, specular reflection being relatively higher for horizontal leaves than for those of other orientations. Since polarization of the reflected light is thought to be produced mainly by specular reflection, horizontal leaves normally polarize more strongly than do those of other orientation. Of course, plants in general have leaves of all possible orientations, although there may be a statistically preferred orientation in each case.

The angular distribution of light reflected in the principal plane from a rapidly growing bluegrass lawn is shown for an angle of incidence of 53° in Fig. 10.4. A turf of this type has leaves of all orientations. The factor π on the ordinate scale gives for each point in the diagram a value relative to what it would have for the reflection from a perfectly reflecting diffuse reflector. The strong change of reflectance with wavelength is evident in the diagram, but so also is a change of reflectance with the angle at which the surface is viewed. This quantity is often termed the bidirectional reflectance of the surface, the origin of the term indicating the two directions of the incident and reflected radiation. The broad minimum of the curves is at roughly 10° from the nadir direction in the antisolar azimuth, with significant increases in the direction backward toward the

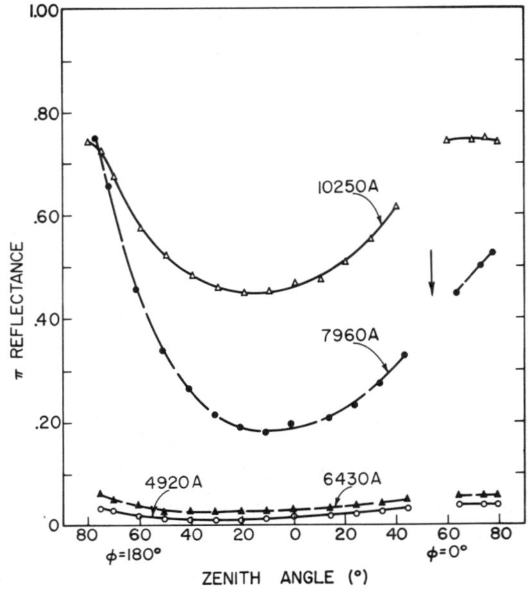

Figure 10.4 Bidirectional reflectance of green bluegrass turf at four different wavelengths in the principal plane for an angle of incidence of 53°. (After Coulson et al., 1965.)

sun and toward the antisolar horizon. These directional characteristics carry over to other angles of incidence as well. Other data show that the total reflectance changes generally by less than a factor of 2 with changing angle of incidence. Measurements by Coulson and Reynolds (1971) on growing stands of alfalfa, sugar beets, bluegrass, rice, and sorghum showed a slight maximum of reflectance at sun elevations of about 10°, with gradually decreasing reflectance with increasing sun elevations up to the highest sun (about 70°) at which measurements were made.

10.3.2 Reflection by mineral surfaces

The reflection spectra of soils, sands, and other mineral surfaces are very different from those of vegetation. In the visible wavelength range, the reflectance of such materials increases with increasing wavelength, as shown by the data of Fig. 10.5. The curves in this diagram are averages for four different types

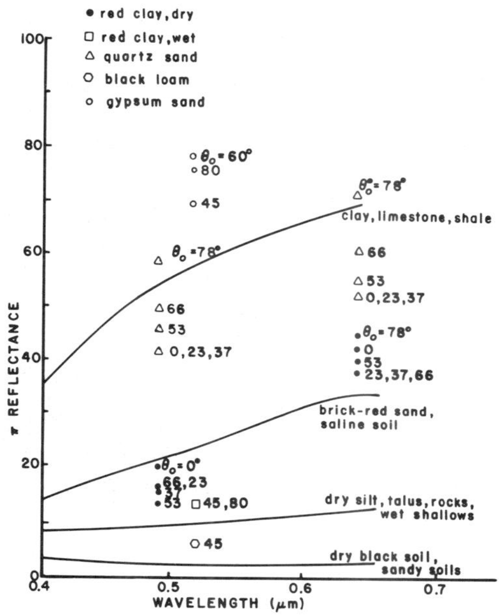

Figure 10.5 Average curve of spectral reflectance for four types of bare areas and soils, according to the measurements of Krinov (1947), and values of reflectance for five surfaces as determined by Coulson et al. (1965) at the indicated angles of incidence. The data are all for the nadir direction.

of bare areas and soils given by Krinov (1953), and the plotted points are from measurements on five different surfaces of soil and sands by Coulson et al. (1965). All of these data apply to the nadir direction of views; associated with the points are the angles of incidence at which the surfaces were illuminated. The increase of reflectance with increasing wavelength is very regular, with the difference between violet and red being generally less than a factor of 2. The total reflectance, however, can be very high for light-colored clay, limestone, and shale, and very low for the dark-colored soils and basaltic minerals.

The smooth character of mineral spectra extends into the near infrared at about 1 μm, but beyond that the spectra of hydrated minerals show strong absorption features due to water of hydration (Hovis and Callahan, 1966). These features are particularly pronounced in the vibration fundamentals of water at about 6.3 and 2.7 μm, with lesser absorptions at 1.4 and 2.2 μm. The total reflectance at wavelengths below about 8.0 μm increases significantly with decreasing particle size for submillimeter size particles, but this dependence is reversed at the longer infrared wavelengths.

As with vegetation, there is a strong dependence of reflectance of mineral-type materials on the angle at which the surface is viewed. This dependence is shown for the principal plane at $\lambda=0.643$ μm and angle of incidence 53° in Fig. 10.6. In addition to the larger differences in magnitude of reflectance among the various materials, a distinctive feature of the distributions is the reflectance maximum in the direction backward toward the source. This feature, which is a maximum for dark-colored materials, is understood to be due to mutual shadowing of the particles. The shadows are densest for strongly absorbing particles, as radiation entering the particles is mostly absorbed before it can emerge in all directions and soften the shadows. The only direction in which the shadows are not seen, of course, is the direction parallel to the incident radiation. For translucent particles, on the other hand, the shadows are softened by light entering the particles and being scattered outward in all directions, thereby illuminating

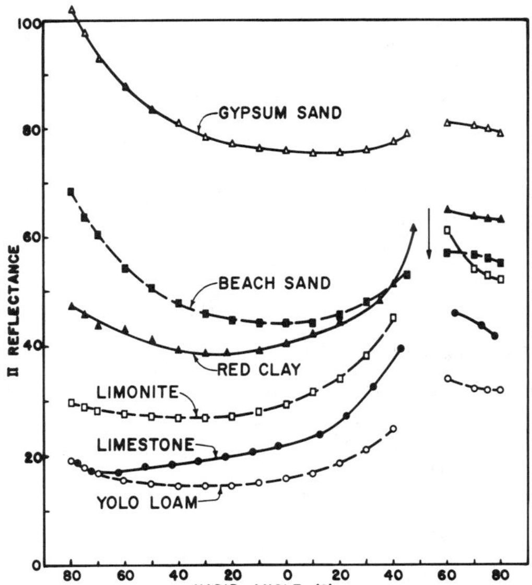

Figure 10.6 Bidirectional reflectance of various types of mineral surfaces in the principal plane for a wavelength of 0.643 μm and angle of incidence 53°. (After Coulson et al., 1965.)

the shadowed areas. This internally scattered light is likewise largely responsible for the generally high reflectance of light-colored surfaces.

10.3.3 Reflection by the sea

The reflection of light by water surfaces (oceans, lakes, bays, estuaries, etc.) is of interest in remote sensing of chlorophyll and phytoplankton in ocean waters, location of schools of fish and mapping of the bottom near beaches, detection of the wakes of ships and other disturbances in the water, as well as in energy balance computations, specification of the background light field for remote sensing of the atmosphere from space, and numerous other applications. The fact that reflection by a plane water surface follows Fresnel's laws of reflection makes the theory more tractable for water surfaces than for vegetated or mineral surfaces in general, although surface waves and bubbles, plankton, and other particles in the water are complicating factors.

The theory of sea surface reflection and the optical properties of sea water have been given in considerable detail in monographs (both in Russian) by Mullamaa (1964) and Shifrin (1983). In particular, Mullamaa's computed reflectances are given in over 250 diagrams and a larger number of tables for different combinations of solar elevation, zenith angle and azimuth of viewed direction, wind speed, and wavelength. A number of computations include the polarization of reflected light. Unfortunately, no observational material is given for validation of the computational results.

Perhaps the most complete set of measurements of total albedo of the sea is that of Payne (1972) taken with upward- and downward-looking pyranometers mounted on a fixed platform in Buzzards Bay, Massachusetts, in water of 20 m depth. The instruments, which were operated continuously over a 4-month period, measured the upward and downward global fluxes over the spectral interval 0.28 to 1.80 μm. The data were averaged over 15-minute intervals, albedo values were computed as the ratio between global fluxes in upward and downward directions, and expressed as a function of sun elevation and atmospheric transmittence. This latter quantity is the ratio of the downward global flux at the surface to that at the top of the atmosphere.

Typical albedo values for the sea, as tabulated by Payne, are shown as a function of sun elevation for atmospheric transmittances of 0.85, 0.65, 0.40, and 0.10 in Fig. 10.7. The data were smoothed and a certain amount of extrapolation was applied to cover all sun elevations and atmopsheric transmittances. The particular transmittances chosen for Fig. 10.7 represent a very clear atmosphere, a moderately turbid atmosphere, a heavily polluted atmosphere, and a case of heavy overcast. The inset in the diagram shows albedo for sun elevations above 40° on an expanded scale. Albedo was found to be a strong function of sun elevation in early morning or late evening for all cases except under overcast conditions. Payne did not observe the sun elevation dependence in cloudy conditions reported by Neiburger (1948) from measurements under stratus clouds off the California coast. Likewise, the average albedo of approximately 6% as observed by Payne under overcast conditions is much below the 10.5% observed by Neiburger, but probably within measurement error of the 6.6% computed by Burt (1954) for diffuse radiation incident on a smooth water surface. In addition, a decrease of reflectance of about 0.5% is brought about by wind roughening of the water surface. The discrepancy of more than 4% in albedo between Neiburger's and Payne's measurements under overcast conditons for the wind-roughened sea cannot be ascribed to light backscattered by bubbles and particles below the surface, as postulated by Neiburger. A previous measurement by Payne (1971) showed that such emergent radiation constitutes at most

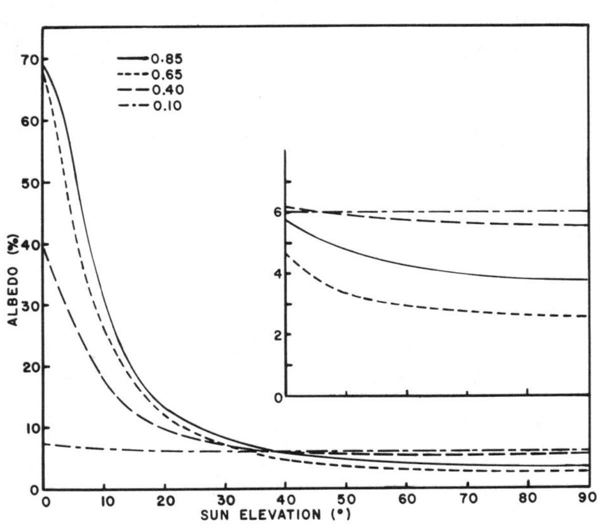

Figure 10.7 Albedo of the sea surface as a function of sun elevation for four different values of atmospheric transparency. Portions of the curves are plotted on an expanded scale in the inset. (Data from Payne, 1972.)

15% of the upward flux, which means that the albedo of 6.6% for diffuse incident flux would be about 7.6% if the submarine backscatter were included. Again, this is much below the 10.5% observed by Neiburger. Payne's (1972) results are close to the observed and computed values of 6% and 7% reported by Ter-Markariants (1961).

Most authors have found that the sea surface albedo decreases slightly with increasing wind speed, but the effect is small. Payne's (1972) data show an essentially negligible dependence for sun elevations of 50° or more, whereas for lower sun elevation the decrease was about 1% of the normalized albedo value for each knot (0.515 m/s) increase of wind speed, at least up to wind speeds at which whitecaps are formed. Monahan (1971) analyzed a large number of photographs of the sea surface under different wind conditions and found that whitecaps are totally absent for wind speeds less than 6 knots (3.1 m/s). For a range of wind speeds of 4 to 10 m/s the maximum whitecap coverage W was given by the relation

$$W = 0.00135\ V$$

whereas V is in m/s. For a 30 knot wind (15.5 m/s), this would give a whitecap coverage of 15%. Generally, however, the coverage was much less than this, observations in the Atlantic Ocean giving only 7.61% coverage for wind speeds between 17.15 and 17.7 m/s. Payne (1972) observed no change of albedo due to whitecaps for wind speeds up to 30 knots (15.5 m/s). Gordon and Jacobs (1977) found in a theoretical study, however, that wind speeds between 9 and 13 m/s could produce a 10% change of albedo by the formation of whitecaps.

The sunlight glitter pattern, a feature readily visible on satellite photos of the ocean surface, has been analyzed by several authors as an indicator of the wave structure of the wind ruffled surface. A unique orientation of a wave surface is required to cause the direct, quasi-parallel light from the sun to be specularly reflected into the aperture of a sensor viewing the ocean surface. For a calm sea the image of the sun on the surface would appear at the specular reflection point as in a mirror, but for a ruffled sea the solar image appears as small glints in numerous water facets of proper orientations surrounding the specular point. The slope of a facet necessary to reflect the solar image toward the sensor is a function of the displacement of the facet from the specular point, the larger the displacement the larger the required slope. This same feature is responsible for the bright path of light across the sea surface in the direction of the setting sun which everyone has seen at the beach. The path may be narrow or wide depending on the state of the sea. Thus the distribution of the glitter facets is an indicator of the wave pattern on the surface. The glitter is superimposed on a natural background of light from the sky which is reflected by the surface and light which enters the water and is backscattered by the submarine particles and bubbles in the water.

The most complete study of the sun's glitter on the sea surface is the classic investigation of Cox and Munk (1954) in which photographs of the wind-roughened sea taken from an airplane near Hawaii were analyzed by fitting a statistical Gram-Charlier series to the data. As shown in Fig. 10.8, the slopes

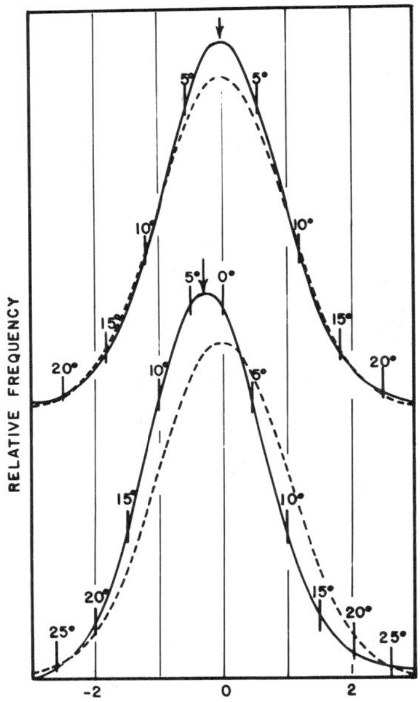

Figure 10.8 The probability distribution of slopes of waves causing the sun's glitter on the sea surface, according to the analysis of aerial photographs by Cox and Munk (1954). The solid lines show the observed distribution in cross-wind (top diagram) and upwind (bottom diagram) components, and dashed lines are for a Gaussian distribution of equal mean square slope components on an abscissa of standard deviation. The degree marks show wave slopes for a wind speed of 10 m/s. (After Cox and Munk, 1954.)

were arranged in an approximately Gaussian distribution around the upwind-downwind and crosswind area. The mean square slopes are larger for the upwind-downwind component (lower curves) than for the crosswind component (upper curves) by a ratio of 1.34 for all photographs. Likewise the distribution is somewhat broader along the upwind-downwind direction than crosswind, two standard deviations being within $\pm$ 20° for the former and $\pm$ 16° for the latter. The mean square slope along both axes was found to increase linearly with wind speed.

Computations of the time-averaged intensities in the glitter pattern, based on the slope probability distribution of Cox and Munk (1954), have been carried out by Plass et al. (1977) and Guinn et al. (1979). In these computations the observer has been assumed to be at altitudes of 5 m and 150 km, the wind direction at 0°, 45°, and 90° azimuths from the sun, and wind speeds varying from 1 to 15 m/s. Typical results of the computations are shown for the case of unit incident intensity, angle of incidence 30°, observer altitude 150 km, and a 10 m/s wind from 90° azimuth in Fig. 10.9. The dashed and solid curves are for an uncontaminated water surface and a water surface covered by oil film, respectively. The top curve is the horizon in both cases. The patterns are symmetric around the principal plane, but a change of the wave slope distribution for the case of an oil slick of refractive index 1.5 on the water shifts the brightest region in the direction away from the horizon and causes a change of the configuration near the horizon. The intensity of the central bright spot is over twice as high for the clean surface as the oil-covered surface, being 5.79×10^{-4} units for the former and 2.32×10^{-4} for the latter. The intensity drops off by more than four orders of magnitude from center to edge of the diagrams, and it becomes so low that the enhancement of the glitter pattern could not be seen against the background light from the surface. Other combinations of parameters utilized by the authors vary the patterns somewhat from those in Fig. 10.9, but asymmetry is not pronounced, even for a wind direction of 45° azimuth, and the general patterns are similar to those shown here.

10.4 Polarizing properties of natural surface

Although it is generally recognized that reflection, including that from soils, sands, water, and vegetation, is an effective polarizing mechanism, the polarization of reflected light has not been pursued to the same extent as the intensity.

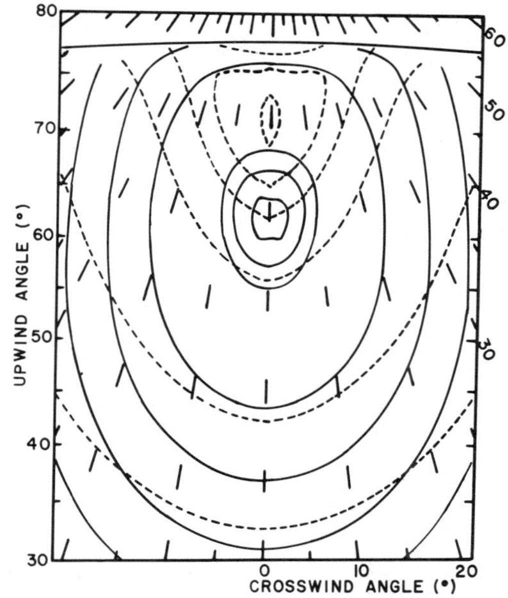

Figure 10.9 Isopleths of intensity on a log 10 scale in the sun's glitter pattern, as computed for the Cox and Munk statistical slope distribution by Guinn, et al. (1979). (Elevation angle of source 30°, observer height 150 km, wind of 10 m/s from 90° azimuth.) The grid points are projected on an inclined plane in front of the observer. The vertical scale, labeled on the left, represents lines normal to the principal plane, and the sloping coordinate lines represent lines parallel to the principal plane after projection to the inclined plane of the diagram.

The reflectance of particular types of mineral surfaces, especially those with complex structures, have been extensively studied in connection with lunar and planetary programs (Hapke and Van Horn, 1963; Salisbury and Glaser, 1964; Oetking, 1966; Orlova, 1952). Attempts to describe reflectance of complex surfaces by theoretical means have been made by Shifrin (1953), Hapke (1963), and Barkas (1939), among others. Several studies (e.g., Egan, 1970; Fitch et al., 1984; Curran, 1978; Vanderbilt and Grant, 1985) show that natural surfaces have polarization signatures which contain information not contained in the intensity field alone, and that the additional information is valuable in remote sensing of agricultural crops, soil moisture, geological formations, and similar purposes. In spite of such useful aspects, however, the polarization fields over natural surfaces have received only minimal attention for practical puposes. For instance, up until a pair of polarizing cameras was carried on a flight of the Space Shuttle in October 1984, no polarization observations of the earth's surface had been made from space. Polarimeters had been flown on space probes to some of the other planets, but the polarization of light reflected to space from our own planet had been investigated only on the basis of theory. Since that pioneer flight of 1984, the polarizing cameras have been on three other flights of the Shuttle, and some interesting results have been obtained (Coulson et al., 1986). This dearth of measurements does not extend, however, to the polarization of the moon and planets.

10.4.1 Polarization of the moon and planets

The early work on the polarization of celestial bodies has been summarized by Lyot (1929). Arago used his newly constructed polariscope for making eight polarization observations of the moon in 1811. He found the lunar seas to be more strongly polarized than the continents, and assumed a light reflecting atmosphere on the moon similar to that of our own planet. In a continuation of Arago's measurements in 1859 and 1860, Secchi observed the same phenomena, and ascribed the polarization to mirror-type surfaces of rock on the moon. Landerer, observing the moon in 1889, determined the phase angle and magnitude of the polarization and compared them to similar data for ice and polished volcanic rocks. From this study he concluded that the lunar surface was not covered by ice, but it might have a large component of glassy substance vitrophyre. Finally, from a comparison of his 1927 measurements of the polarization signatures of the lunar seas and continents with those of numerous terrestrial materials, Barabaschiff concluded that the moon must be covered with a sandy material. In summarizing the lunar observations, Lyot (1929) stated that it was premature to use the polarization data to establish the nature of the rocks which form the lunar ground, and proceeded to concentrate on increasing the precision of the measurements and including polarization observations of the planets.

For these new observations, Lyot constructed a new type of polarimeter (see Chapter 11) by which degrees of polarization very much smaller than 1% could be detected and which had the capability of determining the orientation of the plane of polarization. He mounted the polarimeter on various telescopes at the Meudon Observatory for observing the moon, Mercury, Venus, Mars, Jupiter, and the rings of Saturn between 1922 and 1926. Laboratory measurements during the same period were made on many terrestrial materials, including powdered crystals of different substances, glass grains, snow, sedimentary and igneous rocks, and clouds of water and ice particles. Lyot's conclusions were that the moon and the continents of Mars are both covered with similar materials, which is most probably powders having a constitution similar to that of volcanic ashes on the earth. Equivalent conclusions on the Martian seas were made very difficult by the clouds which are observed to exist in the atmosphere of the planet.

In a continuation of Lyot's work, Dollfus (1955, 1961) and Dollfus and Focas (1966) collected and analyzed a total of 5200 measurements of the polarization of Mars covering the nine apparitions of Mars between 1948 and 1966. Data for the bright regions located at the center of the disk during the first four of these apparitions, as given by Dollfus (1961), are shown by Fig. 10.10. The dashed curve represents the polarization of the surface only, after corrections for effects of the Martian atmosphere had been applied. From a comparison of

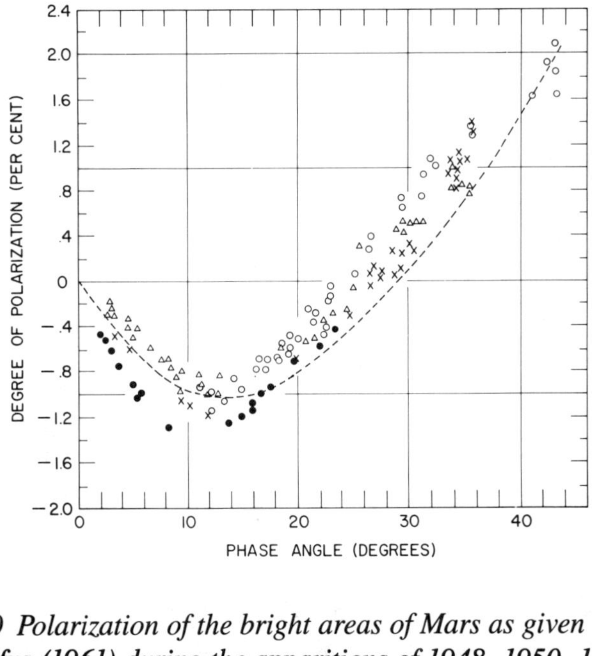

Figure 10.10 Polarization of the bright areas of Mars as given by the observations of Dollfus (1961) during the apparitions of 1948, 1950, 1952, and 1954. The dashed curve represents the polarization of the surface only, after corrections for atmospheric polarization has been introduced.

this curve with laboratory measurements of the polarization of light reflected from the hydrated ferric oxide limonite ($Fe_2O_3 \cdot n\ H_2O$), a yellow powder, Dollfus concluded that the surface of Mars is covered by that material. As summarized by Coulson (1969), however, there were several questions about Dollfus' method, not the least of which was the assumption of a too dense atmosphere. A lesser atmospheric correction would give better agreement between points and curve in Fig. 10.5, but the surface polarization would not then correspond so well to that of limonite. Be that as it may, the planetary observations provided a motivation for measuring the polarizing properties of many terrestrial materials, and the photos obtained from the Viking landers show the Martian surface to have a color which is not incompatible with that of ferric oxides such as limonite.

10.4.2 Polarization by other mineral surfaces

Since these early works, the polarizing properties of many types of minerals have been measured by several authors, both in the laboratory and in the natural environment. Among such measurements are those of Wright (1927), Hapke and Van Horn (1963), Coulson et al. (1965), Egan and Hallock (1966), Egan (1967), Chen and Rao (1968), Stockhoff and Frost (1971), and others. Optical systems for the acquisition of polarized images have been developed by Walraven (1981) and Curran (1978, 1979), the latter of whom interpreted the polarization of light reflected from soils in terms of soil surface moisture.

Two general characteristics emerge from the measurements of the polarizing properties of mineral-type surfaces. First, highly reflecting surfaces polarize less strongly than do dark surfaces. This seems to be a general law which extends to all wavelengths and all geometrical configurations. In fact, this effect was observed in the polarization of light reflected by colored powders by the Russian physicist Umow (1905), and has come to be termed the Umow effect. The physical explanation of the Umow effect is the differing extent to which the incident light enters the particles, as opposed to being reflected at the immediate surface. Light-colored materials having a high reflectance are composed of translucent particles which admit a considerable fraction of light into their interior. This is the case, for instance, for the quartz and white coral particles of beach sand and the gypsum particles of the White Sands area of New Mexico. The radiation which then emerges from the particles has undergone depolarization by scattering and reflection in all directions, and much of the originally produced polarization has been lost in the process. Of course, the light which is reflected at the surface without entering the particles is not subjected to the depolarizing process and retains its polarization by reflection. Dark surfaces, on the other hand, are dark because their constituent particles are highly absorbing. Much of the light which enters the particles is quickly absorbed, and does not emerge in the depolarized state. Thus most of the light which is backward directed is that which has been reflected directly at the surface, and such reflection is a good polarizing mechanism. The net result of all of this is that, in general, surfaces of low reflectance produce high polarization, and vice versa. Of course this explanation is an over simplification, as reflectance of a surface depends as well on index of refraction of the material, birefringence, crystalline structure of the particles, particle size, and other parameters, but the concept is an intuitive explanation of the Umow effect which appears to be valid.

The second general characteristic of reflection by mineral particles is that reflectance increases and polarization decreases with decreasing particle size. This is shown by the data of Egan (1967) in which the reflectance of volcanic ash increased from 0.12 to 0.15 as the particle size was decreased from 0.088 to 0.21 mm to less than 0.037 mm. On the same samples the polarization maximum was decreased from 16% for the large-sized particles to only 7% for the small ones. There are some obvious limitations to this trend, as, for example for particles of size comparable to the wavelength of light. However, for particles of the more usual mineral surfaces the trend is a commonly observed phenomenon.

Examples of laboratory measurements of the polarization of light reflected from various mineral type surfaces, as given by Coulson et al. (1966) are shown in Fig. 10.11. The Yolo loam sample was a fine-grained loam soil from Yolo County, California, the beach sand was a very white quartz sand from Daytona Beach, Florida, and the gypsum sand was from the White Sands area of New Mexico. The limestone and limonite samples were pulverized from the bulk minerals, and the clay was the subsurface soil of Pennsylvania. The antisource direction is indicated by the arrow. The configuration of the curves, with a region of slight negative polarization surrounding the antisource direction and a generally well-defined maximum of positive polarization ranging from 7% to 20% and located at 95° to 110° from the antisource, is typical for mineral particulates.

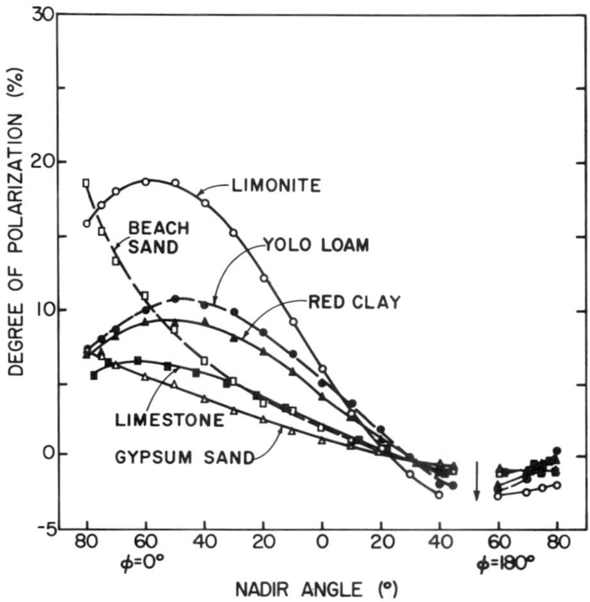

Figure 10.11 Degree of polarization as a function of angle in the principal plane for light reflected from various types of mineral surfaces. Angle of incidence 53°; $\lambda = 0.492$ μm) (After Coulson et al., 1966.)

The anomalous behavior of curves of gypsum and beach sand is produced by the bright glitter of light reflected by crystalline facets of the particles. Aside from these two cases, the magnitudes of both the positive and negative branches of the curves are inversely related to reflectance, as expected from the Umow effect.

The polarization signature of a sample of sand taken from the Mojave Desert of California is shown for the principal plane, an angle of incidence of 53°, and five different wavelengths in Fig. 10.12. A neutral point ($P = 0$), a point of inversion as termed by some authors, is well defined in the curves, located between 20° and 30° above the antisource direction. The curves indicate the possibility of a second neutral point below the antisource. This second neutral point is definitely found for this and other surfaces at smaller angles of incidence, and of course symmetry demands that a neutral point exists at the antisource direction for an angle of incidence of 0°. Similar curves for the case of a black loam soil taken from the rich agricultural lands of Iowa, but for an angle of

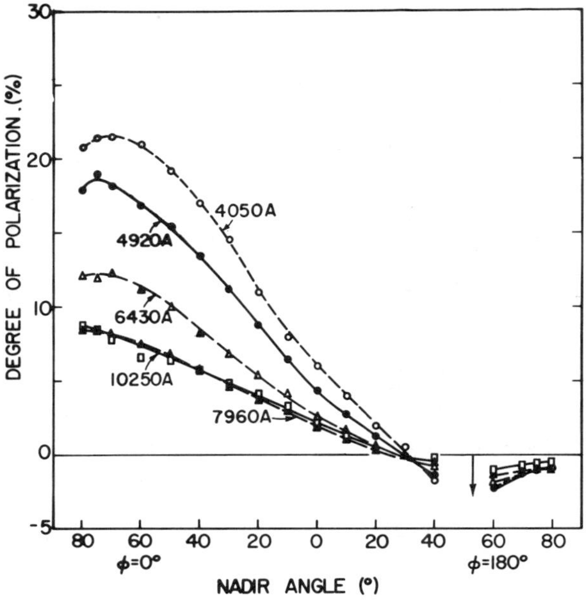

Figure 10.12 Degree of polarization as a function of angle in the principal plane for light of five different wavelengths reflected from Mojave Desert sand for an angle of incidence of 53°. (After Coulson et al., 1965.)

incidence of 78.5°, are shown in Fig. 10.13. A strong wavelength dependence and high polarization at the shorter wavelengths is evident for this strongly absorbing material. Otherwise the pattern is similar to that already seen for desert sand. The hemispheric pattern of polarization of light reflected from desert sand (dashed isopleths) and black loam soil (solid isopleths) for angle of incidence 53° and $\lambda=0.643$ μm is shown in Fig. 10.14. The two patterns are essentially similar, the most obvious feature being the curvature of the isopleths around the antisource direction. This means that the angle between the antisource direction and the direction at which the light is reflected, often termed the phase angle or angle of view, is the dominant geometrical parameter in the polarization field. A small region of negative polarization is evident around the antisource direction, but, contrary to the statement of Egan and Hallock (1966), the only case observed in which the polarization is zero at zero phase angle is for an angle of incidence of zero.

The principal parameters characterizing the polarization field of reflected radiation are the magnitudes of positive and negative polarization maxima, the

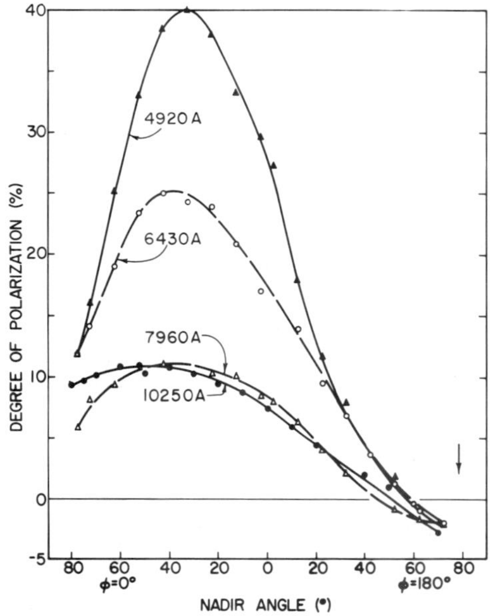

Figure 10.13 Degree of polarization as a function of angle in the principal plane for light of four different wavelengths reflected from black loam soil for an angle of incidence of 78.5°. (After Coulson et al., 1966.)

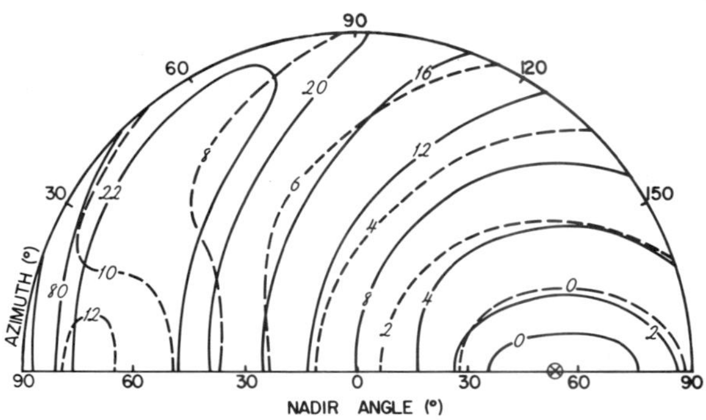

Figure 10.14 Hemispheric patterns of the degree of polarization of radiation of wavelenth λ=0.492 μm reflected from black loam soil (solid isopleths) and desert sand (dashed isopleths). The antisource direction is indicated by the cross at nadir angle 53°. Only one-half of the hemisphere is shown, the other half being a mirror image of this one. (After Coulson et al., 1966.)

phase angle at which the positive maximum is located, and the position or positions of the one or two neutral points. The magnitude and position of the positive maximum is easily determined with acceptable precision, although the broad configuration of the maximum makes its precise location questionable in many cases. The negative branch, together with the neutral points, is difficult to characterize precisely because of low values of polarization, weak gradients, and errors of measurement. The magnitude of negative polarization is only a few percent at most, and the neutral points are in a generally flat polarization field. Furthermore, measurement of the polarization involves a determination of a small difference between two large quantities, a process known to magnify measurement errors. Thus, for practical applications it is better to concentrate on the positive branch of the polarization curve.

A summary of many of the measurements of the polarization of light reflected by different types of mineral surfaces is given in Table 10.1. The type of surface is specified in the first two columns of the table, with angle of incidence and wavelength of the radiation in columns 3 and 4. In some cases, the angle of incidence was not given by the author, and a few of the measurements were in white light. The magnitude of the positive polarization maximum and its phase angle are in columns 5 and 6, with the references from which the data were taken listed in the final column. The polarization maxima are of only modest magnitude, being generally in the 5% to 25% range. Only for the very darkest of surfaces does P_{max} exceed 30%. The phase angle of P_{max} is mainly between 100° and 140°, although there are some exceptions. The very small-grained material (clay, volcanic ash) tends to show the smallest phase angles. The measurements by Egan (1985) were terminated at a phase angle of 90°, which was insufficient to reach the maximum in some cases.

10.4.3 Polarization for indicating soil moisture

Several authors (Curran, 1978, 1979; Stockhoff and Frost, 1971; Frost and Steg, 1970) have suggested use of the polarization of light reflected from soil surfaces as an indicator of soil moisture. The content of moisture in the soil is a critical parameter in the growth of agricultural crops, in the application of fertilizer, in determination of an irrigation cycle, and other aspects of soil management. Both the total amount of available moisture and its distribution in depth

TABLE 10.1 Values of the polarization maximum P_{max} and its angular distance $\theta-\theta_o$ from the antisolar direction for light reflected from various types of surfaces as observed by the indicated authors. θ_o=angle of incidence of the surface illumination, λ=wavelength, A=albedo

Description or category	θ_o(°)	λ(μm)	P_{max}(%)	$(\theta-\theta_o)$(°)	Author(s)
Sands					
Beach Sand A=0.24	—	White	28.0	140	Lyot (1929)
Mixed sand grains	—	"	8.5	105	Dollfus (1961)
Mojave desert sand	78.5	0.492	17.8	133	Coulson et al. (1965)
"	"	0.643	12.4	135	"
"	"	0.796	8.4	133	"
"	"	1.025	8.6	128	"
"	53	0.492	19.0	128	"
"	"	0.643	12.2	123	"
"	"	0.796	8.4	133	"
"	"	1.025	8.6	133	"
Florida beach sand	78	0.643	15.5	148	"
"	53	0.643	14.5(a)	133(a)	"
"	23	0.643	6.0	103	"
"	53	0.492	16.0	133	"
"	53	0.796	10.0	133	"
White gypsum sand	53	0.395	18.0	133	Chen and Rao (1968)
"	53	0.500	15.0	133	"
"	53	0.605	13.0	133	"
Sand treated with HCl	—	White	12.3–13.5	140	Dollfus and Cailleux (1950)
"	—	"	9.0–10.8	120	"
Beach sand, glistening	—	"	22–27.5	138	"
Dune sand from seacoast	—	"	18.0	133	"
Sand, new grains	—	"	18.0	120	"
Sand, dull grains	—	"	12.0	135	"
Minerals					
Quartz, milky, solid	80	White	17.4	109	Dollfus (1961)
Sandstone, rough	80	"	11.5	118	"
Sandstone, A=0.34	—	"	13.8	135	Lyot (1929)
Sandstone, red, A=0.115	—	"	31.5	130	"
Limestone, oolitic	—	"	26		Dollfus and Cailleux (1950)
Limestone, treated with HCl	—	"	8		"
Granite, A=0.14	—	"	17.0	130	Lyot (1929)
Basalt, typical dark	—	"	70–95		Wright (1927)
Basalt, rough cut	—	"	22.2	120	Lyot (1929)
Lava, red-brown, A=0.082	—	"	13.1	110	"
Lava, gray, A=0.15	—	"	24.6	130	"

TABLE 10.1 (Concluded).

Description or category	θ_o (°)	λ (μm)	P_{max}(%)	$(\theta-\theta_o)$ (°)	Author(s)
Lava, gray, $A=0.23$	—	White	22.2	128	Lyot (1929)
Lava, vitrified, $A=0.029$	—	"	70.0	130	"
Volcanic ash, powder, $A=0.166$	—	"	4.0	90	"
" $A=0.206$	—	"	5.4	100	"
Soils					
Clay, white, $A=0.58$	—	White	21.0	150	Lyot (1929)
Clay, gray, $A=0.17$	—	"	16.0	120	"
Clay, red, particles 2mm	53	0.492	9.1	108	Coulson et al. (1965)
"	53	0.643	4.6	103	"
"	78	0.643	3.0	98	"
"	37	0.643	4.8	97	"
"	45	0.520	16.0	105	"
"	80	0.520	12.0	102	"
Clay, red, wet	45	0.520	62.0	120	"
"	80	0.520	68.0	120	"
Silty sand, dry	30	0.350	17.0	96	Egan (1985)
"	30	0.566	9.6	96	"
Silty sand, wet (20.6% water)	30	1.00	4.3	80	"
"	30	0.350	56.8(a)	96(a)	"
"	30	0.566	27.3	96	"
"	30	1.00	13.4	96	"
Loam, black, 20–40 μm	78	0.492	40.0	112	Coulson et al. (1965)
"	78	0.643	25.2	115	"
"	78	0.796	11.3	115	"
"	78	1.025	10.8	125	"
"	53	0.492	36.3	113	"
"	53	0.643	23.8	117	"
"	53	0.796	23.0	116	"
"	53	1.025	10.8	133	"
Miscellaneous					
Glass, pulverized	80	White	8.0	115	Dollfus (1961)
Emery powder, opaque	80	"	9.7	110	"
Sugar, white, granulated	80	"	5.5	122	"
Chalk, $A=0.79$	—	"	5.0	157	Lyot (1929)

(a) No polarization maximum was found within the angles of measurement.

within the soil profile are important parameters. Casual observation shows that a few millimeters of rainfall can make the soil moist or wet at the immediate surfaces, while the root zone of plants remains unaffected and dry. Likewise, the root zone may have moisture amounts near field capacity while the top few millimeters of soil appear perfectly dry. Light reflection takes place in the top few millimeters, which indicates that the polarization of the reflected light would probably not be sensitive to moisture content of the all important root zone. In addition, other soil parameters, such as particle size, extent of agglomeration, tilling status, and absorptive properties, affect surface reflection, so any information on soil moisture contained in the polarization signature must be extracted from this multiparameter domain.

An example of the difficulty is shown in Fig. 10.15, in which the two sets of curves were taken under different regimes of soil moisture. The measurements of Stockhoff and Frost (1971) were for clay and loam soils at a wavelength interval $\lambda = 0.59$–$0.69\ \mu m$ and a phase angle of $110°$. The high sensitivity to soil moisture shown by the data comes about by the moisture content being determined for only the top 5 to 6 mm of soil, with no indication of moisture in the deeper layers. In contrast, when moisture content of the entire soil profile is considered, as was done by Curran (1978) and shown by the second set of curves of

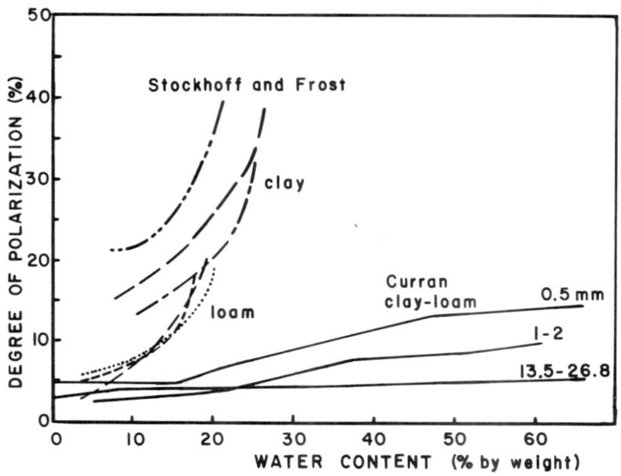

Figure 10.15 Degree of polarization of light reflected from the surface of various types of soil as a function of moisture content of the soil. (Data from Curran, 1978, and Stockhoff and Frost, 1971.)

Fig. 10.15, the polarization of reflected light is very insensitive to total soil moisture. For instance, the addition of an amount of moisture equal in weight to that of the soil itself produced an increase of polarization from 5% for the dry soil to only 14% for the moist soil. The low sensitivity in this latter case comes about because the moisture in the deeper layers, while adding to the total weight of moisture and being most important in plant growth, has little or no effect on the polarization signature. In fact, a change of average size of soil agglomerations from 13.5 to 26.8 mm to less than 0.5-mm diameter changed the polarization signature as much as did the addition of the soil's weight in water. It is interesting that the effect of particle size on polarization operated in this case in reverse from the normal case of polarization increasing with increasing particle size.

It would appear from these and other data that measurements of the polarization of light reflected from soils has only marginal potential in soil moisture determinations for practical use in remote sensing. The polarization signature is not very sensitive to total moisture content, and it is affected by surface roughness, angle of illumination, and, of course, any vegetation or other objects which happen to be in a resolution element of the remotely located sensor. Perhaps further work will indicate more possibilities, but at the present time it appears that other methods (e.g., microwave emission or heat capacity determinations) have more potential for monitoring soil moisture than do polarization measurements.

10.4.4 Polarization by plants

The remote sensing of plants and plant canopies is of great value in assessment of the production potential and areal coverage of agricultural crops, particularly the cereal crops, over the world, as well as in making inventories of forests and forest products as a worldwide resource. As mentioned, the polarization signature of vegetated surfaces may be used to advantage for not only discriminating among the different types of crops but also for indicating the stage of development and possible stress factors (water deficiency, disease, excessive salinity, etc.) affecting overall production. While the use of polarization for such remote sensing applications is still in its infancy, a realization of the value of polarization in remote sensing in agriculture is advancing rapidly and the newer sensing and data-handling techniques are being applied to the problem.

Parametric studies of plant reflection based on laboratory and field measurements are indicating new avenues of approach to a better understanding of reflection and polarization by plants. For instance, Egan (1970) has discussed the use of both intensity and state of polarization of reflected light to characterize crops of alfalfa, potatoes, sweet corn, rye, and wheat. He concluded from the study that the Stokes parameter $Q=I_{\perp}-I_{\parallel}$ may be a more sensitive indicator for discrimination between plant species or indicating soil moisture than either the intensity $I=I_{\perp}+I_{\parallel}$ or the degree of polarization $P=Q/I$. Fitch et al. (1984) obtained images in polarized light of fields of barley and two varieties of wheat at different stages of development. Their results showed that the degree of linear polarization of light reflected by the canopy decreased very substantially as the grain heads emerged, a fact which has potential for determining growth stage of these important food crops by remote sensing techniques. These authors explained the observed decrease of polarization as being due to shading of the topmost leaves, the flag leaves, by the expanded heads of grain, and a consequent decrease of highly polarized specular reflection by the flag leaves. The heads themselves, reflecting in a more diffuse manner, are not good polarizers.

The light reflected by a leaf consists of two components. First, a part of the incident light is reflected by the wax cuticle and other material at the surface of the leaf and does not enter the main body of the cellular structure. According to Vanderbilt and Grant (1985), this type of largely specular reflection is primarily responsible for the polarized component of the reflected light. The second component of the incident light enters the leaf structure, and may be either transmitted, absorbed, or reflected. The processes which take place inside the leaf and result in reflection are due to changes of refractive index among epidermal, palisade, and mesophyll cells, cell sap, chloroplasts, and numerous interstices of air. A tracing of a light ray through the astoundingly complex structure of a leaf is given by Kumar and Silva (1973).

Gates et al. (1965) have postulated that the individual cells, chloroplasts, etc., are large enough to scatter light by the laws of geometrical optics, and that Mie and Rayleigh type scattering are unimportant for reflection by leaves. This is probably true generally, although, as pointed out by Kumar and Silva (1973), certain structures such as ribosomes, cell nuclei, starch grains, and chloroplasts are of the same dimensions as the wavelength of visible light, and would thus be Mie scatterers. In addition, the water molecules would be Rayleigh scatterers, but the combined effect of these smaller particles is probably insignificant in comparison to reflection by cell walls making up the main structure of the leaf. The index of refraction of the epicuticular wax covering of leaves which polarizes light by reflection varies from about 1.54 at $\lambda=0.40$ μm to 1.48 at $\lambda=1.0$ μm (Vanderbilt and Grant, 1985), and that of cell walls is typically 1.52 for visible light (Kumar and Silva, 1973). The index of refraction of chloroplasts and cell sap used by these authors was 1.42 and 1.36, respectively, for $\lambda=0.70$ μm, but the complex part of the index, most important at visible wavelengths, is not well known.

Grant (1985) has made laboratory and field measurements of the reflection of the leaves of 18 common plant species at six different wavelengths from $\lambda=0.40$ to 0.73 μm and at angles of incidence and reflectance of 55°. This is at or near the Brewster angle for the wax coating of leaves. The measurements were made with a portable polarizing photometer which permits in situ measurements on growing plants. The Stokes parameters of the reflected light were, aside from calibration constants, given by the component intensities $I_\parallel$ and $I_\perp$ in and normal, respectively, to the plane containing the incident and reflected light. In this case $U=O$ and parameters I and Q are, as usual $I=I_\perp+I_\parallel$ and $Q=I_\perp-I_\parallel$. The total light reflected by the leaf is proportional to I. That which is polarized by specular reflection at the wax surface is proportional to Q, and the unpolarized remainder $I-Q=2I_\parallel$ is due to reflection by cells and other components inside the structure of the leaf. Finally, the degree of polarization of the reflected light is $P=Q/I$.

The polarized and unpolarized components of the light of five different wavelengths reflected from the top surface of leaves of 18 different plant species, as measured by Grant (1985), are given, along with the computed degree of polarization, in Table 10.2. The species which polarize most strongly, and therefore have a high degree of specularity (oaks, elm, sugar maples, corn, sorghum) generally have relatively flat and glabrous (hair-free) surfaces, whereas those with low polarization (velvetleaf, caladium, milkweed, squash) have generally pubescent (hairy) leaves. Purple velvet is an exception, yielding a relatively strong polarized component in spite of a densely pubescent surface. Apparently the hairs are sufficiently dense on the leaves of purple velvet to themselves constitute the main reflecting surface. The degree of polarization is generally higher for those species with low total reflectance, and vice versa.

TABLE 10.2 Percent reflectance of the polarized (Pol.) and unpolarized (Un.) components and degree of polarization (DP) of light reflected at an angle of 55° from the top surface of leaves of various species of plants, according to the measurements of Grant (1985).

	$\lambda = 0.45\ \mu m$			$0.50\ \mu m$			$0.55\ \mu m$			$0.65\ \mu m$			$0.73\ \mu m$		
Plant Name	Pol.	Un.	DP	Pol.	Un.	DP	Pol.	Un.	DP	Pol.	Un.	DP	Pol.	Un.	DP
Velvetleaf (*Abuliton theophrasti*)	16	35	31	15	15	50	16	23	41	14	19	42	12	50	19
Sugar maple, common (*Acer-sacharum*)	21	8	72	22	10	69	21	13	62	21	9	70	20	43	31
Sugar maple, Indiana (*Acer-sacharum*)	27	7	79	26	9	74	27	11	71	27	9	75	26	45	37
Milkweed (*Asclepias syriaca*)	21	27	43	20	16	56	19	20	49	20	15	57	17	61	22
Caladium (*Caladium bicolor*)	10	34	23	11	38	22	11	40	22	13	38	25	13	42	24
Lambsquarters (*Chenopodium album*)	11	36	23	13	17	43	10	25	29	9	14	39	10	53	16
Squash (*Cucurbita pepo*)	15	9	62	14	12	54	13	17	43	12	13	48	12	43	22
Soybean (*Glycine max*)	16	21	43	21	12	64	21	16	57	23	11	70	21	44	32
Purple velvet (*Gynura aurantiaca*)	17	5	77	16	5	76	15	7	68	19	8	68	17	38	31
Common sun-flower (*Helianthus annuus*)	21	36	37	24	25	49	23	25	48	26	19	58	25	49	34
Tomato (*Lycopersicon esculentum*)	16	11	59	17	10	63	15	15	50	16	10	62	14	40	26
Red oak (*Quercus borealis*	34	9	79	36	7	84	36	10	78	36	7	84	35	40	47
Chestnut oak (*Quercus prinus*)	44	8	85	45	8	85	47	10	82	47	7	87	46	35	56

TABLE 10.2 (Concluded.)

Plant Name	λ=0.45 μm Pol.	Un.	DP	0.50 μm Pol.	Un.	DP	0.55 μm Pol.	Un.	DP	0.65 μm Pol.	Un.	DP	0.73 μm Pol.	Un.	DP
Black oak (*Quercus velutina*	55	7	89	56	7	89	57	9	86	56	7	89	56	41	58
Black nightshade (*Solanum pycanthum*)	22	18	55	19	18	51	19	15	56	19	19	50	17	51	25
Sorghum (*Sorghum bicolor*)	65	41	61	66	30	69	67	30	69	70	18	80	67	52	56
American elm (*Ulinus americana*)	47	5	90	47	6	89	46	8	85	48	6	89	46	40	53
Corn (*Zea mays*)	53	35	60	51	20	72	50	20	71	48	15	76	46	43	52

Radiation of λ=0.73 μm is less strongly polarized than that of the shorter wavelengths because the lack of strong absorption in the near infrared permits this radiation to enter the main structure of the leaf. Reflections by cell walls and other components in the leaf are efficient in redirecting the radiation, but produce little polarization in the redirected stream.

The conclusion of Egan (1970) that the second Stokes parameter, Q, is a more sensitive indicator of plant species than is either the total reflectance or the degree of polarization of the reflected light is further confirmed by the data of Vanderbilt et al. (1985). The data of Table 10.2, together with information on leaf structure, indicates that the magnitude of Q is sensitive to pubescence of leaves, and this characteristic varies strongly among species. The amount and configuration of epicuticular wax on the surface of leaves is likewise species dependent, and this too affects the polarized component of reflection. It is to be expected, therefore, that Q would be the most sensitive indicator of both plant species and status of the leaf surface. This is an oversimplification, of course, as the polarization is also affected by leaf orientation, stage of growth, and a number of other parameters.

The unpolarized component of the reflected light should indicate the status of cells, chloroplasts, and other structures in the interior of the leaf. Vanderbilt et al. (1985) found the unpolarized component to be sensitive to relative water content of corn leaves, the magnitude of this component decreasing with increasing water content. Gausman et al. (1969) found that leaves of cotton plants grown under high salinity conditions were considerably thicker than normal leaves, the additional thickness increasing total reflectance by 6% to 8% in the near infrared where internal structure is most important. These authors did not discriminate between polarized and unpolarized components.

In an extension of measurements of the effects of pigmentation on leaf reflectance, Grant (1985) measured the total, unpolarized, and polarized reflectances and computed the degree of polarization for light of five different wavelengths reflected from variously pigmented sections of leaves of the variegated plant *coleus blumei.* Two hybrids of this plant provided leaves with white, green, and reddish-purple pigmentation for the measurements. The results of the measurements for light at angles of incidence and reflectance of 55° on the top surface of these leaves are shown in Fig. 10.16. Two observations on each of six leaf samples

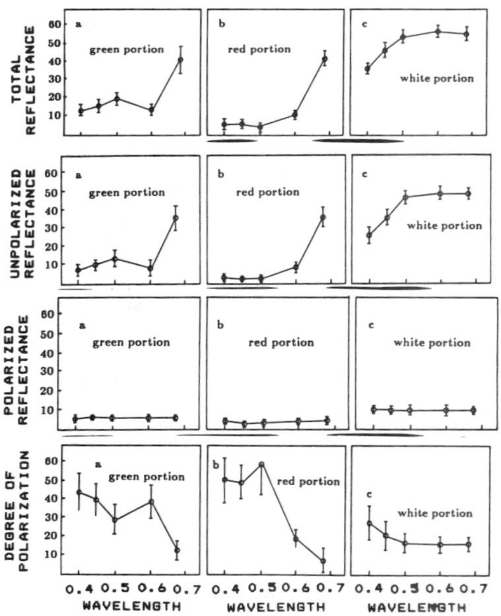

Figure 10.16 Total reflectance, unpolarized reflectance, polarized reflectance, and degree of polarization of light of five different wavelengths reflected from green, red, and white portions of leaves of the variegated plant Coleus blumei (adapted from Grant, 1985). Wavelengths are given in μm and the ordinate scales are in %.

yielded the means and standard deviations shown in the diagram. The data for the green portions are considered to be typical of that for green plants in general. Chlorophyll absorption of light entering the leaf structure makes total and unpolarized reflectance low in blue and red wavelengths, with higher values at green and near-infrared wavelengths. Polarized reflectance, while of low values, is essentially independent of wavelength, a fact which Grant and her colleagues interpret as indicating specular reflection from the wax cuticle of the leaf. The degree of polarization of this weakly polarized radiation is influenced mainly by total reflectance, and the configuration of the curves of degree of polarization is very much a mirror image of that of total reflectance for all three colors. This tendency would be less pronounced, of course, for highly polarizing species, but it is a general characteristic that low reflectance is accompanied by high polarization, and vice versa.

The reddish-purple portions of the leaves have low reflectance throughout the visible due to strong absorption by a combination of chlorophyl and anthocyanin, the latter of which is responsible for the reddish color. Both pigments are effective in promotion of photosynthesis. The white portions of the leaves have generally high reflectance throughout the visible, but carotenoids have some absorption at blue and green wavelengths. Although the magnitude of polarized reflectance is higher in the white than in other wavelengths, the very strong total reflectance of white portions keeps the degree of polarization low. The highest reflectance and lowest degree of polarization for all three colors occurs in the near infrared, a fact which indicates that wavelengths which are not strongly absorbed enter the body of the leaf and are backscattered by the cellular structure. As mentioned, the lack of spectral dependence of the polarized reflectance is indicative of reflection at the surface of the leaf. Thus the polarized and unpolarized components carry different types of information, either or both of which are of potential value in remote sensing applications. Interpretation of the information in terms of agronomic parameters, however, is a hurdle which has not been fully negotiated.

In an effort to better understand the polarizing properties of plants, Vanderbilt and Grant (1985) have developed a simplified mathematical model of light specularly reflected from a plant canopy. The model, which is too detailed for a full discussion here, is based on the assumption of specular reflection by small facets of leaves which happen to be in the field of view. The facets are covered by a wax surface which reflects radiation according to Fresnel's laws of reflection, the only modification being that the Fresnel components are multiplied by constant K to take account of the roughness of the wax surface. A finite number of the facets have the orientation such as to direct some of the reflected light into the aperture of the sensor. These uniquely oriented facets have a probability density function f_a, which is dependent on angles of incident and reflected radiation and on the architecture of the canopy. The authors assumed two different types of average leaf orientation, one being an isotropic angular distribution through an entire 4π sr and the other having uniform orientation in vertical planes but with a variation in azimuth.

Once the angular distribution of facets is appropriately presented, the shading among the leaves has to be considered. There is a finite probability P_i that a given facet will be illuminated directly by the sun, as opposed to being shaded by intervening foliage, and a different probability P_r that the reflected radiation will not be intercepted by overlying foliage before emerging from the canopy. These probabilities, being dependent on the existencec of gaps in the canopy, are functions of type of plants, stage of growth, density and pattern of planting, and other characteristics of the plant stand. The probabilities are related to the leaf area index; for a solid canopy of top leaves P_i and P_r would both be unity.

In summarizing these various dependencies, the portion of incident flux L_i which is specularly reflected and at least partially polarized by the canopy may be expressed as

$$F_p = P_i \, P_r \, f_a \, L_i \, KS_Q \, A_T \cos\gamma \, \Delta\omega_a \, \Delta\omega_r \qquad (10.11)$$

Here γ is phase angle and $A_T \cos\gamma$ is the projected area of leaves per unit horizontal area seen by the sensor. Solid angles ω_a and ω_r are those subtended by source and sensor, respectively, and KS_Q is the polarized flux computed from Fresnel's equations. The agronomic variables which influence these parameters are leaf area index, plant species, plant development stage, cultivation practices, moisture or salinity stress, and plant diseases. They are also functions of wavelength, geometry of illumination and view, wind direction and speed, and other environmental parameters.

From comparisons of model calculations with measurements made on a field of growing wheat in the preheaded stage of development, Vanderbilt and Grant

(1985) concluded that the agreement was favorable, particularly at angles of view 45° and 60° from the zenith. The primary variables determining the amount of flux polarized by reflection from plants are angles of illumination and reflection. Although variation with angle of illumination contains no information about the scene, the variation with direction of view is a function of plant architecture. Geometric configuration of the canopy is intimately related to plant species, and it changes with growth stage, plant stress or plant vigor, possible leaf damage, and other factors. These authors join Egan (1970), Fitch et al. (1984), and others in concluding that remote sensing of the polarizing properties of vegetated surfaces has high potential for assessing the areal extent and production potential of agricultural crops.

10.4.5 Polarization of light by sea surface reflection

The polarization of light reflected by the surface of the sea is of interest in remote sensing of oceanic properties, detection of oil slicks on the surface, assessment of production potential of oceanic waters, and other problems in oceanic optics. In some respects the problem of the polarization of light at sea is simpler than that of polarization from vegetated or mineral surfaces because reflection just at the air-water interface is specular-type reflection and therefore follows Fresnel's laws of reflection. This is not true, of course, for most other types of natural surfaces. However, the problem of sea surface reflection is complicated by the emergence of light from below the air-water interface due to backscatter of incident radiation by particles and bubbles in the water. We shall discuss these topics separately.

The simplest aspect of sea surface polarization is that in the region of the sun glint. For a perfectly flat ocean surface and a solar zenith angle of θ_o, the image of the sun would appear at the specular point located in the principal plane at nadir angle θ_o. The light in that direction would consist of components $I_\perp$ and $I_\parallel$, normal and parallel, respectively, to the principal plane. Fresnel's laws of reflection give the reflected intensity components

$$I_\parallel = \frac{1}{2}\,\frac{\sin^2(\tau - \tau')}{\sin^2(\tau + \tau')}$$

$$I_\perp = \frac{1}{2}\,\frac{\tan^2(\tau - \tau')}{\tan^2(\tau + \tau')} \qquad (10.12)$$

for unpolarized incident light of unit intensity. Angles τ and τ' are angles of incidence and refraction, θ_o having been replaced by its equivalent τ in order to be consistent with the notation of Eqs. (3.37) and (3.39). Snell's law gives

$$\sin\tau = m \sin\tau' \tag{10.13}$$

where m is index of refraction of sea water. To a good approximation the refractive index of air is unity. The light in the specular direction has the intensity components $I_\perp$ and $I_\parallel$ shown as a function of angle of incidence by the solid curves in Fig. 10.17. The dashed curves are for reflection by oil on the surface, to be mentioned further below. For the sun at the zenith, approximately 1% of each $I_\perp$ and $I_\parallel$ is reflected, making the total reflectance in that case approximately 2%. This low value is of interest for remote sensing of the atmosphere from space, as the sea surface makes a very dark background at high sun elevations against which to view the radiation backscattered to space by atmospheric

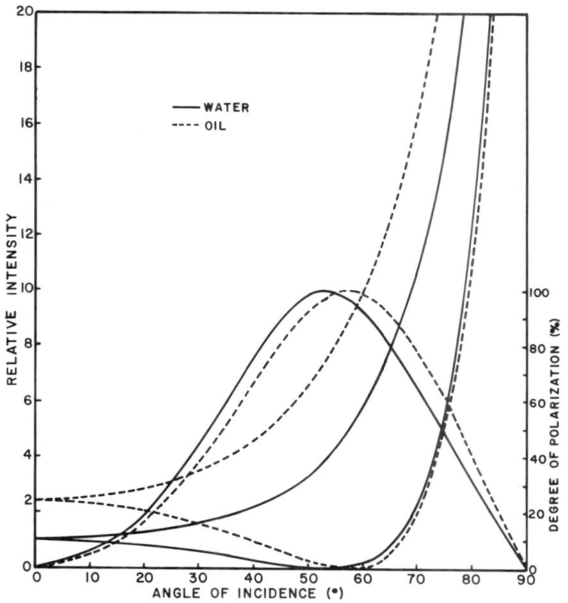

Figure 10.17 Relative intensity of polarization $I_\perp$, and $I_\parallel$ and of degree of polarization as functions of the angle of incidence for light reflected from smooth surfaces of water ($m=1.34$; solid curves) and oil ($m=1.57$; dashed curves) as computed from Fresnel's laws of reflection.

particulates. Even out to $\theta_o=30°$ the reflectance of a calm sea at the specular point is only about 2.2%. At larger values of θ_o the reflectance rises sharply due to the rapid increase of $I_\perp$, even though $I_\parallel$ continues to decrease until it becomes identically zero at the Brewster angle of 53.3° for the index of refraction of 1.34 for sea water.

The Brewster angle is an interesting singular point for the polarization field, as the degree of polarization P reaches its maximum value of 1.0 there. As shown by the polarization curve of Fig. 10.17, $P=0$ at the specular point for $\theta=0°$ and 90°. This, of course, is for the calm sea. In most real cases, the surface waves would change the polarization pattern somewhat, as it changes the sun's glitter pattern of intensity, but P would have values over the glitter region somewhat above or below its smooth water value, depending on the slopes of the waves. Since the solar incident angle is 53.3° for complete polarization, the angle of observation must likewise be 53.3°, making a phase angle of 106.6.° Thus the value $P=1.0$ may be produced by any wave making the phase angle 106.6°. Plass et al. (1977) have computed the distribution of P over the glitter region for the two cases of sun elevations of 5° and 30°. As would be expected from Fig. 10.17, the degree of polarization depends only on angle of incidence. It is a function of wind speed only insofar as the wave slopes depend on wind speed. The statistics of wave slopes, while all important for intensity distribution, is immaterial for polarization. As long as a given slope exists, the polarization is uniquely determined by the combination of sun elevation and wave slope, regardless of how many waves have that particular orientation. For instance, the computations of Plass et al. (1977) show that for a sun elevation of 30° or angle of incidence of 60°, the polarization reaches $P=1.0$ at a nadir angle of roughly 47°. From geometry, it is easy to see that for a phase angle of reflection of 106.6°, the reflected light is directed outward from a nadir angle of 46.6°, and that the responsible wave must be 13.4° below the sun's image with a slope of 6.7°.

Since the degree of polarization of the reflected light depends only on the angles involved, it is of limited value in remote sensing. It carries little information about the sea itself. This does not mean, however, that polarizing optics of an instrument for remote sensing of the sea has no value. As shown by Fig. 10.17, the magnitude of $I_\parallel$ is 1% or less throughout the range of angles of incidence from 0° to 60°, becoming identically zero at 53.3°. At that angle $I_\parallel$ is due only to reflected skylight and light scattered by the water beneath the surface. By knowing the distribution of the intensity and state of polarization over the sky, the skylight contribution to $I_\perp$ could be computed and subtracted out. Likewise the emergent radiation due to Rayleigh scattering in the water could be evaluated and deducted from the background, as it is in the atmosphere. This would leave the bulk of the emergent radiation due to scattering by underwater particles such as plankton, suspended minerals, turbidity in general, and the bottom in shallow water. Needless to say, the numbers and properties of submarine particles are of great interest in marine biology, fisheries and seafood production, sediment transport, and a host of other problems of similar nature. Although much work remains to be done in this aspect, the use of component $I_\perp$ in remote sensing of the sea is a logical first step.

A second use of polarizing optics in remote sensing of the sea is for the detection of oil on the surface. Computations of the reflecting and polarizing properties of an oil film of appropriate thickness (0.2 μm according to Millard and Arvesen, 1973) yield the dashed curves of Fig. 10.17. The index of refraction of the oil was set at 1.57. The reflectance of oil is slightly more than twice as great as that for water at small solar zenith angles, which is probably sufficient to give a good contrast for remote sensing purposes. The contrast can be improved, however, by sensing only the perpendicular component $I_\perp$, as shown for detecting an oil spill by Millard and Arvesen (1973). The magnitude of the parallel component $I_\parallel$ for oil approaches that of water as the solar zenith angle increases, both becoming very small in the region of the Brewster angle. In fact the Brewster angles for the two surfaces differ by only 4.2°, being 53.3° for water and 57.5° for oil. Beyond the Brewster angles, $I_\parallel$ is not much different for the two types of surfaces. Component $I_\perp$ for oil, however, remains well above that for water throughout the range of incidence angles 0° to at least 80°. The difference should give a good contrast for detection of oil spills, as it was found to do by Millard and Arvesen (1973). From aircraft photographs of oil on the sea surface, these authors found that a good contrast between oil and water was observed throughout the extent of oil in component $I_\perp$, but only in the very thickest part of the oil in component $I_\parallel$. Thus $I_\perp$ should be emphasized in the oil detection problem. For $I_\perp$ there is no particular solar angle that is much better than another, but if low intensity were a problem in the instrumentation, an angle of incidence between 40° and 75° would be most appropriate.

The simplest polarization distribution on the sea surface occurs under overcast skies. In case the incident radiation is unpolarized and quasi-isotropic, as it is under a thick overcast, the polarization curves of Fig. 10.17 are applicable at all azimuths. The polarization maximum would be located in a band centered at a nadir angle of about 53° at all azimuths around the point of observation. In fact, the polarization field would be everywhere symmetric in azimuth, with the magnitudes shown by the diagram. For a clear atmosphere, the polarization and nonisotropic nature of skylight could be expected to modify the picture somewhat, but, as far as is known, no observations of the changes have been made.

A word is in order about the contribution of upward radiation from the sea surface due to scattering within the water itself. Estimates of the magnitude of this contribution to the total upward flux vary from 15% (Payne, 1972) to as much as 50% (Neiburger, 1948). The actual fraction certainly varies with altitude of the sun, turbidity of the water, roughness of the surface, and probably with direction at which the surface is viewed. For instance, Waterman (1955), Ivanoff (1956), and Timofeyova (1964) all found the light inside the water to be partially plane polarized, the degree of polarization according to Ivanoff varying from 6% to 34% with direction of view in the water body. Waterman's observations showed the plane of polarization to be horizontal at azimuths of 0° and 180° with respect to the sun, and being tilted somewhat as it would be for Rayleigh scattering of the incident sunlight at other azimuths. Even at depths of 100 m, the effect of the sun's position was visible in the underwater polarization field. From Monte Carlo calculations based on three ocean models of different turbidities, Kattawar et al. (1973) computed the polarization of reflected light to reach a maximum of 71% to 98%, depending on sun elevation. Of particular interest in the calculations is the field of elliptic polarization for downward radiation just underneath the ocean surface. The ellipticity was found to reach 10% for the range of angles in which total internal reflection of the upward radiation occurs at the water surface. The ellipticity is introduced by a phase change of the orthogonal intensity components on total internal reflection. This type of reflection occurs in a cone of angles of incidence larger than the critical angle of θ_o=48.3 given by the relation

$$\sin\theta_o = m_2 / m_1 \tag{10.14}$$

where m_1 and m_2 are the indices of refraction of sea water and air, respectively. Ivanoff and Waterman (1958) measured the ellipticity of underwater illumination, and found appreciable values only in the totally reflected downward radiation just underneath the surface. The ellipticity decreases rapidly with depth, becoming quite small at unit optical depth in the water. The upward radiation, both above and below the surface, has essentially zero elliptic polarization, and multiple scattering in case of the turbid ocean rapidly depletes the elliptic polarization.

References

Allen, W.A., and A.J. Richardson (1968): Interaction of light with a plant canopy, *J. Opt. Soc. Am., 58,* 1023–1028.

Barkas, W.W. (1939): Analysis of light scattered from a surface of low gloss into its specular and diffuse components, *Proc. Phys. Soc. London, 51,* 274–289.

Bunnick, N.J.J. (1978): *The Multispectral Reflectance of Shortwave Radiation by Agricultural Crops in Relation With Their Morphological and Optical Properties,* H. Veenman and Zonen, Wageningen, Holland, 165 pp.

Burt, W.V. (1954): Albedo over wind-roughened water, *J. Meteor., 11,* 283–290.

Chen, H.S. (1985): *Space Remote Sensing Systems—An Introduction,* Academic Press, Orlando, Florida.

Chen, H.S., and C.R.N. Rao (1968): Polarization of light on reflection by some natural surfaces, *Brit. J. Appl. Phys., 1, Ser. 2,* 1191–1200.

Coulson, K.L. (1966): Effects of reflection properties of natural surfaces in aerial reconnaissance, *Appl. Opt., 5,* 905–917.

Coulson, K.L. (1969): Polarimetry of Mars, *Appl. Opt., 8,* 1287–1294.

Coulson, K.L., G.M.B. Bouricius, and E.L. Gray (1965): Optical reflection properties of natural surfaces, *J. Geophys. Res., 70,* 4601–4611.

Coulson, K.L., E.L. Gray, and G.M. Bouricius (1965): A study of the reflection and polarization characteristics of selected natural and artificial surfaces, Report R65SD4, General Electric Space Sci. Lab., Philadelphia, Pennsylvania.

Coulson, K.L., and D.W. Reynolds (1971): The spectral reflectance of natural surfaces, *J. Appl. Meteor., 10,* 1285–1295.

Coulson, K.L., V.S. Whitehead, and C. Campbell (1986): Polarized views of the earth from orbital altitude, *Proc. S.P.I.E., Vol 637-Ocean Optics VIII,* 35–41.

Cox, C., and W. Munk (1954): Measurement of the roughness of the sea surface from photographs of the sun's glitter, *J. Opt. Soc. Am.*, *44*, 838–850.

Curran, P.J. (1978): A photographic method for the recording of polarized visible light for soil surface moisture indications, *Rem. Sens. Environ.*, *7*, 305–322.

Curran, P.J. (1979): The use of polarized panchromatic and false-color infrared film for the monitoring of soil surface moisture, *Rem. Sens. Environ.*, *8*, 249–266.

Dollfus, A. (1955): Etude des planetes par la polarisation de leur lumiere, Thesis, Univ. of Paris; also Technical Translation NASA TT F-188, July 1964.

Dollfus, A. (1961): Polarization studies of the planets, Chap. 9 in G.P. Kuiper and B.M. Middlehurst (eds.) *Planets and Satellites, Vol. III, The Solar System,* Univ. of Chicago Press, pp. 343–399.

Dollfus, A., and J.H. Focas (1966): On the purity of the atmosphere of Mars, *Comp. Rend. Acad. Sci. Paris, B, 262,* 1024–1027.

Dollfus, A., and A. Cailleux (1950): Etude polarmetrique de la lumiere renvoyee par quelques sables et limons, *Comptes Rendus, 230,* 1411–1413.

Earing, D.G., and J.A. Smith (1966): Target Signature Analysis Center: Data Compilation, Inst. of Sci. and Tech., Univ. of Michigan, Ann Arbor.

Egan, W.G. (1967): Polarimetric measurements of simulated lunar surfaces, *J. Geophys. Res.*, *72,* 3233–3245.

Egan, W.G. (1968): Aircraft polarimetric and photometric observations, *Proc. 5th Symp. on Rem. Sens. Environ.*, Univ. of Michigan, Ann Arbor, pp. 169–189.

Egan, W.G. (1970): Optical Stokes parameters for farm crop identification, *Rem. Sens. Environ.*, *1,* 165–180.

Egan, W.G. (1985): *Photometry and Polarization in Remote Sensing,* Elsevier, New York.

Egan, W.G., and H.B. Hallock (1966): Polarimetry signature of terrestrial and planetary materials, *Proc. 4th Symp. Remote Sens. of Environ.* Univ. of Michigan, Ann Arbor, pp. 671–689.

Fitch, B.W., R.L. Walraven, and D.E. Bradley (1984): Polarization of light reflected from grain crops during the heading growth stage, *Rem. Sens. Environ.*, *15,* 263–268.

Frost, R.T., and L. Steg (1970): Visible polarization signature for remote sensing of soil surface moisture, A.3.4, COSPAR, 13th Plenary Meeting, Leningrad, USSR.

Gates, D.M., H.J. Keegan, J.C. Schleter, and V.R. Weidner (1965): Spectral properties of plants, *Appl. Opt.*, *4,* 11–20.

Gausman, H.W., W.A. Allen, and R. Cardenas (1969): Reflectance of cotton leaves and their structure, *Rem. Sens. Environ.*, *1,* 19–22.

Gordon, H.R., and M.M. Jacobs (1977): Albedo of ocean-atmosphere system: influence of sea foam, *Appl. Opt.*, *16,* 2257–2260.

Goudriaan, J. (1977): Crop micrometeorology: A simulation study, Thesis, Centre for Agricultural Publishing and Documentation, Wageningin, Holland.

Grant, L. (1985): Polarized and non-polarized components of leaf reflectance, Doctoral Dissertation, Purdue University, West Lafayette, Indiana.

Guinn, J.A., G.N. Plass, and G.W. Kattawar (1979): Sunlight glitter on a wind-ruffled sea: further studies, *Appl. Opt.*, *18,* 842–849.

Hapke, B.W. (1963): A theoretical photometric function for the lunar surface, *J. Geophys. Res.*, *68,* 4571–4586.

Hapke, B., and H. VanHorn (1963): Photometric studies of complex surfaces, with applications to the moon. *J. Geophys. Rev.*, *68,* 4545–4570.

Hovis, W.A., and W.R. Callahan (1966): Infrared reflectance spectra of igneous rocks, tuffs, and red sandstone from 0.5 to 11 μm, *J. Opt. Soc. Am.*, *56,* 639–643.

Ivanoff A. (1956): Degree of polarization of submarine illumination, *J. Opt. Soc. Am.*, *46,* 362.

Ivanoff, A., and T.H. Waterman (1958): Elliptical polarization of submarine illumination, *J. Marine Res.*, *16,* 255–282.

Justus, C.G., M.V. Paris, and J.D. Tarpley (1986): Satellite-measured insolation in the United States, Mexico, and South America, *Rem. Sens. of Environ.*, *20,* 57–83.

Kattawar, G.W., G.N. Plass, and J.A. Guinn (1973): Monte Carlo calculation of the polarization of radiation in the earth's atmosphere-ocean system, *J. Phys. Oceanography*, *3,* 353–372.

Krinov, E.L. (1953): Spectral Reflectance Properties of Natural Formations, Tech. Transl. TT-439, National Research Council of Canada, Ottawa, 268 pp.

Kubelka, P., and F. Munk (1931): Ein Beitrag zur Optik der Farbanstriche, *A. Techn. Physik, 11,* 593–601.

Kumar, R., and L. Silva (1973): Light ray tracing through a leaf cross section, *Appl. Opt., 12,* 2950–2954.

Lyot, B. (1929): Recherches sur la polarisation de la lumiere des planetes et de quelque substances terrestres, Annales de l'Observatoire de Paris, Section de Meudon, 8, No. 1 (Translation NASA TT F-187, 1964, 144 pp.).

Millard, J.P., and J.C. Arvesen (1973): Polarization: A key to an airborne optical system for the detection of oil on water, *Science, 180,* 1170–1171.

Monahan, E.C. (1971): Oceanic whitecaps, *J. Phys. Oceanography, 1,* 139–144.

Mullamaa, YU.-A.R. (1964): *Atlas of Optical Characteristics of the Ruffled Surface of the Ocean.* Academii Nauk, Estonian SSR, Tartu, Estonia, 384 pp. (in Russian)

Neiburger, M. (1948): The reflection of diffuse radiation by the sea surface, *Trans. Am. Geophys. Union, 29,* 647–652.

Oetking, P. (1966): Photometric studies of diffusely reflecting surfaces with applications to the brightness of the moon, *J. Geophys. Res., 71,* 2502–2513.

Orlova, N.S. (1952): Radial diagrams of scattering for several materials, Trudy, Astronomical Observatory, Leningrad State Univ., 16, 166–193.

Payne, R.E. (1972): Albedo of the sea surface, *J. Atm. Sci., 29,* 959–970.

Plass, G.N., G.W. Kattawar, and J.A. Guinn (1977): Isophotes of sunglint on a wind-ruffled sea, *Appl. Opt., 16,* 643–653.

Rao, C.R.N., and H.S. Chen (1969): An Atlas of Polarization Features of light Reflected by Desert Sand, White Sand, and Soil, Sci. Rept. No. 3, Contr. F19628-67-C-0196, Univ. of California, Los Angeles, 269 pp.

Ross, J. (1981): *The Radiation Regime and Architecture of Plant Stands,* Dr. W. Junk Publishers, The Hague, 391 pages.

Ross, J., and T. Nilson (1974): Radiation exchange in plant canopies, *Proc. Int. Centre for Heat and Mass Transfer.*

Salisbury, J.W., and P.E. Glaser (1964): Studies of the characteristics of probable lunar surface materials, ADCRL Rept. 64–970, Bedford, MA, 309 pp.

Shifrin, K.S. (Ed.) (1983): *Optics of the Sea*, Nauka Pub. House, Moscow.

Smith, J.A., and R.E. Oliver (1974): Effects of changing canopy directional reflectance for feature selection, *Appl. Opt., 13,* 1599–1604.

Smith, W.L. (1986): Meteorological satellite: overview of 25 years of operation, *Science, 231 (4737),* 455–462.

Stockhoff, E.H., and R.T. Frost (1971): Polarization of light scattered from moist soils, *Proc. 7th Inst. Symp. on Rem. Sens. of Environment, 1,* 345–365.

Suits, G.H. (1972): The calculation of the directional reflectance of a vegetative canopy, *Rem. Sens. Environ., 2,* 117–125.

Ter-Markariants, N.E. (1961): Reflection of radiation by the sea, pp. 231–236 in Shifrin, K.S. (ed.) *Actinometry and Atmospheric Optics,* Hydrometeorological Pub. House, Leningrad (in Russian).

Timofeyeva, V.A. (1964): The degree of polarization of natural light in the sea, Doklady, Akad. Nauk, USSR, *Earth Science Sect., 147,* 1–6, 9–12.

Umov, N. (1905): Chromatsche depolarisation durch lichtzerstreuung, *Physikalische Zeit., 6,* 674–676.

Vanderbilt, V.C., and L. Grant (1985): Plant canopy specular reflectance model, *Trans. IEEE, Geoscience and Remote Sensing, GE-23,* 722–730.

Walraven, R.L. (1981): Polarization imagery, *Optical Engineering, 20,* 14–18.

Waterman, T.H. (1955) :Polarization of scattered sunlight in deep water, Papers in Marine Biol. and Oceanography, Suppl. to Deep Sea Res., Vol. 3, 426–434.

Willstatter, R., and A. Stoll (1918): Untersuchungen uber die Assimilation der Kohlensaure, Springer Verlag, Berlin.

Wright, F.E. (1927): Polarization of light reflected from rough surfaces with special reference to light reflected from the moon, *Proc. Natl. Acad. Sci., U.S., 13,* 535–540.

Yates, H., et al. (1986): Terrestrial observations from NOAA operational satellites, *Science, 231 (4737),* 463–470.

CHAPTER 11

POLARIZERS AND POLARIMETERS

11.1 Early work on polarization

A Danish physician and naturalist, Erasmus Bartholinus (1625–1698) published in 1670 the 60-page book *Experimenta Crystalli Islandic Disdiaclastics*, an account of 17 experiments dealing with the birefringent crystal Iceland spar (a variety of calcite). Birefringence is a property of crystals that have the index of refraction different along two optical axes of the crystals. The account of Bartholinus' Experiment No. 7, the most notable of the 17, has been translated into English from the original Latin by Swindell (1975). Bartholinus observed that the crystal of Iceland spar produced a double image in transmitted light, but he did not discover the differences in the two rays of light constituting the images. This discovery was reserved for the Dutch physicist Christian Huygens (1629–1695), who published his brilliant work in 1690, *Traite de la Lumiere*. Huygens showed that the two rays transmitted by Iceland spar in any except one specific direction could be further subdivided by passage through a second crystal, and that the two rays possessed fundamentally different properties. From these observations history has credited Huygens with the discovery of the polarization of light. In spite of his brilliant observations, Huygens was still unable to explain the differences in the two rays, since he accepted the popular view of the time that if light consisted of waves, the waves were longitudinal in nature. The vehement rejection of the wave theory of light by the eminent authority Isaac Newton (1642–1727) was a damper on further investigation of light waves, and the wave theories of Thomas Young (1773–1829) and particularly Augustine Fresnel (1788–1827) were still in the future.

After the work of Huygens, a century passed before the next significant advance was made in the understanding of polarization. In 1808, the Paris Academie des Sciences set up a contest for the best mathematical description of double refraction. The prize was won 2 years later by E.T. Malus (1775–1812) for several important discoveries on polarization. First, while using a crystal of Iceland spar for observing the light of the setting sun reflected from the windows of the Luxemburg palace, Malus observed the light to be strongly polarized at a certain angle, the polarizing angle, and concluded that Newton was correct in thinking that polarization was an intrinsic property of light itself. His careful and detailed observations of reflection led him to the conclusion that "all transparent solids or liquids and opaque objects are able to impress on the molecules of light that strange property that appears to be one of the effects of double refraction" (Malus, 1809). From these observations, Malus formulated the law stating that when a ray of light polarized by reflection falls on a second reflecting surface at the polarizing angle, the intensity of the twice reflected ray varies as the square of the cosine of the angle between the two planes of reflection. Finally, he showed that transmission of light through transparent media also introduces polarization into the transmitted ray. (It should be remembered that Arago discovered the polarization of skylight in 1809. His polarimeter was based on a stack of inclined glass plates.) Malus believed in the corpuscular nature of light, and coined the term polarization to describe the different polarities he conceived for the two different types of corpuscles constituting the different rays of light. An excellent summary of the work of Malus and his predecessors from the Greco-Roman age is given in a monograph on the nature of light by Vasco Ronchi (1970).

The laws of polarization by reflection, so brilliantly initiated by Malus, were extended by the observations of the polarizing angle for many different materials by the Scottish physicist Sir David Brewster. His work led to the enunciation of Brewster's law of polarization. This law states that the tangent of the angle of polarization is the index of refraction of the reflecting material (Brewster, 1815). From this relation, it follows that at the polarizing angle the reflected and refracted rays are perpendicular to each other. Further, when light traveling inside a medium is reflected from the bounding air interface, the relative refractive index is the cotangent of the polarizing angle. In modern usage, the term polarizing angle has been generally supplanted by "Brewster's angle." Measurement of Brewster's angle is a method currently in use for determining the index of refraction of a bulk material.

11.2 Polarizing prisms

Although polarization by reflection is a well-known phenomenon, most modern polarizers and analyzers utilize birefringent crystals. (It is pertinent to note that the distinction between polarizer and analyzer is entirely one of usage of the device. A polarizer is used to introduce polarization into a light beam, whereas an analyzer is used to analyze the properties of a light beam which already contains polarization. Thus a given device may be a polarizer in one application and an analyzer in a different application.)

Some optical materials exhibit selective absorption of the transmitted light, and therefore have limited usefulness as a polarizer or analyzer. This problem is obviated by the use of calcite ($CaO \cdot CO_2$). Calcite is a uniaxial birefringent crystal of the hexagonal crystallographic system. It is transparent from about 0.24 to 1.8 μm. It has two principal refractive indices, the ordinary ω and extraordinary ϵ indices. For wavelength $\lambda=0.5893$ μm, $\omega=1.658$ and $\epsilon=1.486$, the difference of which, termed the coefficient of birefringence, is $J=0.172$. For $\lambda=0.34$ and 0.76 μm, J for calcite is 0.195 and 0.167, respectively (Shurcliff, 1962). These values are so large that the acceptance angle for calcite polarizers is similarly large. As a comparison, the value of J for quartz is only 0.009, a fact which limits the usefulness of quartz for polarizers. Quartz, however, finds widespread use in retarders.

A number of different designs of calcite polarizers have been developed, the differences among them being the angles and numbers of sections of the calcite crystal and the methods by which the sections are reassembled and secured. The principal designs are summarized below, and several additional designs are listed by Glazebrook (1923) and Shurcliff (1962), among others.

11.2.1 Nicol prism

The Scottish physicist William Nicol (1828) was one of the first to make a study of calcite as a polarizer. Although now largely supplanted by other types, the Nicol prism was for many decades the epitome of crystal polarizers. Its general design is shown by the simplified diagram of Fig. 11.1. A rhombohendron of calcite is sectioned, the cut surfaces are polished, and the two parts are reassembled with a film of Canada balsam between them. The index of refraction of Canada balsam, being 1.55, is about midway between those for the O-ray and E-ray of calcite. The O-ray, being deviated more than the E-ray, strikes the

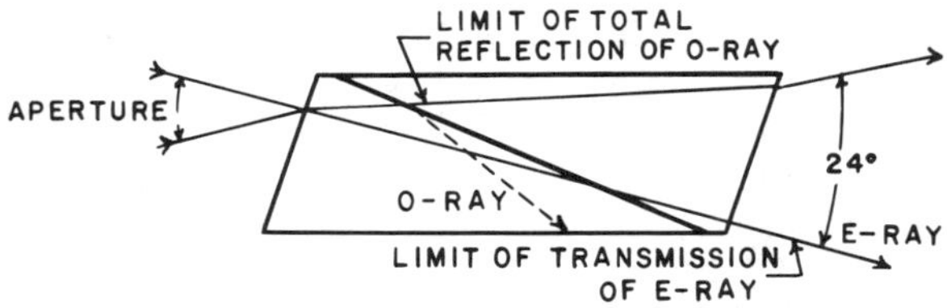

Figure 11.1 Simplified diagram of the paths of ordinary and extraordinary rays in a Nicol prism.

film of balsam at such an angle as to undergo total reflection, while the E-ray is transmitted. This separates the two differently polarized rays. The totally reflected 0-ray proceeds to the side of the prism, where it is absorbed by a coating of black paint, leaving the completely polarized E-ray as the only transmitted part of the original unpolarized radiation.

Undesirable features of the Nicol prism are that the transmitted E-ray undergoes a lateral displacement, the middle line of the incident field is inclined to the longitudinal axis by about 10°, and the external angular field is confined to about 24°. These deficiencies have been largely overcome by later modifications in design of calcite polarizers of which over a dozen have been developed (Glazebrook, 1923).

11.2.2 Glan-Thompson prism

This form consists of two identical calcite prisms with their optic axes parallel to each other and to the input and output faces. A schematic diagram of the arrangement is shown in Fig. 11.2. The gap between the prisms is filled with an isotropic material, such as Canada balsam, whose refractive index of 1.55 is about midway between the two indices of calcite. Although this arrangement is an efficient polarizer, considerable attenuation is produced by reflection at the Canada balsam interface. Better transmission for the normally utilized E-ray is obtained by the use of butyl methacrylate as the cement, the refractive index of which, at 1.49, is close to the index of calcite and thus produces minimal reflectance of this ray at the interface. The totally reflected 0-ray is normally

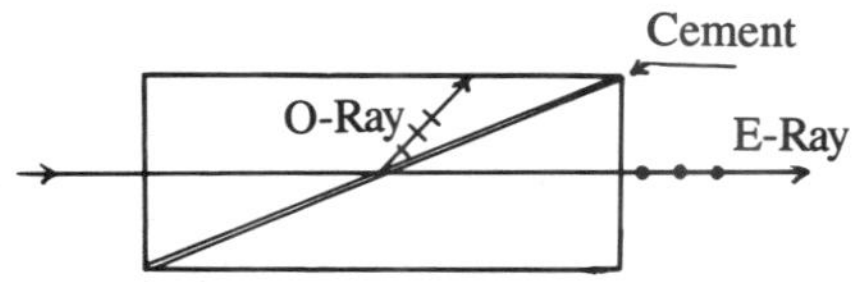

Figure 11.2 Diagram of Glan-Thompson prism. (Points indicate vibrations normal to plane of paper; dashes indicate vibrations in plane of paper.)

absorbed by black paint on the ground side of the prism. The acceptance angle of the Glan-Thompson prism is unusually large (about 40°). Examples of polarimeters utilizing the Glan-Thompson prism are the stellar polarimeter of Hall and Mikesell (1950) and the skylight polarimeters of Seaman and Sekera (1955), Hariharan (1967), Hannemann (1975), and Coulson et al. (1974).

11.2.3 Glan-Foucault prism

The layer of cement between the prisms of most polarizers has a significant amount of absorption in the ultraviolet. For use in this wavelength range, polarizers in which the cement is replaced by an air space are often preferred. In the Glan-Foucault polarizer, a diagram of which is shown in Fig. 11.3, the optic axes of the prisms are normal to the plane of incidence. Disadvantages of the Glan-Foucault prism are that considerable reflection losses occur at the double air interface, and the acceptance angle is small (about 7°).

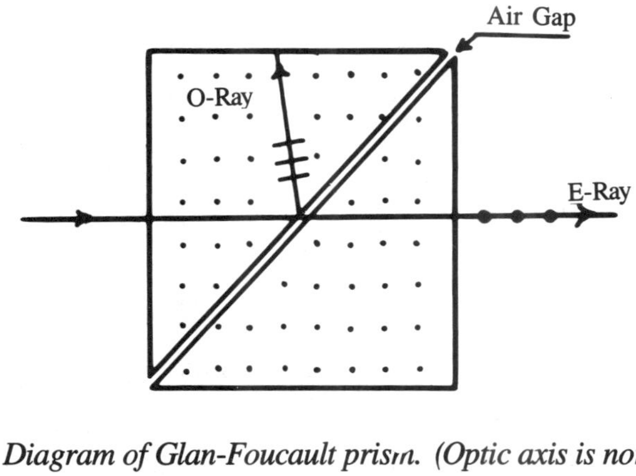

Figure 11.3 Diagram of Glan-Foucault prism. (Optic axis is normal to the plane of incidence.)

11.2.4 Glan-Taylor Prism

This frequently used type of polarizer has the angle at which the transmitted ray strikes the interface near the Brewster angle, which gives a high transmissivity of the prism. The optic axes of the two components are parallel to the plane of incidence, as indicated in Fig. 11.4. In the configuration often used for dumping the rejected portion of a high power laser beam, the Glan-Taylor prism has a polished side window oriented normal to the O-ray through which the excess energy emerges (Miles, 1982).

11.2.5 Wollaston prism

The Wollaston prism produces two orthogonally polarized beams, the two emerging at different angles from the same face of the prism, as shown in Fig. 11.5. One-half of the prism has the optic axis parallel to the plane of incidence, and the other perpendicular to this plane. The two halves are joined with Canada balsam. The transmitted beams are both deviated from the incident direction by amounts which vary with choice of wedge angle and with wavelength. Examples of polarimeters utilizing the Wollaston prism are the telescope photometer of Gehrels and Teska (1960) and the cloud photopolarimeter

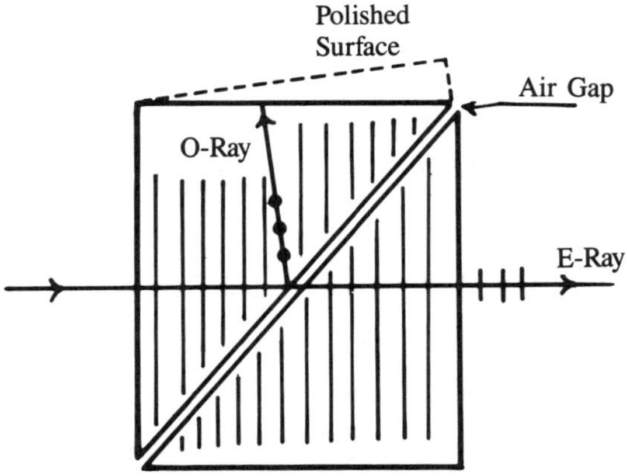

Figure 11.4 Glan-Taylor prism. (Optic axis is parallel to the plane of incidence.) The dashed outline shows a configuration in which the energy of the O-ray is dumped, instead of being absorbed at the side of the prism.

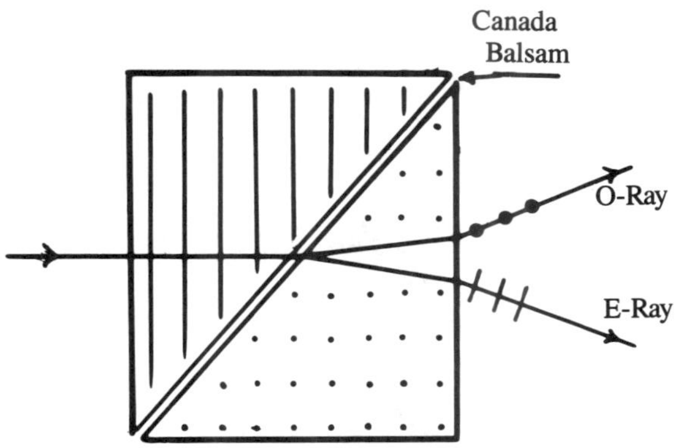

Figure 11.5 Wollaston prism.

of Russell et al. (1977) developed for use on the National Aeronautics and Space Administration (NASA) Pioneer mission to Venus. The method of using a Wollaston prism in a photographic polarimeter is discussed by Hiltner (1962), and a number of other Wollaston polarimeters used in astronomy are listed by Serkowski (1974).

11.2.6 Rochon prism

The Rochon prism is similar to the Wollaston in that it yields two orthogonally polarized beams from the same face of the prism, as shown in Fig. 11.6. In contrast to the Wollaston, however, the O-ray is not deviated from the incident direction. The E-ray is deviated by a relatively large angle, the size of which depends on wedge angle and wavelength. The colinearity of the O-ray with incident direction is an attractive feature of the Rochon prism for various applications.

11.3 Dichroic polarizers

A very convenient and inexpensive type of polarizer is produced by use of the property of some materials that the amount of absorption is dependent on the vibration direction of the incident beam. Such dichroic materials are most often used in linear polarizers, although dichroic circular polarizers are possible

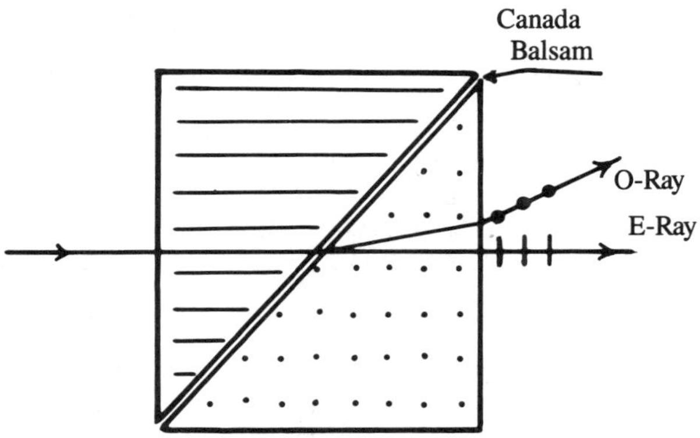

Figure 11.6 Rochon prism.

(Shurcliff, 1962). Tourmaline cyrstals exhibit strong dichroism, and in former days tourmaline polariscopes, consisting of a pair of tourmaline crystals appropriately mounted, were widely used for measuring optical rotation or detecting strains in a glass plate. Because of spectral selectivity, small aperture, and other shortcomings, however, tourmaline polarizers have mostly been supplanted by other types.

Thanks to the work of E.H. Land and his colleagues over a 30-year period beginning in the 1920s, the widely used sheet polarizers of various types were developed to a high degree of efficiency in polarizing light. The methods have been summarized by Shurcliff (1962) and particularly by Land (1951). It is of interest to note that well over a hundred United States patents have been awarded to Land and his colleagues for the development of sheet polarizers and their use in numerous practical applications. The first of Land's sheet polarizers utilized a statistical array of long thin dichroic microcrystals which were embedded in a plastic sheet of nitrocellulose or polyvinyl alcohol. The microcrystals were brought into approximate alignment by stretching the plastic along a single direction, the array of microcrystals then acting similar to a single crystal but without the limitations of the single crystal. By this method sheet polarizers of almost any size can be made. Later methods utilize long thin dichroic molecules instead of microcrystals, but the resulting polarizers have similar properties and manufacturing methods as those with microcrystals. Since the Polaroid Corporation, Cambridge, Massachusetts, has been the foremost manufacturer of these types of sheet polarizers, they are usually termed polaroid sheets or simply polaroid.

A number of different types of polaroid have been developed, as discussed by Shurcliff (1962). We concentrate here on the three most important, the H-sheet, the K-sheet, and the HR-sheet, all of which are molecular-type polarizers.

11.3.1 H-sheet polarizer

The H-sheet polarizer consists mainly of a sheet of polyvinyl alcohol that has been stretched in one direction to align the molecules and laminated to a support sheet of cellulose acetate butyrate. The polyvinyl alcohol is then brought into contact with an ink that contains iodine, and a certain amount of the iodine is absorbed into the polyvinyl alcohol. The three grades of H-sheet (HN-22, HN-32, and HN-38) differ only in the amount of iodine present in the material. The letter *H* designates the general types of polarizer, *N* indicates that the sheet is neutral in color, and the appended number is the approximate total luminous transmittance of a single sheet, expressed as a percentage. In some configurations the laminated sheet is further protected by being placed between cover glasses, but this is normally done only for special applications such as in polarizers for cameras and other optical devices. As normally purchased, the sheet is unprotected. Sheets may be at least 18 inches (45.7 cm) wide and almost any desired length, 12 inches (30.5 cm) square being a popular size. The material is easily cut with scissors, so practically any desired configuration of the polarizer may be achieved.

The useful spectral range of the HN-type polaroid is from about 0.375 to 0.750 μm. Data on the transmittance of the parallel and normal polarization components for the three types of HN polaroid, as given by Shurcliff (1962), are shown for selected wavelengths in Table 11.1. Included also is the degree of polarization of light transmitted by a single sheet of polaroid for unpolarized incident light. Within the precision of the original data, complete polarization is achieved for all listed wavelengths for HN-22, for all wavelengths above $\lambda = 0.450$ μm for HN-32, and for the range $\lambda = 0.550$ to 0.700 μm for HN-38. More complete polarization may be achieved by employing two or more properly aligned sheets, and the useful spectral range may be extended somewhat by multiple sheets. The cost, of course, is a concomitant decrease of total trans-

TABLE 11.1 Parallel (∥) and normal (⊥) transmittances and degree of polarization *P* of transmitted light for unpolarized incident light as functions of wavelength for single sheets of the three types of HN polaroid and of K-sheet polaroid. (Transmittance data from Shurcliff, 1962.)

λ	HN-22			HN-32			HN-38			K-Sheet		
(μm)	∥	⊥	*P*	∥	⊥	*P*	∥	⊥	*P*	∥	⊥	*P*
0.375	0.11	5×10^{-6}	0.999	0.33	1×10^{-3}	0.994	0.54	2×10^{-2}	0.929	0.42	2×10^{-3}	0.991
0.400	0.21	1×10^{-5}	0.999	0.47	3×10^{-3}	0.987	0.67	4×10^{-2}	0.887	0.51	1×10^{-3}	0.996
0.450	0.45	3×10^{-6}	0.999	0.68	5×10^{-4}	0.998	0.81	2×10^{-2}	0.952	0.65	3×10^{-4}	0.999
0.500	0.55	2×10^{-6}	0.999	0.75	5×10^{-5}	0.999	0.86	5×10^{-3}	0.988	0.71	5×10^{-5}	0.999
0.550	0.48	2×10^{-6}	0.999	0.70	2×10^{-5}	0.999	0.82	7×10^{-4}	0.998	0.74	4×10^{-5}	0.999
0.600	0.43	2×10^{-6}	0.999	0.67	2×10^{-5}	0.999	0.79	3×10^{-4}	0.999	0.79	3×10^{-5}	0.999
0.650	0.47	2×10^{-6}	0.999	0.70	2×10^{-5}	0.999	0.82	3×10^{-4}	0.999	0.83	8×10^{-5}	0.999
0.700	0.59	3×10^{-6}	0.999	0.77	3×10^{-5}	0.999	0.86	7×10^{-4}	0.998	0.88	2×10^{-2}	0.956
0.750	0.69	1×10^{-5}	0.999	0.84	2×10^{-4}	0.999	0.90	4×10^{-3}	0.991	0.92	0.57	0.235

mission, but the high transmissivity of HN-38 makes multiple sheets a feasible alternative for increasing the polarization. According to Shurcliff (1962), the useful wavelength range may be extended to cover the range from $\lambda=0.30$ to 0.90 μm by the use of multiple sheets.

11.3.2 K-sheet polarizer

The polarizer in K-sheet is polyvinylene, a small part of the basic polyvinyl alcohol being converted to this long chain molecule. The K-sheet is more resistant to high temperature and high humidity than are the HN polarizers, and, as shown by the date of Table 11.1, it is a generally satisfactory alternative for $0.45\leq\lambda\leq0.65$ μm. Even in the ultraviolet, $P=0.99$, and the transmissivity is good throughout the visible. Only at $\lambda=0.70$ μm and beyond does the K-sheet performance deteriorate badly. According to Shurcliff (1962), glass laminated K-sheet can withstand temperatures of 190 °F (87.8 °C) for long periods, and plastic lamination extends this temperature to as much as 215 °F (101.7 °C).

11.3.3 HR-sheet polarizer

This polarizer is useful for the wavelength range from 0.70 to about 2.2 μm (Shurcliff, 1962), but has a band at $\lambda=0.9$ to 1.0 μm of generally degraded performance. The HR polarizers have good stability at temperatures of 65 °C, and may be subject to temperatures of 93 °C for short periods without appreciable deterioration.

The stability of all of the polyvinyl alcohol polarizers for high temperature and humidity may be enhanced by treating the material with solutions of boric acid. The details of the method and its effect on polarizer efficiency are given by Kostyla and Abcunas (1981).

11.3.4 Polacor glass polarizing filters

A new type of polarizer for wavelengths in the $\lambda=0.8$ to 1.8 μm region has recently been introduced by the Corning Glass Works (Gritz, 1987). This type of polarizer, termed Polacor by the company, consists of elongated silver particles of uniform orientation embedded in a glass matrix. The polarizing effect is due to resonant absorption of one polarization component over the other by the sub-microscopic silver particles in the glass. According to company specifications, the degree of polarization available from a single sheet of the polarizer normally exceeds 99.6% at the specified wavelength, and, by special techniques, 99.8% may be achieved. High polarization efficiency is achieved over a bandwidth of 0.3 μm, the central wavelength of which may be selected in the range 0.8 μm to 1.8 μm, and at angles of incidence up to 30°. Being made of glass with embedded silver particles, the Polacor glass polarizers are stable at temperatures to at least 425 °C for a period of 1 hour, and are essentially unaffected by humidity.

Potential problems in using the Polacor filters are that the particles cause a significant amount of light scattering in the near-forward direction, a factor which may cause blurring of images, and, being of the dichroic type, they are not suitable for use in high power lasers. In addition, some wavefront distortion may occur. Within these constraints, however, it is likely that the Polacor filters will find wide application in optical systems operating at near-infrared wavelengths.

11.4 Reflection polarizers

Reflection of radiation from dielectric surfaces such as glass is a convenient method of either introducing polarization into the reflected beam or analyzing a beam for its inherent polarization. If a beam of unpolarized light is incident at angle θ on a plane dielectric surface such as glass, the reflected light, also at angle θ but at the 180° azimuth, is in general partially plane polarized. If $\rho_\parallel$ is the reflection coefficient for vibrations of the electric vector which are parallel to the surface and $\rho_\perp$ is that for vibrations normal to the surface, Fresnel's equations yield

$$\rho_\parallel = \frac{\tan^2(\theta - r)}{\tan^2(\theta + r)}$$
$$\rho_\perp = \frac{\sin^2(\theta - r)}{\sin^2(\theta + r)} \qquad (11.1)$$

where r is the angle of refraction of the transmitted light. The more interesting of the two relations is that of $\rho_\parallel$, which becomes zero at the angle where $\tan(\theta + r) = \infty$ or $\tan \theta = m$, where m is the index of refraction of the dielectric material. For $m = 1.50$, the polarizing angle θ_ρ (also termed Brewster's angle) is 56.3°. At other angles the degree of polarization of the reflected light is

$$P = \frac{\rho_\perp - \rho_\parallel}{\rho_\perp + \rho_\parallel} \qquad (11.2)$$

for the case of unpolarized incident light.

11.5 Transmission polarizers

Although reflection polarizers may be used for calibration of polarimeters operating in any wavelength range, it is normally more convenient for the near-visible range to utilize a transmission polarizer made from one or more tilted glass plates of known index of refraction and coefficient of absorption. A single glass plate is sufficient to yield a degree of polarization from 0% to about 25%. Walraven (1978) has computed, by the relations

$$T_\parallel = \frac{1 - \rho_\parallel}{1 - \rho_\perp} \qquad (11.3)$$

the transmitted orthogonal components for a nonabsorbing glass plate of index of refraction 1.513 as a function of tilt angle of the plate. Reflection coefficients $\rho_{\parallel}$ and $\rho_{\perp}$ were given in Eq. (11.1). The degree of polarization of the transmitted light is given as a function of tilt angle in Fig. 11.7. The thickness of the plate in relation to the beam radius and detector radius affects the polarization because of multiple reflections inside the plate. Two limiting cases are shown, the (A) and (B) curves being for the beam and detector radii being much smaller and much larger, respectively, than the thickness of the plate. For other relative sizes, the polarization will be between the two cases shown. The precise behavior for tilt angle 60° is given for beam and detector radii up to 0.7 cm by Walraven's computations. The values lie between 0.17949 and 0.19555 for the extremes. For some combinations the polarization is very sensitive to relative dimensions of beam, detector, and plate, whereas for some combinations it is not. In combinations of high sensitivity the polarization values would be sensitive also to the cross-sectional nonuniformity of beam and detector, so these combinations should be avoided for precise calibrations.

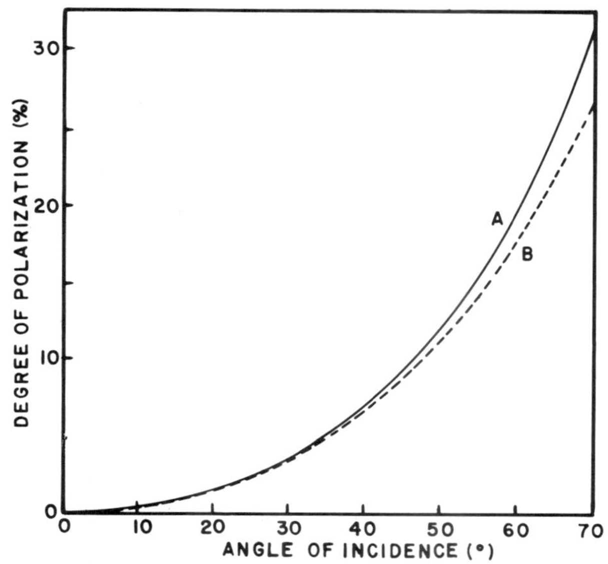

Figure 11.7 Polarization introduced by a tilted glass plate of refractive index 1.513 for (A) both beam and detector radii are much less than the thickness of the plate, and (B) either beam or detector radius much greater than the thickness of the plate. (After Walraven, 1978.)

Higher polarization values may be achieved by stacking glass plates, but the problem of internal reflections is compounded and beam displacement is of concern in this configuration. Shurcliff (1962) has given the designs of three glass plate polarizers, the plates of which are arranged to minimize effects of internal reflections and beam displacement, but there is no real substitute for suitable diaphragms in the system. Weinberg (1964) has tabulated the polarization of the light transmitted by 1, 2, 5, and 10 plates as a function of tilt angle for the limiting cases of no internal reflections and an infinite number of such reflections. By proper design of beam and detector dimensions and use of appropriate diaphragms, the method can yield polarimeter calibrations of accuracy sufficient for all but the most demanding requirements.

Other types of polarizers applicable to special problems are the wire grid polarizer and thin metal film polarizer. The first of these was originally applied to the polarization of microwaves, but as discussed by Yeh (1981), the method has been extended to many infrared devices. A type of polarizer somewhat similar to the wire grid results from the fabrication of thin metal films by vacuum evaporation. According to Slocum (1981), the method is appropriate for making polarizers for the near-infrared and visible wavelength regions. The film consists of an array of discontinuous, parallel metal whiskers distributed uniformly over a transparent surface. Light with the electric vector parallel to the whisker direction is reflected by the whiskers, and that with the electric vector normal to the whiskers is transmitted. The principal requirements for the whiskers, as in wire grid polarizers, are that the whiskers must be long compared to the wavelength and their individual diameters and spacing must be small compared to the wavelength. If these conditions are met, the metal film can be both a stable and relatively efficient polarizer.

11.6 Depolarizers

In instruments used for measuring only the scalar intensity of light, the existence of polarization may cause significant errors in the measurement because of polarization sensitivity of some of the optical components. Many a measurement of intensity made from aircraft has been rendered invalid by the use of mirrors for introducing light into the measuring device, and designers of satellite radiometers have gone to great lengths to take proper account of the polarization of light emerging to space from the top of the atmosphere to be measured by the radiometer. The polarization problem is a particular nuisance in some types of spectrophotometers.

The problem would be alleviated by use of a highly efficient optical component capable of depolarizing the incident light as it is introduced into the radiometer, but, unfortunately, no such device is known. It is much more difficult to eliminate polarization from a radiation stream than it is to introduce polarization into the stream. Some success can be achieved by scattering the light from a cloud of randomly arranged particles, such as those in opal glass diffusers, or by use of an integrating sphere. In both of these cases, however, the direction and general character of the light beam is drastically altered. A better approach for decreasing the effects of linear polarization is by use of a pseudodepolarizer or polarization scrambler of some type. For instance, the optical systems of current satellite instruments, the solar backscatter ultraviolet (SBUV) spectrometer, the total ozone mapping spectrometer (TOMS), and the coastal zone color scanner (CZCS) all have pseudodepolarizers to decrease measurement errors arising from polarization of the incident light (Chen, 1985).

A typical design of a pseudodepolarizer is that used for SBUV, the configuration of which is shown in Fig. 11.8. (Beckman Instruments Technical Report on Contract NAS 5-20970–1980.) It is normally termed a Welford-Lyot depolarizer and consists of four wedges of optical grade cultured crystalline quartz cut at the angles shown. The four wedges constitute two Wollaston prisms, each of which includes two wedges with crossed axes. As light is transmitted through the device, the initial plane polarization is changed to elliptic or circular polarization by the retardations introduced by the varying thicknesses of quartz. Thus instead of being truly depolarized, the state of polarization is scrambled, the amount and type of scrambling dependent on the location and direction of the particular ray as it enters the scrambler. In order to take advantage of this change of polarization state, the instrument must be affected less by elliptic or circular polarization than by the original plane polarization, which is a normal characteristic of this type of radiometer.

Performance specifications for the SBUV depolarizer originally given by NASA, and achieved by the manufacturer, were that the response of the instrument should not be affected by more than 2% for the case of complete polariza-

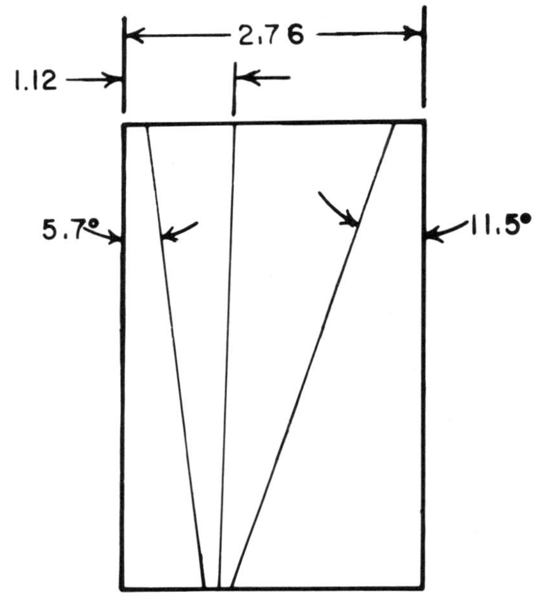

Figure 11.8 Design of the pseudodeopolarizer of the Solar Backscatter Ultraviolet radiometer of NASA. Linear dimensions are in centimeters. (Adapted from June 1980 Technical Report, Beckman Instruments, Inc.)

tion of the incident radiation. In view of the fact that radiation emerging from the top of the atmosphere may have plane polarization of 50% or more in some directions, it is to be expected that errors of 1% or more in intensity measurements could result from the failure to achieve complete depolarization of the ultraviolet light entering the SBUV instrument in orbit. The 2% residual polarization remaining in a completely polarized incident beam represents well the present state of development of optical depolarizers.

11.7 Mueller matrices of optical devices

In order to design an instrument for polarization measurements, one must obtain an analytical representation of the action of the various optical components of the device. A convenient method of representing the action of a polarizer, retarder, or other optical device on a beam of light is by use of Mueller matrices. These are 4×4 matrices that describe how the Stokes vector of radiation incident on the device is transformed by its passage through the device to emerge with different characteristics. For instance, its state of polarization may be changed, such as with a polarizer, the phase of one component may be changed with respect to that of its orthogonal companion, as with a retarder. Furthermore, the action of a train of optical components can be represented very simply by an equivalent product of matrices, one for each element of the train. In the following we shall write the Mueller matrices for polarizer and retardation plate in forms appropriate to perfect devices, and leave the more general representation to be discussed in Appendix A.

If we have a vector intensity $\mathbf{I}'$ incident on a train of components whose individual matrices are $\mathbf{M}_1$, $\mathbf{M}_2$, $\mathbf{M}_3$ . . ., $\mathbf{M}_i$, then the emergent intensity is the product

$$\mathbf{I} = (\mathbf{M}_i)\ .\ .\ .\ (\mathbf{M}_3)(\mathbf{M}_2)(\mathbf{M}_1)(\mathbf{I}') \tag{11.4}$$

It is important that the matrices be arranged in the order shown, $\mathbf{M}_1$, being for the first device encountered by the beam, $\mathbf{M}_2$ for the second, and so on, with the original vector intensity $\mathbf{I}'$ in the indicated position. Then the multiplications are carried out by standard matrix algebra.

As an example, the Mueller matrix of a perfect linear polarizer with no absorption (see Appendix A) which is oriented with the principal axes rotated as shown in Fig. 11.9 is given by

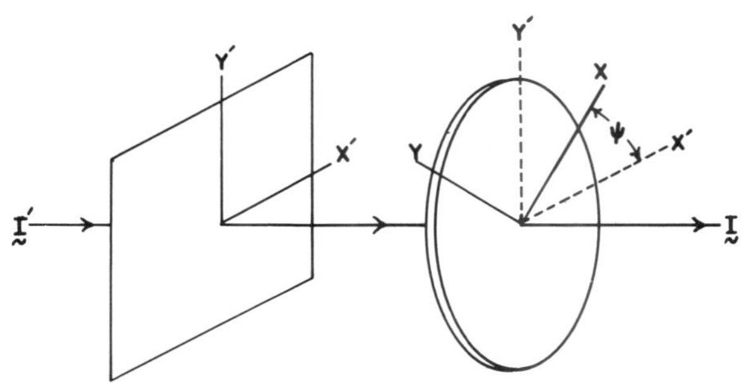

Figure 11.9 Radiation of incident intensity $\mathbf{I}'$ *being transformed to* $\mathbf{I}$ *by passage through an optical device whose principal axes are rotated through angle* ψ *from those of the original beam.*

$$\mathbf{M}_{pp} = \tfrac{1}{2} \begin{vmatrix} 1 & \cos 2\psi & \sin 2\psi & 0 \\ \cos 2\psi & \cos^2 2\psi & \cos 2\psi \sin 2\psi & 0 \\ \sin 2\psi & \cos 2\psi \sin 2\psi & \sin^2 2\psi & 0 \\ 0 & 0 & & 0 \end{vmatrix} \tag{11.5}$$

Now if the incident light is unpolarized and $\psi = 0°$, the emergent light, given as $\mathbf{I} = \mathbf{M}_{pp}\,\mathbf{I}'$ where

$$\mathbf{I}' = \begin{vmatrix} 1 \\ 0 \\ 0 \\ 0 \end{vmatrix} \tag{11.6}$$

is

$$\mathbf{I} = \tfrac{1}{2} \begin{vmatrix} 1+0+0+0 \\ 1+0+0+0 \\ 0+0+0+0 \\ 0+0+0+0 \end{vmatrix} = \begin{vmatrix} \tfrac{1}{2} \\ \tfrac{1}{2} \\ 0 \\ 0 \end{vmatrix} \tag{11.7}$$

This is the vector of a completely horizontally polarized beam.

We take for a second example a nonabsorbing retardation plate with $\delta = 90°$ (i.e., a perfect quarter wave plate) with principal axes at $\psi = 45°$. From Appendix A we have

$$\mathbf{M}_{90} = \begin{vmatrix} 1 & 0 & 0 & 0 \\ 0 & \cos^2 2\psi + \cos\delta \sin^2 2\psi & (1-\cos\delta)\cos 2\psi \sin 2\psi & -\sin\delta \sin 2\psi \\ 0 & (1-\cos\delta)\cos 2\psi \sin 2\psi & \sin^2 2\psi + \cos\delta \cos^2 2\psi & \sin\delta \cos 2\psi \\ 0 & \sin\delta \sin 2\psi & -\sin\delta \cos 2\psi & \cos\delta \end{vmatrix} \tag{11.8}$$

which reduces to

$$
\mathbf{M}_{90} = \begin{vmatrix} 1 & 0 & 0 & 0 \\ 0 & 0 & 0 & -1 \\ 0 & 0 & 1 & 0 \\ 0 & 1 & 0 & 0 \end{vmatrix} \tag{11.9}
$$

If linearly polarized light with plane of polarization horizontal, the vector intensity of which is shown in the previous example, is incident on this quarter wave plate, the resulting intensity is

$$
\mathbf{I} = \begin{vmatrix} 1/2 \\ 0 \\ 0 \\ 1/2 \end{vmatrix} \tag{11.10}
$$

This is a circularly polarized beam of total intensity 1|2.

A number of other special cases are given by Shurcliff (1962), but they are all readily obtained by use of the matrices in Appendix A. Likewise the vector intensity emerging from a train of optical devices is obtained from the product of matrices, one for each optical element of the train. It is necessary, however, that the matrix multiplications be taken in the same order as the elements of the train, as mentioned previously.

11.8 Polarimeters developed by various authors

A large number of optical systems have been developed for observations of polarized light in astronomy, for observations of the planets and other celestial bodies from spacecraft, and for observations of the earth from near the surface to orbital altitude. A list of polarimeters used in astronomy has been prepared by Serkowski (1974), as has a list of polarimeters in space by Coffeen (1974). The various types of polarimeters which have been developed for observations of the earth's atmosphere and surface are listed in Table 11.2. While there is

TABLE 11.2 Types of polarimeters used for skylight polarization measurements or measurements of properties of atmospheric aerosols by various authors. (PM: photomultiplier tube)

Optical system	Detector(s)	References
Rotating $\lambda/4$ plate, fixed Glan-Thompson prisms, 2 channels	End window PM	Seaman and Sekera (1955)
Stepped Wollaston prism with and without depolarizer, 2 channels	End window PM	Gehrels (1962)
Four $\lambda/4$ plates, fixed Glan-Thompson prism, 1 channel	End window PM	Sekera et al. (1963)
Rotating polaroid, 1 channel	Side window PM	Bullrich (1964)
Rotating polaroid, 1 channel	Germanium photocell	Bullrich (1964)
Rotating $\lambda/2$ plate, fixed Glan-Thompson prism, 3 channels	End window PM	Holzworth and Rao (1965)
Rotating polaroid, 1 channel	PM or PbS cell	Egan and Hallock (1966)
Rotating $\lambda/2$ plate, fixed polaroid, 1 channel	End window PM	Gorchakov (1966)
Rotating $\lambda/2$ plate, fixed polaroid, 1 channel	End window PM	Eiden (1966)
Rotating $\lambda/2$ plate, fixed Glan-Thompson prism, 3 channels	End window PM	Rao and Sekera (1967)
Four fixed Glan-Thompson prisms, 4 channels	One end-window PM	Hariharan (1967)
Stepped polaroid, 1 camera	Photographic film	Borofka and O'Brien (1968)
Rotating polaroid, 1 channel	End window PM	Dietze and Seidel (1968)
Rotating Glan-Thompson prism, 1 channel	End window PM	Rao (1969)
Rotating polaroid, 1 channel	End window PM	Fernald et al. (1969)
Rotating polaroid, 2 channels	End window PM	Dandekar and Turtle (1971)
Frank-Ritter quartz prism, 1 channel	End window PM	Boiko et al. (1971)
Six fixed Glan-Thompson prisms, 2 $\lambda/4$ plates, 6 channels	6 side window PM	Hannemann (1975)
Stepped Glan-Thompson prism, 2 channels	End window PM	Coulson et al. (1974)
Rotating polaroid, 1 channel	End window PM	Steinhorst (1977)
Fixed Glan-Thompson prism, 2 channels	End window PM	Sassen et al. (1977)

TABLE 11.2 (Concluded).

Optical system	Detector(s)	References
Savart plate with fixed polaroid	Eye	Gerhaz (1977)
Stepped Glan-Taylor prism	Silicon photodiode	Stowe (1977)
Stepped polaroid, 1 camera	Photographic film	Walraven (1977a)
Fixed polaroids	3 TV cameras	Walraven (1977b)
Rotating polaroid	PM or PbS cell	Reshetnikova and Federova (1978)
Stepped polaroid, 1 camera	Photographic film	Curran (1978)
Rotating $\lambda/2$ plate, fixed polaroid	End window PM	Gorchakov and Siderov (1980)
Fixed polaroids	3 TV cameras	Prosch et al. (1983)
Savart polariscope	Eye	Volz (1983)
Fixed polaroid	Commercial radiometer	Vanderbilt et al. (1985)
Fixed polaroid, 2 cameras	Photographic film	Coulson et al. (1986)
Proposed modification of Landsat Multispectral scanner	18 PM and 6 photodiodes	Egan (1986)

wide diversity of configurations, most of them fall in two main categories: (1) rotating retardation plate with fixed analyzer, and (2) continuously rotating or stepped analyzer or combination of analyzers. Most of the systems use a photomultiplier as the detector, although lead sulfide, germanium, and silicon cells have been used for special devices. The need for polarized images in remote sensing has resulted in a return to photographic film, but also the newer types of television tubes have been used, albeit infrequently so far, for the acquisition of polarized images. Egan (1986) has proposed a modification to an existing satellite radiometer to permit the buildup of polarized images.

In view of the importance of the rotating or stepped analyzer and of the combination retardation plate/fixed analyzer, we discuss in some detail the intensity variations resulting from their use in polarimeters.

11.8.1 Rotating or stepped analyzer

Since Fesenkov (1935) showed that the complete field of linear polarization, including intensity, degree of polarization, and orientation of the plane of polarization, can be determined by measurements of intensity transmitted by a polarizer at three different angular orientations, this has been a commonly used device in polarimeters. In fact, Pyaskovskaya-Fesenkova (Fesenkov's wife) utilized the method extensively for visual determination of polarization, but many electronic polarimeters utilize the same method of analysis. Briefly, if I_1, I_2, and I_3 are the intensities transmitted by the analyzer at angular positions ψ_1, ψ_2, and ψ_3, then the degree of polarization is simply

$$P = \frac{2[I_1(I_1 - I_2) + I_2(I_2 - I_3) + I_3(I_3 - I_1)]^{1/2}}{I_1 + I_2 + I_3} \tag{11.11}$$

and the mean intensity is

$$I = (I_1 + I_2 + I_3)/3 \tag{11.12}$$

Although in principle any three unequal settings of ψ would suffice, the highest precision can be obtained by settings at intervals of 45°. Of course, still higher precision results from a larger number of measurements. The larger number may be achieved by a continuous recording of the output intensity as the analyzer is rotated, or by rotating the analyzer in a discrete number of steps. This latter alternative is the one chosen by Coulson et al. (1974) for the University of California, Davis (UCD) polarimeter, the method of data analysis for which is described below.

In general, the intensity I of light transmitted by the analyzer set at angle ψ with respect to the principal axis of the polarizer for an incident light of Stokes parameters (I', Q', U', V') is given by the relation

$$I = \tfrac{1}{2}[I' + Q'\cos 2\psi + U'\sin 2\psi] \tag{11.13}$$

Thus the output intensity is a composite of a mean intensity plus a superimposed second harmonic. The amplitude of the harmonic is proportional to the degree of polarization, and its phase yields the angle of the plane of polarization. The fact that parameter V does not appear in Eq. 11.13 shows that such an arrangement is not suitable for measurements of elliptic or circular polarization. This is not a serious deficiency for skylight measurements, however, as the elliptic component of polarization in the atmosphere is negligibly small.

If measurements of I are made for several values of ψ, the data can be analyzed by means of Eq. (11.13), and values of I', Q', and U' determined. Walraven (1977c) has developed a method for a very rapid analysis of the data from the UCD polarimeter, in which the analyzer is set at intervals of 45° in the 360° rotation. The resulting eight signal values V_i, $i=1$, 8, after correction for dark count and dead time of the photomultiplier, are described by the functional form

$$V_i = C_0 + C_1 \cos\psi_i + C_2 \sin\psi_i + C_3 \cos(2\psi_i) + C_4 \sin(2\psi_i) + C_5 [(\theta_i - K) / K] \tag{11.14}$$

where $K = 180° (1 - 1/N)$ and $N=8$. The coefficient C_0 is the average intensity of the incident radiation, C_2 and C_4 correspond to Q' and U', respectively in Eq. (11.13), and C_1 and C_3 are first harmonic components of error; C_5 represents linear drift of either intensity or system noise over the period of measurements. The speed with which the measurements can be analyzed was greatly increased by the development of a number of matrices to enter into the computer.

For $N=8$, Eq. (11.14) gives the following coefficients

$$\begin{aligned}
C_0 &= (V_0 + V_1 + V_2 + V_3 + V_4 + V_5 + V_6 + V_7) / 8 \sim I' \\
C_1 &= (V_0 - V_4) / 2 \\
C_2 &= (V_2 - V_6) / 2 \sim Q' \\
C_3 &= (V_0 - V_2 + V_4 - V_6) / 4 \\
C_4 &= (V_1 - V_3 + V_5 - V_7) / 4 \sim U'
\end{aligned} \tag{11.15}$$

The coefficient C_5 due to drift was found to be negligibly small over the approximately 12 seconds necessary for one measurement cycle.

A schematic diagram of the optics and electronics of each of the two channels of the UCD polarimeter is shown in Fig. 11.10. The incoming light enters the optical chain through the well-baffled collimator tube C, which has a total acceptance angle of 1.0°. Shutter S is normally in the closed position to protect the sensor, but is opened by a signal from the computer for a measurement. The first optical element encountered by the incoming light is the Glan-Thompson

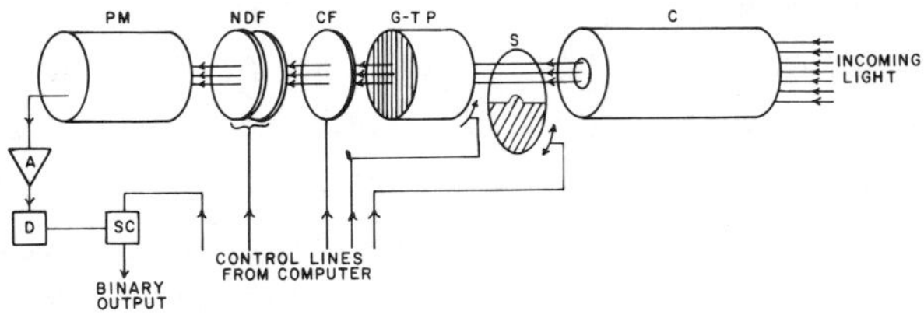

Figure 11.10 Schematic diagram of the stepped analyzer-type polarimeter developed at the University of California, Davis.

C:	*collimator tube*	*PM:*	*photomultiplier tube*
S:	*shutter*	*A:*	*amplifier*
G-T:	*Glan-Thompson prism*	*D:*	*discriminator*
CF:	*color filters*	*SC:*	*scaler*
NDF:	*neutral density filters*		

prism (G-TP) which serves as the analyzer. No lenses or other optical elements are superimposed ahead of the analyzer to avoid parasitic polarization by the instrument itself. Following the analyzer is the interference-type color filter (CF), four of which are available in each channel, and a bank of four selectable neutral density filters (NDF) for controlling the intensity incident on the photomultiplier (PM). The eight color filters permit measurements over the wavelength range $0.32 \leq \lambda \leq 0.90$ μm. The photomultiplier is used in the photon counting mode, in which case a statistical fraction of the photons incident on the cathode produce pulses on the output cable of the tube. These pulses are amplified by amplifier A and separated from electronic noise in the system by discriminator D. The pulses are then counted for a specified period of time by scaler SC, the total number of which is sent in binary form to the computer. Control signals from the computer are used to actuate the shutter, rotate the analyzer, change color filters, and select the correct combination of neutral density filters. Other computer signals, not indicated in the diagram, are used to control the goniometer on which the instrument is mounted, and angular shaft encoders are used to check the goniometer orientation before each measurement.

In normal operation, the analyzer is stopped for a 1-second count period at 45° intervals in the 360° rotation. The resulting eight photon counts, plus a dark count taken with the shutter closed, constitute the basic data of one measurement. The count rate is maintained at values of the order of 10^6/sec by an appropriate selection from the four neutral density filters. After correction for dark counts, the eight photon counts are analyzed by the method described above, and values of the physical quantities of relative intensity, degree of linear polarization, and angle of the plane of polarization, together with the statistical errors of these quantities, are printed out on a real-time basis. A measurement at one goniometer position for the two wavelengths (one in each of the two channels) requires approximately 12 seconds. A scan of the sky in 5° steps from horizon to horizon can be accomplished in about 7 minutes.

This dual-channel UCD polarimeter was used extensively by the present author and his colleagues over a period of several years in Hawaii, at Davis, California, and elsewhere. A single-channel instrument of similar design, but with six interference filters, was used for observations of skylight and of light reflected from the snow surface at the South Pole (Fitch and Coulson, 1977). Both instruments operated generally satisfactorily under various environmental conditions, and, as seen in previous sections of this book, the data were reliable and of sufficient precision for defining the skylight polarization regime of the atmosphere.

One potential problem with the rotating analyzer method which should be mentioned is that the output of the detector may depend on the angle of the plane of polarization of the radiation striking the detector. This feature is particularly troublesome for side-window photomultiplier tubes, and in fact it strongly discouraged use of the rotating analyzer method in the older instruments. However, the newer types of end-window photomultipliers are much less subject to this feature, and while it still has to be considered, a careful selection of the tube to be used can all but eliminate the problem. If the effect is still significant it may be necessary to resort to a depolarizer, such as a Lyot depolarizer, just in front of the photomultiplier. This is not a panacea, however, as depolarizers tend to be wavelength specific and introduce complexity into the system. If a depolarizer is used, its action must be well understood through analysis of the Mueller matrices of all of the elements of the system.

11.8.2 Rotating retardation plate with fixed analyzer

This configuration has been widely used in polarimeters for skylight and other types of measurements, as indicated in Table 11.2 The system has been analyzed in detail by Seaman and Sekera (1955), whose discussion, with modified notation, we follow here. As a preliminary we define four angles. The angle between the plane of polarization of the incident light and an arbitrary system zero is called χ. Conceptually, we designate this as the e axis and consider it to be in the vertical. Axis r is horizontal. Angle ψ is the angle of the fast axis of the retardation plate, and ζ is that of the plane of transmission of the analyzer, both with respect to the vertical. Finally δ is the retardation of the plate.

Seaman and Sekera (1955) showed that the Stokes vector of the radiation leaving the analyzer and falling on the detector for this optical system, and incident radiation $\mathbf{I}'$, is given by the product

$$\mathbf{I} = \mathbf{R}(0) \cdot \mathbf{L}(\zeta - \psi) \cdot \mathbf{R}(\delta) \cdot \mathbf{L}(\psi) \cdot \mathbf{I}' \tag{11.16}$$

Here the rotation matrices $\mathbf{L}(\psi)$ and $\mathbf{L}(\zeta - \psi)$ have the form

$$\mathbf{L}(\psi) = \begin{vmatrix} 1 & 0 & 0 & 0 \\ 0 & \cos 2\psi & \sin 2\psi & 0 \\ 0 & -\sin 2\psi & \cos 2\psi & 0 \\ 0 & 0 & 0 & 1 \end{vmatrix} \tag{11.17}$$

Since they represent a pure rotation of axes, Stokes parameters I and V are unaffected, whereas Q and U have different values after a rotation. Matrices $\mathbf{R}(0)$ and $\mathbf{R}(\delta)$ are of the form

$$\mathbf{R}(\delta) = T_r \begin{vmatrix} 1 & P_r & 0 & 0 \\ P_r & 1 & 0 & 0 \\ 0 & 0 & e_r \cos\delta & -e_r \sin\delta \\ 0 & 0 & e_r \sin\delta & -e_r \cos\delta \end{vmatrix} \tag{11.18}$$

where the total transmission T_r is given in terms of the transmissions along the fast and slow axes t_f and t_s as

$$T_r = \tfrac{1}{2}(t_f^2 + t_s^2) \tag{11.19}$$

Further,

$$P_r = (t_f^2 - t_s^2) \; / \; (t_f^2 + t_s^2) \tag{11.20}$$

and

$$e_r = (2t_f t_s) \; / \; (t_f^2 + t_s^2) = (1 - P_r^2)^{1/2} \tag{11.21}$$

For a perfect retardation plate $t_f = t_s$, $P_r = 0$ and no polarization is introduced by the plate. In this case $e_r = 1$.

Experience with the Seaman-Sekera instrument, however, showed that the retardation plates were not perfect. They were made of mica, and appeared to have a certain amount of dichroism. The amplitude and phase of the second harmonic, which should be dependent only on the ellipticity of the light, varied slightly with the magnitude and direction of the linear polarization. This feature, which resulted in P_r being on the order of 0.01, largely obviated measurements of the small amount of elliptic polarization in the natural skylight. Later information indicates that if the plates had been made of quartz or calcite the problem would have been avoided.

The Mueller matrix of the analyzers in the Seaman-Sekera polarimeter, both of which were high quality Glan-Thompson prisms and represented in Eq. (11.14) by **R**(0), is given by

$$\mathbf{R}(0) = T_a \begin{vmatrix} 1 & 1 & 0 & 0 \\ 1 & 1 & 0 & 0 \\ 0 & 0 & 0 & 0 \\ 0 & 0 & 0 & 0 \end{vmatrix} \tag{11.22}$$

for the plane of transmission parallel to the reference plane. For these prisms $\delta = 0$, $t_f = 1$, $t_s = 0$, $T_a = 1/2$, $P_a = 1$, and $e_r = 0$.

If the retardation plate is rotated around the optical axis with frequency w and the analyzer remains fixed, the intensity leaving the analyzer and falling on the detector is of the nature of a composite signal given by the relation (Sekera, 1956)

$$I = I'[1 + P\cos^2(\delta/2)\cos 2(\chi - \psi)] + V\sin\delta\sin 2(\omega t - 2\psi)$$

$$+ I'P\sin^2(\delta/2)\cos(4\omega t - 2\psi - 2\chi) \quad (11.23)$$

Here P is the degree of linear polarization of the incident light and V is the Stokes parameter representing the ellipticity of the incident light. This shows that all of the information on the ellipticity is contained in the second harmonic of the signal. The degree of linear polarization P appears in both the second and fourth terms of Eq. (11.21), which shows a major limitation of this type of optical system for skylight measurements. The sum of these two terms may be rewritten

$$P[\cos^2\delta / 2 \cos 2(\psi - \chi) + I\sin^2\delta / 2 \cos(4\omega t - 2\psi - 2\chi)] \quad (11.24)$$

This shows that the value of P to be deduced from the measurement is a function of the angle χ of the plane of polarization of the incident light. The degree of polarization of the incident light in nature, however, is not a function of χ. If χ were to have a known value any variation could be accounted for in the data reduction process, but in general χ is not known. It varies over the dome of the sky, and is one of the parameters which should be determined by the measurement. Only in the plane of the sun's vertical χ is known, having a value of either 0° or 90°. In either case the relation is valid, and the value of P deduced from the measurement is valid. Outside the sun's vertical, however, χ is not constant, and erroneous values of P would result from the measurement. For this reason, the Seaman-Sekera polarimeter was confined to measurement in the sun's vertical, and any instrument of the rotating retardation plate/fixed analyzer type would suffer from the same limitation.

A photograph of the Seaman-Sekera polarimeter is shown in Fig. 11.11. It was a two-channel device, each channel fitted with three absorption filters which, taken together, covered the spectrum from the ultraviolet to the far red. The yoke mounting provided a possibility of making scans of the sky in vertical planes at any azimuth, but, as mentioned, the use of a fixed analyzer in the optical train effectively confined its use to the plane of the sun's vertical. The sun shield attached to the dual photocell sun-follower shown in the middle part of the photo protected the photomultiplier tube from damage from the direct solar beam as the scan crossed the sun. Scanning of the instrument across the sky was actuated

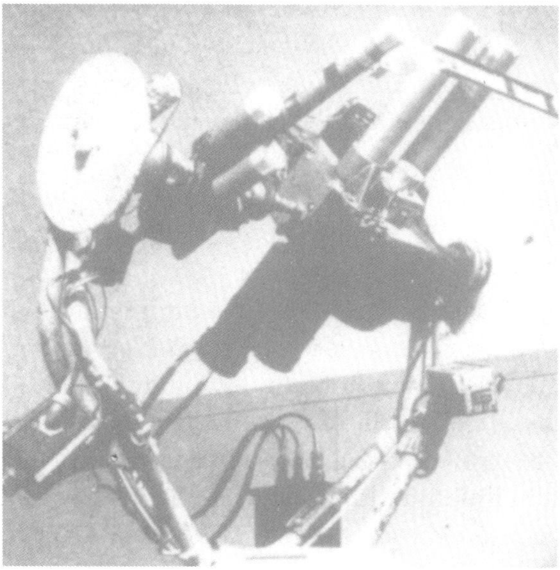

Figure 11.11 Photograph of the Seaman-Sekera dual-channel polarimeter used in various locations in Southern California for measuring polarization and intensity of light from the sunlit sky. (After Sekera, 1956.)

by a small remotely controlled electric motor, the elevation angle of the device being indicated by coded contacts on the protractor mounted at the left in the photo. The retardation plates were rotated by a synchronous motor at 10 Hz, and the signals from the photomultiplier were introduced into a pair of amplifiers tuned to 20 and 40 Hz. The signals from the amplifiers were recorded on strip-chart recorders. The phases of the two harmonics, taken with respect to a fixed zero, were also recorded.

One troublesome feature of the instrument was the method by which the large changes of intensity of the skylight were accommodated. The dynode voltage of the photomultipliers was regulated by a control circuit which held the photomultiplier current at a low and approximately constant value. This arrangement helped prevent fatigue of the tubes, but it introduced spurious noise into the system. The output current of the tubes was approximately proportional to the eighth power of the dynode voltage. This means that changes of intensity on the tube produced very small changes of dynode voltage, with concomitantly large relative errors in the measurement. The time constant of the voltage control circuit was sufficiently long that the dynode voltage did not follow the second and fourth harmonics, so the amplitudes of these harmonics could be extracted from the basic output signal from the photomultiplier tubes by means of tuned amplifiers.

The Seaman-Sekera polarimeter was used for many skylight measurements over a period of several years. The data for UCLA, Pasadena, and Cactus Peak shown in Chapters 5 and 6 were all taken with this instrument. It was particularly well adapted to measurements of the neutral points. The same basic optical system has been used in polarimeters described by Holzworth and Rao (1965), Gorchakov (1966), Hannemann (1975), and others, but various modifications have improved the operation of the instruments and precision of the data. For instance, Holzworth and Rao (1965) increased the number of channels from two to three, utilized quartz retardation plates instead of the original mica plates, and introduced interference filters instead of the older absorption filters. They still varied the dynode voltage for achieving the required dynamic range, however. In an improved version used for observations from a balloon, Rao and Sekera (1967) controlled the dynamic range by means of a calibrated optical wedge. Hannemann (1975) used a complicated combination of six Glan-Thompson prisms with two retardation plates, and Gorchakov (1966) used the basic rotating retardation plate-fixed analyzer configuration in a polarizing nephelometer for measurements of light scattering in the layer of air near the ground.

11.9 Photographic images in polarized light

The use of a polarizer in front of a camera lens has long been a method of controlling the effects of reflection in photographic images. McFarlane (1937) discussed the possibility of enhancing image quality by using a Pola-screen (a sheet polarizer) to minimize reflections by windows, water surfaces, and other reflecting and polarizing surfaces, and to obtain dramatic photos of landscapes by regulation of the highly polarized skylight by means of a polarizer mounted on the camera. Even more impressive effects in photographs may be obtained by illuminating a subject with polarized light and using polarization optics on the camera. Borofka (1967) and Borofka and O'Brien (1968) utilized the photographic method for evaluating the effects of polarization in illumination of a room used for reading magazines and newspapers. By this method, a polaroid is mounted on the camera in such manner that the polaroid may be set at three angles, horizontal, vertical, and at 45°, while the camera is held stationary. The negatives from three exposures, one for each of the polaroid settings, are scanned by a microdensitometer. The signals are then used to compute the degree of linear polarization and the orientation of the plane of polarization, by the method Fesenkov (1935) developed over 30 years previously. The general technique of analysis of polarized images has been discussed by Walraven (1981).

The photographic method was used by Curran (1978, 1979) for indicating soil moisture by means of the polarization signature in landscapes. Only two photos of a scene were taken, the polarizing filter set visually to the angles giving the maximum and minimum intensities of the scene. The well-known difficulties of film nonlinearities in exposure and processing were partially circumvented by positioning black and white reflection panels in each image. The negative density was digitized by a spot densitometer, and the degree of polarization computed from the density values. A similar method was employed by Fitch et al. (1984) for obtaining polarized images of four different cereal crops at various stages of development. In this latter study, however, four photographs of a scene were taken with a polaroid set at angles of 0°, 45°, 90°, and 135°. A nine-level reference gray scale was a part of each scene, permitting corrections for film nonlinearities in radiation response and nonuniformities of the film strip. The four images were digitized by a microdensitometer and brought into registration to ± 0.5 pixel by means of a pattern of dots on the gray scale target. The Stokes parameters were computed from the digital values of each pixel according to the relations

$$\begin{aligned} I &= \tfrac{1}{2}[I(0) + I(45) + I(90) + I(135)] \\ Q &= I(0) - I(90) \\ U &= I(45) - I(135) \end{aligned} \tag{11.25}$$

and the degree of polarization by

$$P = (Q^2 + U^2)^{1/2}/I \tag{11.26}$$

For these computations only relative intensities were used, as no absolute calibration was performed.

As mentioned, an exploratory program for obtaining polarized images from the Space Shuttle has been carried out by Coulson et al. (1986), and more than 400 pairs of photographs of various areas of the earth have been obtained on four Shuttle missions. This is an ongoing program, the objective of which is to obtain space-based images in polarized light for characterizing agricultural surfaces, hydrologic and oceanographic features, and for determination of atmospheric properties. Before the program was temporarily interrupted by the Space Shuttle Challenger disaster in January 1986, a large number of pairs of polarized images, some in black and white and some in color, were obtained on the four missions. Several image pairs were selected for detailed analysis, the basis of the selection being variety of surface characteristics within and between pairs, and different sun-scene-observer geometries.

The cameras for the Shuttle missions are a pair of boresighted and synchronized 70-mm Hassleblad cameras fitted with polarizing filters oriented normal and parallel to the camera axis. For taking photos, the camera pair is rotated so the camera axis is normal to the plane of polarization of the incident light, an operation performed by the astronaut holding and pointing the camera to the feature of interest. Two significant sources of error, in addition to the well-known problems of obtaining high precision data from film, are evident in this mode of operation. First, the astronaut may be unable, by a visual judgment of intensity, to orient the cameras precisely with respect to the plane of polarization. The second source of error is the polarization introduced by spacecraft windows through which the photos are taken. This source is difficult to evaluate because of the change of stress on the window material as the spacecraft goes from atmospheric pressure on the ground to the vacuum of space.

In spite of these difficulties however, some significant results have already been obtained in this exploratory program (Coulson et al., 1986), and more are anticipated. Future plans are for the development of a three-camera television system which will be mounted in the instrument bay and remotely controlled from the cabin, thereby eliminating the problem of viewing through spacecraft windows.

11.10 Video polarimetry

The video polarimetric technique is based on the same principles as that of photographic polarimetry, but the normal photographic cameras are replaced by television cameras. Several advantages are gained by the video method. First, the images are generated instantaneously, and thereby provide the possibility of a quick-look analysis. Second, the rapid repetition rate permits analysis of moving systems or the acquisition of images from a moving platform. Finally, the many problems of photographic film calibration and handling are obviated by use of an electronic system, but, of course, electronic systems have their own limitations arising from nonlinear components, electronic noise, and limited intensity resolution. For remote sensing applications in which high precision is not a requirement, however, video polarimetry has the potential for widespread application.

As mentioned, the intensity and complete state of linear polarization can be determined by measurements of intensity transmitted by an analyzer at three different orientations, or by three analyzers of different orientations according to Eq. (11.11). Thus a system of three television cameras fitted with polarizers oriented, for instance, at $\psi = 0°$, $45°$, and $90°$ should yield all of the information necessary to deduce the intensity I, degree of linear polarization P, and angle of the plane of polarization χ of the incident light. The signals from the three cameras can be processed instantaneously and used to control a color television monitor. In general, the information contained in a color television frame is the spatial position of the elements, along with their intensity, hue, and saturation of color. It should be possible to utilize the signals generated by scans of a scene by the three cameras to control the monitor for displaying these quantities. For instance, scene intensity could control element intensity, the degree of polarization could be represented in terms of color saturation, and the angle of the plane of polarization could correspond to the hue. Thus all of the information on a light field in a scene, either in a given spectral range or in white light, could be displayed in terms of intensity, saturation, and hue on the face of a color monitor. This would make it easy to see at a glance areas of high or low polarization, and the change of the plane of polarization over the scene.

Actually the same information can be obtained by three television cameras, only two of which are fitted with polarizing filters. If camera No. 1 has a polaroid set at $\psi=0°$, camera No. 2 has one at $\psi=45°$, and camera No. 3 has no polaroid, then the intensities for the three cameras are

$$
\begin{aligned}
I_1 &= \tfrac{1}{2} I_0 \, [1 + P \cos 2\chi] \\
I_2 &= \tfrac{1}{2} I_0 \, [1 + P \sin 2\chi] \qquad (11.27) \\
I_3 &= I_0
\end{aligned}
$$

Thus the three signals can be processed to yield I_0, P, and χ. An experimental system of this type was developed at the University of California, Davis, in the 1970s, but it was never used in field studies.

A videopolarimeter utilizing three television cameras with polarizers at 0°, 60°, and 120° and spectral filters in the $\lambda=0.53$ to $0.58\ \mu$m range has been developed by Prosch et al. (1983). The signals from the cameras are digitized for computer analysis, as well as used for the generation of false color images on a color monitor. A color ladder is superimposed on the screen showing specific colors corresponding to predefined polarization properties. The authors estimated that by proper calibration the degree of polarization could be determined to $\pm 3\%$ and the orientation of the plane of polarization to within the range of 1° to about 20°, depending on the precision of the imaging tubes. They used the system for observations of surface features from an aircraft, and concluded that many features were made more clearly discernable by use of polarized light than with unpolarized intensities alone.

An ingenious design for an imaging polarimeter for satellite application has recently been proposed by Egan (1986). The proposed system is based on the currently operating multispectral scanner (MSS) of the Landsat satellites, and would build up images in the same spectral bands ($\lambda=0.5$–$0.6\ \mu$m, 0.6–0.7 μm, 0.7–0.8 μm, and 0.8–1.1 μm) in the solar radiation regime as the MSS. Band 5 ($\lambda=1.04$–$12.6\ \mu$m) was not included in Egan's design. The most difficult problem in such an instrument, aside from the doubling of the data rate, is to avoid parasitic polarization introduced by the system optics. For introduction of radiation from the scene to the MSS telescope, Egan has proposed the techniques of two orthogonally oriented reflecting prisms, the polarization introduced by the second reflection just cancelling that from the first. He estimates the parasitic polarization from the entrance optics to be less than 0.1%. The radiation from the telescope, which still contains the original polarization from the scene, would be divided by means of a beam splitter into the orthogonal polarization components and sensed by separate detectors. The normal component $I_\perp$ would be separated out by reflection at the Brewster angle at the first surface of the beam splitter and sent through a converging lens to a light fiber array which feeds an array detector. The component which is refracted at the first surface of the beam splitter would be a composite of both $I_\perp$ and $I_\parallel$. The $I_\perp$ component would be eliminated by means of a properly oriented polaroid, the remaining $I_\parallel$ being similarly sent to the detector through a converging lens and light fiber array. The signals from the two detectors would build up images as the oscillating mirror scanned in the cross-track direction, as is done now on the MSS.

Regardless of whether or not such an instrument is ever built, one can only applaud the concept of finally getting a satellite polarimeter into earth orbit for remote sensing of natural surfaces. While the polarimeter proposed by Egan appears to be sound, and the fact that the design is a modification of a currently operating radiometer is an attractive feature, there are other simpler and equally attractive methods, such as by use of a video polarizer system, to acquire polarized images from space.

11.11 Calibration of polarimeters

The production of a light beam of known and variable degrees of polarization for use in the calibration of polarimeters is a difficult task. The only point on a calibration curve which is readily defined to high precision is that of complete polarization. The prism polarizers discussed in section 11.2 have the capability of achieving at least $P=0.9999$ for the visible and near visible range of wavelengths, and sheets of polaroid reach $P=0.999$ or better at the wavelengths of their greatest efficiency. In contrast, it is difficult to produce a completely unpolarized beam of light. Incandescent lights, particularly those with coiled filaments, introduce P from one to several percent, and even ribbon filaments are not devoid of the problem. Either of two methods may be used to compensate for or eliminate small amounts of polarization. First, transmission of the beam by opal glass is a generally good depolarizing mechanism. However, not all opal glasses are the same and some care in the selection of the glass is advisable. Another method of compensating for a small amount of linear polarization is by the use of a tilted glass plate interposed in the beam. By rotating the plate around the optical axis to find the plane of polarization of the original beam and then varying its tilt in the orthogonal plane, the polarization introduced by the plate can be made to cancel that of the original beam. In fact, the use of such a tilted plate is a method of getting a precise measurement of polarization from weakly polarized sources such as stars and nebulae.

To a high degree of approximation, the light from the sun, and particularly from the center of the solar disk, is unpolarized. Because of the atmospheric effects, however, the solar intensity, reaching the ground has considerable variability, and the atmosphere may introduce spurious polarization effects. In addition, since polarimeter calibration is usually a laboratory operation, production of the unpolarized beam in the laboratory is normally preferred over use of the direct sunlight.

The production of a calibration source with polarization below about 30% is most conveniently accomplished by a single tilted glass plate of known index of refraction. In this case the effects of internal reflections of the plate are not severe, as shown previously in Fig. 11.9. For higher polarizations requiring a larger number of plates, however, internal reflections cause strong depolarization. The five or more plates required to achieve complete polarization of the transmitted light have numerous internal reflections, the result of which is to cause the polarization from the stack to deviate widely from the theoretical for a single reflection. The case for two and five plates of refractive index $m = 1.5725$ typical of borosilicate glass, as computed by Weinberg (1964), is shown in Fig. 11.12. If one could construct an optical system for which no internally reflected energy reaches the detector, then a stack of either two or five plates would constitute a good polarization calibrator. However, it is very difficult to baffle a system to eliminate all internally reflected light, and even a small fraction of the total can cause significant errors. Furthermore, the displacements of the components are a function of the angle of incidence on the plates, so in any real system the fraction of the internally reflected light reaching the detector, and the error produced thereby, will similarly be a function of angle of incidence. At the other extreme, in which the light from all internal reflections reaches the detector, the degree of polarization reaches only about 60% even for five plates, which is hardly high enough for an adequate calibration. Here also the problem of being sure that all light from internal reflections reaches the detector

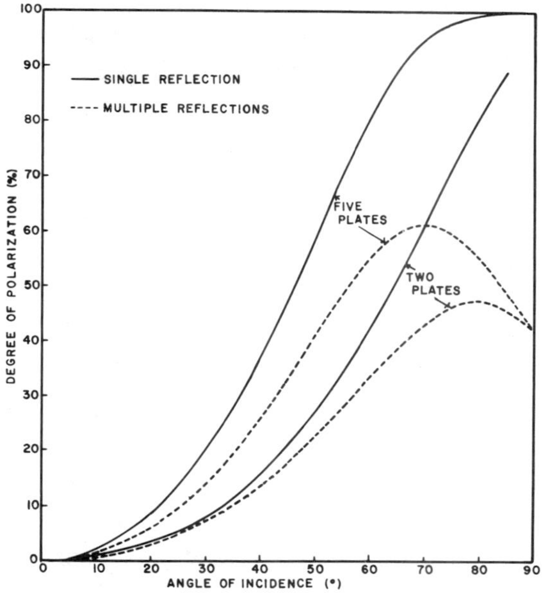

Figure 11.12 Degree of polarization of light transmitted by piles of two and five glass plates for singly reflected and multiply reflected rays for m=1.5725. (Data from Weinberg, 1964.)

is a serious one. The beam displacement is particularly large at the high angles of incidence, and if the optical system accepts light from all internal reflections at low incidence angles but not at higher angles, the curves of Fig. 11.11 would be distorted and unsuitable for calibration purposes. Because of these difficulties, an alternative to the stack of plates is needed for the higher polarizations. The use of a single plate, however, is feasible for the low polarization, as mentioned above.

Two alternatives to the simple pile of plates are given by Shurcliff (1962). In one design, the plates are arranged in two groups of opposite tilt so that the deviation of beam position produced by the first group is counteracted by the second group. The second design has the two groups not only oppositely tilted, but they are arranged in a fanned configuration and the individual plates are wedge shaped. According to the author, virtually none of the reflected rays in these configurations follows the main beam, but the problems of baffling, and the uncertainty of its effectiveness in eliminating reflected rays, are still of serious concern.

A different, and preferred, approach to construction of a polarization calibrator is the reflection polarizer of Rao (1966). Although the calibrator was constructed for field calibration of a polarimeter, and used the sun as the source of unpolarized light, it could readily be adapted for laboratory application. The linear polarization is introduced by reflecting the unpolarized beam from the hypotenuse face of an isosceles right angle glass prism which can be rotated in angle to achieve angles of incidence of from $\theta = 0°$ to $\theta = 90°$. The two polarization components $I_\perp$ and $I_\parallel$ of the light reflected from the prism are shown along with the degree of linear polarization of the reflected light, in Fig. 11.13. The incident light was assumed to be unpolarized and of unit intensity for the computations. The degree of polarization of the reflected light is 0% at $\theta = 0°$, rises to 100% at the Brewster angle, and again decreases to 0% at $\theta = 90°$. The first branch of the curve, which has easily achieved values varying from $\theta = 0°$ to the Brewster angle $\theta = \tan^{-1} m = 57.55°$ for an index of refraction of the prism of $m = 1.5725$, is adequate for calibration over the complete range of polarization values. From geometric ray tracing of the internal reflections of the prism,

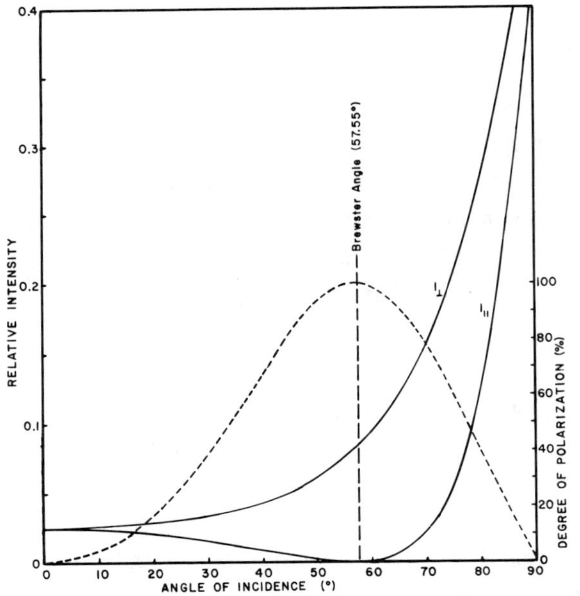

Figure 11.13 Intensity components $I_\perp$ and $I_\parallel$ and degree of linear polarization of light reflected from the plane face of a glass prism of index of refraction $m = 1.5725$.

Rao (1966) has shown that energy in only one-half of the incident beam has the possibility of emerging in the direction of the reflected beam, and that a simple diaphragm centered on the optic axis can effectively eliminate this energy from the reflected beam.

The Rubenson degree of polarization of the light reflected from the face of the prism is given by the expression

$$P = \frac{\sin\theta \, \sin 2\theta \, (m^2 - \sin^2\theta)^{1/2}}{m^2 \cos^2\theta - \sin^2\theta \cos 2\theta} \tag{11.28}$$

From an error analysis of this relation, Rao (1966) has shown that a 1° error in θ gives less than 1% error in P through $20° \leq \theta \leq 57.55°$, and less than a 2% error for $10° \leq \theta \leq 20°$. An error of θ less than 1° reduces the error of P proportionately. An error of 0.1°, or even 0.01°, can readily be achieved in a laboratory setting, in which case the reflection polarizer can provide a light beam of precisely known polarization for the calibration of polarimeters.

References

Boiko, P.N. (1971): Spectropolarimetry of the daytime sky for the ultraviolet part of the spectrum, *Scattering and Absorption of Light in the Atmosphere*, Nauka Pub. House, Alma Ata, Kazakhstan SSR (in Russian).

Borofka, R.P. (1967): Scene polarization measurements by photographic photometry, Ph.D. Thesis, Univ. of California, Los Angeles.

Borofka, R.P., and P.F. O'Brien (1968): Scene polarization measurements for photographic photometry, *Illum. Eng.*, *63*, 415–422.

Brewster, Sir David (1815): On the laws which regulate the polarisation of light by reflexion from transparent bodies, *Phil. Trans. Roy. Soc. London*, *105*, 125–130, 158–159.

Bullrich, K. (1964): Scattered radiation in the atmosphere and the natural aerosol, *Adv. in Geophpys.*, *10*, 99–260, Academic Press.

Chen, H.S. (1985): *Space Remote Sensing Systems*, Academic Press, New York.

Coffeen, D.L. (1974): Optical polarimeters in space, in T. Gehrels (ed.), *Planets, Stars, and Nebulae Studied With Photopolarimetry*, Univ. of Arizona Press, pp. 189–222.

Coulson, K.L., R.L. Walraven, G.I. Weight, and L.B. Soohoo (1974): Photon-counting polarizing radiometer, *Appl. Opt.*, *13*, 497–498.

Coulson, K.L., V.S. Whitehead, and C. Campbell (1986): Polarized views of the earth from orbital altitude, *Proc. S.P.I.E.*, *637—Ocean Optics VIII*, 35–41.

Curran, P.J. (1978): A photographic method for the recording of polarized visible light for soil surface moisture indications, *Rem. Sens. Environ.*, *7*, 305–322.

Curran, P.J. (1979): The use of polarized panchromatic and false-color infrared film for the monitoring of soil surface moisture, *Rem. Sens. Environ.*, *8*, 249–266.

Dandekar, B.D., and J.P. Turtle (1971): Day sky brightness and polarization during the total solar eclipse of 7 March 1970, *Appl. Opt.*, *10*, 1220–1224.

Dietze, G., and H. Seidel (1968): Ein Teilautomatisches gerat zum messen der himmelslicht-polarisation beitage und in der dammerung, *Zeit. fur Met.*, *20*, 9/10. 253–258.

Egan, W.G. (1986): Proposed design of an imaging spectropolarimeter/photometer for remote sensing of earth resources, *Opt. Eng.*, *25*, 1155–1159.

Egan, W.G., and H.B. Hallock (1966): Polarimetery signature of terrestrial and planetary materials, *Proc. 4th Symp. Remote Sensing of Environ.*, Univ. of Michigan, Ann Arbor 671–689.

Eiden, R. (1966): The elliptical polarization of light scattered by a volume of atmospheric air, *Appl. Opt.*, *5*, 569–575.

Fernald, F.G., B.M. Herman, and R.J. Curran (1969): Some polarization measurements of the reflected sunlight from desert terrain near Tucson, Ariz., *J. Appl. Met.*, *8*, 604–609.

Fesenkov, V.G. (1935): Determination de la polarization de la couronne solaire, *Astr. J.*, *12(4)*, 309–321.

Fitch, B.W., and K.L. Coulson (1977): Polarizing radiometer measurements of skylight at South Pole Station, Antarctica, *Proc. S.P.I.E., 112—Optical Polarimetry*, 184–190.

Gehrels, T. (1962): Wavelength dependence of the polarization of the sunlit sky, *J. Opt. Soc. Am.*, *52*, 1164–1173.

Gehrels, T., and T.M. Teska (1960): A Wollaston photometer, *Pub. Astron. Soc. Pac.*, *72*, 115–122.

Gerhaz, R. (1977): Polarization of scattered horizon light in inclement weather, *Arch. met. Geoph. Bioklim.*, *26*, 2/3, 265–273.

Glazebrook, Sir Richard (1923): *A Dictionary of Applied Physics*, Vol. IV, 490–511.

Gorchakov, G.I. (1966): Light scattering matrices in the atmospheric surface layer, *Izv., Atm. and Oceanic Phys.*, *2(6)*, 595–605.

Gorchakov, G.I., and V.N. Siderov (1980): On the spectral dependence of the components of the matrix of light scattering by the atmospheric haze, *Izv., Atm. and Oceanic Phys.*, *16(9)*, 716–720 (English trans.)

Gritz, D.N. (1987): Near-infrared (IR) polarizing glass, Paper No. 750–03 S.P.I.E. Symposium, 11–16 January 1987, Los Angeles, California.

Hall, J.S., and A.H. Mikesell (1950): Polarization of light in the galaxy, *Pub. U.S. Naval Obs.*, *17/1*, 7–62.

Hannemann, D. (1975): A polarimeter to measure the complete state of polarization of scattered solar radiation, *Beit. Physik Atmos.*, *48*, 76–83.

Hariharan, T.A. (1967): A modified skylight polarimeter, *J. Sci. Inst.*, *44*, 478–479.

Hiltner, W.A. (1962): Polarization measurements, Ch. 10 in Hiltner (ed.), *Astronomical Techniques*, Univ. of Chicago Press, pp. 229–246.

Holzworth, G.C., and C.R.N. Rao (1965): Studies of skylight polarization, *J. Opt. Soc. Am.*, *55*, 403–408.

Kostyla, R.J., and K.J. Abcunas (1981): Effect of boration upon the stability of polyvinyl alcohol (PVA) iodine polarizers, *Proc. S.P.I.E., 307—Polarizers and Applications*, 5–12.

Land, E.H. (1961): Some aspects of the development of sheet polarizers, *J. Opt. Soc. Am.*, *41*, 957–963.

Malus, E.L. (1809): Sur une propriete de la lumiere reflechie, *Mem. Phys. Chem. Soc. D'Arcueil*, *2*, 143–158.

McFarlane, J.W. (1937): Photography by polarized light, Ann. Rept. of Smithsonian Inst., 225–233.

Miles, G.M. (1982): Practical prism polarizers, *Proc. S.P.I.E.*, *307—Polarizers and applications*, 2–4.

Nicol, W. (1828): On a method of so far increasing the divergence of the two rays in calcereous-spar that only one image may be seen at a time, *Edinburgh New Phil. J.*, *6*, 83–86.

Prosch, T., D. Hennings, and E. Raschke (1983): Video polarimetry: A new imaging technique in atmospheric science, *Appl. Opt.*, *22*, 1360–1363.

Rao, C.R.N. (1969): Balloon measurements of the polarization of light diffusely reflected by the earth's atmosphere, *Planet. Space Sci.*, *17*, 1307–1309.

Rao, C.R.N., and Z. Sekera (1967): A research program aimed at high altitude balloon-borne measurements of radiation emerging from the earth's atmosphere, *Appl. Opt.*, *6*, 221–225.

Reshetnikova, I.S., and Ye. O. Federova (1978): A study of the optical characteristics of the troposphere and lower stratosphere, *Izv.*, *Atm. and Oceanic Phys.*, *14(11)*, 813–819 (English trans.).

Ronchi, V. (1970): *The Nature of Light*, Harvard University Press.

Russell, E.E., L.A. Watts, S.F. Pellicori, and D.L. Coffeen (1977): Orbiter cloud photopolarimeter for the Pioneer Venus mission, *Proc. S.P.I.E.*, *112—Optical Polarimetry*, 28–44.

Sassen, K., K-N. Liou, and S. Hunter (1977): Scattering of polarized laser light by water droplet and ice crystal clouds, Sci. Rept. No. 1, Dept. of Meteorology, Univ. of Utah, Salt Lake City, Utah.

Seaman, C.H., and Z. Sekera (1955): Photoelectric polarimeter for measurement of skylight polarization, Appendix A, Final Rept., *Contr. A.F. 19*, 122–239, University of California, Los Angeles.

Sekera, Z. (1956): Recent developments in the study of the polarization of skylight, *Adv. in Geophys., Vol. III*, Academic Press, pp. 43–104.

Sekera, Z., C.R.N. Rao, and D. Dibble (1963): Photoelectric skylight polarimeter, *Rev. Sci. Inst., 34*, 764–768.

Serkowski, K. (1974): Polarimeters for optical astronomy, in T. Gehrels (ed.), *Planets, Stars, and Nebulae Studied with Photopolarimetry*, University of Arizona Press, pp. 135–174.

Shurcliff, W.A. (1962): *Polarized Light: Production and Use*, Harvard University Press.

Slocum, R.E. (1981): Evaporative thin metal films as polarizers, *Proc. S.P.I.E., 307—Polarizers and applications*, 25–30.

Steinhorst, G. (1977): Stratospheric aerosol concentration determined by an iterative method from twilight polarization measurements, *Contrib. Atm. Phys., 50*, 508–523.

Stowe, L. (1977): Polarization of reflected sunlight as measured from a high altitude balloon, *Proc. S.P.I.E., 112—Optical Polarimetry*, 176–181.

Swindell, W. (ed) (1975): *Polarized Light*, Dowden, Hutchinson, and Ross, Stroudsburg, Pennsylvania.

Vanderbilt, V.C., L. Grant, L.L. Biehl, and B.F. Robinson (1985): Specular, diffuse, and polarized light scattered by two wheat canopies, *Appl. Opt., 24*, 2408–2418.

Volz, F.E. (1983): Volcanic turbidity, skylight scattering function, and neutral points in New England 1982/83, Preprints, Fifth Conf. on Atm. Rad., Baltimore, Maryland, pp. 240–242.

Walraven, R.L. (1977a): Polarization imagery, *Proc. S.P.I.E., 112—Optical Polarimetry*, 164–167.

Walraven, R.L. (1977b): Video polarimeter, Unpublished.

Walraven, R.L. (1977c): Reduction of polarizing radiometer data, *Rev. Sci. Inst., 48*, 613–616.

Walraven, R.L. (1978): Polarization by a tilted absorbing glass plate, *Rev. Sci. Inst., 49(4)*, 537–540.

Walraven, R.L. (1981): Polarization imagery, *Opt. Eng.*, *20*, 14–18.

Weinberg, J.L. (1964): On the use of a pile-of-plates polarizer—The transmitted component, *Appl. Opt.*, *3*, 1057–1061.

Yeh, P. (1981): Generalized model for wire grid polarizers, *Proc. S.P.I.E.*, *307—Polarizers and Applications*, 13–21.

APPENDIX A

REPRESENTATION OF MUELLER MATRICES

In order to represent the Mueller matrices, we rewrite the definitions of the Stokes parameters from Chapter 1 (Eq. 1.34).

Thus we have

$$
\begin{aligned}
I &= A_x A_x^* + A_y A_y^* \\
Q &= A_x A_x^* - A_y A_y^* \\
U &= A_x A_y^* + A_y A_x^* \\
V &= i\,(A_x A_y^* - A_y A_x^*)
\end{aligned}
\tag{A.1}
$$

where the A_{ij} are complex amplitudes of the electric vector.

If we pass radiation of vector intensity $\mathbf{I}_1$ through some optical device, the transmitted radiation will have intensity $\mathbf{I}$, as was shown above in Fig. 11.8. The amplitudes of the two intensity components can in general be related as follows:

$$
\begin{aligned}
A_x &= a_{11} A_{x1} + a_{12} A_{y1} \\
A_y &= a_{21} A_{x1} + a_{22} A_{y1}
\end{aligned}
\tag{A.2}
$$

The transformation of the radiation passing through the device is conveniently represented by use of the Mueller matrix $\mathbf{M}$ as

$$\begin{vmatrix} I \\ Q \\ U \\ V \end{vmatrix} = \begin{vmatrix} M_{11} & M_{12} & M_{13} & M_{14} \\ M_{21} & M_{22} & M_{23} & M_{24} \\ M_{31} & M_{32} & M_{33} & M_{34} \\ M_{41} & M_{42} & M_{43} & M_{44} \end{vmatrix} \begin{vmatrix} I_1 \\ Q_1 \\ U_1 \\ V_1 \end{vmatrix} \tag{A.3}$$

where the elements M_{ij} are to be determined in terms of the complex amplitudes a_{ij}.

From Eq. (A.1) we have the relations

$$\begin{aligned} A_{x1}A_{x1}^* &= \frac{1}{2}(I_1 + Q_1) \qquad & A_{x1}A_{y1}^* &= \frac{1}{2}(U_1 - i V_1) \\ A_{y1}A_{y1}^* &= \frac{1}{2}(I_1 - Q_1) \qquad & A_{y1}A_{x1}^* &= \frac{1}{2}(U_1 + i V_1) \end{aligned} \tag{A.4}$$

On combining these relations with Eq. (A.2) we obtain, after some algebra,

$$\begin{aligned} I = {} & \frac{1}{2}(|a_{11}|^2 + |a_{21}|^2 + |a_{12}|^2 + |a_{22}|^2)\, I_1 \\ & + \frac{1}{2}(|a_{11}|^2 + |a_{21}|^2 - |a_{12}|^2 - |a_{22}|^2)\, Q_1 \\ & + \frac{1}{2}(a_{11}a_{12}^* + a_{21}a_{22}^* + a_{12}a_{11}^* + a_{22}a_{21}^*)\, U_1 \\ & + \frac{i}{2}(a_{12}a_{11}^* + a_{22}a_{21}^* - a_{11}a_{21}^* - a_{21}a_{22}^*)\, V_1 \end{aligned} \tag{A.5}$$

From this we see

$$
\begin{aligned}
M_{11} &= \frac{1}{2}(|a_{11}|^2 + |a_{12}|^2 + |a_{21}|^2 + |a_{22}|^2) \\
M_{12} &= \frac{1}{2}(|a_{11}|^2 + |a_{21}|^2 - |a_{12}|^2 - |a_{22}|^2) \\
M_{13} &= Re(a_{11}a_{12}^* + a_{21}a_{22}^*) \\
M_{14} &= Im(a_{11}a_{12}^* + a_{21}a_{22}^*)
\end{aligned} \tag{A.6}
$$

We have used the identities $Re(a_{12}a_{11}^*) = Re(a_{11}a_{12}^*)$ and $Im(a_{11}a_{12}^*) = -Im(a_{12}a_{11}^*)$.

By representing the other three Stokes parameters in a manner similar to Eq. (A.5), we obtain the remaining matrix elements as

$$M_{21} = \frac{1}{2}(|a_{11}|^2 + |a_{12}|^2 - |a_{21}|^2 - |a_{22}|^2)$$

$$M_{22} = \frac{1}{2}(|a_{11}|^2 + |a_{22}|^2 - |a_{12}|^2 - |a_{21}|^2)$$

$$M_{23} = Re(a_{11}a_{12}^* - a_{21}a_{22}^*)$$

$$M_{24} = Im(a_{11}a_{12}^* - a_{21}a_{22}^*)$$

$$M_{31} = Re(a_{11}a_{21}^* + a_{12}a_{22}^*)$$

$$M_{32} = Re(a_{11}a_{21}^* - a_{12}a_{22}^*)$$

$$M_{33} = Re(a_{11}a_{22}^* + a_{21}a_{12}^*)$$

$$M_{34} = Im(a_{11}a_{22}^* + a_{21}a_{12}^*)$$

$$M_{41} = Im(a_{21}a_{11}^* + a_{22}a_{12}^*)$$

$$M_{42} = Im(a_{21}a_{11}^* - a_{22}a_{12}^*)$$

$$M_{43} = Im(a_{22}a_{11}^* + a_{21}a_{12}^*)$$

$$M_{44} = Re(a_{22}a_{11}^* - a_{21}a_{12}^*)$$

For optical components such as linear polarizers or retardation plates, some of the elements reduce to zero, as indicated in section 11.3.

Consider a flat plate with principal axes x' and y' oriented at angle ψ with respect to axes x and y of the original beam of radiation. On passage through the plate, the radiation is transformed to components

$$A'_{11} = \exp[-i(2\pi d\ /\ \lambda)(n_{x'} - ik_{x'})]$$

$$A'_{22} = \exp[-i(2\pi d\ /\ \lambda)(n_{y'} - ik_{y'})] \tag{A.7}$$

$$A'_{12} = A'_{21} = 0$$

where d is the thickness of the plate, λ is a wavelength in a vacuum, n_{ij} are the principal indices of refraction and k_{ij} is the principal absorption coefficients of the plate. In the case of a nonabsorbing retardation plate $k_{x'} = k_{y'} = 0$, and for a perfect dichroic polarizer $n_{x'} = n_{y'} = 0$. For a polarizer such as those of calcite, $n_{x'} \neq n_{y'}$, and the absorption coefficients are not necessarily the same but are very small.

The electric field components in the (x,y) coordinates are related to those in the (x',y') system by the relations

$$A'_x = A_x \cos\psi + A_y \sin\psi$$

$$\tag{A.8}$$

$$A'_y = -A_x \sin\psi + A_y \cos\psi$$

This gives

$$
\begin{aligned}
A_{11} &= A'_{11} \cos^2\psi + A'_{22} + \sin^2\psi \\
A_{12} &= A_{21} = (A'_{11} - A'_{22}) \sin\psi \cos\psi \qquad \text{(A.9)} \\
A_{22} &= A'_{11} \sin^2\psi + A'_{22} \cos^2\psi
\end{aligned}
$$

On substituting these into Eq. (A.9), we obtain the matrix elements

$$M_{11} = \frac{1}{2}(|A'_{11}|^2 + |A'_{22}|^2)$$

$$M_{12} = \frac{1}{2}(|A'_{11}|^2 - |A'_{22}|^2) \cos2\psi$$

$$M_{13} = \frac{1}{2}(|A'_{11}|^2 - |A'_{22}|^2) \sin2\psi$$

$$M_{14} = 0$$

$$M_{21} = M_{12}$$

$$M_{22} = \frac{1}{2}[|A'_{11}|^2 + |A'_{22}|^2 \cos^2 2\psi + Re(A'_{11}A'^{*}_{22}) \sin^2 2\psi]$$

$$M_{23} = \frac{1}{2}[|A'_{11}|^2 + |A'_{21}|^2 - 2Re(A'_{11}A'_{22}) \cos2\psi \sin2\psi]$$

$$M_{24} = Im(A'_{22}A'^{*}_{11}) \sin2\psi$$

$$M_{31} = M_{13}$$

$$M_{32} = M_{23}$$

$$M_{33} = \frac{1}{2}[|A'_{11}|^2 + |A'_{22}|^2] \sin^2 2\psi + Re(A'_{11}A'^*_{22}) \cos^2 2\psi$$

$$M_{34} = Im(A'_{11}A'^*_{22}) \cos 2\psi$$

$$M_{41} = 0$$

$$M_{42} = -M_{24}$$

$$M_{43} = -M_{34}$$

$$M_{44} = Re(A'_{11}A'^*_{22}) \tag{A.10}$$

We shall now apply this formalism to some specific optical devices.

1. Dichroic linear polarizer with no retardation

In this case $n'_x = n'_y$, so $|A_{11}|^2 = e^{-2\pi d k_x/\lambda^2} \equiv k_1$; $|A_{22}|^2 = e^{-2\pi d k_y/\lambda^2} \equiv k_2$; $A_{11}A^*_{22} = e^{-2\pi d k_x/\lambda}\, e^{-2\pi d k_y/\lambda} = \sqrt{k_1 k_2}$; and the Mueller matrix is

$$\mathbf{M}_p = \begin{vmatrix} (k_1 + k_2) & (k_1 - k_2)\cos 2\psi & (k_1 - k_2)\sin 2\psi & 0 \\ (k_1 - k_2)\cos 2\psi & (k_1 + k_2)\cos^2 2\psi + 2\sqrt{k_1 k_2}\sin^2\psi & (k_1 + k_2 - 2\sqrt{k_1 k_2})\cos 2\psi \sin 2\psi & 0 \\ (k_1 - k_2)\sin 2\psi & (k_1 + k_2 - 2\sqrt{k_1 k_2})\cos 2\psi \sin 2\psi & (k_1 + k_2)\sin^2 2\psi + 2\sqrt{k_1 k_2}\cos^2 2\psi & 0 \\ 0 & 0 & 0 & 2\sqrt{k_1 k_2} \end{vmatrix} \tag{A.11}$$

For a perfect dichroic polarizer, $k_1 = 1$ and $k_2 = 0$, in which case

$$\mathbf{M}_{pp} = \begin{vmatrix} 1 & \cos 2\psi & \sin 2\psi & 0 \\ \cos 2\psi & \cos^2 2\psi & \cos 2\psi \sin 2\psi & 0 \\ \sin 2\psi & \cos 2\psi \sin 2\psi & \sin^2 2\psi & 0 \\ 0 & 0 & 0 & 0 \end{vmatrix} \tag{A.12}$$

2. Retarder with no absorption

In this case $k_1 = k_2 = 0$, phase shift δ is given by $\delta = (2\pi d/\lambda)(n_{x'} - n_{y'})$ and we have $|A_{11}|^2 = |A_{22}|^2 = 1$ and $A_{11}A_{22}^* = e^{-i\delta}$. Then

$$\mathbf{M}_R = \begin{vmatrix} 1 & 0 & 0 & 0 \\ 0 & \cos^2 2\psi + \cos\delta \sin^2 2\psi & (1 - \cos\delta) \cos 2\psi \sin 2\psi & -\sin\delta \sin 2\psi \\ 0 & (1 - \cos\delta) \cos 2\psi \sin 2\psi & \sin^2 2\psi + \cos\delta \cos^2 2\psi & \sin\delta \cos 2\psi \\ 0 & \sin\delta \sin 2\psi & -\sin\delta \cos 2\psi & \cos\delta \end{vmatrix} \tag{A.13}$$

For a quarter-wave retardation plate, $\delta = 90°$ and the Mueller matrix is

$$\mathbf{M}_{90°} = \begin{vmatrix} 1 & 0 & 0 & 0 \\ 0 & \cos^2 2\psi & \cos 2\psi \sin 2\psi & -\sin 2\psi \\ 0 & \cos 2\psi \sin 2\psi & \sin^2 2\psi & \cos 2\psi \\ 0 & \sin 2\psi & -\cos 2\psi & 0 \end{vmatrix} \tag{A.14}$$

Similarly for a half-wave plate, $\delta = 180°$ and the matrix is

$$\mathbf{M}_{180°} = \begin{vmatrix} 1 & 0 & 0 & 0 \\ 0 & \cos 4\psi & \sin 4\psi & 0 \\ 0 & \sin 4\psi & -\cos 4\psi & 0 \\ 0 & 0 & 0 & 1 \end{vmatrix} \tag{A.15}$$

SUBJECT INDEX

AUTHOR INDEX